Chemotherapy in Aquaculture:

from theory to reality

Symposium

Paris, 12 - 15 March 1991

Editors:
C. Michel and D.J. Alderman

ISBN 92-9044-301-4

CONTENTS

Session 1 - Chemotherapy in aquaculture today

Session 2 - Toxicity and side effects of chemotherapeutants

Session 3 - Problems of resistance

Session 4 - Residues

FOREWORD

Through its permanent Fish Diseases Commission, created in 1960, the International Office of Epizootics deals with questions related to the health of fish, and since 1990 now also deals with all of the health products of aquaculture.

The health of aquaculture animals is continually jeopardized by bioagressors such as viruses, bacteria, parasites and fungi which play their noxious role either spontaneously or after induction by aquatic animal husbandry practices, and often both. Indeed, it is generally known that disease problems follow the techniques developed for animal production.

If one considers the fish health problems in aquaculture according to their control methods, the pathogens of aquaculture animals can be divided into two groups. The first group encompasses the agents that are not treatable (viruses, myxosporeans), or are drug resistant or difficult-to-treat bacteria such as *Renibacterium salmoninarum*, and are also of socioeconomic importance. The second group encompasses the other causes of diseases, some of which are susceptible to the improving effects of therapy.

For those diseases which do not respond to therapy, the most common approach is to prevent their dissemination with the international transfer of living animals or their products. The OIE has played a role in this approach to the control of diseases associated with aquaculture, as dealt with in the provisions put forward in the *International Animal Health Code*. The measures in the *Code* were designed to deal with the more common serious diseases in intensive aquaculture and in developed countries where the techniques and the logistics for their control are available.

However, most often the major difficulty for aquaculturists is dealing with each disease outbreak on a case by case basis, and the need to have their problem solved immediately. The urgency of the need to respond to these outbreaks of diseases has resulted in an uncontrolled approach to both the applications of drugs and to the proper diagnosis and prescription of the needed medications to control the disease. Nevertheless, this problem must not hide the fact that therapy for animal production sites to achieve day-by-day control of pathogens will help to ensure the success of this industry.

The use of therapy leads obviously to the transit of drugs and to their more or less long lasting persistence in goods intended for human consumption. It also leads to the release of drugs or their by-products in to the aquatic environment. Hence the criticisms raised in the media against the use of therapy in aquaculture, and the restrictive legislations set up in many countries under the pressure of public opinion. Some people almost seem to want the public to believe that the drug resistances of bacteria responsible for human infections, originate exclusively or almost exclusively from the consumption of animal products such as those provided by aquaculture!

Therefore, in the near future, the major threats for the health of aquaculture animals could lie in restrictions placed on the use of therapy as well as in the causative agents of diseases. Such a situation is becoming a particular matter of concern, especially at a time when aquaculture is acknowledged by both the media and politicians as an essential resource for humankind in the next century.

Besides its role in health control in international transfers of living animals, the OIE also provide rules for the hygiene of animal products intended for human consumption. It therefore became mandatory for the Fish Diseases Commission to take into consideration the problems of therapeutic control of the aquaculture animal diseases insofar as they concern both animal and human health.

The need was to gather pieces of well established data, make an inventory of gaps in knowledge, reach a working spirit enabling reasonable debate, and demystify the use of therapy. The latter is widely presented as a permanent threat to the health of consumer populations, of whom the average life span is continuously increasing, despite the poor living conditions they are said to undergo.

The endeavour took the form of an international meeting drawing together fish pathologists, pharmacologists, toxicologists and therapeutists. Data on treatments in aquaculture and implications in food quality or even in aquatic environment properties and thus in human health, were presented and discussed over a three and-a-half day meeting held at the OIE headquarters, Paris, in March 1991.

However, the impact of this meeting would have remained limited if the diffusion and release of the resultant information had not been achieved. Thanks to the outstanding editorial work done jointly by Drs Christian Michel and David Alderman, the content of the communications presented and the resulting discussions is presented here in a clear and attractive form. I can hardly acknowledge the editors enough for the remarkable performance that they have achieved simultaneously with their normal workload. Their work will broaden the success of a meeting which had already been successful *per se* due to the quality of the communications presented and the discussions.

The data presented and considerations they raised, will set the basis for the reasonable use of effective therapies in a context of respect for the safety of the human consumer, the environment and animal welfare. This constitutes another aspect of the involvement of the OIE in the field of health in aquaculture.

P. de Kinkelin
President Fish Diseases Commission
(1988-1991)

PREFACE

The remarkable development of aquaculture and the growing interest of numerous Member Countries of the Office International des Epizooties (OIE) in this type of production encouraged the OIE Fish Diseases Commission to organise an international conference on a particularly important subject: therapeutic interventions in the aquatic medium.

Such interventions have become increasingly common because of the multiplication of infections associated with the intensive farming of fish, molluscs or crustaceans. The need to resort to chemotherapy against infections in these species is accompanied by the problems already encountered in mammals and birds: correct dosage, toxicity, resistance, residues, etc.

It was therefore useful to review these problems and to open discussions on possible solutions and their eventual international harmonisation.

It fell to Dr P. de Kinkelin (President of the Commission), Dr C. Michel and Dr D.J. Alderman to prepare this conference, and to invite the leading specialists from 22 countries.

The conference, held from 12 to 15 March 1991 at the OIE Headquarters in Paris, was a great success, and its usefulness will be extended by the publication of this book.

The reader will find a collection of very detailed documents concerning the principal topics of the conference:

- Current conditions of chemotherapy: drugs used, dosage, methods of administration, constraints, regulations, etc.;
- Toxicity and potential side effects of chemotherapy: evaluation of direct and indirect toxicity, and toxicity after release into the aquatic medium;
- Resistance of micro-organisms to antibiotics: its importance in aquaculture and consequences for public health;
- Drug residues, their assay, technical procedures and pharmacokinetic studies.

The editors of this book and all the authors are warmly congratulated and acknowledged. Their effort will contribute towards the awareness of animal health and public health authorities to future problems, and will constitute an important step in any harmonisation of international regulations which may be put forward by the OIE.

J. Blancou
Director General of the
Office International des Epizooties

A NOTE FROM THE EDITORS

The organisation of a conference on "Problems of chemotherapy in aquaculture: theory to reality" in Paris in March 1991, and the publication of this collection of contributions to the meeting, has been a most stimulating challenge, particularly in view of the new arrangements applicable throughout the European Community from 1992. Both events have enabled new light to be shed on these most important issues. Despite unfavourable international events at the time of the conference, attendance was high. Very few invited specialists were unable to participate. Even though the balance of certain sections suffered slightly, notably the one devoted to the effect of drug residues on the environment, the 56 contributions compiled in this book provide a rich source of information and reflect the nature of the conference. We wish to thank sincerely all the authors, whose goodwill and patient cooperation has greatly facilitated our task.

As with every enterprise of a collective nature, it has been impossible to avoid gaps and duplication. In order to reduce these to a minimum, we have provided short yet comprehensive accounts of the discussions which followed the presentations. The poster presentations, initially grouped together, have been inserted in the different chapters at appropriate places. These were not, of course, followed by discussions, but they illustrate points raised during the lectures. The paper by Dr F. Meyer, who was unable to attend and was replaced by Mrs Rosalie Schnick, has been included because of the originality of the ideas presented. Finally, special attention has been paid to the summaries in order to compensate for the inconvenience of bilingualism; the size of the book precludes full translation.

The problems associated with the use of drugs in acquaculture do not merely represent topics of current interest. They illustrate the different points of view that exist and indeed the controversies. The reader must not be surprised to encounter irreconcilable opinions expressed in the pages which follow. It was not the intention of the OIE meeting to resolve these questions and to express ideas that cannot be supported by existing knowledge, but to bring together different ideas and experiences. The aim was to present, even if this gave rise to rather lively dialogue on occasions, the maximum number of arguments and facts to support the reflections and ideas of observers or readers not necessarily specialised in this field. We hope that this modest aim has been achieved particularly for those whose customs and responsibilities may be put to the test by the adoption of many new regulations.

C. Michel & D. J. Alderman

SESSION 1

Chemotherapy in aquaculture today

CHEMOTHERAPY IN AQUACULTURE TODAY

D.J. Alderman[*] & C. Michel[**]

[*] MAFF, Fish Diseases Laboratory, Weymouth,Dorset, DT4 8UB, UK.
[**] INRA, Lab.de Virologie et d'Immunologie Moléculaires, 78350, Jouy en Josas, France

***Summary** :Although chemotherapeutants have been used in fish farming over many years, the expansion in aquaculture together with greater public awareness of environmental issues has emphasised many of the potential problems associated with the use of increasing quantities of potent chemical compounds in the aquatic environment. Whilst many of these problems do exist, they are often, as yet, ill defined and may, in some limited cases, be more imagined than real. This paper forms an initial introduction to many of the problems and constraints involved in the use of medicines in aquaculture which will be considered in greater detail elsewhere in these proceedings.*

Résumé : **Les thérapeutiques dans l'aquaculture d'aujourd'hui.**

Le développement de méthodes intensives en élevage aquicole a accru l'impact des maladies infectieuses et parasitaires. Les mesures préventives, fondées sur l'hygiène, la lutte sanitaire et l'immunoprophylaxie ne s'avèrent pas toujours efficaces, et en dépit de ses inconvénients la thérapeutique demeure souvent incontournable. Le recours à des traitements massifs et fréquents suscite cependant des interrogations quant aux effets pouvant résulter pour le consommateur et l'environnement aquatique.

Les principales indications de traitements (Tableau I) portent sur la pathologie bactérienne, souvent causée par des germes à Gram négatif dont le pouvoir pathogène est modulé par des facteurs externes, ainsi que sur les mycoses et parasitoses externes (protozooses, infestations à monogènes et à crustacés copépodes) que l'on peut contrôler par des bains thérapeutiques curatifs ou préventifs. Les bactérioses chroniques et les infections parasitaires internes, parmi lesquelles dominent des bactéries à Gram positif, des mycoses profondes, des microsporidioses et des myxozooses, sont en revanche rebelles à la plupart des agents thérapeutiques. Les substances les plus employées sont les antibiotiques (Tableau II) et des produits antiseptiques à usage externe comme le sel, le formol, le sulfate de cuivre, le permanganate de potassium, le vert malachite, les ammoniums quaternaires, auxquels s'ajoutent des composés organochlorés.

Cette apparente richesse de moyens est en fait limitée par des considérations de coût, par des méthodes d'administration spécifiques (bain et voie orale), et par la mise en place de réglementations souvent disparates selon les pays, dont les plus restrictives fondent les autorisations d'emploi sur des dossiers techniques sévèrement codifiés. Le handicap le plus sérieux provient de ce que les critères techniques, généralement extrapolés des élevages terrestres, s'appliquent mal au contexte des exploitations aquicoles.

Parmi les problèmes préoccupants figurent les résistances, souvent transmissibles, dont on peut redouter la propagation à des agents pathogènes pour l'homme, et qui surtout rendent vite caducs les traitements spécifiques lorsque une panoplie réduite de produits interdit d'en alterner l'emploi. Le recours aux antibiotiques à titre prophylactique ou comme activateurs de croissance devraient être proscrits, bien que certains éleveurs y souscrivent systématiquement. La question des résidus, actuellement très débattue, ne pourra être résolue que par la multiplication des travaux scientifiques. En effet, la variété des espèces et l'influence de la température sur les mécanismes pharmacocinétiques multiplient les cas de figures. En principe les délais d'attente légaux devraient tenir compte des températures d'élevage, et se référer à la notion de degré-jour. Mais la définition de limites maximales de résidus (LMR) reste difficile faute de critères objectifs, bien que la sensibilité des techniques de détection n'ait cessé de s'améliorer. Enfin, l'impact des traitements sur l'habitat aquatique, sans doute réel, justifie qu'on s'interroge, mais aussi bien les méthodes d'étude que les données établies n'en sont pour l'instant qu'à leurs balbutiements!

Introduction

The increasing intensity of fish farming combined with the introduction of new species has inevitably been paralleled by increases in disease. Whilst inadequate understanding of physiological, biological and nutritional requirements of newly domesticated species is undoubtedly a major contributor to losses, the high stocking densities and stress of intensive farming favour the occurrence and spread of infectious disease (25, 33).

Whilst optimisation of culture conditions will allow fish to express natural defences and thus reduce losses, such ideal conditions are often not practicable commercially and infectious disease agents will still persist in the farm environment. The most effective disease control method is, therefore, to prevent infectious agents from reaching a farm by means of health control systems policed by national authorities. However, such controls need both industrial and international co-operation and the acceptance of some restrictions of commercial trade to be effective (23). In the case of endemic disease, attempts at eradication will be expensive both in time and money once the disease is established.

In theory, vaccination is the next most valuable approach, but, so far, effective vaccines are available only for a limited range of pathogens (9, 24).

The limitations of more general control methods and of vaccines have led to chemotherapy being considered as the most effective and flexible weapon against infectious disease (11, 18, 35, 40). Although chemotherapeutants are available against a wider range of pathogens than vaccines, they too have limitations in terms of the pathogen groups against which they are effective. Chemotherapy is also a palliative rather than a final solution to problems of disease and its use in the aquatic environment is not free from practical problems. For example, attention is being drawn increasingly to the potential hazards that intensive use of drugs could represent for the environment and for human health (2, 6, 28, 30).

Disease Groups

Fish diseases are caused by a wide range of infectious organisms including viruses, bacteria, fungi, protozoan and metazoan parasites. The main groups involved in fish mortalities are listed in Table I, illustrated by species of major economic significance for the aquaculture industry. Viral diseases are widespread and have a major adverse effect on the production of shrimp larvae and oysters as well as of carp and salmonids. Apart from spring viraemia of carp (SVC) no successful vaccine is yet available against a viral disease in aquaculture. Unfortunately, viral diseases are not susceptible to control by chemotherapy, therefore, despite their considerable significance as pathogens, viruses will not be considered further here.

In contrast, bacteria, whilst of equal or greater importance in aquaculture , present a larger variety of pathogenic agents, many acting as secondary invaders which can take opportunity of any cause of weakness in fish to overcome the natural host defences. Such opportunistic bacteria are widespread in all environments and are a threat every time fish handling or other husbandry operations are undertaken. However, their harmful effects rarely persist and generally cease with the removal of the original source of stress.

Many bacteria are true pathogens and all fish culture systems suffer outbreaks of primary bacterial infections as do invertebrate hatchery systems which are particularly susceptible. In fish, gram- negative bacteria usually produce septicaemic infections characterized by haemorrhagic and/or necrotic lesions. Chronic infections, which progress more slowly and result in granulomatous internal lesions, are more commonly associated with Gram positive bacteria. Whilst it is often possible to control acute diseases with antimicrobial therapy, chronic conditions are more difficult. In vertically transmitted (i.e. via the ova) infections (Renibacteriosis for instance) and where infection may be intracellular, it is difficult to ensure that effective drug concentrations are achieved within the affected cells or tissues, and only partially effective treatments exist.

Table I Etiology of the most important diseases of commercial aquatic species

PATHOGEN	HOST	DISEASE	ENVIRONMENT
BACTERIA			
Enterobacteriacae			
Edwardsiella tarda	Eel	Red pest disease	Japan, warm water
E. ictaluri	Channel catfish	Enteric septicaemia	USA,warm fresh water
Yersinia ruckeri	Salmonids	Red mouth	N America, Europe
Vibrionacae			
Vibrio anguillarum	All fish	Vibriosis	Ubiquitous, marine
V.salmonicida	Atlantic salmon	Cold water vibriosis	North Sea
Vibrio spp	Molluscs & crustacea	Larval infections	Ubiquitous, marine
Aeromonas salmonicida	Salmonids	Furunculosis	Ubiquitous
A. hydrophila	All fish	Septicaemia	Freshwater
Pasteurellae			
Pasteurella pisicida	Yellowtail	Pseudo-tuberculosis	Japan, marine
Pseudomonadacae			
Pseudomonas anguilliseptica	Eel	Red spot disease	Japan
Flexibacteriaceae			
Flexibacter columnaris	All fish	Columnaris disease	Freshwater
Flexibacter psychrophilus	Salmonids	Cold water disease	Freshwater
Flavobacteriaceae			
Flavobacter branchiophilum	All fish	Gill disease	Freshwater
Mycobacteriaceae			
Mycobacterium spp	All fish	Tuberculosis	Marine, aquaria
Nocardiaceae			
Nocardia kampachi	Yellowtail	Nocardiosis	Japan, marine
Nocardia spp	Crustaceans	Nocardiosis	Ubiquitous
Coryneforms			
Renibacterium salmoninarum	Salmonids	Bacterial kidney disease	Freshwater
Cocci			
Streptococcus spp.	All marine fish	Septicaemia	Marine
Aerococcus viridans	Lobster	Gaffkaemia	Marine
EUMYCOTA			
Oomycetes			
Saprolegnia spp	Fish and crayfish	Topical mycosis	Freshwater
Achlya spp	Fish and crayfish	Topical mycosis	Freshwater
Aphanomyces astaci	Crayfish	Crayfish plague	Europe,N America, freshwater
Lagenidium marinum	Crustaceans	Larval mycoses	Marine
Haliphthoros milfordensis	Crustaceans	Larval mycoses	Marine
Hyphomycetes			
Fusarium solani	Crustaceans	Burn spot disease	Ubiquitous
Exophiala spp	Salmonids	Visceral mycosis	Ubiquitous
Ochroconis spp	Salmonids	Visceral mycosis	Ubiquitous
Ascomycetes			
Trichomaris invadens	Snow crab	Black mat disease	Alaskan waters
Ichthophonus spp	All fish	Visceral mycosis	Marine, esp flatfish

PATHOGEN	HOST	DISEASE	ENVIRONMENT
PROTOZOA			
Flagellates			
Oodinium spp	All fish	Velvet disease	Ubiquitous, aquaria
Ichthyobodo	All fish	Skin and gills	Freshwater
Ciliates			
Trichodina spp	Fish, molluscs	Skin and gills	Freshwater
Ichthyophthirius sp	All fish	Skin	Freshwater
Cryptocaryon	Marine fish	Skin	
Epistylis	Fish & crustaceans	Skin & exoskeleton	Ubiquitous
Zoothamnium	Crustaceans	Exoskeleton	Marine
Apicomplexa			
Eimeria	All fish	Gut, systemic	Ubiquitous
Perkinsus	Oysters, clams	Systemic	N Atlantic, Australia
Microsporida			
Nosema	Fish & crustaceans	Systemic	Ubiquitous
Pleistophora spp	Fish & crustaceans	Systemic	Ubiquitous, aquaria
Thelohania contejeanii	Crayfish	Porcellain or cotton dis	Ubiquitous
Bonamia spp	Oysters	Haemocytes	Europe,US(W),N. Zealand
Microcytos spp	Oysters	Haemocytes	USA W. coast, Australia
MYXOZOA			
Myxobolus spp	All fish	Systemic	Freshwater
M. cerebralis	Salmonids	Whirling disease	Freshwater
Ceratomyxa shasta	Salmonids	Systemic	Freshwater
Henneguya spp	All fish	Systemic	Ubiquitous
PKX agent	Salmonids	Proliferative kidney dis	Europe, N America
Marteilia spp	Oysters	Gut and digestive gland	Europe, Australia
MONOGENEA			
Dactylogyrus spp	All fish	Gills	Freshwater
Gyrodactylus spp	All fish	Skin and gills	Freshwater
TREMATODA			
Diplostomum spp	All fish	Eyes	Freshwater
Bucephalus spp	Cyprinids	Internal tissues	Europe, freshwater
Ligula intestinalis	Cyprinids	Peritoneal cavity	Freshwater
Bothriocephalus sp	Predator fish	Intestine	Freshwater
NEMATODA			
Camallanus	All fish	Digestive tract	Ubiquitous, aquaria
ANNELIDA			
Leeches	All fish	Skin and gills	Freshwater
CRUSTACEA			
Branchiura			
Argulus	All fish	Skin	Freshwater
Copepoda			
Lernea spp	All fish	Skin	Ubiquitous
Ergasilus	All fish	Gills	Freshwater
Lepeophtheirus	Marine fish	Skin	Marine
Isopoda	Marine fish	Skin	Marine

Mycoses are significant in all aquaculture species. Oomycetes produce superficial infections in fresh water fish, particularly when stress or superficial wounds occur. Such infections are susceptible to chemotherapy As with bacterial infections, mycoses are primarily diseases of hatchery operations in invertebrate culture and can be serious for crustacean larval production. Higher fungi, principally Hyphomycetes, result in systemic infections in fish and crustaceans. No effective therapy is available for this type of deep seated infection.

Protozoans are responsible for a range of infections from topical, often opportunistic, to systemic. Again, as with fungi, topical infections are controllable by chemotherapy although the range of available treatments is limited and have distinct environmental problems associated with their use.

Systemic infections, largely associated with Myxosporida and Microsporida are responsible for some of the most serious losses in fish and shellfish, are much more refractory to treatment and their practical significance is enhanced by the lack of effective chemotherapy.

Although a great many metazoan parasites of fish have been described, most have a multiple host life cycle which cannot easily be completed in intensive farm conditions. Some cause problems in extensive and cage culture and external parasites can proliferate in all environments where convenient fish species favour their development, and represent a constant threat in all kinds of facilities. Direct development and propagation allow quick spread of monogenetic trematodes. Some parasitic crustaceans can also find suitable conditions for their development in fish farms, particularly when naturally infected waters exist nearby. A number of treatments have been proved effective against fish ectoparasites, but their use is limited by many environmental factors.

Drug Groups

The use of chemicals in fish culture has a long history, commencing with attempts to control ectoparasites with salt baths. The first recorded use of formalin was in 1909 to control *Costia* infections (26). Other topical "disinfectants" including copper and malachite green came on the scene in the 1920's and 1930's and quaternary ammonium compounds in the 1940's (2).

Attempts at therapy of systemic bacterial diseases commenced in the late 1930's and suphamerazine was introduced in the USA in 1948 (15). As new drugs were introduced into human and veterinary medicine, suitable representatives were soon investigated for application to fish farming. Despite the enormous expansion of the aquaculture industry throughout the world in the last 30 years, and despite the wide range of drugs suggested in the scientific literature as being suitable for use in aquaculture, the range of veterinary medicines which are legitimately available to fish farmers is still very limited.

Table II Major antimicrobial drugs used in aquaculture.

	PRODUCT	ROUTE	DOSE	INDICATION
ANTIBIOTICS				
ß lactams	Ampicillin Amoxycillin	oral oral	50-80mg/kg 10days 50-80mg/kg 10 days	Gram -ve bacteria Gram -ve bacteria
Aminoglycosides	Neomycin Kanamycin	(oral ((bath	50-80mg/kg 10 days 20mg/l	Gram -ve bacteria
Tetracyclines	Tetracycline Oxytetracycline Doxycycline	(oral ((bath	50-80mg/kg10 days 20mg/l	Gram -ve bacteria
Macrolides	Erythromycin	(oral	50mg/kg 10 days (bath (eggs)	Bacterial kidney 2mg/l 1 h
Non-classifiable	Chloramphenicol	(oral ((bath	50-80mg/kg10 days 20mg/l	Gram -ve bacteria
SYNTHETIC ANTIBACTERIAL AGENTS				
Sulphonamides	Sulphamethazine Sulphadimethoxine Sulphaguanidine	((oral ((	200mg/kg 10 days	Gram -ve bacteria
Potentiated sulphonamides	Trimethoprim + sulphadiazine	oral	50mg/kg 10 days	Gram -ve bacteria
Nitrofurans	Furazolidone Furaltadone Nifurpirinol	(oral ((oral (bath	50-80mg/kg 10 days 10-50mg/kg 10 days	Gram -ve bacteria Gram -ve bacteria
Quinolones	Oxolinic acid Flumequine	((oral	12mg/kg, 10 days	Gram -ve bacteria

The legal constraints on the availability of fish medicines which are considered below, inevitably are of increasing importance for the future, but are only partly responsible for the current limited range of fish medicines in widespread use. If the range of potential fisheries therapeutants which have been proposed in the scientific literature were listed, rather than those actually in use, a book rather than a short list would result (20, 21) . There are therefore only a few major drug groups to consider. The major antimicrobial groups are listed in Table II. The numerous "topical disinfectants" and miscellaneous antiparasitics may be found in detail in Herwig (20) and Hoffman and Meyer (21).

1. "Topical disinfectants": A disparate group including salt, formalin, malachite green, copper sulphate, quarternary ammonium compounds etc. Of various degrees of efficacy, these are partly interchangeable products employed to eliminate topical parasites such as opportunistic bacteria, protozoans and fungi. Some, such as malachite green, have potentially adverse environmental impacts.

2. Organo-phosphates: Primarily dichlorvos used to control sea lice parasites in sea cages. Although essential to production of salmon, the use of dichlorvos is regarded by many as having potentially serious environmental effects and as being in urgent need of replacement.

3. Antimicrobials:-

a. Sulphonamides: A group of antimicrobial agents that act by blocking the folic acid pathways of susceptible micro-organisms. Sequential blocking of that pathway can produce synergism, hence the use of potentiated sulphonamides which have been widely used in the last 15 years in many countries.

b. Tetracyclines: A group of broad-spectrum antibiotics whose action is mainly bacteriostatic and which interfere with bacterial protein synthesis by binding to the bacterial 30S ribosomal subunit. Effective against a wide range of gram-negative and gram-positive bacteria, the most widely used tetracycline in aquaculture is oxytetracycline, although doxycline and chlortetracycline have been used.

c. 4-Quinolones: First introduced into aquaculture in Japan, this group of antimicrobials is very active against gram-negative bacteria. Oxolinic acid and flumequine have achieved wide usage in Europe and there is considerable interest in the newer 4-quinolones which are now beginning to become available. Antimicrobial action is both bacteriostatic and bacteriocidal, by interference with bacterial DNA gyrase. Cross-resistance between the different 4-quinolones can occur easily.

d. Nitrofurans: Nitrofurans such as nitrofurazolidone and nifurpirinol were once thought to hold considerable promise for use in aquaculture. In many countries their use is now discouraged.

e. Erythromycin: A macrolide antibiotic, erythromycin is the only drug with reported effect against Renibacteriosis. Its ability to prevent vertical transmission is still in dispute, although recent results are encouraging (8). The related drug, vancomycin, has been used in Canada to control gaffkaemia in lobsters.

f. Chloramphenicol: Achieved wide usage in a number of countries including the USA and France. Its use is now strongly discouraged in most countries because of its potential value as a human medicine.

4. Others: A range of anthelminthics (e.g. metronidazole) have been proposed for use with fish, but in general these have not achieved wide acceptance or use.

Regulatory Constraints

These undoubtedly represent the most basic limitation on the availability and use of any medicinal product in aquaculture. If effective controls prevent a fish farmer from obtaining a medicine, it cannot be used, no matter how effective a therapeutant it may be.

Legal constraints on the supply and use of veterinary medicines now exist in most countries. However, in many cases, these regulations were formulated with 'conventional' veterinary medicine (i.e. medicines for farm mammals) in mind and either do not cover fish (e.g. The Netherlands) or have some minor limitations in their application to fish medicines (e.g. the UK Medicines Act 1968 (13) which does not apply to fish eggs or to fish fry before they feed independently). Such regulations vary from strict to lax, although it must be borne in mind that nominally strict regulations may not be enforced rigourously. Often, some form of derogation (temporary or otherwise) may be applied for aquaculture where the effect of new regulations not strictly designed for that market would be too restrictive.

The closest control on the use of medication in aquaculture appears to be that applied in Norway, where all medicines are supplied under the control of the Norwegian Medical Control Authority which maintains a centralised record of the amount of drugs prescribed (12). Norwegian workers are currently exploring the practicabilities of retrieving uneaten medicated feed from farm cages in order to recycle it and reduce the potential environmental impact.

In the UK, all products for which a medicinal claim is made or which are used for a medicinal purpose must hold a veterinary product license. Some compounds such as antibiotics may not be supplied for any purpose (except research) except under prescription. Other compounds which have non-medicinal use (e.g. formalin) require licensing only when supplied or used for a veterinary medicinal use. Not all licensed medicines require veterinary prescriptions before being used for a medicinal purpose. Requirements of efficacy and safety (including consumer and environment) must be satisfied before any medicine may be licensed and the requirements to be satisfied have adapted to reflect changing and increasing usage of aquaculture medicines over the years. Further legislation exists (14) which controls discharges from fish farms and imposes additional constraints on the presence of any medicines in that water.

In the USA, the Food and Drug Agency exercises a similar regulatory control over the registration of veterinary medicines for aquaculture (38). Products such as salt have been registered with the support of the US Fish and Wildlife Service in a category known as Generally Regarded as Safe (GRAS). Others such as malachite green are currently permitted only under strict control for limited use as Investigational New Animal Drugs (INAD). In the case of malachite green, the permit requires that the malachite green be removed from the effluent water of hatcheries permitted to use it by means of carbon filters before the water may be discharged. More details are given elsewhere in these proceedings (Schnick). Canadian authorities are introducing similar regulations for aquacultural medicines and the possibility of harmonising regulations to match those of the USA is under consideration.

In Europe, the harmonisation of veterinary medicines regulations is underway in preparation for the open market at the end of 1992. Such harmonisation may lead to a greater uniformity in the use of aquaculture medicines, but only if the harmonised regulations are applied and policed properly. More details are given elsewhere in these proceedings (Hankin).

The object of all such medicines licensing procedures, harmonised or otherwise, is to ensure that veterinary medicines are effective and safe in use, that their use will not result in unacceptable residues in fish meat or damage to the environment. Many of the aspects which present special problems for the licensing of aquacultural medicines are considered below.

Constraints on Route of Application

In addition to legal controls, the use of chemotherapy in aquaculture is limited by numerous practical constraints. In theory almost all treatments can be administered by parenteral injection, as is usual in veterinary practice. However, with fish this is costly in time and labour, requiring that fish be caught and anaesthetized with inevitable stress associated. Therefore, parenteral treatments can only be applied to fish of high individual value, such as salmon and sturgeon broodstock, or ornamental fish. Since most disease problems of invertebrate culture which are readily susceptible to chemotherapy involve larval stages in the hatchery, the parenteral route is not relevant, with the exception of valuable lobster stocks which can repay treatment by the injection route in the presence of gaffkaemia.

Most fish treatments are carried out on farms producing table fish and on animals of low inherent individual worth. Under such conditions, not only must the route of application be economic to apply, but the cost of the drug itself will often be of overriding importance.

Ideally, the selection of a therapeutic regime should follow upon the identification of the specific pathogen responsible by reliable diagnostic procedures, and by reference to information on the efficacy of available drugs against that organism. Selection should also take into account the bioavailability of the drug, the concentrations at which it accumulates in target tissues, and the rate at which it is eliminated. Finally, in an ideal world, the environmental conditions prevailing at and near the farm site should be considered, in order to take into account any effects of water chemistry on the efficacy and toxicity of the drug.

In the real world of fish production; however, most of these conditions are difficult if not impossible to comply with, particularly during an active fish mortality, when speed and cost of treatment is paramount.

Bath treatments, although easy to carry out with products of high solubility by immersing fish in concentrated solutions or adding the drugs directly to the fish's environment, are mainly restricted to recirculating units or to tanks of limited size for reasons of cost as much as for potential environmental impact. As the ways in which drugs are absorbed from water by fish are only partly understood (and that for a few drugs only), it is difficult to determine dose to ensure appropriate tissue concentrations of active product with the bath method. When a bath treatment is carried out by means of flush treatment (i.e. in flowing water) the dose determination becomes even more difficult. Bath treatments have therefore found their most widespread application in the treatment of topical infections and in particular very small animals (shellfish, shrimp, fish fry) or fish suffering skin and gill infections are likely to be treated with success using this route of administration.

Perhaps the best way to ensure good uptake of controlled amounts of chemicals is by the oral route, particularly for antibacterials. It permits treatment of large numbers of fish in intensive culture conditions with minimal waste of the product, so that even high-value compounds can be employed in cases of emergency. Inevitably, there are limitations to the oral route of administration. Thus absorption from the gut must be adequate to achieve therapeutic levels in target organs. Some compounds can render medicated feed unpalatable and palatability can vary from species to species. Thus chinook salmon accept the addition of nifurpirinol to their diet, whilst coho salmon refuse to take food treated with it (5). Oral administration of drugs has become the prime route of medication for finfish, but is rarely applied to invertebrates. This is chiefly due to the feeding behaviour of Decapoda, which results in incomplete consumption of the diet and renders any treatment rather expensive. With fish, diseased animals loose appetite rapidly and therefore any delay in applying treatment reduces efficacy and increases cost.

Different factors combine to make dose control more difficult in fish than in homeothermic species. Compounds added directly to water may be bound or inactivated by organic matter and inorganic ions (e.g. oxytetracycline (22)). The efficacy and toxicity of certain products is known to change according to the water quality, so much that in complex environmental systems the fish farmer may be unable to predict the full effects of a treatment. Feeding habits of fish also introduce degrees of uncertainty in the dosage of medicinal compounds. Many species exhibit hierarchical behaviour which results in competition for food and thus in uneven distribution of the drug within stock. To take this into consideration, and to compensate additional losses of active molecule by dissolving in surrounding water, it is often normal to supply higher doses than those recommended for mammals, even though such practice has the disadvantage of increasing the potential impact of drug use on the environment and the risks of persistent residues.

Despite such difficulties, it is important that therapeutic regimes be designed to minimalise the risks of the appearance of resistant pathogens. The best approach, in conventional medical and veterinary practice, is to carry out therapy only when the target pathogen and its drug sensitivities have been properly identified. In particular, speculative chemoprophylaxis, to prevent the possible development of disease, should never be employed in the aquatic environment which is an ideal one for the development and spread of drug resistance.

The fact that it is rarely possible to determine the probable efficacy of any proposed therapy before it is carried out is one of the most difficult aspects of chemotherapy in aquaculture. Even if knowledge from previous outbreaks or from rapid identification of the causative organism can sometimes provide pointers to the selection of a therapeutant, there still remains a significant risk of failure, and the requirement to change the drug when the laboratory result is known cannot be eliminated. Such a procedure does not conform to good practice for antibiotic therapy, but on the other hand, it is unrealistic to expect a fish farmer to remain inactive whilst his fish are dying. It must be borne in mind that practical and economic factors mean that fish treatments cannot in practice always adhere to the same rules as those which should prevail in terrestrial veterinary medicine.

Resistance, Environmental and Residue Constraints

Despite the long history of the use of medicines in aquaculture, it was only after 1960, when the rapid expansion of fish farming coincided with an increasingly enthusiastic application of antibiotics, that some observers began to point out the potential hazards and side effects of drug use in the aquatic environment. As in the case of other farmed animals, attention was drawn at first to the increasing occurrence of drug resistance. Such resistance can rapidly reduce the ability to control disease and, it was considered, hold the potential for transmission into human pathogenic bacteria (28, 44). More directly, there were concerns arising from the possibility of unacceptable drug residues in the flesh of marketable fish (10, 17, 19, 27). The possible environmental consequences of repeated discharges of chemotherapeutants in fish farm effluents, have been considered more recently (23, 31, 36, 40).

Drug resistance problems have become increasingly evident in the last ten years and the reasons for this are not difficult to find. The availability of only a few compounds really suited to control of aquatic pathogens does not encourage fish farmers to vary successful treatments. The approach used is always likely to be "if it works don't change it", rather than rotating the available therapeutants in an attempt to prevent single drug resistance becoming established. Antimicrobials are susceptible to the rapid emer-gence of bacterial resistance, and there is now sufficient information to confirm that the development of resistance to antibiotics in the aquatic environment correlates with intensive repeated therapeutic use (3, 4, 42).

One of the most serious examples of this problem has occurred recently in marine salmon farms in Scotland. Only two antibiotics were licensed for use to control furunculosis on these sites : oxytetracycline, for which transmissible plasmid mediated resistance is well known, and oxolinic acid. The latter was used extensively during the rapid expansion of the salmon farming industry in Scotland in the 1980's and, as a result the occurrence of chromosomal resistance emerged as a major problem (16, 42). Repeated use of single therapeutants and insufficient effort to prevent the spread of resistance now means that none of the licensed drugs for control of furunculosis is effective.

When new drugs become available, there is a strong tendency for them to be taken up by industry as a fad or fashion and no other will do. This again is the route to single drug use, drug resistance and failure. Until 1980, French fish farmers frequently used chloramphenicol against bacterial fish diseases. In 1981, this practice was severely attacked in the press, producing a consumer reaction which resulted in a marked drop in trout consumption. At the same time, the 4-quinolones antimicrobials began to become available (particularly flumequine in France) to control gram-negative bacterial disease in fish. In less than a year, the use of chloramphenicol had been given up by fish farmers, and flumequine had become the most popular treatment in France.

The rapid introduction of the 4-quinolones resulted from newly published scientific observations on their efficacy. This example shows that scientific investigations have a positive role to play in the development of fisheries chemotherapy. However, misinterpretation of scientific data may have very different consequences. A striking illustration of this point has been recorded, again in France and again with choramphenicol, where the recent demonstration (and perhaps new occurrence) of 'cold water disease' caused by *Cytophaga psychrophila* is accompanied by significant losses of fry in intensive trout farms. The special requirements of *C. psychrophila* in artificial culture media make antibiotic sensitivity studies difficult and, indeed, few compounds seem to control the disease *in vivo*. Experimental trials reported in a recent paper (7) seemed to indicate that only tetracyclines and chloramphenicol at high doses were likely to be effective. The main consequence of this paper was to restore old malpractices in the form of the use of chloramphenicol at high doses on fish farms.

Unrestricted use of drugs has also become established in many shrimp farms in tropical regions. Infection by *Vibrio* spp can be a major limiting factor in the production of shrimp larvae. Repeated prophylactic use of antimicrobials, to prevent bacterial infections, was adopted early on and is now a permanent part of production. Regular addition of tetracyclines to the culture water is also thought to provide better growth of shrimp and this aspect has encouraged the abusive spread of such practice. The inevitable result has been the selection of resistant *Vibrio* spp.

The use of antimicrobials, as growth promoters in fish feed, has been a matter of debate. Although some workers have demonstrated positive growth effects with tetracyclines (1, 34, 41), others have not (39, 43). There is now reason to believe that fish farmers in some countries may be employing antimicrobials as growth promoters, but there is currently no means of estimating the degree of such usage. Bearing in mind the problems which have occurred with therapeutic use, it is certain that a widespread and regular supplementation of diet with drugs will result in serious problems in the areas concerned, particularly if drugs such as tetracyclines, which have major use as aquaculture therapeutants, are used in this way. Workers investigating the growth promotion potential of antimicrobials in aquaculture should bear in mind the major responsibility bearing on those who advocate such use in the face of the damage which may result.

Similarly, the prophylactic use of drugs, which is regarded as being unacceptable practice in other fields of veterinary medicine, cannot be regarded as an acceptable practice in aquaculture. The fact that most of the specific fish pathogens survive in infected areas, in the environment or in carrier fish, leads fish farmers to anticipate outbreaks of disease at critical times of year and at times of stressful farm operations. Where the pattern of disease on a site is well understood and can be predicted reliably, then prophylactic treatments may be acceptable. In Scotland, such treatments have been employed on fresh-water salmon rearing sites with endemic furunculosis. Fish are treated with a therapeutic dose of antibiotic immediately before transfer to the marine environment. Such treatment in fresh water is less expensive in terms of the amount of drug required there, as opposed to the amount required in sea water to achieve the same serum levels. However, such treatments have not prevented the development of drug resistant strains of *Aeromonas salmonicida* which are now capable of transmitting infections in sea water

The need for prophylactic treatment is perhaps better justified for external parasites, which cannot be eliminated completely from the farm environment. In these cases, treatments mainly aim to control the parasite population level and must be repeated at convenient intervals of time to avoid excessive proliferation and ensuing lesions. Infestation of salmon with sea lice is perhaps the prime example of this problem. The periodicity will be a function of the active multiplication of the parasites, and of course, of other environmental parameters. In this instance, the limits of acceptable practice and unacceptable practice are less easy to determine!

Whilst the problem of drug residues in fish meat must be addressed in assessing acceptability of compounds as aquaculture medicines, the subsequent monitoring of residues lies more in the area of food safety authorities, which of course represent a further regulatory area which impinges on aquaculture.

There appears to be no reason why residue tolerances in fish meat should differ from those for other meat. The specific problem relating to aquaculture is that the animals concerned are poikilotherms whose metabolic rates are largely controlled by the temperature of the water environment. Since an increase of 1°C in water temperature can increase metabolic rates by 10%, the pharmacokinetic parameters of uptake, distribution and elimination may be expected to vary markedly with water temperature. It is normal in licensing veterinary medicines to impose a minimum time, a with-holding period, between last drug treatment and slaughter for human consumption, expressed in terms of days. Whilst this is acceptable for homeotherms in which metabolic rates and therefore drug residue elimination rates are not environmentally determined, it is not for poikilotherms, where environmental temperatures will play a critical role.

Undoubtedly, many of the early aquaculture drugs were permitted on the basis of fixed number of days with-holding periods. However, initial pharmacokinetic studies clearly demonstrated that drug residues were more persistant in colder water, with a demonstration of a near 10% per 1°C for oxytetracycline (37). These authors recommended a with-holding period of 60 days at temperatures above 10°C for oxytetracycline and potentiated sulphonamide and 100 days at water temperatures below 10°C for oxytetracycline. They proposed that potentiated sulphonamides should not be used below 10°C, because of residue persistence at lower temperatures.

Whilst fixed but temperature linked with-holding periods may be suitable for relatively constant temperature waters, rapidly fluctuating water temperatures are found on many fresh-water fish farms in temperate climates where a drop in water temperature from 16°C to 8° in less than 60 days occurs every autumn. Which procedure should be adopted with sulphonamide therapy in such a case - maintain the 60 day period or 'stop the clock' until water temperatures increase again? To answer this problem in the UK, a tentative move was made towards the introduction of the degree day with-holding period. This is a concept familiar to fish farmers where the age of developing fish eggs is frequently expressed in degree days, the figure being attained by adding the water temperature for each day cumulatively during the developmental period. The degree day approach enables a single with-holding period to be set which can be applied to any farm at any time of year with a reasonable expectation that it may more accurately reflect the varying residue excretion rates being imposed by varying water temperatures.

The degree day concept requires pharmaceutical manufacturers to provide data on the performance of their products at more than one water temperature in order to demonstrate that there is a clear temperature related effect on the elimination rate for that product. Although there is at the moment only limited scientific information to support the approach, the concept has attracted considerable interest and is currently proposed as one approach to the harmonisation of licensing requirements for aquaculture medicines within the EC. Ensuring that regulations on with-holding periods remain simple and easy to apply is essential both to the progress of aquaculture medicines and to consumer protection. Complex and impractical requirements are an invitation to evasion.

The protection of the environment against adverse effects of the use of veterinary medicines on fish farms is a less easy matter, not so much because there is clear evidence of any adverse effect, but more because of the difficulty investigating the problem. Many of the more obvious potentially adverse effects (e.g. carcinogenicity, teratogenicity) will already form part of the standard requirements for evaluation for any veterinary medicine. Longer-term effects on flora and fauna will require long-term evaluation and may be difficult if not impossible to separate from the other impacts of the fish farm on the aquatic environment. Also difficult to separate will be any effects from other sources of similar or identical drugs from human and land farming sources via sewage discharges or run-off. For these reasons, no good study has yet been published which adequately addresses this topic, although some approaches to the problem are considered in contributions to this conference.

Conclusion

The increasing recognition of the need to regulate the use of aquaculture medicines is reflected in the increase in the number of scientific publications on the subject describing the associated problems and risks. It is also reflected in a change in the nature and the orientation of those publications. There is now an increasing emphasis on drug kinetics and particularly on drug residue elimination. Attention is also being paid to the environmental aspects of the use of medicines because of a general feeling that some adverse effects, beyond the simple resistance and residue problems, must exist. Whether this really is so is a matter for the future. The conference which this paper introduces is, in itself, an indication of the increasing importance of the problems real or perceived in aquacultural medicine, in that it is the first international conference on the topic sponsored by the OIE.

References

1). AHMAD, T.S. & MATTY, A.J. 1989 - The effects of feeding antibiotics on growth and body composition of carp (*Cyprinus carpio*). *Aquacult.*, **77**, 211-220

2). ALDERMAN, D.J. 1988. - Fisheries chemotherapy. In R.J. Roberts. *Recent advances in Aquaculture*. 3, Croom Helm., London, 1-61.

3). AOKI, T., KITAO, T. & KAWANO, K. 1981 - Changes in resistance of *Vibrio anguillarum* in cultured ayu, *Plecoglossus altivelis* Temminck and Schlegel, in Japan. *J. Fish Dis.*, **4**, 223-230.

4). AOKI, T., KITAO, T., IEMURA, N., MITOMA, Y. & NOMURA T. 1983. - The susceptibility of *Aeromonas salmonicida* strains isolated in cultured and wild salmonids to various chemotherapeutics. *Bull. Jap. Soc. Sci. Fish.*, **49**, 17-22.

5). AMEND, D.F. 1972. - Efficacy, toxicity and residues of nifurpirinol in salmonids. *U.S. Dept. Int., Bureau Sport Fish. Wildl., Tech. Pap.* **Nº 62**, 13 pp.

6). AUSTIN, B. 1985. - Chemotherapy of bacterial fish diseases. In A.E. Ellis, *Fish and shellfish pathology*, Academic Press, London, pp. 19-26.

7). BAUDIN-LAURENCIN, F., VIGNEULLE, M., CASTRIC, J. & TIXERANT, G. 1988. Myxobacteriose viscérale de la truite arc-en-ciel, essais thérapeutiques. *La Piscic. fr*, **94**, 5-7.

8). BROWN, L.L., ALBRIGHT, L.J. & EVELYN, T.P.T., 1990. - Control of vertical transmission of *Renibacterium salmoninarum* by injection of antibiotics into maturing female coho salmon *Oncorhynchus kisutch*. *Dis. Aquat. Org.*, **9**, 127-131

9). ELLIS, A.E. 1988. - *Fish vaccination*. Academic Press, London, 255pp

10). FRIBOURGH, J.H., ROBINSON J.A. & MEYER F.P. 1969. - Oxytetracycline residues in tissues of blue and channel catfishes. *U.S. Dept. Int., Bur. Sport Fish. Wildl., Tech. Pap.* **Nº 38**, 7 pp.

11). GRATZEK, J.B. 1983. - Control and therapy of fish diseases. In C.E. Cornelius & C.F. Simpson, *Adv. Vet. sci. comp. med.*. **27**, 297-324.

12). GRAVE, K., ENGELSTAD, M., SØLI, N.E. & HASTEIN, T. 1990. - Utilization of antibacterial drugs in salmonid farming in Norway during 1980-1988. *Aquacult.*, **86**, 347-358

13). GREAT BRITAIN - Parliament 1968. - The Medicines Act 1968, 165pp.

14). GREAT BRITAIN - Parliament 1989. - The Water Act 1989, 419pp.

15). GUTSELL, J.S. 1948. - The value of certain drugs, especially sulfa drugs, in the treatment of furunculosis in brook trout (*Salvelinus fontinalis*). *Trans. Am. Fish. Soc.*, **75**, 186-199

16). HASTINGS, T.S. & McKAY, A. 1987. - Resistance of *Aeromonas salmonicida* to oxolinic acid. *Aquacult.* **61**, 165-171.

17). HEATON, L.H. & POST, G. 1968. - Tissue residues and oral safety of furazolidone in four species of trout. *Progve. Fish Cult.*, **30**, 208-215.

18). HERMAN, R.L. 1970. - Chemotherapy of fish diseases: a review. *J. Wildl. Dis.*, **6**, 31-34.

19). HERMAN, R.L., COLLIS, D. & BULLOCK, G.L. 1969. - Oxytetracycline residues in different tissues of trout. *U.S. Dept. Int., Bur. Sport Fish. Wildl., Tech. Pap.* Nº **37**, 6 pp.

20). HERWIG, N. 1979. - *Handbook of drugs and chemicals used in the treatment of fish diseases*. Charles C. Thomas, Springfield, Ill., 272pp

21). HOFFMAN, G.L. & MEYER, F.P. 1974. - *Parasites of freshwater fishes.* TFH Publications, Neptune City, New Jersey, 224pp.

22). JACOBSEN, F. & BERGLIND, L. 1988. - Persistence of tetracycline in sediments from fish farms. *Aquacult.*, **70**, 375-370.

23). DE KINKELIN, P. & HEDRICK, P. 1991. - International veterinary guidelines for the transport of live fish or fish eggs. *Ann. Rev. Fish Health,* (In press)

24). DE KINKELIN, P. & MICHEL, C. 1984. - *Symposium on fish vaccination*. O.I.E., Paris, 248 pp.

25). DE KINKELIN, P., MICHEL C. & GHITTINO P. 1985. - *Précis de pathologie des poissons*. I.N.R.A., O.I.E., Paris, 348 pp.

26). LEGER, P. 1909. - La costiase et son traitement chez les jeunes alevins de truite. *C. R. Seances Acad. Sci. Paris, Series D*, **148**, 1284-1286.

27). McCRACKEN, A., FIDGEON, S., O'BRIEN, J.J. & ANDERSON, D. 1976. - An investigation of antibiotic and drug residues in fish. *J. Appl. Bact.*, **40**, 61-66.

28). MICHEL, C. 1986. - Practical value, potential dangers and methods of using antibacterial drugs in fish. *Rev. Sci. Tech. Off. int. Epiz.*, **5**, 659-675.

29). POPOFF, M. & DAVAINE, Y. 1971. - Facteurs de résistance transférables chez *Aeromonas salmonicida*. *Ann. Inst. Pasteur*, **121**, 337-342.

30). RASMUSSEN, F. 1988. - Therapeutics used in fish production. Pharmacokinetics, residues, withdrawal periods. *EIFAC/XV/88*, Inf.**13**, 1-25.

31). RASMUSSEN, F. & FIJAN, N. 1990. - Report of the working party on the withdrawal periods for drugs. *EIFAC/XVI/89*, Inf.**13**, 1-6.

32). REYNOLDS, J.E.F. 1989. - *Martindale, the extra pharmacopoiea.* The Pharmaceutical Press, London, 1896pp.

33). ROBERTS, R.J. 1978. - *Fish pathology*. Baillère-Tindall, London, 318 pp.

34). RIJKERS, G.T. TEUNISSEN, A.G. VAN OOSTEROM, & R. VAN MUISIWINKEL, W.B. 1980. - The immune system of cyprinid fish. The immunosuppressive effect of the antibiotic oxytetracycline in carp (*Cyprinus carpio*). *Aquacult.*, **19**, 177-189.

35). RUCKER, R.R. 1972. - Fish disease therapy, past, present and future. In R.W. Moore, *Progress in fishery and food science. Univ. Wash. Publ. Fish., New series* **5**, 135-140.

36). SAMUELSEN, O.B. 1989. - Degradation of oxytetracycline in sea-water at two different temperatures and light densities and the persistence of oxytetracycline in the sediment from a fish farm. *Aquacult.*. **83**, 7-16.

37). SALTE, R. & LIESTØL,, K. (1983).- Drug withdrawal from farmed fish: depletion of oxytetracycline, sulfadiazine and trimethoprim from muscular tissue of rainbow trout (*Salmo gairdneri*). *Acta Vet. Scand.*, **24**, 418-430

38). SCHNICK, R.A. 1988. - The impetus to register new therapeutants for aquaculture. *Progve. Fish Cult.*, **50**, 190-196.

39). SNIESZKO, S.F. 1957. - Use of antibiotics in the diet of salmonid fishes. *Progve. Fish Cult.*, **18**, 81-84.

40). STUART, N.C. 1983. - Treatment of fish disease. *Vet. Rec.*, **112**, 173-177.

41). SUKHOVERKHOV, F.M. 1967. - The effect of cobalt, vitamins, tissue preparations and antibiotics on carp production. Proc. World Symposium. on Warm water pond fish culture. *FAO Fish. Rep.*, **44**, 401-407

42). TSOUMAS, A., ALDERMAN, D.J. & RODGERS, C.J. 1989 - *Aeromonas salmonicida*: development of resistance to 4-quinolone antimicrobials. *J. Fish Dis*. **12**, 492-507.

43). WAGNER, E.D. 1954. - The effects of antibiotics and arsanilic acid on the growth of rainbow trout fingerlings. *Progve. Fish Cult.*, **16**, 36-38.

44). WATANABE, T., OGATA, Y., AOKI, T. & EGUSA, S. 1969. - Studies on the drug resistance of fish-pathogenic bacteria. I. Detection of R factors. *Jap. J. Bacteriol.*, **25**, 42-43.

Discussion

FIJAN What do you think about the influence of dosage on the emergence of resistance in bacteria?

MICHEL I am not sure that the influence of dosage is very important, because it has been demonstrated that even low residues of antimicrobial drugs in water are sufficient to produce resistance. It is very difficult to visualise because as I said, you have different behaviour of fish which prevents you from knowing exactly what is the dose of product which is absorbed and according to the different conditions under which fish are kept, you can also have difference in the metabolism, so that you cannot avoid some leakage of products into the water.

ALDERMAN One of the main problems in terms of dosage is the question of prophylactic use of drugs. In other words, you expect a certain problem to occur on a particular farm site at any one time and therefore you treat before the disease appears. That approach is dangerous in the extreme and is certainly likely to increase resistance. Nonetheless, one can understand why it may happen. If a farm has a number of very valuable fish that they cannot afford to loose, they will worry about being able to get the drug into the fish before they start to loose appetite and tend to use prophylaxis. The second difficult area relating to dosage is in regard to resistance and the need to ensure rotation of drugs used. We have relatively few drugs available, but if prescribers were to use different groups of drugs alternately, then they would reduce the chance for resistant strains to develop and to survive. Encouraging this to is none to easy and the general tendancy is for a farm has to use an effective drug until resistance problems eliminate it as a useful drug.

LAFONT A question and a comment. I was rather surprised to see your slide about the MICs of the various quinolone antibiotics against *Aeromonas salmonicida*. Baytril is a fluorinated quinolone and it appeared that as having higher MICs than oxolinic acid which is a first generation quinolone. Is it usual to have such results with a quinolone?

ALDERMAN I think you may have misinterpreted the slide. Baytril is a fluorinated quinolone which is much more active against *Aeromonas salmonicida*. Although there were a few strains with resistance to baytril, those strains were much more resistant to oxolinic acid. MICs for baytril were always much lower.

LAFONT I agree that shifting antibiotics is the best way avoiding resistance problems as you suggested, even if it is very difficult procedure to apply.

LEWIN I think that Dr Lafont has made an interesting point. If you look at the activity of the fluoroquinolones compared to oxolinic acid against Enterobacteriaceae and other gram negatives, you usually get a thousand fold increase in activity. In the case of baytril and other fluoroquinolones such as sarafloxacin, you get something like a ten fold to fifteen fold increase in MICs against fish pathogens.. They are somewhat less active than one would expect and perhaps oxolinic acid is somewhat more active at first than one would expect in the aquatic environment.

CURRENT PRACTICES OF CHEMOTHERAPY IN FISH CULTURE

H.J. Schlotfeldt

State Fish Epidemics Control Service of Lower Saxony and Fish Health Service
Eintrachtweg 17, D-3000 Hannover 1, Federal Republic of Germany

Summary: *Until recently the use of chemotherapeutants by bath or medicated feed was characterized in the European Community (EEC) by an amazingly broad and in some ways heterogenous spectrum. This situation was not very different in fish producing countries outside the EEC. The presently used drugs for fish in the EEC areas as well as in Norway, Iceland, Canada and the USA are outlined below.*

The manner of the therapeutical efficacy of drugs in fish is often poorly understood. The legal aspects of drug application, registration and licensing is also surprisingly heterogenous both in-and-outside the EEC.

The main methods of practice of the application of chemotherapeutants in fish culture are outlined with particular reference to practices in the Federal Republic of Germany (FRG). The efficacy of chemotherapy in fish is briefly considered and its efficacy under practical conditions is discussed. The main needs for standardisation and safety of drugs used in aquaculture are itemized and the over all situation in view of the EEC market without national boundaries outlined. The possible future requirements in the EEC area of drugs used in fish and protection for, and compatibility with the environment are discussed.

***Résumé*: Applications de la thérapeutique aux élevages de poissons**

Les applications thérapeutiques chez les poissons sont dictées par le coût des traitements et par la commodité d'administration à de grands effectifs. Seules la voie orale et la balnéation sont couramment utilisées, l'injection parentérale étant réservée à des cas particuliers. Selon les pays de grandes variations existent dans les listes de produits autorisés et dans les procédures d'autorisation, ceci aussi bien dans la CEE qu'à l'extérieur.

En Allemagne, on distingue des produits ayant reçu leur autorisation de mise sur le marché (1 seule AMM à ce jour!), des produits bénéficiant d'une autorisation provisoire, et des produits traditionnels d'usage externe, dont la validité sera directement entérinée par une commision fédérale si aucune remise en question n'est explicitement formulée à leur égard. En Norvège et en Irlande les procédures d'homologation sont très rigoureuses, mais les vétérinaires peuvent être autorisés à prescrire des médicaments destinés aux espèces terrestres. Les législations les plus restrictives restent celles des Etats-Unis et du Canada, où les autorisations ne sont acquise que pour des types de poissons et des maladies clairement spécifiés. Semblable hétérogénéité de situations se retrouve en matière de prescription vétérinaire et de formulation des délais d'attente après traitement.

La création du grand marché européen représente une opportunité pour atténuer ces disparités. De fait plusieurs organismes internationaux, en tête desquels la FAO, ont commencé à réfléchir à des normes et à des codes de conduite pratique qui pourraient prendre valeur de référence. L'urgence en est d'autant plus ressentie que la complexité et le coût des études nécessaires à de multiples AMM dissuadent les industriels d'investir dans le marché forcément limité de l'aquaculture, alors que le besoin de médications nouvelles se fait au contraire pressant. A ces appréhensions s'ajoutent les contraintes prévisibles que risque d'imposer le souci de protèger l'environnement aquatique. C'est donc un véritable défi que devront relever les experts de la CEE.

Current Situation in the Federal Republic of Germany

Within the routine practice of an ordinary Fish Health Service the routes of administration of chemotherapeutants in fish culture are focused on:

- bath treatments
- oral treatments, mainly through medicated feed and
- (occasionally) injection treatments.

Bath treatments are carried out mainly by

- dipping
- short term bath of a few minutes until half an hour
- medium term bath up to one hour
- long term bath for eight hours or more - in some cases until chemical decomposition in the (not flowing) water of the drugs involved.

In the Federal Republic of Germany (FRG) the dosage of drugs used orally through medicated feed treatments are "standardized" in the sense that 1.5% of the body weight is fed per day for one week. This is changed, respectively extended for 0.8% for 2 weeks in cases of reduced feed intake because of low water temperatures.

In Northern Germany two fish feed mills have the status of official approval to produce medicated feed in advance, to store it and achieve a quick delivery once a precription has been written. Mills without this official approval can manufacture medicated feed only for a single case, one fish population with one prescription, but in practice this is results in excessive delay.

Occasional injection treatments are carried out, mainly intraperitoneally.

The amount of chemotherapeutant used, the legal aspects of its application as well as registration and licensing factors are presently characterized by its heterogenous spectrum in the European Community (EEC) as well as in countries outside the EEC which devote efforts in aquaculture.

Taking as a basis example the situation in the FRG, Table I shows the drugs licensed, temporary authorised, proposed for a standard marketing authorisation, and drugs additionally used in the practice of the Fish Health Service of Lower Saxony.

Table I Federal Republic of Germany

Drugs licensed for use in fish (temp. marketing authorisation)

5-nitro-2-thiazolylamine	Dibutyl tin oxide
Furazolidone	Ampicillin trihydrate
Sulphadimidine	Trimethoprim
Chloramphenicol	Oxytetracycline
Chlortetracycline HCl	Paromomycin sulphate
Dimetridazole	Sulphadoxine
Estomycine disulphate	Metrifonate
Nifurprazine HCl	

The only one fully licensed = Trimethroprim/Sulph.

Chemicals proposed for a standard marketing authorisation :

NaCl	Malachite green
CaO	Formaldehyde
Chloramine T	$KMnO_4$
Niclosamide	

Additionally used in the practice of the Fish Health Service of Lower Saxony :
* - gill swelling : ActomarB-100^{R}, Mefarol^{R} (quarternary ammonium products)
* - ecto- and endoparasites : Mebendazole, Praziquantel^{R}, Fumagillin, Methylene Blue, Trypaflavin, Acriflavin.
* - anaesthesia : Benzocaine, Ethyleneglycolmonophenylether.
* - pond disinfection : Calcium cyanide, Calcium chloride

Table II adds drugs presently tested in the routine antibiogrammes of the Fish Health Service of Lower Saxony and summarizes the drugs licensed or temporary licensed <u>really</u> used in practice, which are for the moment no more than seven and among them only one fully (long term) licensed (trimethoprim/sulfadimethoxine). It also includes the current withdrawal periods in use.

Table II Antibiotics and other drugs presently tested in routine antibiogrammes.

Name	Licenced for fish	Remarks	Use as	Withdrawal period (days)
Chloramphenicol (palmitate)	Yes	TL	Premix, bath, injection	30
Chlortetracycline HCl	Yes	TL	Premix, bath	30
Furazolidone	Yes	TL	Premix, bath	90
Trimethoprim/sulphadiazine	Yes	LL	Premix	49
Nifurprazine HCl	Yes	TL*	Premix	30
Oxytetracycline HCl	Yes	TL**		30
Erythromycin	No	LN		
Ampicillin $2H_2O$	No	LN		
Lincomycin	No	LN		
Oxolinic acid	No	LA		
Flumequine	No	LW		
Enrofloxacin (Baytril)	No	LA		
Ectoparasiticides				
Masoten (Trichlophon, Metrifonate, Neguvon))	Yes	LL	Bath	30
Endoparasiticides				
Paramonycinsulfate	Yes	TL	Premix	90
Di-n-butyl-tin-oxide	Yes	TL	Premix	30

* not used in practice due to resistance.
** licence not utilised inpractice.

Key
TL : temporary licence
LL : long term licence
LN : licence presently not sought
LA : licence applied for
LW : application withdrawn

The "Standard Marketing Authorisation" refers to those mainly simple drugs like e.g common salt (NaCl) for which it is to be expected that pharmaceutical enterprises will never apply for licensing because of obvious reasons, and means that certain drugs can be exempted from the obligation to obtain a marketing authorisation, if there is no fear of any hazard to human or animal health (1). Such an exemption is the decision of the Federal Minister, to whom the Federal Health Office (Bundesgesundheitsamt/BGA, Berlin) is scientifically and technically responsible and carry out the whole and very time consuming work up to compilation and publication of monographs for each of these substances. Finally, the Standard Marketing Authorisation Commision, consisting of representatives of human and animal medicine, pharmaceutical industry, pharmacologists, etc... is consulted by the Federal Minister and a decision wether to exempt the drug from authorisation is made (1,4).

The present paradoxical situation of "not allowed but not forbidden" status for those chemicals proposed for standard marketing authorisation in the FRG (Table I) is an unpleasant fact for the prescribing veterinarians of the Fish Health Services. It has lasted now for decades and now needs a rapid solution.

Only the use of bath-applied chemicals in net cage culture is officially forbidden. The main reason for it are the unknown subterranean currents in such waters. Also the use of e.g. Masoten[R] -licensed for use in fish (Table II)- is not allowed in areas from which potable water is obtained, drinking places for large animals (cattle and others) and waters used for bath purposes for human populations.

This is -very briefly- a general overview of the situation in the FRG.

Current Situation in the EEC and Elsewhere

Table III shows the chemicals used in the 12 member states of the EEC and again the heterogeneity mentioned arises as the common denominator. The majority of the rather simple chemicals have been used for decades (or even a century) in fish culture but a minority -or none- are officially licensed for this use.

Table III European Economic Community

Chemicals used in fish farming in the EEC - 1986

Chemical	Country
Malachite green	10
Copper sulphate	9
Formalin	10
Potassium permanganate	7
Chloramines	8
Trichlorphon	6
Dichorvos	2
Sodium chloride	7
Benzalkonium	3
Iodophor	4
Methylene blue	2
Hypochlorite	1
Acriflavine	1

Only dichlorvos has been fully licensed in the UK.

(After report of Round Table Consultations on Documentation of Drugs for use in Aquaculture, Oslo, Norway, June 25-28, 1990).

Within this frame and as an example the situation is interesting in the European countries not members of the EEC but in which fish farming represents a main national effort and income, as for example, in Norway and Iceland. This is shown in Tables IV and V.

Table IV Norway

Drugs used in Salmonid farming in Norway

- Chlorbutanol
- Iodophor
- Malachite green oxalate
- Formaldehyde
- Praziquantel and enbendazole
- Nitrofurazolidone, oxolinic acid,
- Oxytetracycline chloride,
- Trimethoprim-sulphadiazine and flumequine
- Trichlorphon and dichlorvos

All pharmaceutical specialities must be registered, but veterinarians can apply for a registration exemption for unregistered specialities.

It is intended to produce a definitive list of medicines to be allowed for use in aquaculture.

(After report of Round Table Consultations on Documentation of Drugs for use in Aquaculture, Oslo, Norway, June 25-28, 1990).

The use of chemotherapeutants for fish in Norway is widely known. Sometimes not in a very positive sense in the today's audiovisual media which is not always objective. As an example, towards the end of 1990, a TV statement led by ecologists and transmitted atpeak broadcasting time in the FRG,outlined the dangers for the environment and consumers of the use of dichlorvos in salmon culture in Norway. The main property of the compound as a specific "nerve poison" was highlighted - but not a word for what (for the salmon lice). Thus for the uninformed TV-watcher the idea left that salmon - and therefore perhaps the consumer - would have been the target of such a dangerous "nerve poison".

The registration exemption for unregistered drugs to be used in fish farming by veterinarians tends now to be reduced (9). In Iceland the barriers are higher in view of the fact that the registration procedure is the same for both human and veterinary drugs.

Table V Iceland

The following drugs and chemicals are in use		
Drugs	*	Oxytetracycline
	*	Trimethoprim/Sulphadiazine
	*	Trimethoprim/Sulphamethoxazole
	*	Erythromycin
	*	Oxolinic acid
	*	Metrifonate
	*	Dichorvos
	*	Benzocain
Chemicals	*	Malachite green
	*	Benzalkonium chloride
	*	Chloramine
	*	Iodophor
	*	Formalin

In Iceland the same registration proceedure is used for both human and veterinary drugs.
(After report of Round Table Consultations on Documentation of Drugs for use in Aquaculture, Oslo, Norway, June 25-28, 1990).

Looking overseas, the rather restrictive policies in Canada and USA are reflected in Tables VI and VII. Thus, in Canada only one drug - Oxytetracycline - is fully registered and approved for use in fish farming. In the USA the situation is heavily complicated by the fact that each drug is approved only for the treatment of certain diseases of certain fish species.

Table VI Canada

The following drugs and chemicals are in use

* Trimethoprim/Sulphadiazine
* Oxolinic acid
* Sulphadimehtoxine/Ormetoprim
* Oxytetracycline
* Ovaprim
* Pituitary extract
* Estradiol
* MS222
* Marinil
* 2-Phenoxyethanol
* Quinaldine sulphate

Only oxytetracycline is registered and approved

(After report of Round Table Consultations on Documentation of Drugs for use in Aquaculture, Oslo, Norway, June 25-28, 1990).

Table VII USA

The following drugs are FDA approved (each drug for certain species and diseases only).

* Sulphadimethoxine and ormetoprim combination ("Romet 30)
* Oxytetracycline
* Sulphamerazine
* Formalin
* Tricaine methanesulphonate (MS222)

The following drugs are being approved (each drug for certain specoes of fish and diseases only).

* Choramine T
* Oxytetracycline
* Formalin
* Erythromycin
* Virginiamycin
* Benzocaine

There is a long list of approved chemicals for controlling insects, algae, plants and micro-organisms in ponds and raceways.

(After report of Round Table Consultations on Documentation of Drugs for use in Aquaculture, Oslo, Norway, June 25-28, 1990).

The heterogeneity of the handling of drugs used for fish farming within the EEC is demonstrated by the very different situation regarding to

a) the need for a prescription (mainly by veterinarians) and

b) the existence of withdrawal periods.

This is shown in Table VIII.

Table VIII Legislation concerning fish drugs in European countries

	Need for a prescription	Withdrawal periods
Greece*, Poland	-	+
Belgium, Cyprus, Switzerland	+	-
Austria, Czechoslovakia, Denmark, Fed. Rep. Germany, Finland, France, Hungary, Ireland, Israel, Italy, Netherlands, Norway, Portugal, Sweden, United Kingdom, Yugoslavia	+	+

* From Greece it is pointed out that drugs for use in animals are generally sold by veterinarians themselves - and thus without specific prescription. Withdrawal times are in preparation.

EIFAC questionaire, 1988 (after Bernoth 1990).

It is beyond the scope of this presentation to look into details within the countries mentioned in Table VIII, but again a rather heterogenous picture is clearly shown. At least in the FRG each prescription must be reported and there is an at least a yearly check on them by the authority.

Another crucial point is the effectivity of chemotherapy in fish which in many cases is far from being properly known. Available pharmacokinetic data tend more to residual behaviour of drugs as main target than to efficacy as such. This is daily finding in the practice of a Fish Health Service. As a simple example- Fig.1 shows the antibiogramme sensitivity of *Yersinia ruckeri* against 11 drugs within a time of two and a half years of practice of the Fish Health Service of Lower Saxony.

Of course these are *in vitro* trials - but 15 years of Fish Health Service field experience showed that on average there are no substantial differences between those *in vitro* results and field treatment success. In general, drugs which show highly sensitive (and in some cases only less sensitive) in the antibiogramme, show also succesful in field therapies (9). As it is to be expected, the best results are shown by those drugs not yet (officially) used for fish in the FRG like the quinolones enrofloxacine, flumequin, oxolonic acid, nalidixic acid, piromidic acid and (partially) the antibiotics lincomycin and spectinomycin.

Fig 1 Antibiogramme sensitivity of Yersinia ruckeri : Fish Health Service of Lower Saxony : 1986 to mid 1988

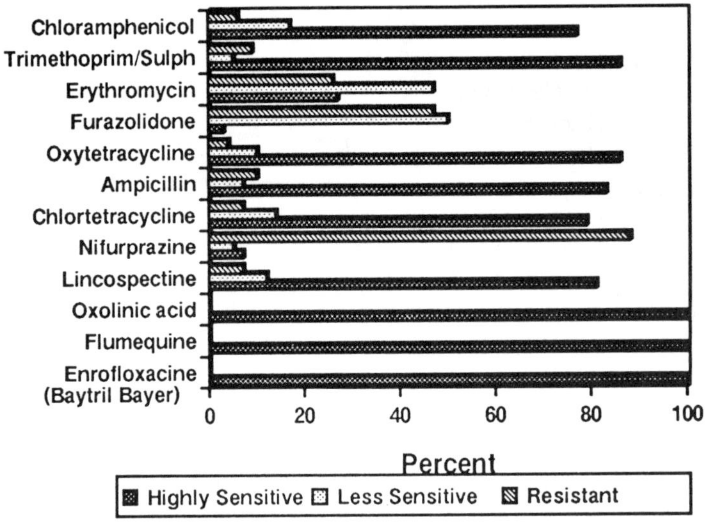

(after Schlotfeldt et al 1991)

The establishment of the EEC market without national boundaries after January 1 1993 offers an unique opportunity to overcome the heterogeneity deplored in this presentation and to bring a certain order into the field of chemotherapy in fish culture. This refers essentially to the need to find common pathways in legislation. Table VIII is a very limited reflection of this and refers only to prescription and withdrawal periods. There are so many other aspects for which guidelines must be provided, even beginning with the primary question of whether the fish is to be regarded as a domestic animal or not. This point has a fundamental implication for the policies of fish epidemic control and last but not least therapy and use of drugs.

In the meantime some efforts on this topic have been made by the EEC Committee on Veterinary Medical Products (CVMP) (5), the European Inland Fisheries Advisory Commission (EIFAC) (6) and the Codex Alimentarius Commission (CAC) (2) of FAO as to elaborate a "Code of Practice for the Safe Use of Drugs in Aquaculture".

The main needs realized are summarized in Table IX.

Table IX Main needs for standardisation and safety of drugs used in aquaculture

1.- more and better established pharmacokinetic data on drugs used in fish
2.- standarized rules for use of chemicals and therapeutants in fish farming: 2.1 licensing of drugs 2.2 prescription modalities 2.3 possibilities to prescribe for fish drugs approved for other animals 2.4 production of medicated feeds
3.- standarized withdrawal periods, based on relevant data of residual behaviour for each drug and in dependance of at least three different water temperatures
4.- elaboration of guidelines for 4.1 tolerance tests 4.2 microbiological studies of the drug ingredients 4.3 metabolism (and residues) of active substances involved and influence of water temperature, pH, salinity, day light, fish age, fish species and growth rate 4.4 excreted products 4.5 effects on the aquatic ecosystems 4.6 mutagenicity 4.7 subchronic and chronic toxicity 4.8 carcinogenicity 4.9 reproductive toxicity
5.- rules to protect the aquatic environment: possible impact of drugs used for fish in the environment and compatibility with the environment

This needs established represent an ideal approach- but perhaps a bit far from reality even in countries in which fish farming is a main industry. In general it is to be feared that the fish segment of the whole animal health market is too small as to support such heavy investments. Many pharmaceutical companies will desist from apply for licensing of drugs for use in fish and the detriment will be for the fish farming industry. The reduced size of the market for later sales of drugs for use in fish is a crucial problem which the Fish Health Services in the FRG are daily confronted with. No substantial innovations in this field have been made for many years.

Especially point 5 of Table IX -protection of-and compatibility with the aquatic environment- represents a serious threat to the whole business if really carried out to perfection.

Each drug e.g used in bath treatment should be then diluted down to therapeutical inefficacy, as to avoid any impact on the environment.

For a thorough protection of the aquatic environment techniques mainly based on filter systems of activated carbon were recently developed as to guarantee a practically total removal of- for example- malachite green (8) or benzocaine (7) from effluents waters. But again the aspect costs for such a system working throughout the year represent both a threat and a limitation for a "fish farming in complete compatibility with the environment".

It is a challenge for the responsible EEC body involved (and its experts) to find out within the remaining months to January 1993 suitable ways to simplify and standardize administrative procedures, to reduce costs and adapt the needs to the economic reality and practical feasability of aquaculture industry, at least in the EEC area.

Graphs by D.W. Kleingeld, State Fish Epidemics Control Service of Lower Saxony and Fish Health Service, Hannover, FRG.

References

1). BERNOTH E.M.,1990. Marketing authorisation of fish drugs-Current status and future intentions. *J.Vet.Med.* **B 37**, 401-406.

2). CODEX ALIMENTARIUS COMMISSION/CAC, 1989. Questionnaire on the content and scope of the proposed Code of Hygiene Practice for Aquaculture (CHPA). CAC. CL 1989/13-FFP.

3). CODEX ALIMENTARIUS COMMISSION, 1990. Report of the fifth Session of the Codex Committee on Residues of Veterinary Drugs in Foods. *Food and Agriculture Organization and World Health Organization*, Rome, Italy. ALINORM 91/31A, 53pp.

4). (German) DRUG LAW (Arzneimittelgesetz) of August 24,1976, (*Federal Law Gazette* I p.2445-2448), in the version of the third Law Amending the Drug Law of July 20, 1988 (*Federal Law Gazette* I p.1050).

5). EEC DIRECTIVE 81/852, September 28.,1981. *O.J.*No. L 317 of November 6, 1981.

6). EIFAC, 1988. Therapeutics used in fish production - pharmacokinetics, residues, withdrawal periods for drugs. *European Inland Fisheries Advisory Commission (EIFAC)*, 15th Session, Göteborg, Sweden. May 31, 1988- June 7,1988. EIFAC/XV/88/Inf. 13.

7). HOWE G.E.,BILLS T.D. & MARKING, L.L., 1990. Removal of benzocaine from waters by filtration with activated carbon. *The Progressive Fish Culturist*, **52**, 32-35.

8). MARKING L.L., LEITH, D. & DAVIS, J., 1990. Development of a carbon filter system for removing malachite green fron hatchery effluents. *The Progressive Fish Culturist*, **52**, 92-99.

9). NORWEGIAN MEDICINES CONTROL AUTHORITY, 1990. Report of the round table consultations on documentation of drugs for use in aquaculture Oslo Norway,June 25-28.

10). SCHLOTFELDT, H-J., HERBST, J., ROSENTHAL, H.,ALVARADO,V., STANISLAWSKI, D. & BOEHM, K.H., 1991. Freshwater fish production and fish diseases in Lower Saxony (Federal Republic of Germany): Fish Health Service experiences. *J. Appl. Ichthyol.*, **7**, 26-35.

Discussion

TREVES-BROWN The speaker mentioned that, in the USA, licences are given for individual species of fish. The same applies in the United Kingdom, because virtually our only farmed species are the rainbow trout, which is kept in fresh water and the Atlantic salmon which is kept in sea water; and we do not believe that you can do residue studies and come up with a withdrawal period for either of those species and assume that the same will apply to the other, because their metabolisms are bound to be so very different.

DE KINKELIN [To Alderman/Michel] Do you want to answer?

ALDERMAN I agree with Mr Treves-Brown. You cannot set a single withdrawal period, whether it be a degree day based or a fixed day based withdrawal period for salmonids unless those salmonids are at least in the same environment. Once you start consider sea water and fresh water, you have to have data from both environments. It may be possible that the withdrawal period may be identical under both conditions, but until you have the data, you must assume that it is not. Certainly now that in the EC, fish are coming under the Meat Residues Directive for the presence of antibiotics and other drug residues in tissues, we are going to have to be more certain about withdrawal periods than in the past.

SCHLOTFELDT As far as I know three temperature ranges will be considered?

ALDERMAN That is one proposal. I think probably two would be adequate. The ideal (for salmonids) would be to take temperatures something like 16°C and 8°C and if those show a reasonable compatibility with the effect of temperature on fish metabolism for a drug, then that would be sufficient. However if the residue studies at those two temperatures show difficulties, then additional temperature studies may be necessary. Selected temperatures must be appropriate for the target disease and fish species concerned.

SCHLOTFELDT Another problem that we have in practice in our service is that there are additional drugs not even being considered for standard marketing authorisation. For instance for anaesthesia, there is an anaesthetic called 2-phenoxyethanol which has been shown to be very very effective, is very well tolerated and cheap (that is a most important factor!). It water soluble so that it is not necessary a to make a stock solution in acetone as with benzocaine. Of course we have no idea of a withdrawal period for this chemical, but it is so cheap that our farmers are using it more and more for stripping of trout.

FIJAN In Yugoslavia we do require registering of drugs and there is a withdrawal period which is stated within the registration procedure.

CHEMOTHERAPY IN THE CONTROL OF MOLLUSCAN DISEASES

D.J. Alderman

MAFF, Fish Diseases Laboratory, Weymouth, Dorset, DT4 8UB, UK.

Summary : *Molluscs are susceptible to a wide range of diseases, many of which produce very severe mortalities. Few of these diseases are susceptible to chemotherapy, either because of environmental constraints - the impossibility of applying a chemotherapeutant on extensive shellfish beds covering many hectares in area, or because the pathogen is not susceptible to any available therapeutant. Effective chemotherapy in mollusc culture is, so far, only applicable to bacterial and protozoan/protistan diseases under controlled hatchery conditions where the volume of water to be treated is very limited.*

***Résumé* : Contrôle thérapeutique des maladies des mollusques**

Les mollusques sont prédisposés à de nombreux types de maladies dont les effets se traduisent par de sévères mortalités. Peu de cas se prêtent vraiment à un contrôle thérapeutique, soit que les agents pathogènes responsables ne répondent pas aux traitements disponibles, soit que le contexte d'exploitation en parcs extensifs de plusieurs hectares interdise toute intervention. Jusqu'à présent la démarche thérapeutique ne s'est appliquée qu'à des infections bactériennes et à des parasitoses causées par des champignons ou des protistes, dans des écloseries où les volumes d'eau à traiter demeurent limités. On peut citer le cas du champignon Ostracoblabe implexa, *sensible au chlorure mercurique mais n'ayant de ce fait qu'une valeur historique, et surtout celui des bactérioses à* Vibrio *et* Aeromonas *spp., dont diverses formules antibiothérapiques (pénicilline-streptomycine, chloramphenicol, acide oxolinique, furazolidone parmi les plus courantes) permettent de limiter l'impact. D'autres traitements ont été tentés avec des organophosphorés contre le copépode* Mytilicola intestinalis, *avec de la cycloheximide contre* Perkinsus, *ou avec le chloramphenicol contre les Thraustochytriales.*

Although the culture of molluscs is an old practice, the recognition of the occurrence of serious diseases is relatively recent. Since mollusc culture is, for the most part, extensive rather than intensive, attempts to control disease other than by controls on movement of shellfish have been very limited. Only in the limited area of the mollusc hatchery can chemotherapy be applied with any success. Elsewhere the nature of extensive culture means that cost and environmental constraints preclude the use of chemotherapeutants. Despite this, the first attempt at disease control in molluscan culture was in the Netherlands, where Korringa (8), attempting to grapple with the problem of the neglected Dutch oyster beds in the years immediately after World War II, found that a major problem was the presence of a fungus *Ostracoblabe implexa* growing in the shell of the oysters and causing irritation to the mantle resulting in the laying down of abnormal shell. During the wartime neglect of the oyster beds, much old shell had accumulated, producing a reservoir of infection which quickly resulted in the infection of young oysters setting on the old infected shells. Since shell material ("cultch") is necessary to encourage setting of oyster larvae, it was not practical to clear all the old shell from the beds. Instead, experiments demonstrated that if cleared shell was dipped in mercuric chloride, any fungal hyphae present in the shell would be killed and the shell could be returned to the oyster beds after rinsing to act as cultch once more. Whilst technically practical, the use of mercury compounds cannot today be countenanced and this first attempt at chemical control of a mollusc disease was never adopted extensively.

When problems with reliable production of young oysters increased in the post war period in Europe, research both into new oyster species and into artificial production of larvae was undertaken. Under laboratory conditions, larvae both of traditional species and of new species incapable of breeding naturally in the new environment could be produced readily. Techniques for the production of *Ostrea edulis, Crassostrea gigas* and other bivalves were developed, but it rapidly became evident that a range of gram negative motile bacteria, principally *Vibrio* spp and *Aeromonas* spp presented major problems.. These bacteria, present in the hatchery water and in the algal feed culture stocks caused rapid mortalities of larval cultures. Hidu and Tubiash (6) demonstrated that *Mercenaria mercenaria* and *Crassostrea virginica* larvae could be successfully reared in the presence of mixed bacterial cultures provided antibiotics were also present in the culture water, but not without. The disease concerned has been termed larval bacillary necrosis. Penicillin-streptomycin and chloramphenicol were effective (13) in encouraging the survival of larvae. Also suggested were polymixin B, erythromycin and neomycin and, more recently, oxolinic acid has been used with considerable success. The use of chloramphenicol, apart from being inappropriate for reasons discussed by Alderman and Michel (this conference) proved to be of limited value since, presumably due to toxicity, treated larvae frequently cease feeding (12).

Vibrio spp (principally *Vibrio parahaemolyticus*) are also responsible for mortalities in juvenile oysters and hard clams in hatchery culture and on-growing systems (5). As water temperatures rise surfaces of shell and tank systems become coated with dense layers of bacteria which attack and destroy the conchiolin (protein matrix) of the shell. *Vibrio* spp also produce ciliostatic toxins which prevent the larvae from feeding or swimming. Often secondary invasion by other bacteria will then occur. Various approaches have been tried, from antibiotic treatment with chloramphenicol and penicillin-streptomycin to sodium hypochlorite baths (up to 2h in 18ppm over a 3 to 4 week period) (7, 12). Such problems of biofouling are exacerbated when biological filters are employed in the system. Whilst such filters may aid in controlling the organic loading in the culture water, they also act as a reservoir of infection which will return after chemotherapeutic treatments have ended. The filters themselves must be separated from the system whilst disinfection or chemotherapy is in progress, since any antimicrobial agent will destroy the biofilter.

Indeed, where commercially practicable, the best approach to control of microbial diseases in the molluscan hatchery is likely to be to discard the larval batch and restart from the beginning, employing optimal husbandry techniques. One example of improved husbandry techniques lies in the attractant used to encourage the settling of larval oysters. This usually takes the form of a coating of oyster extract. The temptation is not to sacrifice good oysters, but instead to use the oldest, probably stressed and dying stock. Such animals will harbour populations of *Vibrio*'s and encourage the development of bacteraemia's in the resulting juveniles. A further husbandry improvement is ensure the physical separation of areas used for larval culture, spat culture and broodstock maintenance. Chemotherapy, whilst possible in culture systems is not a good substitute for good husbandry. Unless there is a great need that a particular batch of larvae or juveniles survive, the last resort of chemotherapy should not be employed.

Vibrio spp have also been associated with 'brown ring disease' in the clam *Ruditapes philippinarum* and recent trials (9) have demonstrated that furazolidone is effective in controlling this disease. Administration of furazolidone in the sea water at 10mg/l for three days gave 100% protection against the *Vibrio*. Whilst this may be an effective therapy under laboratory conditions, it is difficult to envisage a situation in which furazolidone therapy could be applied in extensive clam culture.

Similar limitations also apply to the demonstration that an unnamed organophosphate would control the levels of infestation of mussels (*Mytilus edulis*) with the copepod *Mytilicola intestinalis*. Infestation by this copepod has been attributed as being responsible for a 60% reduction in production of mussels in Charente and Mont St Michel Bay between 1982 and 1984 (1). Treatments consisting of two hour exposures to 30mg/l of the unnamed organophosphate were responsible for the expulsion of 90 to 93% of copepods from a trial lot of 2 tonnes of mussels. However, unless the mussels can be treated in an enclosed tank during movement of stock from one area to another, it is difficult to see how this treatment can be applied in extensive mussel culture. Here the low individual value of mussels militates against any labour expensive means of exposing the animals to chemotherapy.

Protozoans are perhaps the most important pathogens of molluscs, but little success has been recorded in attempts to control infection by means of chemotherapy. Ray (10) found that administration of 1mg/ml cycloheximide per week would inhibit the development of *Perkinsus marinus* in laboratory oyster stocks.

In British Columbia Bower (2, 3, 4) demonstrated that a member of the Thraustochytriales, *Labyrithuloides haliotidis* was responsible for a lethal disease of larval and juvenile abalone under 6 months old in an abalone mariculture facility So invasive was this protistan that the hatchery was unable to operate. The Thraustochytriales are a protistan group of uncertain taxonomic affinity, members of which are increasingly being recognised as potentially serious pathogens of molluscs from gastropods to cephalopods (11). As with *Perkinsus marinus*, chloromycetin appeared (3, 4) to offer some hope of control of this lethal infection.

The brevity of this manuscript is a clear reflection of the limitations of chemotherapy in molluscan culture. These limitations are imposed in the main by the nature of molluscan culture which is traditionally extensive. Disease is perhaps the major limitation on the exploitation of molluscs, but the practicality of applying chemotherapeutants on any scale to extensive marine culture systems means that only under intensive hatchery and juvenile on-growing systems can disease be controlled except by prevention and disinfection.

References

1). BLATEAU, D., LE COGUIC, Y., MIAHLE, E., GRIZEL, H. & FLAMION, G. (1990) - Traitement des moules (*M. edulis*) contre le copépode *Mytilcola intestinalis*. *Proc. 4th Internat. Colloq. Pathol. Marine Aquacul.* Pontevedra, Spain, 97-98.

2). BOWER, S.M. (1987) - The life cycle and ultrastructure of a new species of thraustochtrid (Protozoa: Labyrinthomorpha) pathogenic to small abalone. *Aquacult.*, **67**, 269-272.

3). BOWER, S.M. (1987) .- *Labyrinthuloides halotiditis* n.sp. (Protozoa: Labyrinthomorpha), a pathogenic parasite of juvenile abalone in a British Columbia mariculture facility. *Can. J. Zool.*, **65**, 1996-2007.

4). BOWER, S.M. (1987) .- Pathogenicity and host specificity of *Labyrinthuloides halotiditis* (Protozoa: Labyrinthomorpha), a parasite of juvenile abalone. *Can. J. Zool.*, **65**, 2008-2012.

5). ELSTON, R., ELIOTT, E.L. & COLWELL, R.R. (1982) - Conchiolin infection and surface coating *Vibrio*: shell fragility, growth depression and mortalities in cultured oysters and clams, *Crassostrea virginica, Ostrea edulis* and *Mercenaria mercenaria. J. Fish Dis.*, **5**, 265-284.

6). HIDU, H. & TUBIASH, H.S. (1966) - A bacterial basis for the growth of antibiotic treated bivalve larvae. *Proc. natn. Shellfish. Ass.*, **46**, 25-29.

7). KRAEUTER, J.N. & CASTAGNA, M. (1984) - Disease treatment in hard clams, *Mercenaria mercenaria. J. World Maricult. Soc.*, **15**, 310-317.

8). KORRINGA, P. (1951) - Investigation on shell disease in the oyster, *Ostrea edulis* L. *Rapp P.-v Reun. Cons. perm. int. Explor. Mer.*, **128**, 50-54.

9). NOEL, T., AUBREE, E., BLATEAU, D., MIAHLE, E. & GRIZEL, H. (1990) - Treatments against the *Vibrio* sp suspected to be responsible for mortalities in *Ruditapes philippinarum. Proc. 4th Internat. Colloq. Pathol. Marine Aquacul.* Pontevedra, Spain, 22-23.

10). RAY, S.M. (1966) - Cycloheximide: Inhibition of *Dermocystidium marinum* in laboratory stocks of oysters. *Proc. natn. Shellfish. Ass.*, **56**, 31-36.

11). SINDERMANN, C.J. (1990) - *Principal diseases of marine fish and Shellfish.* Volume 2, Academic Press, New York, 516 pp.

12). SINDERMANN, C.J. & LIGHTNER, D.V. (1988) - *Disease diagnosis and control in North American aquaculture.* Developments in Aquaculture and Fisheries Science, **17**, Elsevier, Amsterdam, 431 pp.

13). TUBIASH, H.S., CHANLEY, P.E., & LEIFSON, E. (1965) - Bacillary necrosis, a disease of larval and juvenile bivalve mollusks. I. Etiology and epizootiology. *J. Bacteriol.*, **90**, 1036-1044.

Discussion

MUNRO David, there is one other example perhaps and this concerns the use pesticides on oyster beds to control various invertebrate fauna. I am thinking perhaps of Sevin which has certainly been recorded in the grey literature on the

US west coast. This is an acetylcholinesterase inhibitor and, as I understand it, these materials can be spread over oyster beds to control various pests. So we have the situation, I think of a pesticide, perhaps in modern definition, being used as a medicine.

ALDERMAN Of course Dr Munro is also thinking of another pesticide being used as a medicine which is a problem in the UK. Yes, that sort of use has been suggested, but I don't think that such use is very extensive today. Other, mechanical methods are far preferable, because the dosage that would be necessary is these days unacceptable. It was used quite extensively at one time and there is quite a lot in the literature suggesting various pesticides that could be used, for example, to control crabs, starfish, etc. The method has never been used in the UK and I don't think it would ever have been countenanced there.

FIGUERAS There are two more treatments used on shellfish in Spain. One for the brown ring disease in Manilla clam caused by *Vibrio* P1 and is treated with flumequine. The other is for *Mytilicola intestinalis* a copepod in mussels. A product has been licensed by a French pharmaceutical company, I don't know if you are familiar with this?

ALDERMAN I was not aware of the *Mytilicola intestinalis* treatment. Of course, quinolones are appropriate treatments for gram negatives and we have tried oxolinic acid in our research hatchery at Conwy, so I am not surprised to hear that they are being used elsewhere to control bacterial disease.

FIGUERAS This is a bacterial disease of adults.

ALDERMAN Is use under intensive conditions or under extensive conditions.

FIGUERAS Extensive.

[***Note : the details and references mentioned by Dr Figueras have been added to the preceding text***]

CHEMOTHERAPY IN AQUACULTURE TODAY - CURRENT PRACTICES IN SHRIMP CULTURE: AVAILABLE TREATMENTS AND THEIR EFFICACY

T.A. Bell & D.V. Lightner

University of Arizona, Department of Veterinary Science
Tucson, AZ 85721, USA

Summary: *Shrimp culture throughout the world has grown rapidly in the past decade. Concomitant with this growth has been an increase in the intensification of culture techniques and a need to combat ever increasing disease outbreaks. Disease treatment techniques have significantly lagged behind the growth of the culture industry and might best be considered as imprecise. Currently important shrimp disease syndromes which are either potentially treatable or are presently being treated, irrespective of the treatment efficacy, are reviewed.*

Résumé: **Applications de la thérapeutique aux élevages de crevettes: traitements et efficacité.**

La production des crevettes d'élevage s'est développée de façon spectaculaire en l'absence de réel support logistique, et cette évolution s'est accompagnée de pathologies nouvelles, dont l'étiologie est souvent complexe, et difficile à interpréter. En matière thérapeutique l'empirisme prévaut, et les résultats des quelques tentatives menées à l'échelon local sont rarement publiés. Très peu de produits sont légalement autorisés pour le traitement des crevettes, le plus répandu étant le formol.

Aux stades de production larvaire les élevages sont régulièrement décimés par des mycoses attribuées aux genres Lagedinium *et* Sirolpidium, *et par des infections microbiennes d'origine et d'expression clinique variées : syndromes bactériens non spécifiques, syndrome "bolitas", syndrome à bactéries luminescentes, d'où l'on isole le plus souvent des* Pseudomonas, *des* Aeromonas *et des* Vibrio. *Dans le premier cas le vert malachite et la trifluraline ont apporté des résultats intéressants à condition d'être employés précocement. Dans le deuxième cas les antibiotiques sont requis, et des modalités d'administration originales, comme la bioencapsulation dans les larves d'*Artemia *destinées à l'alimentation, sont à l'étude.*

En phase de croissance les troubles d'origine bactérienne ne sont guère mieux définis: nécrose septique de l'hépatopancréas, syndrome des mouettes, syndrome des étangs du Texas. Là encore, l'antibiothérapie, surtout à base de sulfamides ou de tétracyclines administrés par bain ou par voie orale, est régulièrement mise en oeuvre, parfois à titre préventif. Il existe également des infections à grégarines contre lesquelles des agents anticoccidiens ont récemment été expérimentés. Enfin la prolifération d'organismes épibiontes sur les branchies (fouling disease), aux conséquences parfois redoutables, est classiquement contrôlée par le permanganate de potassium, le sulfate de cuivre et le formol, employés seuls ou en combinaison.

Introduction

Within the past decade, shrimp culture throughout the world has progressed at a very rapid pace. Unfortunately, some of the requisite logistic support has lagged significantly behind the commercial "culture" *per se*. One area in which the technical base is lacking is the therapeutic treatment of shrimp diseases. In general, pond and tank-side treatments, as practiced today, have been derived by a less than scientific means. Typically the treatment of choice, if one exists, has been established from a "trial and error" process of compounds which were most readily available at that particular time. Often no particular compound has been positively established as being significantly better than any other compound.

Of equal concern is the fact that at the time of treatment, for most of the disease phenomena encountered, the etiology is seldom known. Unfortunately, there are very few shrimp disease syndromes which are fully understood, and most appear to have rather complex etiologies. Many of the known disease syndromes appear to have a bacterial component. However, the application of an efficacious antibacterial treatment may, in some cases, have no positive impact on halting the course of the disease. Often, in the long run, the converse may even be true. Bacterial "pathogens" associated with shrimp disease syndromes differ little from other bacteria; the non-judicious use of antibiotics inherently has the potential to promote the formation of resistant strains, further reducing the efficacy of the compound.

Without proper scientific investigations into treatment regimes, there has been a tendency for individual farms to do their own research. More often than not, the information which is generated (albeit informal, non-controlled studies) is considered as proprietary and is not shared, especially if it would appear to management that its use would provide market advantage. Therefore, very little documented information exists. Information provided in the following text, which addresses specific chemotherapeutants, should be assumed to be based primarily upon second hand knowledge (at best), unless otherwise specified. Even information which has been properly documented may often be dated, especially in such a dynamic field.

There presently exists only one compound approved for use in shrimp culture in the United States and at least two compounds approved for use in

Japan. To our knowledge, no other shrimp-producing countries have formally approved the use of any other chemothera- peutants. Our laboratory is actively involved in the approval process for several chemotherapeutants. We are presently working, in conjunction with the USDA and/or pharmaceutical companies on the approval process for: oxytetracycline, florfenicol, sulfadimethoxine plus ormetoprim, sarafloxacin, enrofloxacin and several other as of yet un-named products. In addition to these, we have been instrumental in obtaining formal US FDA approval for formalin use in shrimp culture; formal announcement is all that is lacking.

Disease syndromes of larval culture

Larval mycosis

Larval mycosis (LM) has been a recognized disease of larval penaeid shrimp since the early days of shrimp culture. The disease is well understood. The treatment of choice is universally accepted and effective.

Clinical signs: Affected larvae are typically white and very opaque. In those severely infected individuals, movement is highly restricted and thus they are usually found near or on the tank bottoms. Affected larvae contain highly branched pale yellowish-green fungal mycelium. Individual hyphae contain nondescript refractile globules and relatively few septa and are approximately 10-14 μm in diameter. During fungal sporulation, specialized hyphae exit the cuticle, producing large spherical discharge vesicles containing multiple biflagellated zoospores which are ultimately discharged and capable of infecting a new larval host. Tanks affected with this disease typically experience a rapid increase in daily mortalities. If unchecked, mortalities can reach 100% within a few days.

Pathology: After a zoospore has germinated on a larvae and sends its first vegetative hyphae through the cuticle, the hyphae rapidly branch while they consume the surrounding tissue. Within a short period it can completely fill the cuticle and in the process consume the entire larva, excluding the cuticle. Histological examination reveals no host response to the invasion.

Geographic and host distribution: LM appears to be a worldwide problem of cultured shrimp. Although certain shrimp species may be less susceptible than others, it still represents probably the most widespread identifiable disease of penaeid shrimp. Although it has been noted in those stages up to and including very early post-larvae, it more commonly infects the egg, nauplius and protozoea stages. It is not only recognized from shrimp, but it has also been noted on lobsters, blue crabs, dungeness crabs and barnacles.

Etiology: LM in penaeid shrimp is most often caused by the phycomycete fungus *Lagedinium callinectes*. Less often a similar fungus, *Sirolpidium* sp., has been implicated. The clinical signs are nearly identical, only the fungal morphology differs somewhat.

Potential treatment: Numerous antifungal compounds have been tried against LM, usually with very little success. Two compounds, however, have proven to be useful. Malachite green has been used at levels of 2 to 10 ppb with varying success. Its effectiveness being contingent upon early treatment. The use of malachite green is not without serious human health dangers and thus stands little chance of ever being approved for use in US aquaculture and should be avoided whenever possible. Fortunately another compound has been used with even more success and with apparently few, if any, down-sides. The compound is called trifluralin and marketed by at least one major industrial chemical / pharmaceutical firm. It is normally sold for use as an agricultural herbicide.

Trifluralin is typically applied every so many hours or continuously. The desired working concentration in the tanks is approximately 10 to 20 ppb. Its effectiveness is likewise dependent upon its being administered early enough in the course of the disease. Trifluralin has little effect on the vegetative hyphae enclosed within the shrimp's exoskeleton, but instead is very toxic to the motile infective zoospore stage.

The key to the success of treatment is to provide trifluralin before the zoospores are able to infect the larvae. Hence, the treatment is not considered as one being applied to the shrimp, but instead the target is the water-borne zoospores. Nearly all, if not all, major shrimp growing regions routinely use trifluralin prophylactically, most often beginning with the fertilized eggs in the spawning tank water and ending in the late nauplius or early protozoea stages.

Trifluralin has an extremely wide margin of safety and is inherently unstable in the larval-rearing tank environment. The latter characteristic minimizes chances of unwanted residues in harvested shrimp or in the discharged effluent of treated tanks.

Non-specific bacterial syndromes

Non-specific bacterial syndromes (NSBS) might best be considered as a "catch-all" category. There may be a multitude of unique maladies which fall into this general category, but they do not differ enough from one another, at a clinical level, to be separated. This is in no way meant to suggest that each separate malady responds in the same way to a given treatment. In fact, this may in part be the cause for the wide variation seen in response to treatment.

Clinical signs: Typically the afflicted populations will display a dramatic reduction in feed intake. Larvae will often be noted to stall-out in a particular molt or life-stage. Significant population mortalities are most commonly noted following a molt.

Although characteristic clinical signs have been noted in all larval stages, the nauplius and protozoea stages may be somewhat more predisposed than mysis and postlarval stages.

Pathology: The most consistent pathology noted is hepatopancreas (HP) or muscle atrophy, accompanied by some degree of lethargy. The HP tubules, midgut and/or stomach will often contain high numbers of bacteria. Quite often there will be deformities and/or melanized erosion of the appendages and sub-appendages (setae). The intestinal tract will often be empty, with no fecal strands being noted.

Geographic and host distribution: NSBS-caused mortalities have been reported from *Penaeus vannamei* and *P. stylirostris* hatcheries throughout Ecuador, and possibly other Central and South American countries. Similar reports have been received from the major shrimp culture regions of Asia. Because of the nearly all-inclusive nature of this category, it is very likely that similar disease syndromes have been noted for all cultured penaeid shrimp species.

Etiology: Bacteria and/or bacterial toxins appear to be playing some role in these related diseases. This assumption is based primarily upon the consistent presence of significant levels of bacteria in the HP, stomach and midgut, and by the apparent efficacy of some antibiotics in stopping the mortalities if provided early in the disease. Bacteria noted typically fall into the genera: *Vibrio*, *Pseudomonas*, *Aeromonas*, and other related Gram-negative rods. On rare occasions Gram-positive bacteria have also been isolated.

Potential treatment: Numerous antibiotics have been tried against this "catch-all" disease phenomenon. It would appear that only a select group are effective, although this may vary from facility to facility. The antibiotics are applied via the tank water, but must be applied very early in the course of the disease. No particular compound can be singled out as being most efficacious, which is understandable considering that, within the NSBS category, we may be actually dealing with more bacterial species or strains than there are antibacterials. Some of the compounds added to the tank water, with varying success, have been: formalin, EDTA, oxytetracycline, furazolidone, erythromycin, oxolinic acid, nalidixic acid, sodium nifurstyrenate, flumequin, enrofloxacin, chloramphenicol, florfenicol, nitrofurazone, neomycin, virginiamycin, acriflavine, sulfonamides, potentiated sulfonamides such as sulfamethoxine+ormetoprim and sulfamethoxazole +trimethoprim. This list should not be considered as definitive, but instead representative. It should be noted here that application of chemotherapeutants via the water is not the only technique to have been used for treatment of this syndrome. A relatively new procedure of bioencapsulating a drug has been used in a few instances with success. Very briefly, the drug to be administered (assuming it is relatively water insoluble) is fed to *Artemia* (a normal living feed for shrimp larvae), which after filling their intestinal tract, are in turn fed to the shrimp larvae.

"Bolitas" syndrome

This syndrome should normally be classified under the NSBS problems if it were not for a rather unique clinical signs, the "bolitas" or small dark spheres common to the syndrome.

Clinical signs: This disease appears to be a distinctive variation of the NSBS noted above. The major distinction being the presence of a dark hepatopancreas or dark spherical objects ("bolitas") in the hepatopancreas and/or anterior midgut. Significant mortalities seem to always be associated with the appearance of the "bolitas". The feeding rate likewise is significantly reduced.

Pathology: Although HP and muscle atrophy is often associated with this disease, it is not unique. The distinctive gross signs of the hepatopancreas translate directly to distinctive pathology. The hepatopancreas tubules typically will contain high levels of sloughed epithelial cells, and often numerous small to large dark staining spherical bodies. These spherical bodies often appear as being expelled from B-cell vacuoles.

Geographic and host distribution: Like that of the NSBS, this disease has been noted in *Penaeus vannamei* and *P. stylirostris* hatcheries throughout Ecuador and possibly other Central and South American countries.

Etiology: Here again, bacterial toxins are suspect. This assumption is based upon the consistent presence of significant levels of chitinophilic bacteria on the cuticle of the mouth appendages, esophagus and stomach and in the HP and midgut. The assumption is also based on the apparent efficacy of some antibiotics in stopping the mortalities if provided early in the disease.

Potential treatment: The same variety of antibiotics tried against non-specific hatchery diseases have also been used against this disease. The only possible difference being that there might be even fewer compounds which have shown some degree of efficacy. The lack of controlled trials has made the evaluation process very difficult. The use of bioencapsulated chemotherapeutants may be equally applicable here as it is for the NSBS.

Luminescent bacteria syndromes

Luminescent bacteria syndrome (LBS) is another syndrome, or syndromes, which are generally very similar to the NSBS. They differ primarily in having a more distinctive, possibly smaller group of associated disease agents.

Clinical signs: This likewise is a unique variation of the NSBS. A major distinction being the presence of brown or melanized mouth appendages, esophagus and stomach region. In addition, and probably the most distinctive characteristic of afflicted larvae, is their inherent luminescence (best noted in the dark). Significant mortalities seem to always be associated with the appearance of this discoloration. The feeding rate is significantly reduced.

Pathology: For the most part, the pathology, apart from the dense bacterial colonization and melanization of the mouth/stomach region, is similar to that seen in the NSBS. Muscle and HP atrophy, likewise, are commonly associated with this syndrome.

Geographic and host distribution: It would appear from documented and undocumented reports that the occurrence of this problem has become worldwide. The predominance of reports come from *P. vannamei* and *P. monodon* hatcheries, although it is assumed that it may as well be occurring in the less popular species and in *P. chinensis* hatcheries of China.

Etiology: Bacteria have been experimentally confirmed to play a significant role in the mortalities observed in LBS. LBS-associated isolates fall within the same general groups found to be associated with NSBS, however they differ most often by being luminescent. This is not to say that luminescent bacteria can not be found in the other related syndromes, it is just that they have not been noted to date.

Potential treatment: The same variety of antibiotics tried against NSBS have also been used against this disease. However, one possible difference being that there might be even fewer compounds which show some degree of efficacy. The lack of controlled trials has made the evaluation process very difficult. The bioencapsulation procedures may be equally applicable here as it is for the two previous bacterial syndromes.

Disease syndromes of nursery and/or growout phases

Septic hepatopancreatic necrosis and/or "red disease"

Septic Hepatopancreatic Necrosis (SHPN) is a relatively recent disease, that to date has been found most commonly in cultured shrimps of Asia. It may in fact be the same disease as that previously referred to as "red disease" from *P. monodon* cultured in Asia.

Clinical signs: In this disease shrimp (juveniles or subadults) are characterized as displaying extreme lethargy and grossly opaque muscles. Such shrimp are typically found swimming near the surface and the edges of the ponds. Many of the shrimp exhibit a coloration change, ranging from very light to very dark, with actual colors ranging from red to nearly black. Afflicted populations will experience reduced feed intake and growth. These will normally be accompanied by a rapid increase in daily mortalities.

Pathology: The primary pathology noted is within the hepatopancreas. It is usually significantly atrophied. Atrophied hepatopancreata are either very pale and soft, or quite hard and very dark in color. Hepatopancreatic bacteria and/or viruses may or may not be present.

Geographic and host distribution: To date the disease has been noted primarily in *P. monodon* from Indonesia, Malaysia, Thailand, the Philippines and Taiwan. Occasionally SHPN has been observed in cultured *P. stylirostris* and *P. vannamei* in the Americas.

Etiology: Presently the etiology is unknown, as is the case of "red disease". However, in both cases the clinical signs, epizootiology and pathology suggest a bacterium is involved.

Potential treatment: Non-controlled antibiotic feed trials have been conducted in Indonesia. The results of these are scattered and inconclusive. At least one antibiotic preparation (actual incorporated antibiotic unknown) is commercially available in Asia. This and other similar antibiotic preparations are presumably available in all areas where SHPN has been noted. It has been reported that many of the compounds listed in this publication under NSBS may also have been used. In most instances, the chemotherapeutant is administered via the feed. In some instances, in particular high to super-high density systems, the chemotherapeutant may be applied directly in the water as an indefinite bath treatment.

Seagull syndrome or "síndrome de gaviotas"

Seagull syndrome (SGS) appears to be a rather unique phenomenon. In addition to distinctive pathology, SGS has a somewhat limited geographic and temporal distribution.

Clinical signs: The afflicted shrimp exhibit extreme lethargy with grossly opaque and often luminescent muscles. They are found typically swimming near the surface and the edges of the ponds. Waterbirds such as gulls or terns ("gaviotas" in Spanish) are often seen feeding on dying shrimp at the surface. Afflicted populations will often display reduced feeding rates, reduced growth and dramatic increases in mortality rates.

Pathology: Typically the shrimp display a bacteremia and multiple small melanized hemocytic nodules containing bacterial colonies. These nodules occur primarily in the lymphoid organ, heart, gills and less often scattered throughout nearly all tissues.

Geographic and host distribution: So far the disease has been limited to *Penaeus vannamei* nursery ponds in Ecuador and Texas.

Etiology: The etiology of this disease is still unclear, although a degradation (eg. eutrophication and elevated salinity) of the seawater source appears to be an important predisposing factor. Characteristic bacteria, tentatively identified as *Vibrio parahemolyticus*, *V. alginolyticus*, *V. vulnificus*, *V. damsela* and other *Vibrio* spp., have been routinely isolated from the hemolymph of afflicted shrimp and readily from the pond bottom and water. These bacteria are usually chloramphenicol sensitive, urease-positive and/or luminescent.

Potential treatment: Numerous antibiotics have been administered in the feed. These however must be provided before full scale epizootics are underway, therefore most treatment regimes have been prophylactic, as opposed to therapeutic. Sulfamethoxine+ormetoprim, oxytetracycline (and possibly other chemotherapeutants) have been used with varying success.

Texas pond mortality syndrome

Texas pond mortality syndrome (TPMS) appears to be a relatively unique shrimp disease, primarily with respect to its pathology. However, it does share many clinical signs with other shrimp pond diseases.

Clinical signs: This disease is typically characterized by any or all of the following: a significantly increased mortality rate, a decreased feeding level, very poor growth rate, soft shells, black gills, surface fouling, and/or lethargy. The shrimp are often found near the pond surface, edges and outlet.

Pathology: Severe HP atrophy is the primary pathology noted. This is usually accompanied by focal to generalized necrosis and inflammation, along with septic hepatopancreatic abscesses and granulomas. Intracellular bacteria are typically noted within hepatopancreatocytes.

Geographic and host distribution: This disease has so far been restricted to *P. vannamei* from the central and south-central Texas (USA) coast.

Etiology: Like many of the previous diseases, the etiology of TPMS is unclear. There does appear to be several agents associated with the disease. *Vibrio parahemolyticus*, *V. damsela* and *Pseudomonas* spp., all of which are non-luminescent and urease-negative have been consistently isolated from diseased shrimp. Additionally an unidentified intracellular organism or organisms is present in the HP lesions. This organism has tentatively been categorized as a probable rickettsia, chlamydia or mycoplasma-like organism.

Potential treatment: Experimental trials with oxytetracycline medicated feeds have shown some degree of efficacy and are quite encouraging. Similar new controlled field trials with oxytetracycline, and possibly several other antibiotics, are tentatively scheduled for the 1991 growing season.

Gregarines

Shrimp culturists have recently become aware of this group of protozoan parasites (Phylum: Apicomplexa, Class: Sporozoa, Order: Eugregarinida). Gregarines may have a negative impact on shrimp growth and survival, both directly as a parasite and indirectly as a predisposing factor for other diseases.

Clinical signs: Typically, gregarine-infected shrimp are very difficult to discern from non-infected shrimp, especially in those with relatively low gregarine counts; the exception being those with extreme infections which may show a yellowish coloration to the midgut. Populations with a high prevalence of gregarine infestation may often show any or all of the following: reduced growth rate, increased surface fouling, decreased feed intake and/or slightly to moderately elevated mortality rates.

Pathology: Severely infected shrimp may display significant pathology of the midgut mucosa. The gregarines are primarily found within the lumen of the midgut, and are intimately associated with the gut epithelium. On occasions, they have been found within the lumen of the anterior midgut cecum, posterior stomach, primary ducts of the HP and anterior region of the hindgut. Most frequently the large trophozoite stage is found in all but the hindgut. The hindgut typically is the site where the gametocyte stage is found. Two types of gametocytes have been noted, one containing microgametes and a second with macrogametes.

Geographic and host distribution: Gregarines have been noted in American and Asian penaeids.

Etiology: The parasitic organism may be somewhat host-specific. It has been suggested that the gregarines found in South American penaeids belong to the genera *Nematopsis* and *Cephalobus*, as may that found in the Gulf of

Mexico penaeids. The gregarine *Caridohabitans setnai* has been documented in the Asian penaeid *P. semisulcatus*.

Potential treatment: Initial reaction by shrimp culturists to the presence of gregarines in the midgut was ambivalent, for there appeared to be no direct host response. As the prevalence of the disease increased, along with the actual levels of infestation in individual shrimp, it became more apparent that the gregarines were quite probably an indirect, or possibly direct, source of growth and survival problems. Recently, individual farms and feed companies began experimenting with medicated feeds. The majority of this information has remained confidential, but it appears that most of the anti-gregarine feeds include the addition of poultry or cattle dose-levels of commercially available anti-coccidian compounds. Results of some of these medicated feeding trials would seem to indicate that some efficacy is being provided. To our knowledge, the results of controlled studies establishing efficacy has yet to be made public.

Fouling diseases

Fouling diseases (FD) is a broad category of agents (epibionts) which have the ability to infest the external cuticular surfaces of shrimp, including the gills. Problems arise typically when high numbers are present and restrict respiratory function.

Clinical signs: In spite of the diversity of epibionts potentially involved in FD, the clinical signs are nearly the same for all types. Principally, shrimp can experience reduced feeding rates, reduced growth rates and elevated mortality rates as a direct result of low to moderate levels. Higher levels of epibionts can often be directly seen, macroscopically, on gills and other cuticular surfaces. These higher levels can be directly responsible for extremely elevated mortality rates, especially when associated with marginal to poor water quality.

Pathology: Shrimp with low to moderate fouling levels are seldom noted with gross or microscopic pathology. However, high fouling levels can result in necrosis of gill epithelium (presumably due to microbial toxins) which in turn is manifested in inflammation and melanization of the gill epithelium.

Geographic and host distribution: All cultured penaeid species in all geographic regions. Some penaeid species may be more or less sensitive to specific fouling organisms than other penaeid species, and the same may hold true for sensitivity relative to the shrimp's age.

Etiology: Several major groups of organisms can foul the surface of shrimp. Among the protozoa are representative ciliates (eg. *Zoothamnium*, *Epistylis*) and flagellates (eg. *Bodo*-like organisms, choanoflagellates). Several bacterial groups or genera are commonly found, including *Leucothrix* and *Vibrio* spp. or *Vibrio*-like spp. Pennate diatoms, certain oscillatoriciae, and other algae are also frequently encountered. Metazoan fouling parasites have been noted on penaeid shrimp, but seldom pose problems, especially in cultured shrimp. In addition to the actual epibionts that actively attach to the shrimp surface, FD can also comprise various forms of debris and incidental organisms (eg. phytoplankton) trapped in the mesh of attached protozoans and filamentous organisms.

Potential treatment: Several water quality compounds have been used successfully against FD. Included would be: potassium permanganate, formalin and chelated copper compounds. The latter two, in combination, are used routinely (and very successfully) in the periodic treatment of penaeids in super-intensive ponds, tanks and raceways.

Discussion

DE KINKELIN Thankyou for your very interesting talk, because most of us are not familiar with shrimps, prawns and prawn pathology.

FLEGEL That vibriosis syndrome with the gregarine parasitism, did it come in at any particular age?

LIGHTNER It occurred in juveniles and sub-adults. There seemed to be a density effect rather than an age effect, so in nursery ponds it would occur at a density threshold and the same seemed to be true in grow out.

ALDERMAN You have indicated the use of antibiotics quite extensively to control many of these diseases. Do you have any thoughts on problems that areas such as the EC should be considering in regard to potential drug residues in shrimps as a result of the use of antibiotics to control these bacterial diseases?

LIGHTNER Clearly it is in the interests of the shrimp farming industry to use these substances responsibly and I think that the first step is to know what withdrawal times should be, particularly in the juvenile and grow out stages. In larvae, of course, there is an inherent withdrawal time which makes the chances of residues fairly remote. There are studies being done in cooperation with some of the pharmaceutical companies that are interested in shrimp culture to determine what residues should be under typical temperature regimes of culture conditions. So there is some data available now and there is more being accumulated.

FLEGEL For the gregarines, do they use those anticoccidial drugs for treatment or for prevention?

LIGHTNER Again, my experience is in Ecuador. I am not exactly sure how they do it. Many of the companies consider this to be proprietary, but I believe what happens is that when the gregarine level at a given farm on the average of several of the ponds begins to exceed about 1000 individual trophozoites per shrimp intestine, then they begin to use the anticoccidials. They try to keep the level below say 100 per shrimp. I think they believe that 1000 or so starts to hurt them economically in terms of growth rate, perhaps even some mortalities.

JOHNSTON In your non specific bacterial syndrome in hatcheries you mentioned chloramphenicol and nitrofurans as the chemotherapeutants. What is the mode of application to the shrimps?

LIGHTNER They are added as bath, directly to the water.

STRUB Could you tell us something about the shrimp's own defence mechanisms, specific or unspecific defence mechanism against all these kinds of infections. Is there any, what would be the normal regulation of such a disease?

LIGHTNER That is kind of a complicated question, but shrimps belong to the Arthropod phylum and arthropods lack the ability to produce specific antibodies, so their responses are non specific generally, with cellular response perhaps the first order of protection with agglutinins also playing important roles. So resistance by shrimp to a given pathogen is going to be defined genetically rather than as inducible kinds of responses.

EFFECT OF OXYTETRACYCLINE AND SULPHAMETHAZINE ON WEIGHT GAIN AND SURVIVAL OF *PENAEUS MONODON* UNDER STRESS

I. Higuera-Ciapara, J. H. Brown & K. Jauncey

Institute of Aquaculture, Stirling University, Stirling, Scotland FK9 4LA.
United Kingdom.

Summary: *Two size groups (postlarvae and adults) of black tiger shrimp* P. monodon, *were fed a medicated pellet containing 250 mg* kg^{-1} *of oxytetracycline hydrochloride and 250 mg* kg^{-1} *of sulphamethazine sodium salt for four weeks. Both groups were exposed to severe stress due to infection by* Vibrio *spp. Weight gain, survival and antibiotic residues were measured weekly in representative samples.* Vibrio *counts on the abdominal muscle were performed at the start and end of the experiment. No difference in weight gain was evident from the use of the medicated feed in any of the experimental groups according to statistical comparison of the regression lines* ($p < 0.05$) *but survival of postlarvae was significantly higher in the antibiotic-fed shrimp. Residues of oxytetracyline at the end of the trial were 150-200 µg* kg^{-1} *while those of sulphamethazine were 200-250 µg* kg^{-1} *in the adult shrimp.*

***Résumé*: Effets de l'oxytétracycline et de la sulfaméthazine sur la croissance et la survie de *Penaeus monodon* soumis à des agressions microbiennes.**

Deux groupes (postlarves et adultes) de crevette tigre Penaeus monodon *ont été nourris pendant 4 semaines avec des granulés contenant 250 mg/kg de chlorhydrate d'oxytétracycline et 250 mg/kg de sulfaméthazine sodique. Les deux groupes ont subi des atteintes microbiennes sévères causées par* Vibrio *spp. La survie, le gain pondéral, et la teneur en résidus thérapeutiques étaient établis chaque semaine sur des échantillons représentatifs. La numération des* Vibrio *dans les muscles abdominaux a également été effectuée d'un bout à l'autre de l'expérience. Aucune différence significative n'est apparue dans le gain pondéral des groupes témoins et des groupes antibiosupplémentés lors de l'analyse des courbes de régression* ($p < 0{,}05$)*. Par contre la survie des postlarves a été significativement plus élevée chez les crevettes ayant reçu les antibiotiques. Les résidus d'oxytétracycline mesuré chez les crevettes adultes à la fin de l'expérience étaient de 150 à 200 µg/kg contre 200 à 250 µg/kg pour la sulfaméthazine.*

An urgent need exists to generate quantitative information on various aspects of antibiotic use in shrimp aquaculture i.e. residue accumulation under stress conditions and effect of sub-therapeutic and therapeutic treatments on overall performance. In spite of this, there has been very little research on the issue of withdrawal times and therapeutic effectiveness of common treatments (3,4,10). Recent evidence has suggested that the role played by opportunistic bacteria, particularly *Vibrio* spp., may be essential in understanding mass mortality incidents (7). This investigation was designed to study the combined effect of oxytetracyline and sulphamethazine on the performance of two size groups of shrimp under severe stress due to *Vibrio* spp. The combined effect of these and other antibiotics, could help to decrease the high concentrations (> 5,000mg kg^{-1}) currently used in commercial operations which cause adverse impact on the environment due to pellet leaching.

Materials and Methods

Experimental design

Shrimp were maintained in plastic aquaria with 1 m^2 surface area containing approximately 50 l of seawater. These aquaria were kept at a temperature of 28 ±2°C. A recirculating system comprising a biological filter and a protein skimmer was used to maintain water quality parameters under normal conditions for *P. monodon* culture. Salinity was maintained at 33°/$_{oo}$ throughout the experiment. All the aquaria had their lids removed for approximately six hours every day.

Each aquarium was stocked with three shrimp in the case of the adult group, and with thirty-five shrimp in the case of the postlarvae group. Postlarvae were from a single spawning from Frippak Feeds (Findon, Scotland) and grown in the Tropical Prawn Unit (TPU) of the Institute of Aquaculture until they each weighed 0.15g. The adult shrimp were purchased from a commercial producer in England. The initial mean weight per tank in the adult group was 36.7 g and 4.9 g in the postlarvae group. Both groups were fed the same pellet formulation which was prepared according to a standard formula used in the TPU. Oxytetracycline hydrochloride and sulphamethazine sodium salt (Sigma) were mixed in powdered form with the rest of the ingredients to make a final concentration of 250 mg kg^{-1} of each drug in the pellet. Individual pellets fed to the adult group averaged 0.70 cm in length while those fed to the postlarvae averaged 0.25 cm. Average diameter was 0.25 mm. The amount of feed given to each experimental group was increased or decreased daily, depending on the amount eaten, so that shrimp always had food available and leaching of the pellet was minimized. Growth and survival were determined every week during a four week period and dead shrimp were removed daily.

Antibiotic detection and quantification

The Charm-7000 System was used for the detection and quantification of oxytetracycline and sulphamethazine in the tail of the adult shrimp. Extracts were prepared by high-speed homogenization of equal amounts of shrimp muscle and a buffer provided with the Charm System, incubation at 60°C for 30 min., centrifugation for 10min. at 1600 g and collection of the clear supernatant for the test. When necessary, pH was brought to 7.5-8.5 according to instructions given by the equipment supplier (8). The Charm System uses a radioactively-labelled antibiotic which competes for specific binding sites on a cell or ribosome with the contaminating drug (2). The counts per minute (cpm) are, therefore, inversely proportional to the amount of antibiotic present in the sample. Briefly, the sample is first incubated with the binding substrate for 2 min. at 35°C, then with the radioactive antibiotic tracer at the same temperature and centrifuged to collect a solid pellet containing the bound drugs. This pellet is resuspended in a scintillation cocktail and read for one minute on the ^{3}H channel of a scintillation counter.

Vibrio identification and enumeration

Thiosulphate citrate bile salt sucrose agar (Oxoid) was used to assay shrimp muscle tissue for the presence of *Vibrio* spp. by recommended procedures (1). *V. alginolyticus* was differentiated from other species by yellow colonies and ability to grown on 8% NaCl. *V. parahaemolyticus* was differentiated by green colonies and equal tolerance to salt, i.e. 8%. Also, histochemicalexamination of muscle was carried out to look for evidence of necrotic damage.

Results and discussion

Growth and survival

The addition of oxytetracycline and sulphamethazine to the pellet formulation did not exert any statistically-significant effect on weight gain in either of the experimental groups (Figs. 1 & 2).

In the adult group, the antibiotic-fed shrimp decreased the total amount of feed consumed as compared to the control and only a marginal increase in weight was evident throughout the experiment. During the period of acclimatization, several mortalities occurred unexpectedly. *Vibrio alginolyticus* and *V. parahaemolyticus* were thus identified in the control as well as in the antibiotic-fed shrimp at the start of the experiment. However, no physical signs of ulceration in the body surface could be observed and histological examination showed that the infection had not produced any melanized nodules typical of advanced vibriosis.

A threshold infectious level for the manifestation of disease has been reported for *Vibrio parahaemolyticus* (7) in juvenile shrimp. However, little information is available on pathogenicity of *Vibrio alginolyticus* to penaeid shrimp.

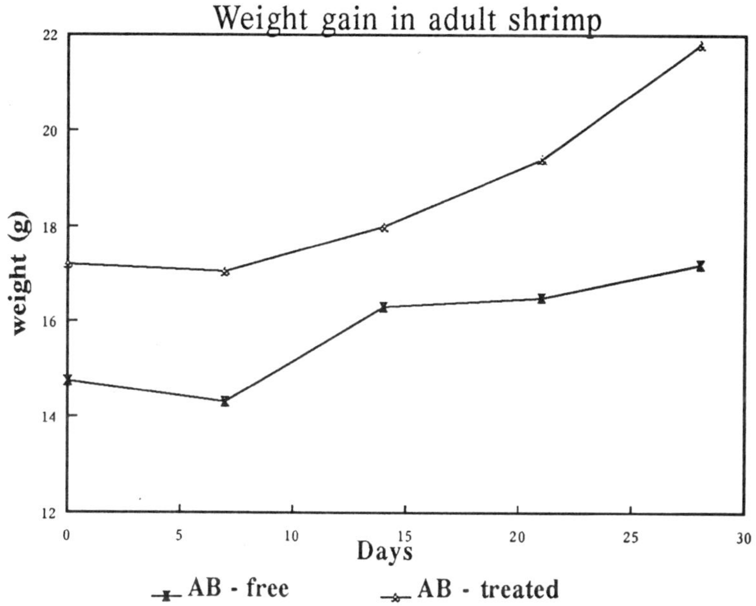

Fig 1 Weight gain in adult shrimp

In this study, weight gain was severely impaired by *Vibrio alginolyticus* in both experimental groups. The negative control in the adult group showed an increase of only 2.6 g during the four-week trial, while the antibiotic-fed shrimp increased only 4.4 g on average. The average mean weight of each group was used to construct the growth response curves. The slopes of these curves were then compared by a two-tailed F test according to standard procedures (9). No significant difference was evident from such analysis (Table I). Moreover, the mortality rate was identical in both groups (16%).

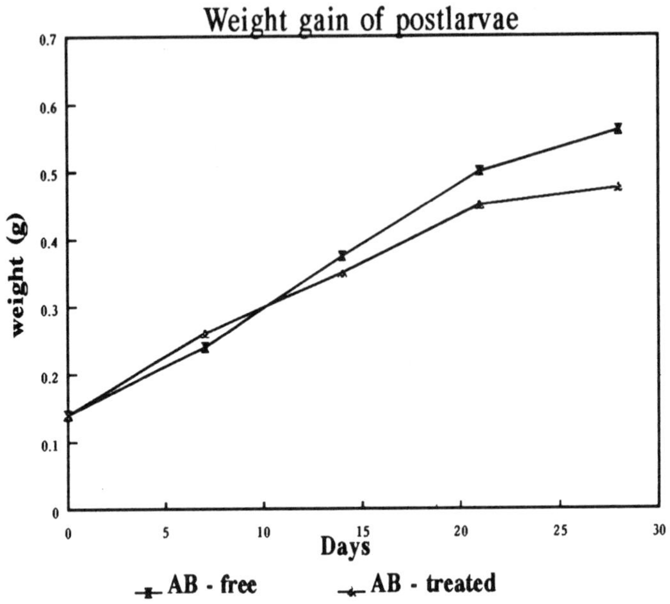

Fig 2 Weight gain of postlarvae

Table I: Comparison of regression lines: Antibiotic-free and antibiotic-treated shrimp: adult group

		df	x2	xy	y2	Reg. coef	Deviations from regression df	SS	MS
	Within								
1	AB-free	4	1470	1166	1265	.79	3	232	77
2	AB-treated	4	1470	1388	1861	.94	3	299	99
3							6	531	176
4	Pooled	8	2940	2555	3126	.87	7	617	88
5	Difference between slopes						1	86	89

F = 0.50 (df 1, 6): not significant.

In the postlarvae group, no statistically significant difference in growth rate was evident either. Hwever mortality was 34% in the antibiotic fed tank and 40% in the control.

The inclusion of oxytetracycline in amounts as low as 100 mg kg^{-1} in shrimp feed has been previously shown to result in a decrease in pellet acceptability but a favourable growth response (4). In this experiment, however, no beneficial effect could be assessed from the combination of that drug and sulphamethazine at the 500 mg kg^{-1} level.

Antibiotic residues

The concentration of oxytetracycline and sulphamethazine was determined weekly in two shrimp tails of the adult group (Table II). After two weeks, the levels of oxytetracyline were 100-150 μg kg^{-1} and those of sulphamethazine were slightly higher (150-20 μg kg^{-1}). They increased to approximately 200 and 250 μg kg^{-1} respectively during the third week and remained at the same concentration until the end of the experiment. These levels are substantially lower than those previously reported in the literature (4). They reflect reduced feeding rates by the shrimp because of low palatability of the diet or possibly, because of the vibriosis detected. The levels of antibiotic in the tissues did not reach the minimum inhibitory concentration (MIC) of oxytetracycline reported, for most *Vibrio* spp., i.e. 200-400 μg kg^{-1} (10) until the third week.

Sulphamethazine was detected after seven days of feeding the medicated pellet, whereas oxytetracycline was detected after two weeks. Oxytetracycline has been reported to be poorly absorbed from the gastrointestinal tract of aquatic animals due to complex formation with calcium ions (5). The Charm System is known to be more sensitive to sulpha drugs than to the tetracylines (8). While this has not been clearly established for shrimp, preliminary work indicated a minimum detectable level of 50 μg kg^{-1} for the sulphas and 100 μg kg^{-1} for oxytetracycline.

It is also important to note that due to the relatively high seawater temperatures needed for shrimp farming, leaching of antibiotics and other soluble compounds from the pellet can be very rapid. Leaching of oxytetracycline at 28°C and pH 7.5 in seawater can be as high as 50% in four hours (6). This is quite relevant to the present investigation because it is known that during vibriosis outbreaks shrimp are less attracted to feed and therefore take a much longer time to ingest it. This results in a large proportion of the antibiotic being lost into the environment.

***Vibrio* identification and enumeration**

Results from *Vibrio* analysis on shrimp muscle from both groups are presented on Table III. *Vibrio alginolyticus* comprised more than 95% of the colony forming units detected. *Vibrio* spp. decreased more rapidly in the antibiotic-fed shrimp than in the negative controls. However, the levels found at the end of the experiment suggest that their detrimental effect was not fully overcome through the use of the oxytetracycline-sulphamethazine combination in this experiment.

Table II: Antibiotic concentration in the muscle of adult shrimp

	Control ^3H CPM		`Positive ^3H CPM		Eq. conc. $\mu g\ kg^{-1}$	
Days	OTC	SMZ	OTC	SMZ	OTC	SMZ
0	1,018	5,600	-	-	-	-
7	1,049	6,693	1,021	1,328	-	40-50
14	1,154	7,352	763	1,361	100-150	150-200
21	1,448	5,942	658	744	150-200	200-250
28	1,266	4,836	702	532	150-200	200-250

Average of three or more readings

Table III: *Vibrio* counts of experimental shrimp (cfu g^{-1})

	Postlarvae	
Days	Control	Antibiotic-treated
0	1.2×10^2	3.0×10^2
28	1.5×10^2	1.3×10^2

	Adult shrimp	
Days	Control	Antibiotic-treated
0	5.2×10^3	7.5×10^3
28	4.2×10^2	2.5×10^3

It is well known that when the antibiotic concentration in the tissues does not reach the MIC for the microorganism of concern, elicitation of resistance may occur quickly due to the positive pressure of the drug which triggers highly specific defense mechanisms in the bacterial cell (11).

It is suggested that a high concentration of antibiotic in the pellet (>5,000 ppm) would be more favourable in providing the appropriate amounts of antibiotics to the shrimp. However, novel ways to protect the environment from antibiotic leaching should be investigated.

Acknowledgements: The support of the Overseas Development Administration of the UK Foreign and Commonwealth Office for the penaeid shrimp research at the Institute of Aquaculture is gratefully acknowledged. The authors are grateful to the EC Directorate for Science and Technology for the fellowship for Dr Higuera-Ciapara and to Ms Sarah Christie for technical assistance. The generosity of Frippak Feeds, Findon in supplying penaeid nauplii is very much appreciated

References

1). AUSTIN J. P. & AUSTIN D. A., 1989. Methods for the Microbiological Examination of Fish and Shellfish. *Ellis Horwood Series in Aquaculture and Fisheries Support*. Ellis Horwood LTD, Chichester, England, pp. 206.

2). CHARM S. E. & CHI, R., 1988. Microbial receptor assay for rapid detection and identification of seven families of antimicrobial drugs in milk: a collaborative study. *J. Assoc. Off. Anal. Chem.*, **7** (2), 304-316.

3). CORLISS J. P., LIGHTNER D. & ZEIN-ELDIN Z. P., 1977. Some effects of oral doses of oxytetracycline on growth, survival and disease in *Penaeus aztecus*. *Aquaculture*, **11**, 355-363.

4). CORLISS J. P., 1979. Accumulation and depletion of oxytetracyline in juvenile white shrimp, *P. setiferus*. *Aquaculture*, **16**, 1-16.

5). CRAVEDI J. P., CHOUBERT G. & DELOUS G., 1987. Digestibility of chloramphenicol, oxolinic acid and oxytetracycline in rainbow trout and influence of these antibiotics on lipid digestibility. *Aquaculture*, **60**, 133-141.

6). HIGUERA-CIAPARA I., BROWN J. H. & JAUNCEY K., 1990. Leaching of oxytetracycline from pelleted shrimp feeds. Poster Paper presented at the Conference on Bacterial Diseases of Fish. Institute of Aquaculture, Stirling University, Stirling, Scotland. June.

7). NASH G. L., 1990. *Penaeus monodon* growout diseases. Paper presented at the Aquatech '90 Conference, Putra World Trade Center, Kuala Lumpur, Malaysia. 35pp.

8). PENICILLIN ASSAYS INC., 1990. The Charm System for Monitoring Antibiotics in Tissue. *Operator's Manual*. Malden MA. USA. 12pp.

9). SNEDECOR G. W. & COCHRAN W. G., 1980. *Statistical Methods*. Iowa State University Press. Ames, Iowa. USA., p. 386-387.

10). TAKAHASHI Y., ITAMI T., NAKAGAWA A., NISHIMURA H. & ABE T., 1985. Therapeutic effects of oxytetracycline trial tablets against

Vibriosis in cultured Kuruma prawns. *Penaeus japonicus* Bate., **51** (10), 1639-1643.

11). VASQUEZ-MORENO L., BERMUDEZ M. C., LANGURE A., HIGUERA-CIAPARA I., DIAZ de AGUAUYO. & FLORES E.,1990. Antibiotic residues and drug resistant bacteria in beef and chicken tissues. *J. Food. Sci.*, **55** (3), 632-634.

LE LEVAMISOLE - UN AGENT IMMUNOSTIMULANT ET THERAPEUTIQUE CONTRE LA SAPROLEGNIOSE DE LA CARPE

M. Prost, & A. Sopinska

Centre de Pathologie des Poissons, Faculté Vétérinaire, Académie d'Agriculture de Lublin, Pologne

Résumé: *On a étudié les réactions cellulaires de carpes atteinte de saprolégniose et traitées au vert malachite et au lévamisole, en comparaison avec des carpes indemnes. L'activité cellulaire était appréciée selon trois méthodes: le test NBT, adapté aux recherches chez les poissons; le test des rosettes E; et le test biochimique ANAE. Les résultats du travail ont demontré que: 1/ l'infection par* Saprolegnia *stimule les défenses cellulaires de la carpe, 2/ le lévamisole est un immunostimulant efficace dans l'évolution et l'involution de la saprolégniose, 3/ les bains thérapeutiques et les injections intraperitonéales de levamisole ont des effets semblables; en pratique on peut recommander la première méthode. 4/ les résultats obtenus ne permettent cependant pas d'établir avec certitude l'existence d'une activité immunosuppressive du vert malachite.*

Summary : **Levamisole : an immunostimulant and therapeutant against Saprolegniasis on the carp.**

Cellular immunoactivity was studied in carp, both healthy and infected with Saprolegniasis treated with malachite green and levamisole. assessment of cellular activity was performed following the following methods : the NBT test modified for fish studies and the rosette E and biochemical test. The following results were obtained : Saprolenia infection stimulates the cellular immunoactivity of carp, levamisole appared to be an effective immunostimulant and therapeutic agent during the course and recovery from saprolegniasis. Levamisole applied by bath immersion and intraperitoneal injection produces similar results in fish so that for practical applications bath treatment can be recomended. The results obtained do not show clearly that malachite green shows imunosuppressive activity.

Introduction

Le traitement de la saprolégniose - maladie fréquente chez les poissons d'élevage - est realisé surtout avec le vert malachite. C'est un remède toxique, mutagénique (4), pourtant employé jusqu'à présent à cause de son exceptionnelle efficacité fongicide.

Des résultats positifs obtenus dans l'immunostimulation du système leucocytaire de la carpe (1, 3) nous ont engagés à tenter l'application de cette méthode à la lutte contre la saprolégniose de la carpe.

Materiel et méthodes

On a utilisé un lot de 50 carpes (K1) d'un poids de 120-150 g qui présentaient des symptômes typiques de saprolégniose, ainsi que 10 carpes saines de poids semblable.

L'activité immunologique du système leucocytaire a été évaluée chez les carpes infectées et les témoins selon trois méthodes: 1/ le test NBT, qui permet de mesurer l'activité métabolique des neutrophiles et s'avère bien adapté aux recherches sur poissons (2), 2/ le test des rosettes E, qui permet de déterminer le nombre de rosettes formées à partir de 500 lymphocytes T dénombrés dans le sang des poissons, et 3/ la méthode cytochimique ANAE, qui permet de différencier les lymphocytes B et E en recherchant l'activité de l'acide-naphthyl-acétate-estérase (ANAE), présente uniquement dans les membranes des lymphocytes T. Les résultats sont exprimés en pourcentage de cellules ANAE positives (ANAE +) dans le sang des carpes. La description précise de ces techniques a été donnée dans un travail précédent (2).

Comme critères supplémentaire on a retenu également, en se référant aux méthodes classiquement employées, le nombre total des leucocytes et le nombre de neutrophiles par mm3 de sang.

Le traitement des carpes atteintes de saprolégniose a été éxécuté par balnéation en solution de vert malachite cristallisé et exempt de zinc, à la concentration de 1 g pour 150 l d'eau et pendant une heure. Ces conditions sont celles appliquées en pratique dans les élevage de carpes. La température de la solution thérapeutique était de 15°C.

Le lévamisole (Jansen) a été employé comme agent immunostimulant. Son efficacité comme vermicide et comme stimulant de l'activité des lymphocytes T et des cellules phagocytaires est bien connue chez les animaux domestiques. Les carpes ont reçu individuellement 0,1 mg de lévamisole en injection intrapéritonéale (IP), ou ont été placées pendant 4 jours dans un bac contenant 75 mg de produit pour 75 l d'eau. Ces modes d'administration étaient définis d'après des expériences préalables.

Les carpes ont été réparties en six groupes de 10 poissons, à savoir:

I - carpes témoins, saines, n'ayant reçu aucun traitement.

II - carpes infectées par la saprolégniose, non traitées

III - carpes malades, traitées au vert malachite

IV - carpes malades traitées au vert malachite et stimulées 24 heures plus tard par injection IP de lévamisole

V - carpes malades, stimulées par lévamisole administré en bain
VI - carpes malades, stimulées par injection IP de lévamisole

Des prises de sang ont été effectuées, 7 jours après le traitement au vert malachite dans le groupe III, 7 jours après l'injection IP de lévamisole dans les groupes IV et VI, et 3 jours après l'exposition en bain de lévamisole dans le groupe V. Cette façon de procéder s'avère la plus intéressante d'après les résultats d'essais précédents.

On a comparé statistiquement les résultats quantitatifs obtenus par le test t de Student, le degré de signification étant établi au risque d'erreur $p \leq 0,05$.

Résultats et discusssion

La comparaison des résultats entre les groupes I et II (Tableau I) montre qu'au cours de la saprolégniose l'activité leucocytaire est augmentée: celà s'exprime par l'augmentation du nombre des leucocytes et des neutrophiles, et par une augmentation d'activité des lymphocytes T se traduisant par la formation d'un plus grand nombre de rosettes E. Pourtant cette activation n'est pas détectée par la methode ANAE, sans qu'on puisse bien en expliquer les raisons. La chute significative de l'activité métabolique des neutrophiles

Tableau I. Résultats des recherches ($x \pm s$)

N° du groupe	Nombre de leucocytes dans 1 mm³ de sang	Nombre de neutrocytes dans 1 mm³ de sang	mg de formazane NBT dans 1 mm³ de sang	mg de formazane NBT dans 10^6 neutrocyte	Nombre de rosettes E pour 500 lymphocytes	Pourcentage de lymphocytes ANAE
I	19300 ± 2055,8[a]	1346,2 ± 433,1[a]	0,077 ± 0,005[a]	0,654 ± 0,306[a]	4,98 ± 1,89[a]	56,38 ± 9,04[a]
II	51600 ± 3657[b]	19573 ± 4007[b]	0,056 ± 0,003[a]	0,029 ± 0,005[b]	36,87 ± 5,95[b]	40,46 ± 6,47[b]
III	31800 ± 4147[a]	8984 ± 2246[a]	0,027 ± 0,003[a]	0,030 ± 0,007[a]	9,1 ± 3,15[a]	54,36 ± 9,92[a]
IV	47530 ± 4792[b]	9192 ± 196,07[b]	0,078 ± 0,005[b]	0,088 ± 0,022[b]	27,87 ± 2,92[b]	64,62 ± 6,95[b]
V	67400 ± 3565[a]	15529 ± 3739[a]	0,078 ± 0,005[a]	0,052 ± 0,013[a]	47,44 ± 3,30[a]	76,07 ± 8,37[a]
VI	58700 ± 5207[a]	6514 ± 2388[b]	0,080 ± 0,009[a]	0,142 ± 0,060[b]	47,11 ± 9,42[a]	73,52 ± 7,53[a]

Explications: dans tous les groupes expérimentaux une différence significative pour $p \pm 0,05$ est exprimée par des lettres différentes (a, b); l'absence de signification est indiquée par des lettres identiques (a, a ou b, b); la comparaison a été exécutée entre le groupes I et II, III et IV, V et VI.

On constate que le vert malachite provoque une réduction du nombre des leucocytes et des neutrophiles, ainsi que du nombre des rosettes E formées par les lymphocytes T (groupes II et III). Ces résultats pourraient signifier que cet agent thérapeutique manifeste des propriétés immunosuppressives. Pourtant, les résultats obtenus par les tests NBT et ANAE ne le confirment pas.

L'immunostimulation par le lévamisole, tant par voie intrapéritonéale que par bain, conduit à des résultats semblables (groupes V et VI). Appliquée après traitement par le vert malachite elle provoque une augmentation significative de tous les paramètres mesurés (groupes III et IV). Pourtant les valeurs enregistrées après immunostimulation des poissons non traités (groupes V et VI) sont plus elevées que celles des poissons traités (groupe IV). On peut penser que chez les poissons du groupe IV l'effet immunostimulant a été contrarié par l'influence du vert malachite.

On a observé chez les carpes malades traitées par le lévamisole une disparition plus rapide du mycelium de *Saprolegnia*, ainsi qu'une amélioration de l'état général plus nette que chez les carpes malades non immunostimulées. Les mycoses, chez les poissons n'ayant reçu aucun traitement, ont été très sévères, affaiblissant considérablement les carpes. Les seuls cas de mortalité sont survenus sur des animaux non soumis à l'immunostimulation.

Les résultats de ces recherches permettent de constater que le lévamisole est un agent immunostimulant efficace, et que cette activité lui confère un réel intérêt dans le traitement des saprolégnioses de la carpe.

Références

1). PROST M., SOPINSKA A., 1988. Ocena stanu zdrowotnego karpi na podstawie bdan hematologicznych podczas inwazji pijawek oraz po leczeniu Neguvonem i zastosowaniu preparatu immunostymulacyjnego. *Medycyna Wet.*, **44**, 455-457.

2). SOPINSKA A., 1985. Przydatnosc testu NBT do badan aktywnosci metabolicznej granulocytow krwi obwodowej karpi. *Medycyna Wet.*, **41**, 738-741.

3). SIWICKI A.K., 1989.Immunostimulating influence of levanisole on nonspecific immunity in carp (*Cyprinus carpio*). *Developmental and Comparative Immunology*, **13**, 87-91.

4). STEFFENS W., LIEDER U., NEHRING D., HATLOP H.W., 1962. Möglichkeiten und Gefahren der Anwendung von Malachitgrün in der Fischerei. *Z.Fisch.*, **10**, 745-771.

AN OVERVIEW OF THE REGULATORY ASPECTS OF CHEMOTHERAPY IN AQUACULTURE

R. A. Schnick

National Fisheries Research Center, 2630 Fanta Reed Road, La Crosse, WI 54603-1223, USA

Summary: *Few drugs are registered worldwide for use in aquaculture. One major obstacle to this registration is the various regulations applied to aquaculture drugs. Another major obstacle to this registration is the great expense to the pharmaceutical industry for completing extensive requirements in relationship to potential markets. Mechanisms must be found to encourage the generation of data that would be acceptable in many countries so that drugs could be marketed worldwide without any repeated studies. With international and united efforts the aquaculture industry could become attractive to the pharmaceutical industry.*

***Résumé*: Aspects de la réglementation en thérapeutique aquacole**

La sécurité du consommateur tient une place privilégiée dans la réglementation des médicaments employés en aquaculture, dont l'innocuité doit impérativement être prouvée. Cette contrainte se heurte cependant au problème de l'inadaptation ou de la rareté des produits actuellement autorisés pour affronter les situations pathologiques (3 produits sont autorisés aux Etats-Unis, 8 en Europe, 24 au Japon). De nombreux pays s'efforcent de développer des règles d'évaluation tenant compte des législations déjà en vigueur pour d'autres médicaments vétérinaires, des protocoles d'étude élaborés chez les mammifères ou pour l'environnement, et de la spécificité des espèces aquatiques. Les difficultés liées au contexte aquicole sont réelles, les modes d'administration par exemple compliquant la distinction entre médicaments et pesticides.

En Afrique, en Asie et en Amérique latine, les réglementations sont rares. De sérieuses complications en résultent pour l'exportation. Aux Etats-Unis, la sécurité, l'efficacité, et la toxicité des substances médicamenteuses pour les organismes vivants et l'environnement sont contrôlées par la Food and Drugs Administration (FDA) et l'Environment Protection Agency (EPA), dont les compétences ne sont pas toujours bien tranchées. La question très aiguë des métabolites et de la dynamique des résidus a conduit à concevoir des codes de bonnes pratiques, tant pour les recherches de laboratoire qu'aux stades de fabrication. La multiplication des études, cas par cas et espèce par espèce, alourdit cependant les coûts financiers. Des dispositions plus souples sont prévues pour des produits dits "d'usage restreint", mais leur mise en oeuvre est lente. Au Canada, le Bureau des Médicaments Vétérinaires se montre encore plus sévère, tandis qu'au Japon les dispositions sont les mêmes que pour les produits destinés aux espèces terrestres.

La CEE a constitué un Comité des Produits de Médecine Vétérinaire qui dès 1992 aura précisé les notions d'efficacité, et de sécurité pour l'espèce cible, de risques pour le consommateur, pour l'utilisateur et pour l'environnement. Selon les produits les résidus seront négligés ou devront obéir à des limites fixées provisoirement ou définitivement. Certains composés seront interdits. L'étude des résidus, également codifiée, doit être menée à plusieurs températures et exprimée en degrés-jours. Dès 1997 tous les produits devront être approuvés. Jusque là le droit de prescription vétérinaire ou le recours à des substances d'emploi traditionnel sont encore reconnus.

Le coût des études d'évaluation, l'hétérogénéité des réglementations, l'étroitesse du marché aquicole, découragent les industriels et expliquent la pénurie en médicaments autorisés. Seules peuvent y pallier, des solutions conçues à l'échelle internationale, avec le concours de tous les acteurs concernés. Les agences de décision, pour développer des règles adaptées à l'aquaculture, doivent se soucier d'en connaître le contexte, se donner les moyens humains et techniques de leur politique, instaurer un dialogue permanent entre elles-mêmes, les chercheurs, les producteurs et les organisations internationales. Ces dernières paraissent bien placées pour promouvoir le soutien des recherches, la coordination des législations et des méthodes de contrôle, l'adoption de codes internationaux, ainsi que pour assurer efficacement la circulation de l'information.

Introduction

The consumer plays a very important role in the use and registration of aquaculture drugs and chemicals. Proof must be given to the consumer that the food products they eat are safe. In 1991, the National Academy of Sciences (NAS) in the United States released a document in which they concluded that chemicals are not effectively controlled in aquaculture. Because of the lack of control, NAS recommended that additional studies be completed on contaminant risks and that all imported sea food must meet U.S. standards for safety (1). A balance needs to be found between the need to protect the consumer and the environment and meeting the needs of the aquaculture industry. The growth of aquaculture is increasing throughout the world with few compounds fully registered to meet the increase. Where products are regulated, losses are increasing because there are not enough effective compounds registered for the diseases, especially bacterial diseases. Currently, the United States has 3 therapeutic drugs registered and available in the registered form, Canada has one, the European countries have 8 (with Germany having an additional 7) and Japan has 24 (11). Each country, or region, as in the case with the 12 countries in the European Economic

Community, is trying to develop regulations and guidelines to protect the human food supply while addressing the unique conditions of raising a variety of species in an aquatic medium. The solutions are not easy. There are really three major factors to consider: a country's regulations, laws, and policies on drugs and chemicals used on all food animals, the existing protocols for required studie to assess mammalian and environmental safety, and the development of protocols to test chemicals on aquatic species.

These factors are complicated by the application methodology used in aquaculture; namely, the use of chemicals applied directly to the water to control disease organisms on the cultured species. No country has completely solved the question whether this constitutes a drug or pesticide use.

Regulation and registration of aquaculture chemicals in United States, Canada, Europe, and Japan

The control of aquaculture drugs varies among countries from no regulation to very restrictive regulations. Generally, few requirements are needed for a therapeutant to be licensed for use in South America, Africa, and most of Asia. Seafood-exporting countries are increasingly concerned because other countries may no longer accept their products without a guarantee that the products contain no chemical residues of concern. At the present time, each country or region sets regulations and data requirements needed for aquaculture drugs. There is a general awareness that regulations need to be adjusted to accommodate the uniqueness of aquaculture.

United States

In the United States, the application of chemicals to organisms or to their environments is regulated by the U.S. Food and Drug Administration (FDA) and the U.S. Environmental Protection Agency (EPA). FDA controls the use of drugs and anesthetics and EPA controls the application of chemicals and pesticides to the environment. In cases that involve water treatments to control pathogens, the jurisdiction becomes unclear and has been changed over time. Decisions on jurisdiction need to be made on pesticides, like copper sulfate, that have therapeutic properties. At present, each agency develops appropriate guidelines and policies to implement the laws for their respective fields of responsibility (10).

Data requirements for therapeutant approval include information on human safety, efficacy against target organisms, toxicity to nontarget organisms, residues in food animals, and effects on the environment (15, 17). For a variety of reasons, both EPA and FDA are becoming more concerned over the potential adverse effects of chemicals released in aquacultural effluents. FDA recommends a tiered approach to testing to determine environmental impact (14). For drugs used on food animals, data are required on metabolites and residue dynamics in order to establish tolerance levels and withdrawal times. Studies to generate the required data must be conducted according to Good Laboratory Practices (GLP) requirements and each drug must be produced and formulated according to Good Manufacturing Practices (GMP) regulations. These GLP and GMP requirements significantly increase the cost of developing a new drug (10, 16).

Since 1983, FDA has attempted to help such minor-use industries as the aquaculture industry by registering drugs under a minor-use policy that (i) permits the extrapolation of data from a major animal species to a minor species, (ii) accepts data on inherent withdrawal times on a case-by-case basis, and (iii) provides guidelines for the preparation of data for minor species (14). Registering drugs for aquaculture is difficult because the minor-use guidelines apply better to terrestrial animals than they do for aquatic species. There are few, if any, registrations of drugs used on major species that are needed for use in aquaculture. Thus, few or no mammalian safety or environmental data are available for those drugs (12).

Ten years of effort was required to establish the minor-use program at FDA. Since the beginning of the minor-use policy, FDA has provided funding for minor-use workshops, encouraged the funding of drug studies, and recognized that available funds are not adequate to do the required studies. FDA's position is that the U.S. Congress will have to fund the studies needed to resolve the minor-use problem.

An Aquaculture Committee has been set up by FDA to discuss the problems faced by aquaculture. Since the aquaculture industry is relatively new in the United States, FDA is being cautious as they are not familiar with cultural practices. FDA needs to know the potential toxicity of veterinary drugs before they register a product, but they need data for evaluation. Without knowledge of the safe level for each drug, FDA has difficulty in determining the level of sensitivity needed for the analytical methodology for detection of the drug in animal tissue. FDA wants additional time to become more knowledgeable about aquaculture. In the future, FDA might reduce data requirements for drugs or chemicals used on early life stages (eggs, fry, and fingerlings) and change their present policies. In the meantime, FDA plans to review the drugs on a case-by-case basis. FDA feels that the Federal Food Drug and Cosmetic Act (FFDCA) allows little or no flexibility to change regulations and that the aquaculture industry would not be served well by increased flexibility because GLP and GMP requirements can not be relaxed. FDA is being careful not to overreact to pressure from the aquaculture industry because of scrutiny by the public to any relaxation of regulations.

Canada

Requirements for registration data in Canada are similar to requirements in the United States. However, Canada has significantly different regulations and approval processes (6). Drugs commonly used in Canadian aquaculture are either licensed for other food animals and prescribed by veterinarians or not specially banned and used in an emergency under the direction of the Canadian Bureau of Veterinary Drugs (4, 6). The agency is concerned about the lack of data on the pharmacokinetics of fishes, especially the differences in uptake of drugs at a range of temperatures. Chloramphenicol and tributyltin compounds are the only two classes of compounds that never can be used in Canadian aquaculture. Canadian regulations also differ from the United States in that they have no minor-use policy or classifications such as Generally Recognized as Safe Compounds.

Europe

The European countries will implement compulsory, uniform regulations on all veterinary drugs in 1992 for the 12 countries of the European Economic Community (EEC). National regulations are in effect until that time. Licensing of drugs will be directed by the EEC Committee on Veterinary Medicinal Products with guidelines on (i) safety to the target species, (ii) efficacy to control target pathogens, (iii) risks to the consumer from exposure to residues in food, (iv) hazards to the workers handling the chemical, and (v) potential for polluting the environment. The EEC has developed a system for classifying compounds subject to regulation. This system consists of four lists: pharmacologically active substances for which maximum residue limits have been fixed; substances for which it appears that a residue limit is not necessary; substances for which a provisional limit is acceptable; and a banned list of substances constituting a positive hazard to consumer health at any limit. Companies have until January 1997 to get their products approved. At that time all unapproved chemicals will be prohibited. Residue and pharmacokinetic studies will be required for aquaculture drugs to determine the pattern of residue depletion at different temperatures and to develop withdrawal times, expressed in degree days, that are based on temperature-related metabolism. Mammalian safety studies are also required to determine chronic toxicity, mutagenicity, carcinogenicity, reproduction, and teratogenicity of the product. If a product is registered in one country, it should be easily registered in other EEC countries, although different environmental conditions may require additional field trials. Most other European countries are interacting with the EEC to develop similar regulations and guidelines [2, 3, 5, 7, 8, 13].

Under current regulations, chemicals allowed for use in various European countries are either fully licensed for aquacultural use (e.g., oxytetracycline, oxolinic acid) or can be prescribed by veterinarians if they are licensed for use on other food animals. In addition, previously unlicensed chemicals that are applied to the water (topicals) may now be granted a so called "grandfather clause" if no one questions their safety. The Ministry of Health, Germany, is evaluating certain chemicals for Standard Marketing Authorization. These chemicals have been used in fish culture for decades without registration and include salt, lime, formalin, copper sulfate, potassium permanganate and malachite green. Discussions have also developed in other countries over whether a chemical is a medicine or a pesticide. For example, dichlorvos (Nuvan 500 EC) was initially designated as a pesticide in the United Kingdom, but was later categorized as a medicine. A similar product, trichlorfon (Masoten), was treated the same way in the United States.

Japan

In Japan, registration of drugs for aquatic species require the same data as those required for drugs used on other food animals. The Ministry of Agriculture, Forests and Fisheries and the Ministry of Welfare control the use of chemicals in aquaculture in Japan [9]. The preclinical data requirements include product chemistry, toxicity (acute, subacute, special) using rats and mice, safety to target animals, and metabolism. The requirements for clinical data include availability and residues.

Potential solutions to the registration situation

The lack of registered drugs for aquacultural use is a worldwide problem the solution to which lies in international cooperation. The cost to fully register drugs for aquaculture far outweighs the market potential. In the United States, for example, a minimum of $3.4 million is needed to complete the registration requirements. Thus, a worldwide market is needed for the pharmaceutical industry to become interested in aquaculture. Other aspects and actions are needed as well to attract the pharmaceutical industry and also to better regulate aquaculture chemicals. Various groups and organizations can play key leadership roles to improve the regulatory aspects of chemotherapy in aquaculture.

Agencies from all aquatic farming countries that regulate the use of chemicals in aquaculture need to (1) become familiar with the cultural practices in aquaculture and to learn more about the aquatic animals themselves, (2) set-up special task forces to address all the issues involved with aquaculture, (3) develop or change regulations, policies, and guidelines that take into consideration the unique cultural conditions of aquaculture, (4) communicate with other regulatory agencies (within and without) the country to remain aware of worldwide developments, (5) review the guidelines for mammalian safety and environmental studies to determine if they are similar to the OECD guidelines or other international guidelines, (6) develop risk assessment techniques to determine whether minute residues are a problem, and (7) develop meaningful tolerances and withdrawal times that protect the consumer but at the same time, are realistic. These agencies need to consider inherent withdrawal times that are associated with early life stages.

Individuals involved in any aspect of research associated with drugs or chemicals that are used, or have the potential for use in aquaculture, need to (i) determine the effects of the various factors (temperature, pH) on the pharmacokinetics of drugs so that models or uniform protocols and data can be developed for worldwide use, (ii) determine the effects that chemicals have on the environment in a variety of settings (seawater, brackish, freshwater, both cold and warm water) so that protocols and data can be developed for worldwide use, (iii) develop analytical methods that are quick, easy, reliable, quantitive, and confirmatory for the compounds currently used in aquaculture, (iv) develop comparative toxicity data that will help regulatory agencies accept crop groupings, and (v) perform research for drug registration according to Good Laboratory Practices and accepted protocols (if available).

Producers of aquatic species need to (i) comply with current regulations and use codes in their country, (ii) educate the regulatory agencies about the cultural conditions in aquaculture, and (iii) use chemicals only when necessary and restrain from using hazardous chemicals.

Various national and international organizations that are associated with aquaculture need to (i) promote the funding of research needed to answer regulatory questions on orphan compounds, (ii) promote and coordinate the development of uniform regulations, policies, guidelines, and protocols on a worldwide basis, (iii) encourage the development of international use codes that promote the safe use of chemicals in aquaculture, (iv) promote the establishment of an international clearinghouse of information such as the use of the newsletter on registration of veterinary drugs published jointly by OIE & ITCVDR, and (v) work with regulatory agencies on any matter that helps the aquaculture industry (education, information, etc.).

Discussions and conclusions

All individuals, organizations, and regulatory agencies involved in aquaculture need to work together to accomplish regulatory balance for all of aquaculture on a worldwide basis. Without international awareness, cooperation, coordination and participation, each country will continue to struggle to meet the chemotherapy needs of the aquaculture industry and at the same time, ensure the safety of the food supply.

References

1). AHMED F.E., 1991. Seafood safety. National Academy Press, Washington, D.C., 446 pp.

2). ANONYMOUS, 1990. EC sets residue levels for drugs. *Fish Farming Int.*, **17** (9), 3.

3). BERNOTH E.-M., 1990. Marketing authorization of fish drugs Current status and future intentions. *J. Vet. Med.*, **37**, 401-406.

4). BROOKS G.E., 1989. The regulations of drugs used in aquaculture. *Aquacult. Assoc. Canada Bull.*, No. **89** (4), 39-42.

5). BROWN L.A., 1987. Aquaculture in Europe. *Vet. Hum.Toxicol.*, **29** (Suppl.1), 54-57.

6). BUREAU OF VETERINARY DRUGS, 1989. *Guide for the preparation of veterinary new drug submissions.* Bureau of Veterinary Drugs, Health and Welfare Canada, Ottawa, Ontario, 62 pp.

7). HOWARTH W., 1990. *The law of aquaculture: the law relating to the farming of fish and shellfish in Britain.* Fishing News Books, London, England. 271 pp.

8). NORDIC COUNCIL ON MEDICINES, 1989. Drug applications: medicines for veterinary use, Nordic guidelines. Nordic Council on Medicines, Uppsala, Norway, NLN Publication No 27., 69 pp.

9). SANO T. & FUKUDA H., 1987. Principal microbial diseases of mariculture in Japan. *Aquaculture*, **67**, 59-69.

10). SCHNICK R.A., 1988. The impetus to register new therapeutants for aquaculture. *Prog. Fish-Cult.*, **50** (4), 190-196.

11). SCHNICK R.A.,1991. Aquaculture, chemicals. *Kirk-Othmer Encyclopedia of Chemical Technology*, 4th edition, New York, in press.

12). SCHNICK R.A., 1991. Major crisis in chemical and drug registration. *Fisheries*, **16** (2), 3.

13). SHEPHERD J., 1990. Regulatory and licensing issues for aquaculture chemotherapeutics and biologicals in Europe. Conference

Proceedings: *Aquaculture feeds and veterinary products:* Worldwide business opportunities for feed, pharmaceutical and chemical companies, Stamford, Connecticut, June 6-8, 1990.

14). U.S. FOOD AND DRUG ADMINISTRATION, 1986. *Guidelines for the preparation of data to satisfy the requirements of Section 512 of the act regarding the use of animal drugs.* U.S. Food and Drug Administration, Center for Veterinary Medicine, Rockville, Maryland.

15). U.S. FOOD AND DRUG ADMINISTRATION, 1986. *General principles for evaluating the safety of compounds used in food-producing animals.* U.S. Food and Drug Administration, Rockville, MD.

16). U.S. FOOD AND DRUG ADMINISTRATION, 1990. *Good laboratory practice for nonclinical laboratory studies.* Code of Federal Regulations, Chap. 21 (Part 58).

17). U.S. FOOD AND DRUG ADMINISTRATION, 1990. *New Animal Drug Applications.* Code of Federal Regulations, Chap. 21 (Part 514).

PROBLEMS OF CHEMOTHERAPY IN AQUACULTURE REGULATORY ASPECTS: AN OVERVIEW

F.P. Meyer

U.S. Fish & Wildlife Service (retired),
518 No. St., La Crescent, MN 55947, USA

Summary : *In the U.S., the aquaculture industry is being restrained by overly cautious, conservative regulatory policies and by excessively costly data requirements of the Food and Drug Administration. The agency continues to disregard differences between production systems for land and aquatic animals, ignores the similarities in physiology among such organisms, and has failed to accept its own guidelines for minor use drugs and for minor species of food animals. Graphic evidence of how the regulatory restraints have affected American aquaculture is provided by the fact that only one new drug (Romet 30) has been approved in the past 12 years. Unless the regulatory agency becomes willing to cooperate with researchers and the aquaculture industry in funding or conducting those studies that must be run, the impasse will continue well into the future.*

Résumé : **Aspects de la réglementation en thérapeutique aquacole.**

Aux Etats-Unis l'industrie aquacole est handicapée par une politique réglementaire conservatrice et excessivement prudente, qui impose de fournir des données d'obtention très coûteuse. La Food and Drug Administration ne tient aucun compte des différences entre les systèmes de production d'animaux aquatiques et terrestres, ignore les points communs physiologiques entre ces organismes, et a même rejeté ses propres directives pour les produits pharmaceutiques à usage restreint ou destinés à des espèces mineures d'animaux. Les effets des contraintes réglementaires sur l'aquaculture américaine se voient à l'évidence dans le fait qu'on n'a approuvé qu'un seul produit pharmaceutique (le Romet 30) en douze ans. A moins que la FDA n'accepte de coopérer avec les chercheurs et les industriels, et de financer ou mener à bien les études nécessaires, on ne peut qu'aboutir à une impasse dans l'avenir.

Introduction

Regulatory controls over the use of drugs and chemicals have the same objectives whether the use is in human medicine or in veterinary practice. In concise terms, there are 4 goals that must be met. They are : 1) to ensure that use of a drug is safe in treated animals and poses no hazard to the user; 2) that use of a drug will leave no harmful or undesired residues that could contaminate the human or animal food supply; 3) that there will be no significant contamination or damage to the environment associated with the use of a drug; and 4) that the product offered for sale is effective for the intended purpose when used as directed.

These objectives are essential the same for every nation but individual governments may place greater emphasis on some objectives than others. In the United States, there has been a continuing increase in regulatory controls over the past 2 decades. Initially, regulations were broad and general and any compound that proved effective was considered good. Concern about safety of the human food supply soon followed and data on residue levels and residue persistence were required before a drug could be approved. Thus change in regulatory control occurred in the late 1960's and early 1970's. These dates coincided with the start of commercial aquaculture of warm water fishes in the United States. Although cold water species, such as trout and salmon, had been cultured for many years, a major new thrust developed in all types of aqua culture during this same period. The data requirements and the cost of studies to generate those data were relatively modest and a number of drugs were approved for use in fish culture. Sulfamerazine and oxytetracycline were approved by the U.S. Food and Drug Administration for use in treating bacterial diseases of trout and catfish. Because of the relatively low cost of meeting regulatory requirements and the favourable growth potential of aqua culture, a number of pharmaceutical companies became actively interested in developing products specifically for use in the aqua culture industry.

However, developments during the 1970's resulted in major increases in the types of studies and in the amount of data that would be required in order to obtain regulatory approval for any drug, whether it was for aqua culture, human medicine, or for veterinary use. The new regulations added tests for carcinogenesis, mutagenesis, teratology, reproduction and other potential impacts. The impact of these added requirements was immediately evident in the escalating costs of studies that had to be conducted. By 1980, the cost of meeting all of the data requirements had become so high that most pharmaceutical companies were no longer interested in aquaculture products. Growth of the aquaculture industry was at a slower rate than the escalation of registration costs and the profit potential declined sharply.

In the early 1980's, regulatory attention turned to metabolic products produced in treated animals and a new generation of added requirements followed. New technology had to be developed to identify the possible metabolites and to quantify them in edible tissues, blood, urine, feces, and the environment. Another major increase in costs accompanied these new requirements. Because industrial chemicals were appearing in ground water and surface waters, regulatory interest was also prompted concerning the environmental fate and effects of pharmaceutical drugs and pesticides. The addition of tests to answer these regulatory concerns represented yet another major cost increase.

Also, in the 1980's, the Food and Drug Administration imposed quality assurance checks and good laboratory practices guidelines on all studies being conducted to develop data needed to apply for a drug registration and approval for use. These requirements added another 30X to the already high cost of obtaining approval for a pharmaceutical or pesticide product.

In 1990, the estimated cost of obtaining Food and Drug Administration approval for a new drug was $3.4 million. While no firm figures are available, pharmaceutical companies estimate that the potential market must be at least 100 x the cost of obtaining a registration if they are to become interested in a new product. At present, no drug need in American aquaculture can meet this criterion.

It has become evident that regulatory encumbrances have priced the development of new drugs beyond the needs of all but human medicine and the major food animal industries.

Assessment Of The Need

The American aquaculture industry has grown dramatically in the last decade. The channel catfish industry continues to grow at a rate of 15 to 20 percent per year (H. Dupree, personal communication, U.S. Fish and Wildlife Service, Stuttgart, Arkansas). Annual production exceeds 160,000 metric tons and has a value of about $280 million. The production of rainbow trout is about 30,000 metric tons worth $73 million. Statistics are not available for the bait fish, ornamental and pet fish, shellfish, or crustacean culture industries.

Losses in American aquaculture due to disease are staggering. Estimates for the catfish industry place 1989 losses at $23 million; over half of this was caused by Gram negative bacterial infections. Losses in the trout industry for the same period were estimated to be at least $2.5 million (3).

Aquaculturists have only a few approved therapeutants available for use in combating infectious and parasitic diseases (5). Only oxytetracycline (Terramycin) and a combination of sulfadimethoxine and ormetoprim (Romet-30) may be used to treat bacterial infections. Formalin has been approved for use as a parasiticide and fungicide. Acetic acid, copper sulfate, sodium chloride, and potassium permanganate may be used for non-specific purposes. Sulfamerazine, nifurpirinol (Furanace), and trichlorofon (Masoten) were once approved for aquacultural uses but these are no longer available to American producers. Malachite green is not allowed for use in aquaculture.

Few therapeutants are fully approved for aquacultural uses in any part of the world. In addition to the United States, Canada, European countries, and Japan are implementing regulatory control over the use of drugs and chemicals in the production of fish and shellfish for human food.

As international trade in farm-raised fisheries products continues to develop, the aquaculture industry will be asked to certify that its products are free of undesirable chemical residues. The U.S. Food and Drug Administration has increased its surveillance for chemical residues in fishery products that originate outside the country. This action has been stimulated by regulatory awareness that drugs currently banned in the U.S. may be allowed in other countries.

There can be no questions about the need for additional therapeutic agents for use in aquaculture. A wide array of potential products is available, most have already demonstrated efficacy against infectious or parasitic agents similar to those that affect fish and shellfish. Excellent discussions of such drugs and chemicals are provided in Alderman (1) and Meyer and Schnick (4).

Problem Areas In The Regulatory Process

During the last decade, the American aquaculture industry and researchers have met regularly with the U.S. Food and Drug Administration and the U.S. Department of Agriculture in a series of workshops. It was hoped that such meetings would educate all agencies concerned about aquaculture systems, the species being cultured, the diseases that plague the industry, and data needs of the regulators. After over 10 years of such meetings, little has been accomplished. Progress seemed likely when a set of minor use guidelines was published by the FDA in 1986 (6). It was hoped that the new guidelines would reduce the cost and amount of work required to obtain regulatory approval for aquacultural therapeutants. Unfortunately, the several Divisions within the agency chose to interpret the guidelines differently. Some officials were overly-conservative, others attempted to apply regulations for land-based culture systems to aquatic situations, and still others refused to consider the unique systems involved in aquaculture. The result has been endless regulatory bickering, conflicting directives, and inconsistent agency policies. Some regulators do not even recognise that fish and shellfish are animals and are concerned that aquaculture species might have unique metabolic systems although a special workshop devoted to this topic clearly showed that fish and shellfish utilise the same enzyme systems and energy pathways as land animals.

Because fish are poikilothermic, some regulators have been concerned because the typical elapsed days concept of withdrawal time cannot be applied. Since the rate of metabolism in fish rises and falls with changes in the water temperature, they feel it may not be possible to draft enforceable withdrawal periods. Such reasoning is further indication of the inflexibility of regulators and suggests why they refuse to even consider degree days as the measure for required withdrawal times.

A problem that must be addressed by both the regulators and the pharmaceutical companies relates to the need for adequate analytical methods. The U.S. Food and Drug Administration will not permit the use of any drug for which analytical methods are not available to check for residues and metabolites. While this attitude may be defendable as a policy decision, it is also an admission that the agency is unable to check for residues in imported products. As a result, domestic producers are placed at a regulatory disadvantage because of the Food and Drug Administration's vigorous enforcement program within the U.S. Producers in many foreign countries are not restrained in their use of needed therapeutic drugs.

The entire aquaculture industry is concerned about the fate of chemicals in the environment. No other type of husbandry is more concerned than aquaculturists. Any drug or chemical that is applied, or that may otherwise enter an aquatic culture system, has the potential to adversely affect the animals being raised. Because fish and shellfish are continually surrounded by water, any compound that is added to an aquatic environment has the potential to be biomagnified in the flesh. This is true whether the substance is added intentionally, as for therapeutic purposes, or unintentionally by industrial, domestic or agricultural practices. Present FDA and EPA concerns about the fate of aquaculture compounds in the environment are reflected in their strong emphasis on forcing industries to comply with pollution guidelines. The regulators fail to realise that the addition of harmful chemicals to an aquatic culture system would immediately place the animals under culture at risk of being killed or rendered unmarketable if persistent residues developed.

Recommendations For Resolving The Problems

The present impasse between governmental regulators and the aquaculture industry cannot be resolved without **significant** changes in attitude and policies. Although aquaculturists generally desire to comply with regulations affecting their use of drugs and chemicals, they cannot be expected to continue to sustain losses to disease if effective drugs are known but not allowed because of policy rather than demonstrated hazard. The solution lies in a compromise between the regulators and those being regulated.

The aquaculture industry is willing to help regulatory agencies become acquainted with their culture systems, the species being reared, and the unique requirements of each. Regulators must recognise that guidelines for terrestrial animals must be modified if they are to be applicable to aquatic species. Officials within the regulatory agencies must make the effort to become knowledgeable about aquaculture and to work with producers so that they will understand why regulations are needed and how they apply to the aquaculture industry.

The issue of needed analytical methods can also be resolved by cooperative efforts among all parties. Data relating to residue levels and residue dynamics are needed by pharmaceutical firms, by regulators, and by the producers. Yet, those data cannot be generated without analytical methods. Because of the high cost of developing original analytical methods, it is imperative that available funds be used wisely and efficiently. If the regulatory agencies would reduce their emphasis on enforcement and reprogram the funds to methods development, all parties would benefit. Likewise, the aquaculture industry should identify its highest priority drug needs and then assist in funding research to develop the needed analytical methods. Success in this effort would also help researchers determine the fate of the drugs in the environment.

Conducting mammalian safety studies needed to answer concerns about potential hazards associated with the use of a drug represents the major portion of the cost of obtaining regulatory approval. It is imperative that such studies be accurately conducted according to the highest professional standards in order to protect the user, the food supply, and the environment. No one will argue against this requirement. However, any study that meets these standards should be acceptable for submission to regulatory agencies in support of a desired registration for use. Consideration should also be given to data generated in other species of animals and from other uses for answering questions related to metabolites, routes of excretion, and sites of retention. Failure to consider data from such studies when dealing with minor use drugs will mean costly duplication of studies, unnecessary expense, and create obstacles to the safe use of drugs urgently needed by the aquaculture industry.

Data related to the persistence of a drug or its metabolites in edible tissues, to possible oncogenicity, teratology and adverse physiological effects should be developed but the full battery of tests required for use in major food animal species should not be necessary.

The Food and Drug Administration should also reexamine its definition of what constitutes a "food animal". In aquaculture, early life stages are subject to a variety of external protozoans, parasites, fungal agents, and bacterial pathogens. In all cases, extended periods after the early life stages are required for the young animals to grow to marketable sizes. In fish, during the first year of life, fin fish are called fingerlings because of their small size. Six months or more of feeding and growth are needed to produce food sized animals so an inherent withdrawal period is provided. In some species, the period may be over one year. Brood fish are valuable animals maintained only for the purpose of providing eggs and replacement stock. Because this is their primary purpose, drugs needed to maintain good physical condition would not affect the food supply. If necessary, a condition of use could include a prohibition of the use of brood fish for human or animal food. Similarly, eggs produced by brood fish that are treated during incubation never enter the food chain. Any fry or larvae that hatch from treated eggs will have to undergo increases in body mass of 10,000 or more times. The period of time before hatchlings would become food animals would be even greater than for fingerling animals.

The U.S. Fish and Wildlife Service and the American aquaculture industry have repeatedly urged the Food and Drug Administration to accept the premise that eggs, fingerlings, and brood stock of fin fishes should be considered non-food stages. As recently as October 1990. the Food and Drug Administration has rejected this proposal based on its conservative attitude toward all aquaculture.

Conclusion

In the U.S., the aquaculture industry is being restrained by overly cautious, conservative regulatory policies and by excessively costly data requirements of the Food and Drug Administration. The agency continues to disregard differences between production systems for land and aquatic animals, ignores the similarities in physiology among such organisms, and has failed to accept its own guidelines for minor use drugs and for minor species of food animals. Graphic evidence of how the regulatory restraints have affected American aquaculture is provided by the fact that only one new drug (Rome 30) has been approved in the past 12 years (2). Unless the regulatory agency becomes willing to cooperate with researchers and the aquaculture industry in funding or conducting those studies that must be run, the impasse will continue well into the future.

References

1). ALDERMAN D.J. (1988). - Fisheries chemotherapy: a review. *Recent Advances in Aquaculture,* Vol. 3: 1-60.

2). MEYER F.P. (1989). - Solutions to the shortage of approved fish therapeutants. J. *of Aquatic Animal Health.* Vol. 1: 78-80.

3). MEYER F.P. (1991). - Aquaculture disease and health management. *J. of Animal Science* (in press).

4). MEYER F.P. & SCHNICK R.A. (1989). - A review of chemicals used for the control of fish diseases. *Critical Reviews in Aquatic Sciences.* Vol. 1: 696-710.

5). SCHNICK R.A., MEYER F.P. & GRAY D.L. (1986). - A guide to approved chemicals in fish production and fishery resource management. U.S. Fish & wildlife Service and University of Arkansas Cooperative Extension Service. Little Rock, Arkansas. MP 241-llM-1-86. 27 pp.

6). U.S. FOOD & DRUG ADMINISTRATION. (1986). - Guidelines for the preparation of data to satisfy the requirements of Section 512 of the act regarding minor use of animal drugs. U.S. FDA, Center for Veterinary Medicine, Rockville, Maryland.

CONSTRAINTS ON DRUG PRESCRIPTION - A VETERINARY VIEWPOINT

R. H. Richards

Institute of Aquaculture, University of Stirling, Stirling, FK9 4LA, Scotland

Summary: *Veterinary control of fish disease in salmon, particularly applies to the treatment of furunculosis and sea-lice infection. The supply of prescription only medicines places a responsibility on the veterinarian to ensure on-farm clinical control, to have complete control of the storage and use of such compounds, and to stipulate a withdrawal period.*

Constraints placed on the veterinarian include a) the limited range of compounds available, b) the inability to treat during extensive withdrawal periods prior to necessary harvest and c) competing regulations preventing discharge of medicaments in effluent from farms without agreement from the relevant River Purification Boards and d) a complete lack of suitable treatments for a number of significant fish diseases.

Decreasing the cost and reducing the time necessary for manufacturers to obtain full product licences for fish treatment would dramatically improve the ability of the veterinary surgeon to control fish disease.

Résumé : **Les contraintes de prescription en aquaculture : point de vue du vétérinaire.**

Le contrôle vétérinaire des maladies des poissons, plus précisément dans le cas des saumons élevés au Royaume Uni, s'applique particulièrement au traitement de la furonculose et de l'infestation par le "pou de mer" Lepeophtheirus salmonis. *La délivrance des médicaments est opérée seulement sur ordonnance et rend le vétérinaire responsable du contrôle complet de l'acquisition et de l'utilisation des produits actifs, des opérations de surveillance clinique, et de l'estimation des périodes d'attente à observer avant la vente.*

Les contraintes imposées en pratique au vétérinaire comprennent a) le choix limité des produits disponibles b) l'impossibilité de mettre en oeuvre un quelconque traitement pendant la période précédant la commercialisation des poissons c) les réglementations antinomiques qui empêchent le relargage de médicaments dans l'effluent des fermes d'élevage sans un accord préalable des Conseils d'Epuration des Rivières , et d) le manque crucial de traitements adéquats pour nombre de maladies importantes chez les poissons.

Enfin la diminution des frais et des délais nécessaires aux fabricants pour obtenir l'autorisation des produits destinés au traitement des maladies des poissons augmenterait sensiblement pour le vétérinaire la possibilité de maîtriser ces maladies.

Introduction

The use of chemotherapy to control fish disease has been practised for decades but such traditional remedies have been few in number, consisting principally of in-water treatment with chemicals such as formalin, malachite green and the quaternary ammonium range of compounds. The rapid development of intensive aquaculture over the last 15 years or so has been associated with increasing problems of bacterial and parasitic disease such that, at the present time, a number of operations are failing because of the high cost of losses and treatment Intensification has resulted in a greater opportunity for the rapid build up of pathogens and created difficulties in treating the large numbers of fish involved in a disease outbreak. The viability of modern aquaculture depends very much on improving current disease control practices aided by the use of appropriate legislation.

Veterinary involvement in the UK salmon industry

The involvement of the veterinary profession in controlling fish disease in the UK varies considerably from farm to farm. In some cases, veterinary surgeons are involved on a continual basis in monitoring and diagnosis, but in others, a fish health team permanently employed by the fish farming company will carry out monitoring, diagnosis and many treatment regimes, only calling upon the veterinary surgeon when the use of a controlled "prescription only medicine" or P.O.M. legally requires such involvement under the 1968 Medicines Act. Veterinary diagnosis and control of animal diseases is stipulated in the Veterinary Surgeons Act (1948) and specifically mentions the individual major farm and companion animals. Fish are neither included nor specifically excluded from this legislation, as indeed is the case with most other "minor" species and this has resulted in non-veterinarians taking a major role in fish disease diagnosis. Fish disease training is currently provided to a limited extent at all veterinary schools and short intensive courses as well as MSc. courses are available for those who wish to specialise. This has resulted in a small pool of "specialist" veterinarians dealing with fish disease.

As veterinary involvement is specifically legally required when using a P.O.M. compound it is this aspect of chemotherapy that I will discuss, in particular applied to the two most serious problems facing the salmon industry, namely furunculosis (caused by *A. salmonicida*) and sea-lice infection (caused by *Lepeophtheirus* or *Caligus* spp).

Prescription only Medicines (P.O.M.'s)

a) The Range

Only four antibacterial substances are currently licensed for use in the chemotherapy of fish bacterial disease, namely: oxytetracycline, oxolinic acid, potentiated sulphonomides and recently, amoxycillin. Multiple resistance to all but the latter compound has been problematical in recent years creating major difficulties in ensuring the rapid selection of the appropriate treatment. Dichlorvos is currently the only P.O.M. licensed for the treatment of sea-lice infestation. Though other molecules are being developed for fish use, the licensing process is both costly and slow, particularly with the increasing, though very necessary, requirements for environmental assessments. The relatively small size of the international market for such products has restricted the development of novel treatments on financial grounds until relatively recently, when environmental constraints in particular are reducing possible development. It should also be noted that there are currently a number of significant infectious diseases for which no treatments are currently available. These include Bacterial Kidney Disease (caused by *Renibacterium salmoninarum*) and certain systemic protozoal diseases.

b) Administration and supply

The vast majority of antibiotic treatments are given orally, incorporated either within, or on the surface of, feed pellets. Because of the poikilothermic nature of the animals concerned, standard feed inclusion rates are not employed, but antibiotics are incorporated at varying rates based on the feeding-rate being used so as to achieve even uptake of the required medicine over the total feeding period. This means that premedicated diets are not available at the food-processing factories and medicated diets only made to order. Mill-mixing will only be considered for quantities of 1 tonne or more of feed and delays are always possible resulting from (a) supply of antibiotic, (b) availability of mixing plant and (c) delivery to the fish farm site. Because of the urgent nature of most treatments, not least because of disease-induced appetite depression, this necessitates a considerable quantity of on-farm mixing, binding the antibiotic to the surface of the pellet with oil or gelatin. Though this is very efficiently carried out, quantities of medication in excess of 20kg/tonne of food will not adhere to the pellet surface making treatment at low feeding rates difficult (e.g. at low temperatures or in broodstock treatment), especially if using medicinal feed additives containing a low percentage of active ingredient.

The necessity for rapid production of medicated feed at the farm site also creates a need for holding stocks of antibiotics at the fish farm. Such stocks are under the direct control of the prescribing veterinary surgeon and by law, only small quantities can be held. In practice, stock holding is usually limited to quantities sufficient for the first two days of a treatment. The remainder is supplied directly from veterinary wholesalers or the veterinary practice itself, necessitating the maintenance of large basic stocks at these sources. Because of the distances involved and the remote nature of many sites the use of expensive overnight delivery services is also necessary.

Dichlorvos is applied in bath form to treat sea-lice infested fish in net pens, using either completely enclosed tarpaulins or "skirts" surrounding only the sides of the pens. No in feed medication is currently available for sea-lice treatment. Further problems arise with stock holding and transport related to the supply of dichlorvos. The currently commercially available licensed preparation is provided in packaging which cannot be legally supplied by air or rail unless it is repackaged into smaller containers and all supplies are currently provided by dedicated road transport. This necessitates the holding of much larger stocks at farm sites based on predicted usage.

c) Veterinary responsibility

(i) Clinical control

The provision of P.O.M. medication places a legal responsibility on the prescribing veterinary surgeon to provide clinical care to the stocks being treated. This is usually interpreted as representing a continuing clinical relationship with the farm concerned through site visits and sample analysis as well as by regular telephone contact.

Because of the distances between veterinary surgeon and farm site, it is not always possible to make on site visits on every occasion that P.O.M. medication will be used - this particularly applies if frequent dichlorvos treatment is necessary. Such medication, however, remains firmly under the control of the prescribing veterinary surgeon and will only be carried out after thorough discussion and agreement on the necessity of treatment. Clinical care also implies a detailed knowledge of the condition being treated. The relatively small numbers of veterinary surgeons specialising in fish health often make the distance between veterinary surgeon and farm site even greater.

(ii) Withdrawal periods

It is also the duty of the prescribing vet to stipulate a "withdrawal period" or minimum time following the end of a treatment period before harvesting of the stock can take place. Though a specific minimum period is stipulated in the licence conditions for a particular product, the veterinary surgeons may insist on an extension of this period if they consider it appropriate. This provides consumer protection in ensuring that any residues following treatment are below the accepted maximum limit. Such periods are long, varying from 30 days for some compounds to 500 degree C days for others at present. Any variation to the licence stipulation in terms of extending a period of treatment or increasing a therapeutic dose-rate, which may legally be carried out in exceptional cases, for instance if a proportion of the stock is feeding very poorly, necessitates the automatic application of a 500 degree C. day withdrawal period. Such very long withdrawal periods create particular problems in the face of repeated attacks of disease requiring treatment, as stocks of fish in their withdrawal period prior to necessary harvesting may suffer extensive losses and also act as a focus for continuing infection. This problem applies particularly to fish because of the inability to eradicate diseases such as furunculosis and sea-lice due to regular reinfection from wild stocks.

Because of the limited range of antibiotics available and in the face of increasing problems with resistance, dispensation has also been given to veterinarians to prescribe in-feed antibiotics licensed for other species, provided that a 500 degree C. day withdrawal is observed. Although this move was welcomed, in practice only one extension to the licensed fish range was possible because the active ingredients in such mixtures were present at low concentrations (commonly 10%), making surface-coating impossible. Though additions to the medicines used to control sea-lice infection are also theoretically possible under this dispensation, the unknown ecological effects of such compounds have prevented their use. Any use of medicament outwith the strict licence conditions also places sole legal responsibility for use directly on the prescribing veterinarian, creating a substantial degree of risk for that individual.

(iii) Effluent Discharge

A further constraint that could affect the veterinarian's ability to control disease is the possibility of the River Purification Board preventing the use of a certain medicament because it or its metabolites would escape from the farm site in the water passing through it (the effluent). The boards concerned ask for a list of type and quantity of medicament to be used and when possible, require 48 hours notice of use to allow sampling to take place during the treatment period. In practice, such advance notice is often not possible but fortunately the boards recognise the necessity for speedy treatment and currently accept immediate notification on treatment. It is also accepted that a medicine may be used without prior notification provided it has been prescribed by a veterinary surgeon, and is a licensed compound. A further consideration is that the welfare of affected fish, and thus prevention of unnecessary suffering, also dictates rapid response to disease outbreaks.

(iv) Necessity of Monitoring

The use of licensed P.O.M. medications in fish necessitates extensive monitoring programmes. In the case of bacterial disease, knowledge of the changing sensitivity patterns of existing pathogens in the farm environment is necessary in order to rapidly commence an appropriate antibiotic treatment at the start of a disease outbreak before full laboratory results from that outbreak are available. In the case of sea-lice infections, continuous monitoring programmes allow treatment to be undertaken at the appropriate stages of the life-cycle, avoiding unnecessary treatment and avoiding the development of serious lesions in fish. Though this comment applies to *Lepeophtheirus* infection where parasites are developing in the host for several weeks before causing damage, it is possible for large numbers of *Caligus* parasites to infest fish overnight, making continual careful fish observation essential.

Future developments

The prevention of animal disease is usually the best form of control. Improvement in general husbandry is already in use as a means of non-specific prevention. Specific control can be achieved either by (a) eradicating the disease agent from the environment - this is unlikely to be possible in the case of furunculosis and sea-liceinfection or (b) by increasing natural resistance through the use of aids such as Vaccines. Vaccines are available currently for furunculosis control but have only given moderate protection. Major research programmes are already underway to improve furunculosis vaccines and to develop novel sea-lice vaccines. However successful these vaccines become, new products will always be needed to combat outbreaks of disease when host defences become overwhelmed by the pathogen.

Provision of legislation which ensures safety, quality and efficacy of the medicaments used but at the same time does not place an excessive financial burden or undue time loss on manufacturers is a necessity in future disease control.

Discussion

SCHLOTFELDT You don't have in the UK an arrangement that approved mills can produce medicated feeds in advance and store it, so when you prescribe it, you can have it within a day? In northern Germany we have two of this kind of mill which are officially approved to produce in advance, so that it goes very quickly.

RICHARDS We do have approved mills for mixing antibiotics. The problem that we have is that we wish to treat each case with what we consider to be the optimum dose rate. So we don't have a fixed incorporation rate. Pre dosage of medicaments would therefore not be appropriate as they would have to hold all sorts of mixtures of different pellet sizes and different incorporation rates of antibiotics. Thus it is simply not practical in our situation.

CONSTRAINTS ON CHEMOTHERAPY: THE FISH FARMING INDUSTRY VIEW

G.H. Rae

Scottish Salmon Growers Association, Drummond House, Scott Street, Perth, PH1 5EJ, Scotland

Summary: *Modern fish farming techniques require the use of a range of chemotherapeutants for control of enzootic fish diseases. Fish farmers are constrained by the present lack of drugs and vaccines. This means that an integrated approach to health management cannot be taken. Current legislation for the licensing of products is complicated, expensive and slow. There is a relatively small market for fish health products and manufacturers have to be coaxed into taking their products to licence. The current development of ever more sensitive techniques for measuring residues of medicants in food products and the ever lengthening withdrawal periods put serious restrictions on the ability of the farmer to carry out his business. There is an urgent need for an understanding of the causative agents of diseases of farmed fish and subsequent development of efficacious vaccines against all current major fish pathogens.*

Résumé : **Les contraintes de traitement en aquaculture: point de vue de l'éleveur.**

Il est impossible à l'aquaculteur de se passer de traitements, même s'il s'impose des règles d'hygiène et de gestion rigoureuses. La conduite de ces traitements s'accompagne de soucis qu'un diagnostic précis ne suffit pas à atténuer. Il faut en effet affronter des situations pathologiques diversifiées par l'intensification des méthodes d'élevage en usant de produits en nombre restreint, que de fréquentes résistances, tant chez les agents bactériens que chez Lepeophtheirus salmonis, rendent toujours plus aléatoires. De fait les autorisations d'emploi ne sont accordées qu'après des procédures complexes où les contrôles écotoxicologiques pèsent lourdement, car la santé du poisson paraît négligeable face aux risques encourus par l'environnement. Les producteurs de saumons écossais ont entrepris de financer les études de molécules nouvelles, mais cela n'atténue en rien le coût et la lenteur des procédures légales.

Les bains thérapeutiques ont l'inconvénient de relarguer les produits actifs dans les effluents. Au Royaume-Uni les Conseils d'Epuration des Rivières (River Purification Boards, ou RPB) doivent être avisés 48 heures à l'avance des intentions de traitement et des modalités envisagées. En théorie ils peuvent s'opposer à la prescription. Des produits comme le formol, le vert malachite, la benzocaïne et le dichlorvos ne sont en général pas contestés, bien que le dernier, très critiqué, figure sur une liste rouge. Les résidus d'antibiotiques administrés par voie orale sont également concernés. L'amoxocilline a été récemment ajoutée à la liste des 3 principes déjà autorisés (oxytétracycline, acide oxolinique, association sulfamide-triméthoprime), trop tard cependant pour éviter la faillite d'un certain nombre de fermes marines ravagées par la furonculose.

Le respect de délais d'attente après traitement oblige à conserver des animaux vulnérables sans pouvoir intervenir en cas de rechute. Ces délais, fixés par référence au niveau 0 de détection des résidus, tendent à s'allonger avec les progrès des techniques d'analyse. En pratique, les traitements sont lents à instaurer, et coûteux : il n'est pas possible de garder des aliments médicamenteux en stock, leur fabrication sur place s'accompagne de pertes de produit actif, le comportement alimentaire hétérogène des animaux oblige à en prolonger l'administration et l'absorption intestinale n'est pas toujours optimale. Enfin, l'effet des traitements n'est jamais définitif.

Des solutions techniques sont bien sûr à l'étude pour pallier ces inconvénients : encapsulation des médicaments, techniques d'injection automatisées, mise au point de vaccins performants font l'objet de recherches active. Il reste que les producteurs souhaiteraient un choix de produits plus large pour pouvoir élaborer de véritables stratégies de traitement, ainsi que davantage de souplesse dans les critères d'établissement des délais d'attente.

Introduction

Although good management and fish husbandry can reduce the impact of fish diseases on farms, opportunistic and obligate pathogens do cause problems and require treatment from time to time. Diagnosis of these diseases is now well established and management strategies, trained staff and treatment equipment are in position on the farm. Once a disease is diagnosed there are a number of questions which the farmer has to answer before treatment can go ahead. He needs to know what to use, how to use it, how long fish will have to be withdrawn before harvest and what are the legal implications of use. There are also the questions of how efficacious the treatment will be and if there is a vaccine against the disease. In this paper it is intended to discuss these points under the following headings.

- Chemotherapeutants
- Administration regimes
- Effluent discharge consents
- Withdrawal periods
- Vaccines

Chemotherapeutants

Limited range available

The farmer requires a reasonable range of treatments to combat different situations as they arise and at present there are not enough licenced products available for planning any control strategy. This is probably due to the youth of the industry and the fact that for many years old problems could be solved using traditional remedies. However,the recent intensification of salmon farming in particular has meant that the limited range of available therapeutants has become inadequate.

Resistance

There is only one chemical licenced to treat *Lepeophtheirus salmonis*,the Sea Louse which parasitises salmon, and work funded by the Scottish Salmon Growers Association (SSGA) has shown that there is now developing resistance in some populations of the parasite. There is also an increasing number of sensitivity patterns being found in fish bacteria and these resistant strains require a range of antimicrobials for control.

Licensing constraints

The requirement for more drugs to allow a strategic approach to be taken has come at the same time that regulations on the licensing of products have become more complicated, and the process is now much more expensive and time consuming than in the past. Some of the limiting factors to licensing are shown below.

- Fish not considered important to the manufacturer
- Market for products is relatively small
- Fish welfare not considered a significant issue
- Environmental issues preeminent

The decision on whether to take a product to licence inevitably lies with the manufacturer who must balance the expense of getting products to licence against the possible sales and returns. Compared to human medicine, the aquaculture market is small and manufacturers cannot expect large sales and subsequent profits. To try to combat any inertia, the SSGA has encouraged companies to go to licence in Great Britain by funding a post-doctoral worker at the Institute of Aquaculture (University of Stirling) to screen candidate antimicrobial drugs for suitability in fish culture. At the same institute another SSGA funded project is assessing additional chemicals which may be suitable for use against sea lice. As a result of this pump-priming by the SSGA these two projects have contributed to the successful licensing of one product and four others are in the process of submission for Animal Test Certificates, leading to Product Licence, in the United Kingdom.

The process of licensing in the United Kingdom is complex, expensive and slow, as mentioned previously, and these factors conspire against the fish farmer who needs products with which to treat his fish. The increasing requirement for ecotoxicological data on potential fish medicines can bewilder farmers who are waiting for products to be licensed. They can easily buy the same products to treat pets or livestock that live on land but cannot use them to treat disease or parasites in fish.

Administration regimes

Herd treatment

Fish are relatively small, shoaling animals and as such farmers tend to hold large numbers in distinct populations in tanks or in net pens. Any attempt to separate individual fish for treatment will result in increased stress for the whole population and the problem will be exacerbated. This means that treatment is usually of the whole population at one time. Given that a suitable drug is available the farmer has three methods of administration from which to choose. These are by bath or dip, via the feed and by injection.

Bath treatment

External parasites are usually treated by adding a diluted chemical to the water in the holding facility. This can either be a flowing treatment or it can be a static bath. The methodology is well established and has been in use for many years. Both methods result in a release of chemical to the environment and for this an Effluent Discharge Consent is required from the relevant River Purification Board (RPB) in Scotland. This requirement can constrain the fish farmer in his attempts to keep his fish healthy for the following reasons.

Effluent Discharge Consent

In his application for a Discharge Consent the farmer must define the chemicals which he expects to use and the likely annual amounts of each. The RPB then has the authority to grant or deny permission to discharge. It is a condition of the licence to discharge that details of treatments must be notified to the authority 48 hours before the event. This in itself is difficult to comply with because treatment is mostly recommended immediately after diagnosis and farmers must act without delay if the treatment is to be effective.

The real difficulty comes if the veterinarian prescribes a drug which has not been applied for in the discharge application. In this case the RPB has authority over the veterinarian in deciding the fate of the fish. However the RPB will refer to the Medicines Act when making any decision and give assurances that it is unlikely that they would ever use their power to refuse permission provided that it is a licenced medicant which is being prescribed by the veterinarian. The licensing of a product involves extensive ecotoxicological assessments so that the granting of a licence is enough to assure the boards that the product does not pose any threat to the environment.

Should the veterinarian wish to prescribe a drug which is licenced for use in other animals but not yet for fish a problem may exist. RPBs do not employ experts on fish diseases and should they disagree with the vets decision and the fish cannot be treated then they clearly have the authority to decide the fate of the fish. Their judgement ought to take into account the welfare of the fish. Perhaps because they live underwater fish welfare is not something to which man has paid much attention in the past.

Discharge Consents currently allow the release of only the following chemicals into rivers or into the sea.

1. Formalin
2. Malachite Green
3. Benzocaine
4. Dichlorvos

Even this limited range is under threat with dichlorvos receiving most media attention to date. So far it has survived because all the ecotoxicological data have shown that its controlled use on salmon farms is safe. However it is on the United Kingdom Government RED LIST and as such its release into coastal waters must be reduced by 50% by 1995.

In-feed treatment

Consents also control the release of metabolites of fish feed and these can include antimicrobials. The most common use of antimicrobials in fish farming is in feed.

Most in-feed treatments are prescribed to combat bacterial diseases. These diseases include furunculosis which I will use as a model in this talk. Furunculosis is caused by the bacterium *Aeromonas salmonicida* which causes septicaemia, morbidity and mortality. The disease is named from its ability to cause large boils or furuncles to appear on the fish. It occurs as chronic, acute and peracute disease with acute mortality being the most common condition in Atlantic salmon. To date there are only four LICENSED PRODUCTS available to combat the disease. viz.

1. Oxytetracycline
2. Oxolinic Acid
3. Trimethoprim/Sulphadiazine
4. Amoxycillin

The fourth drug has only recently become available but its licence came too late to save closure of some salmon farms, premature harvest of many hundreds of tons of our salmon and unnecessary suffering to hundreds of thousands of farmed fish due to the presence of a strain of furunculosis which was only sensitive to amoxycillin. This was particularly galling for the fish farmers who knew that the drug was licenced for use in other food animals but could not be used to save their fish.

Withdrawal periods

Given that the veterinarian has prescribed an efficacious drug the farmer has to get it into his fish. To do this he has to consider a number of criteria. These revolve round the fishes appetite, the recommended dose of drug, the palatability of the medicated feed, duration of treatment, in-feed/on-feed incorporation and withdrawal period to harvest. These are all problems which the modern fish farmer understands and copes with, however, after treatment the fish must go through a period of clearance while the level of medicant in their tissues returns to zero. This period is called the Withdrawal Period during which the fish cannot be sold for human consumption. The continual lengthening of these periods, due to the development of increasingly sophisticated methods of analysis for residues and the application of the results, makes his job much more difficult,not least because the disease can return well before harvest.

This obviously constrains the farmer as summarised below.

- Disease returns in the withdrawal period
- Health management is jeopardised
- Wothdrawal fish are a point source of infection
- Periods too long?

The return of clinical disease in the withdrawal period has led to the development of management strategies to allow the farmer to continue his business. However these strategies are less than ideal for the eradication of the bacterium. They involve selection of fish destined for harvest while they are under treatment and the further treatment of those fish which are to be kept for the next growing period. These growing fish are threatened by the release of bacteria from the "harvest" fish as they wait to pass through the withdrawal period and the farmer suffers more outbreaks of disease.

Injection

Another albeit little used option is to inject each fish with medicant. This method is usually only quoted as being of use when small numbers of valuable fish are to be treated, however mass injection of fish is now well proven.

Automatic machines are being developed and will, one day, out-perform humans. This method is, however, more applicable to vaccines than to antimicrobials.

Efficiency of incorporation and efficacy of medicant

Antimicrobials are expensive and one of the concerns of the farmer is the loss of any of it before it gets into the fish. Once the fish has eaten it he is concerned that it is taken up by the fish and that it is efficacious.

Losses in manufacture

There are two problems associated with in-feed medication an these are the loss of drug during the manufacturing process and the loss of time to the start of treatment while the feed is being made up. It used to be possible to hold stocks of medicated feed ready for predicted disease outbreaks but the increasing variety of bacterial sensitivity has made this impracticable.

Inefficient uptake

Research sponsored by the SSGA has shown that food uptake can be very uneven with an average of 26% of fish eating 65% of the food in any one spell of feeding. This means that the efficacy of antimicrobials can be less than expected. To overcome this, treatments are prolonged so that every fish gets the proper dose and so that non-feeders will die off during the treatment and will not threaten the feeding, treated fish at the end. Despite this the effect of antimicrobial treatments does not last and repeat treatments are necessary for diseases like furunculosis which are not cured.

Once eaten, not all the drug is taken up by the fish. The inefficient uptake of the drug from the feed is something which has concerned farmers from the beginning. The drugs are expensive and any loss of efficacy is important because no-one wants to pay for something which has no effect. Some research in this area could benefit the farmer and the manufacturer. e.g. There may be an opportunity to apply modern encapsulation technology to this question

Vaccines

The use of efficacious vaccines would obviate the need for the repeated use of antimicrobial and antiparasitic agents. Vaccines for fish have now been developed and they are being used to control a number of diseases. e.g. Vibriosis and enteric redmouth. These two are very efficacious and are immersion vaccines. However there are not enough fish vaccines available yet and once again I would like to refer to the case of furunculosis and the lack of a protective vaccine against this disease.

Attempts to produce an efficacious vaccine for furunculosis have been made sporadically since 1939 when Duff demonstrated "a degree of immunity to subsequent infection". Although there are now some licenced furunculosis vaccines on the market it still remains a problem because protection is limited. Better furunculosis vaccines are still being developed and the most successful test preparations are injectable. This may seem to be a constraint on the farmer if he is small and cannot manage mass fish injection, but efficacy can be shown in injectable preparations while other methods of administration are being developed, and actually mass injection is not as hard as might be thought. The ideal vaccine preparation for the farmer to administer is an oral one. This method of administration is still in the early stages of development and no oral preparations are currently available. However it is generally agreed that oral vaccines will be the products of the future. The farmer will benefit because of the ease of application, the ability to boost the vaccines and from the ability to plan administration according to the best research results. Some methods of preparation are being tested now and products can be expected to emerge in this year or in the next. Fish vaccine development has similar problems to the licensing of drugs in some respects. The market for vaccines is still relatively small and profits are limited. Administration regimes have not been fully developed. The pathogens are complex in nature and fish immunology is still in its infancy.

Conclusion: the way ahead

In the short term the farmer requires more licenced drugs so that he can employ a strategic approach to treatments. The continual lengthening of Withdrawal Periods requires review as this is the most serious constraint on treatment. There is a long term need for efficacious vaccines and improved administration regimes. It is felt that the study of fish immunology and the development of genetic engineering techniques for production of antigens will be of benefit to future fish health management.

Discussion

SCHLOTFELDT The small market for products seems to be the overwhelming common denominator in all countries. There is a hope that, if the Community becomes an open market without boundaries after 1993, the market will no longer be so small, especially if, one day, Scandinavia becomes a member of the Community.

RAE That is more of a comment than a question, but we can't get bigger until we get the products and we can't get the products until we get bigger I suppose.

CONSTRAINTS ON CHEMOTHERAPY: A MANUFACTURER'S VIEW

Lydia A. Brown

Abbott Laboratories, Mill Leat Annexe,
West Gomeldon, Salisbury, Wilts SP4 6JZ. UK.

Summary: *Manufacturers of therapeutants for use in aquaculture must be prepared to license medicines on a global basis in order to produce an adequate return on their investment. Factors which hinder this practice are centred on the national differences in product registration. No guidelines are available to manufacturers for use when designing studies and there are no assurances that such studies are acceptable in other countries around the world.*

Résumé **: Les contraintes de traitement en aquaculture : point de vue du fabricant de médicaments.**

Le développement d'un médicament nouveau s'accompagne de coûts dont l'amortissement est conditionné par l'importance du marché. A cet égard, le marché de l'aquaculture apparaît bien étroit, puisqu'on l'estime à 90 millions de dollars, contre 9,7 milliards pour l'ensemble de la santé animale. Pour être rentable, un produit devrait rapporter au moins 100 fois le montant des sommes investies dans son développement. Or le coût des études est aussi élevé sinon davantage, pour les produits destinés au poisson que pour ceux destinés aux espèces homéothermes. La mise au point d'un produit requiert toujours de longues années et coûte 3 à 6 M de dollars!

Les réglementations varient beaucoup selon les pays, mais seuls les produits traditionnels bénéficient parfois de procédures d'autorisation simplifiées. Souvent, le poisson est ignoré par les législations sur le médicament. Les procédures d'autorisation font alors appel à des critères plutôt arbitraires. Certains types de produits sont même carrément interdits. Les études requises pour présenter un produit nouveau doivent donc être répétées selon les normes propres à chaque pays, les différences portant sur les doses et la durée des protocoles d'études, ainsi que sur la limitation des genres ou des espèces pour lesquels l'autorisation sera validée. Les protocoles imposés sont souvent théoriques, voire inapplicables quand les laboratoires et les équipements nécessaires à leur réalisation n'existent pas. Les études cliniques s'avèrent tout aussi incommodes à mener sur le terrain.

L'évaluation des résidus est codifiée selon des critères sévères, bien que les systèmes d'inspection demeurent généralement embryonnaires. Dans beaucoup de pays, les données doivent être établies à plusieurs températures. L'impact des produits sur l'environnement est encore plus compliqué à établir, faute de références précises. On réclame des composés qui concilient une dégradation rapide dans l'environnement avec une certaine stabilité au moment de la préparation de l'aliment médicamenteux.

On comprend dès lors que peu de formules nouvelles aient été mises sur le marché ces dernières années. Pour améliorer cette situation, il a été suggéré de réduire les études, par l'adoption de codes de bonnes pratiques de laboratoire et de protocoles standards agréés par tous les pays. La CEE s'emploie désormais à cette harmonisation, et les autorités norvégiennes se sont déjà penchées sur la mise au point de standards d'expérimentation réalistes.

The size of the market

A manufacturer who is considering the development of a product for the aquaculture market today has to make a rational assessment of the market and consider the potential for recouping the development costs which will be required. Unlike other animal medicines, chemotherapeutants for aquaculture must be licensed globally. In some countries the total aquaculture market is not strong enough to support the development costs of the product. The fee for submitting a product licence application for a novel compound may be greater than the potential antimicrobial market for farmed fish in that country (for example in Australia).

The worldwide chemotherapeutant aquaculture market is a relatively small industry estimated at $90 million (2) compared with the traditional animal health industry which is estimated to be $9,700 million in total sales (1). The cost to register a product in aquaculture is equal to if not greater than registering a product for any other food animal since a number of studies have to be repeated at different water temperatures. This is due to fish being poikilothermic, that is they are cold blooded animals, whose basal metabolic rate varies with water temperature.

To register a novel compound for aquaculture even in one country requires a financial commitment on the part of the pharmaceutical company of anything from $3 million to $6 million. This money must be invested over several years before a product licence even goes into an agency.

Variations in licensing requirements

Currently the requirements of individual countries for registration of aquaculture chemotherapeutants vary enormously and are of concern to those manufacturers wishing to enter the market with novel chemotherapeutants targeted specifically for aquatic animals. Some countries allow traditional generic compounds to be licensed in an abridged form of licence because there is a great shortage of adequate remedies. In some cases withdrawal periods are placed on the drugs by the regulatory authority when no studies on residues have been provided. Such a product has an enormous price advantage over the novel compound entering the market for the first time which has not had to provide any residue data. All of these practices tend to ensure that only traditional `me too' generic compounds will be authorized.

Some countries do not include fish in their medicines legislation as in the Republic of Ireland. Here a veterinarian doesn't have to prescribe chemotherapeutants for fish. The fish farmer can buy drug from agricultural supply companies, supermarkets or directly from a producer. Other countries may provide an arbitrary ban on a particular class of compounds as does Japan which once again reduces the manufacturers chances to make a return on his development costs by operating in a global market.

Studies required

However the constraints are not all on the strictly financial side. One of the main costs is duplication of effort by having to repeat studies in slightly different ways to satisfy different regulatory authorities. Target animal safety studies for example need to be done at 1x, 3x and up to 5x the recommended therapeutic dose for three times the recommended duration in the USA but for 1x, 2x and 4x the recommended therapeutic dose in the UK. Kinetic studies in the USA may require 5 to 6 fish per sample point although 12 is a more usual number in Europe. Even the same animal can be looked on differently in particular countries. Again using the USA as an example Atlantic salmon, Pacific salmon and all species of trout are classified as salmonids. In the UK Atlantic salmon are different species compared with rainbow trout; and separate studies are also required depending on whether they are in fresh water or sea water.

Standards of reporting on studies can vary from country to country. Specific studies requiring Good Laboratory Practice (GLP) procedures can add 50 to 75% over the cost of a non GLP study. The theory of the regulatory requirements in this regard often cannot be adequately met in the practicality of performing the tests.

Some studies are extremely difficult to perform, for example the methodologies for metabolism studies in fish are such that quantitative studies wholly acceptable to regulatory authorities cannot be done. Enclosing the fish in a metabolism chamber stresses the fish and it is no longer a healthy normal animal. Obtaining correct material balance for a radiolabelled study in an aquatic animal for such a study is impossible even under chamber conditions.

One should remember that there are research laboratories in the same governmental agencies, which authorize product licences, that are not able to conform to the GLP requirements laid down by their own departments.

Research facilities for aquatic animals are extremely limited. The manufacturer has difficulty in identifying a research laboratory in any one country which is able to perform adequate aquatic animal studies, especially for salt water target species of the size normally receiving the medication.

Clinical studies

EEC guidelines developed several years ago (Paper III/121/87-EN Rev 2) suggested that positive and negative controls be assessed for field efficacy studies. For welfare and ethical reasons it is impossible to ask a veterinarian who supervises a trial to withhold treatment in a cage of `animals under his care' in order to provide a negative control. The welfare implications of such guidelines do not unfortunately appear to have been addressed by the committee which issued the recommendations.

A major problem with performing field efficacy studies is that one is not able to see the animals under test nor to count every individual fish before the trial starts. In fact on most cage salmonid systems it is possible that the only time the fish have ever been physically counted in a cage is when they were originally transferred from fresh water to sea water as young smolts.

Analytical methods are also a major problem for manufacturers. To what level of sensitivity do we go? Some authorities suggest that no drug should be present in tissue of food fish at slaughter. Some say that the concentrations of drug in the tissues should be at the lowest detectable limit possible while others use the concept of maximum residue levels (MRL) and no observable effect levels (NOEL) obtained from toxicology studies in mammals. These limits are quite rightly imposed to ensure that the consumer is not contaminated by levels of residues in fish tissue. However, I am sure the consumer would be pleased to see more speed in defining residue limits at the international level. Equally they may like to see a system of public health inspection of aquatic animals. This does not appear to be a priority for the majority of countries.

In the USA withdrawal periods are recommended based on data collected at one water temperature. In the UK data is required at more than one temperature. The EEC guidelines cited earlier (which I believe are being redrafted) call for data at three water temperatures. This seems excessive when it is well established that metabolic rates at different water temperatures do not have a directly linear effect.

Environmental studies

We (fish farmers, manufacturers and regulatory officials) are all concerned about environmental impact of our end products (be it fish or chemotherapeutants). However, requirements for studies in this area appear to be particularly ill defined. It is so little understood that constant dialogue with environmental agencies is needed to agree on studies.

In this respect aquaculture remedies have to be uniquely formulated so that they are stable under the high temperatures encountered in feed mill operations, ultimately reach the aquatic environment.

Although this paper highlights a number of concerns on the part of the manufacturer, with goodwill and cooperation between the regulatory authorities and the manufacturers, the process of licensing a product for a global market can be streamlined. Suggestions to help in reducing the load from the manufacturers and easing the paperwork burdens between countries are:

1. Reduce the numbers of studies requiring GLP

2. GLP in each country should be readily acceptable across national boundaries

3. Provide adequate practical protocol guidelines (worked up from practices which are known to work in the field) and which will be accepted by all countries

The EEC are trying to move toward a unified approach to the licensing of chemotherapeutants for animals. However, this is still in the future. Whether or not harmonization will occur by 1992 remains to be seen.

The Norwegian Medicines Control Authority (SLK) are attempting now to provide a standard set of guidelines which would ultimately achieve a practical and realistic approach to licensing an aquaculture chemotherapeutant throughout the world (3). We should unreservedly enter dialogue with such bodies to make this goal a reality.

References

1). ANIMAL PHARM, 1990. Citing Vivash Jones Consultants. Volume 214. October 26th. Page 10.

2). BROWN L. A., 1989. *The Aquaculture Market.* PJB Publications Ltd. 18/20 Hill Rise, Richmond, Surrey. TW10 6UA. UK.

3). STATENS LEGEMIDDELKONTROLL, 1990. Round Table Consultations on documentation of drugs for use in Aquaculture. SLK, Sven Oftedals vei 6, 0950, Oslo 9, Norway.

Discussion

TREVES-BROWN Dr Brown said that we are suggesting overdose at 1x, 2x and 4x and the USA, 1x, 2x and 5x. We are not fussy. If you come and ask us, we will suggest 1x, 2x and 4x, but if you present data with 1x, 2x and 5x, that's quite satisfactory. The important thing with overdose studies is not to try to show that your fish necessarily stay alive at very large doses. The important thing is to know qualitatively what happens to the fish when it is overdosed, so that you can introduce into your labelling satisfactory warnings to enable product to be used safely. Dr Brown listed three things that she would like to see happen to make the pharmaceutical industry's life easier. I would like to add a fourth, which is one of the problems that has been mentioned several times by different speakers. It is the problem of residues: the problem that if you have a long withdrawal period, your fish is going to get ill again before it has gone to market. The problem is, we are working on maximum residue levels which are established for mammalian meat consumption. We need to call on the international authority, insofar as there is one which has any authority in this area, which is the joint committee of FAO and WHO (JECFA). We need to call on them to come up with realistic maximum residue levels for fish meat, which may be different from those for mammalian meat.

LYDIA BROWN I would like to agree with Mr Treves-Brown on the second point, but on the first point I believe that the manufacturers in particular would welcome a standardisation of protocols and guidelines. I believe that this is way overdue, surely all of the countries can get together and agree on the particular studies that they require. You say, quite rightly, that we can come and ask you and you will tell us one thing, but if we come to you with another, you will still look at it. I am afraid as a manufacturer we might like a bit more assurance that what we are doing is the right study before we spend the money.

RESTRICTIONS ON THE USE OF DRUGS IN AQUACULTURE IN JAPAN

A. Okamoto

Research Division, Research Department, Fisheries Agency, Ministry of Agriculture, Forestry and Fisheries, 1-2-1 Kasumigaseki Chiyoda-ku Tokyo, Japan

Summary : *The Japanese Pharmaceutical Affairs Law regulates the use of drugs for farmed fish in Japan. The Ministry of Agriculture, Forestry and Fisheries of Japan is responsible for selecting the Subject Animals and the Subject Drugs. The regulation appliess when farmers use the Subject Drugs for treatment of the Subject Animals. Users are obliged to keep to the standards concerning the use of drugs. We always need to reform the contents of the regulation in accord with development of aquaculture.*

***Résumé*: Restrictions d'emploi des médicaments pour l'aquaculture au Japon.**

Au Japon la sécurité des consommateurs est garantie par une loi sur l'hygiène des denrées alimentaires qui proscrit la présence de résidus thérapeutiques, et par une loi relative aux questions pharmaceutiques. Une ordonnance ministérielle sur le contrôle des médicaments destinés aux animaux a renforcé ces dispositions en 1980.

Conformément à ces textes, les médicaments destinés aux poissons doivent faire l'objet d'une demande d'autorisation auprès du Ministère de l'Agriculture, des Forêts et des Pêches. L'utilité des produits est évaluée espèce par espèce. Dans le cas des antibiotiques, l'estimation de l'efficacité est complétée par celle de l'innocuité et par l'étude du devenir des résidus. Le dossier technique, qui repose sur le travail de 2 laboratoires indépendants, conduit à définir des délais d'attente variables selon les espèces.

Les autorisations, accordées produit par produit et espèce par espèce, imposent que des données techniques soient fournies aux utilisateurss sous forme d'une notice figurant sur l'emballage et précisant les espèces concernées, le mode d'administration, les doses et les délais d'attente. Toutes applications autres que celles prévues par la loi sont prohibée.

Pour étayer ces dispositions le ministère tient à jour des listes-types des espèces de poissons et des agents antimicrobiens les plus répandus en aquaculture. Cette évolution récente de la loi sur les questions pharmaceutiques sera sans doute suivie de nouvelles réformes compte-tenu du développement rapide de l'aquaculture japonaise.

Introduction

The Japanese Food Sanitation Law provides controls to prevent people from having antibacterial agents remain in food in Japan from point of view of public health and sanitation. The Pharmaceutical Affairs Law regulates the manufacture, sale or use of drugs. One of the purpose of the regulation is to prevent drug residues in livestock or farmed fish. I would like to explain the regulation which is relating to users of drugs.

Manufacturers of drugs are required to present drug users with the necessary information to avoid the occurrance of drug residues. That is why the names of the species of fish on which people can use each drug, how to use the product, how much use per day and withdrawal periods before slaughter, on data labels attached to sold drugs. Of course, even if the labels present the necessary information, users may not observe them. Therefore we amended the Pharmaceutical Affairs Law into its present form, which enables us to oblige users to use drug properly. The Ministerial Ordinance Regarding the Control of Use of Drugs For Animals establishes the details of the regulation.

Approval for manufacturing drugs

First, I need to explain the proceedure for approval for the manufacture of drugs, because understanding this is necessary to the understanding of the Ministerial Ordinance. Pharmaceutical companies have to get permission from the Ministry of Agriculture, Forestry and Fishery to manufacture fish drugs and to sell them. Pharmaceutical companies present data showing ability of the drugs. The Ministry of Agriculture, Forestry and Fishery gives permission for the drugs after assessing if the drugs are useful. It is necessary to present data showing of utility of drugs for each species of fish. Therefore there are a lot of types of drugs which have permission for use on one species of fishes and not for any others. Table I shows the status of approval for manufacturing drugs.

The Ministry of Agriculture, Forestry and Fishery determines how long the drugs remain in fish tissues in addition to the efficiency and safety of drugs when the pharmaceutical companies make an application for permission for drugs containing antibacterial agents. Pharmaceutical companies need to establish withdrawal period before shipping by using data concerning residue properties investigated in two laboratories. They often establish different withdrawal periods in species of fish, since they get separate authorisation for each species of fish.

Table I: The status of approved fish drugs

Drug	*a*	*b*	*c*	*d*	*e*	*f*	*g*	h	i	j	k	l	m
Ampicillin	O	X	X	X	X	X	X	X	X	X	X	X	X
Erythromycin	O	X	X	X	X	X	X	X	X	X	X	X	X
Alkyltrimethylammonium calcium oxytetracycline	O	X	X	X	X	X	X	X	X	X	X	X	X
Oxytetracycline HCl	O	O	O	X	O	O	X	O	X	X	X	X	X
Oxolinic acid	O	X	O	O	O	O	O	O	O	O	O	X	X
Oxolinic acid (bath)	O	X	X	X	X	X	X	X	X	X	X	X	X
Oxolinic acid (liquid)	X	X	X	X	O	X	O	X	X	X	X	X	X
Kitasamysin	O	X	X	X	X	X	X	X	X	X	X	X	X
Spiramycin embonate	O	X	X	X	X	X	X	X	X	X	X	X	X
Sulphadimethoxine or its sodium salt	X	X	X	X	X	O	X	X	X	X	X	X	X
Sulphamonomethoxine or its sodium salt	O	X	O	X	O	O	O	X	O	O	X	X	X
Sulphamonomethoxine or its sodium salt (liquid)	X	X	X	X	X	O	X	X	O	X	X	X	X
Thiamphenicol	O	X	X	X	X	X	X	X	X	X	X	X	X
Florfenicol	O	X	X	X	X	X	X	X	X	X	X	X	X
Sulphamonmethoxine & ormetoprim	X	X	X	X	X	X	O	X	X	X	X	X	X
Amoxicillin	O	X	X	X	X	X	X	X	X	X	X	X	X
Colistin sulphate	X	X	X	X	X	X	X	O	X	X	X	X	X
Sulphisoxazole or its sodium salt	O	X	X	O	X	O	O	X	X	X	X	X	X
Tetracycline HCl	O	X	X	X	X	X	X	X	X	X	X	X	X
Doxycycline HCl	O	X	X	X	X	X	X	X	X	X	X	X	X
Lincomycin HCl	O	X	X	X	X	X	X	X	X	X	X	X	X
Josamycin	O	X	X	X	X	X	X	X	X	X	X	X	X
Nalidixic acid	X	X	O	X	X	O	O	X	O	O	O	X	X
Sodium nifursytenate	O	X	X	X	X	X	X	X	X	X	X	X	X
Sodium novobiocin	O	X	X	X	X	X	X	X	X	X	X	X	X
Piromidic acid	X	X	X	X	O	X	X	X	O	X	X	X	X
Flumequine	O	X	X	X	X	X	X	X	X	X	X	X	X
Sodium polystyrene sulphonate	O	X	X	X	X	X	X	X	X	X	X	X	X
Miloxacin	X	X	X	X	O	X	X	X	X	X	X	X	X

Key to Table I

Fish: a: yellowtail b: sea bream c: coho salmon d: carp
e: eel f: rainbow trout g: ayu h: kuruma prawn
i: amago salmon j: yamame salmon k: other salmonids
l: crucian carp m : other fishes
X: there is permission for use in these species.
O: use not permitted in this species.
italic : the Subject Animals and the Subject Drugs.
bold : extent of The Ministerial Ordinance

Designs of the regulation

The Ministry of Agriculture agrees the species of animals (called the Subject Animals) and species of drugs (called the Subject Drugs). The Ministerial Ordinance starts when people use the Subject Drugs for the Subject animals. For example, ampicillin, which is the Subject Drug, has permission for only yellowtail among fish. In this case people can use ampicillin for yellowtail. Additionally, users are obliged to follow the use instructions as to route, dose to use per day, and withdrawal periods before slaughter for shipping, as outlined on the labels attached to sold drugs.

On the other hand the Ministerial Ordinance prohibits people from using ampicillin for the Subject Animals except yellowtail. Those are sea bream, coho salmon, carp, eel, rainbow trout and ayu. Because the Ministerial Ordinance establishes the regulation by using data which are presented to get permission to manufacture the drugs we know how to use the drug, how to use per day and withdrawal period before shipping when we use drugs for fish having permission. However there is no information of this type available if we use drugs on fish having no permission. Ampicillin seems to be effective to treat diseases of fish species other than yellowtail, but a withdrawal period has not been established in case of the use of ampicillin for sea bream or coho salmon. If people were to use it on them, they might ship fish in which drug residues remain. That is why the Ministerial Ordinance prohibits people from using ampicillin for any of the Subject Animals except yellowtail.

The Subject Animals

The Ministerial Ordinance appoints some fish species which are produced in large quantities and which people use drugs for, as the Subject Animals. When the Ministerial Ordinance was made in 1980, six species of fish, yellowtail, sea bream, carp, eel, rainbow trout and ayu, were appointed as the Subject Animals. After that coho salmon was added in 1990, because of increases in its production. We are farming a lot of species of fish recently. Production of some species which are not the Subject Animals, like horse mackerel, is increasing. We are now planning to make new the Subject Animals.

The Subject Drugs

The Ministerial Ordinance appoints some species of antibacterial agents as the Subject Drugs. These are only species which are much used or have to be used with taking special care of their residues in present. We are going to appoint the other species of antibacterial agents as the Subject drugs in the future.

Conclusion

An amendment to Pharmaceutical Affairs Law gave the government greater rights and responsibility to secure nation from danger of drugs than before. Foundation of the Ministerial Ordinance was one of them. We always need to reform the regulation in accord with development of aquaculture.

Discussion

[Editors : Dr Okamoto was unable to present this paper, which was presented by Dr R. Schnick, who therefore attempted to respond to questioners in this discussion session]

BROWN Dr Schnick, do you know if there are any quinolone compounds licensed for use in fish in Japan?

SCHNICK Well I know that oxolinic acid is, are you talking about the fluoroquinolones, the more advanced....?

LYDIA BROWN I was just wondering which ones were licensed in Japan?

SCHNICK Oxolinic acid, nalidixic acid, piromidic acid, miloxacin and flumequine are the quinolones licensed. I know that there is another compound that is being considered for registration in Japan. I think this is an important area of consideration by the pharmaceutical industry. I have been in discussion with a number of pharmaceutical companies within the United States and this is why I talk about a window of opportunity, I am hopeful that we can attract at least a couple more pharmaceutical companies to the aquaculture industry , so that we can do what we are talking about, namely changing the treatment regimes so that we are not always using the same compound and then trying to avoid some of these resistance problems that we have seen.

SCHLOTFELDT Do you know perhaps, how they managed to licence so many drugs for fish. The licensing costs, which may be three to six million dollars in the USA, must be considerably lower there?

SCHNICK As I understand it, according to the paper, the requirements are that they establish efficacy and a withdrawal time based on analysis and I am assuming that there are target animal safety studies, but I am not sure how much is done to determine mammalian safety.

DESPLENTER Is there a global consensus on what is considered edible tissue in fish?

SCHNICK At this point this may not be true, but in the United States for salmonids it is the muscle and the skin, for channel catfish, it is just the muscle. Now I am not sure about the other countries, whether anything else is used, I know that there is some use of livers, but whether this is a universal situation I don't know.

DESPLENTER So that means for the calculation of the ADI, you have or might have differences on a country by country basis.

SCHNICK Yes, but I think that's probably fairly rare. I think that generally you will find that, with an animal in which the skin is left on in the preparation or cooking process, the residues in the skin will be included in withdrawal requirements. When it is not eaten, then you are just dealing with the muscle. I think this is fairly straightforward compared to other food animals where consumers may be exposed to other internal organs of the animal.

VETERINARY MEDICINES REGULATIONS FOR FISH IN THE EEC

R. Hankin

Pharmaceuticals Service, (III/C/3)
Commission of the European Communities, 200 rue de la Loi,
B - 1049 Brussels, Belgium

Summary *: The organisation of the open market within the EEC is accompanied by a coordination of the regulations for veterinary medicined legislation throughout that market. This paper describes the EC directives which impose the coordination and considers how those directives (81/851/EEC, 87/22/EEC, 90/676/EEC and 90/677/EEC) will affect aquaculture medicines.*

Résumé: **Réglementation des médicaments vétérinaires destinés aux poissons dans la CEE.**

L'organisation du grand marché européen doit s'accompagner d'une réglementation commune sur les produits de médecine vétérinaire. Celle-ci repose sur une directive adoptée dès 1981 (n°81-851), complétée jusqu'en 1990 par une série de texte qui entreront pleinement en vigueur en janvier 1992 et codifient les analyses pharmacotoxicologiques, les essais cliniques, le cas des médicaments dérivés des nouvelles technologies, et prévoient la définition de limites maximales de résidus (LMR) dans les denrées d'origine animale. Un Comité des Produits de Médecine vétérinaire (CVMP) a été constitué pour en préciser les applications.

Autorisations de mise sur le marché (AMM): Une définition précise des produits médicamenteux destinés à l'usage médical ou vétérinaire a été formulée. Depuis 1981 le poisson est expressément inclus parmi les animaux concernés par la législation vétérinaire. Tout nouveau produit doit subir une procédure d'AMM. Actuellement les exigences sont fixées par chaque état membre, mais depuis 1990 est envisagée une procédure globale qui s'exercerait sous le contrôle d'une Agence européenne des Médicaments.

Qualité, innocuité, efficacité doivent être vérifiées pour obtenir l'AMM. La première garantit un produit stable, homogène, qui obéit aux règles de production industrielle communément admises. Les principes sont calqués sur l'exemple des médicaments humains, mais des mesures particulières adaptées au cas des poissons pourront être spécifiées. La seconde prévoit l'absence de résidus, la définition des espèces-cibles, et l'innocuité pour l'utilisateur. Les problèmes liés à la protection de l'environnement sont également à l'étude et devront être pris en compte dès janvier 1992.

La protection du consommateur repose sur l'établissement des LMR, sur l'observance de délais d'attente qui les garantissent, ainsi que sur une surveillance par échantillonnage aléatoire. Dès janvier 1992 tout produit nouveau ne pourra être accepté que s'il y a accord unanime sur les LMR, et les LMR de tous les produits existants devront être établies avant 1996. Les principes qui prévalent en la matière sont les mêmes que pour les autres denrées d'origine animale. Néanmoins les difficultés d'étude de la pharmacocinétique chez les poissons imposeront sans doute des mesures spécifiques. Aucun contrôle formel n'est encore appliqué aux poissons, mais un programme de surveillance volontaire est en préparation.

Médicaments sans AMM: du fait du coût des études et de l'étroitesse du marché aquicole peu de produits ont actuellement reçu une AMM. On recourt donc souvent à des traitements non homologués, soit à titre temporaire, soit en puisant dans les formules autorisées pour d'autres espèces. Un amendement à la directive 81-851 devrait préciser les limites de tels recours et les placer sous un contrôle vétérinaire très strict. Dans de tels cas les délais d'attente seront fixés au maximum, et les interventions devront être consignées dans des documents d'enregistrement.

Un travail considérable et de réels progrès ont été accomplis, mais les experts de la CEE ont encore beaucoup à faire pour aboutir à des règles communautaires parfaitement adaptées aux médicaments à usage aquicole.

Introduction

As part of its overall programme for the realisation of the internal market within the 12 Member States, the European Community has adopted a series of Directives and one Regulation governing veterinary medicinal products. The basic directive is Directive 81/851/EEC on the approximation of the laws of the Member States relating to veterinary medicinal products, which entered into force in 1983. This directive has recently been substantially amended by Directive 90/676/EEC which enters into force on 1 January 1992. In addition, Directive 90/677/EEC, which also enters into force in January 1992, has extended the scope of the Community rules to over immunological products.

Directive 81/852/EEC harmonizes the laws of the Member States on the analytical, pharmaco-toxiological and clinical tests and trials of veterinary medicinal products. Directive 87/22/EEC establishes a special Community authorization procedure for products derived from biotechnology and other high-technology medicinal products. Finally, reference should be made to Regulation 2377/90 which lays down a Community procedure for the establishment of maximum residue limits (MRLs) for residues of veterinary medicines in foodstuffs of animal origin.

In accordance with the basic principles of Community law, this legislation is binding on Member States and overrides any conflicting provision of national law. It also confers direct rights on individuals which they may invoke before national courts.

In addition, the Committee for Veterinary Medicinal Products has been established. The Committee provides a forum for the discussion of issues relating to the evaluation of veterinary medicines, and is also responsible for giving opinions on applications submitted through the special Community registration procedures.

Marketing Authorization

In European Community pharmaceutical legislation, a medicinal product is defined as

"Any substance or combination of substances presented for treating or preventing disease in human beings or animals".

"Any substance or combination of substances which may be administered to human beings or animals with a view to making a medical diagnosis or to restoring, correcting or modifying physiological functions in human beings or in animals is likewise considered a medicinal product."

A veterinary medicinal product is defined as any medicinal product intended for animals.

Although there might appear to be some scope for arguments of a legal nature about whether the reference to animals in the definitions includes fish, in practice all Member States interpret the Community directives as applying to fish, and indeed recent amendments to Directive 81/851/EEC have for the first time expressly mentioned fish. It therefore appears clear that Community legislation governing veterinary medicinal products is fully applicable in the case of medicinal products intended for use in fish.

It follows that no medicinal product intended for use in fish may be placed on the market of the Member States unless a marketing authorization has been granted by the competent authorities of the Member State concerned. At the present time authorization for each Member State must be obtained separately. However, in September 1990, the Commission issued a comprehensive package of proposals for the future system for the free movement of medicinal products in the Community, including the establishment of a new European Medicines Agency and new binding Community authorization procedures.

Demonstration of quality, safety and efficacy

In order to obtain authorization, the applicant must demonstrate that the product satisfies the three traditional criteria of quality, safety and efficacy.

Quality covers the verification of the ability of the applicant to produce a stable and homogenous product, in accordance with its approved specifications, on an industrial scale. In principle, the quality requirements for veterinary medicines are the same as those laid down for human medicines, although it is recognised that there are specific pharmaceutical forms used in veterinary medicine which may require the development of specific quality guidelines, including possibly certain pharmaceutical forms used in fish. This matter is under review, but no decisions have yet been taken.

Safety currently covers three aspects:

- the safety of any residues of the product for human consumers of foodstuffs of animal origin;
- the safety of the product in the target species of animal;
- the safety of the product for the animal holder and the veterinarian.

Moreover, from 1 January 1992 onwards, consideration will also have to be given to the safety of the product for the environment.

The protection of the consumer against risks resulting from the presence of residues in foods is based upon three factors:

- the safety evaluation of the compound, leading to the establishment of an acceptable daily intake and maximum residue limits in individual food commodities;
- the establishment of withdrawal periods to allow time for residues to deplete to levels below the MRL;
- residue monitoring and surveillance, with the random testing of food commodities for the presence of residues.

From 1 January 1992 onwards no new active substance will be permitted to be used in food animals, including fish, unless a maximum residue limit has been established by the Community for residues of the compound concerned. Moreover, the Community is required to establish MRLs for all existing compounds over a five year period from 1992-96. Of course, the basic principles of the safety assessment of a compound are the same, whether the compound is used in mammals, poultry or fish and in principle MRLs which are established for meat commodities are also applicable to fish. However, where a compound is used in fish, and residue data are available, this will of course be taken into consideration during the establishment of MRLs.

There are specific difficulties in the conduct of pharmacokinetic and residue depletion studies designed to enable the determination of withdrawal periods. Once the work on general guidelines on the conduct of residue depletion studies is completed, consideration will be given to additional Community guidelines dealing specifically with the conduct of residue depletion studies in fish.

So far as residue monitoring is concerned, the formal Community legislation requiring the establishment of national surveillance plans does not apply to fish. However, preparations are at an advanced stage for the establishment of a voluntary surveillance programme for residues in fish, which will involve screening for residues of antibiotics, in particular chloramphenicol, for malachite green and possibly for residues of hormones. The objective of this programme, in the first phase, will be to determine whether there are in fact potential residue problems in fish at present.

The other safety issue of potential concern from the regulatory point of view concerns the effects of the use of medication on the environment. As noted above, new legislation will require specific consideration to be given to this aspect from 1 January 1992 onwards. At the present time the Commission is working with Member States in order to define the circumstances in which an environmental assessment will be required, and the criteria to be taken into consideration during that assessment.

Efficacy covers the verification that the product does indeed have the effects claimed by the manufacturer, for example, for the prevention or treatment of disease, or for the increase in growth or yields from the treated animal. In addition, in the case of a product intended to increase growth or yields, the efficacy criterion extends to the verification of the quality of the foodstuffs obtained from treated animals. At the present time, the efficacy working party of the CVMP is preparing specific guidelines on the demonstration of efficacy of veterinary medicinal products in farmed fish, which will be released for consultation in the course of 1991.

Extra-label use in fish

For various reasons, linked with the cost and uncertainties of obtaining authorization and the limited size of the market for fish medication, there are at present only a limited number of products specifically authorized for use in fish, and much medication is undertaken through the extra-label use of compounds authorized for use in other animal species, or possible the use of non-authorized compounds in extemporaneous preparations. Such use is inevitably rather problematic, and may give rise to difficulties for the farmer, particularly if residues are detected.

A new amendment to Directive 81/851/EEC is intended to clarify the legal rules applicable to this type of use and to define the conditions under which it may be necessary. In principle, recourse to non-approved therapy will be acceptable only if undertaken by a veterinarian in respect of animals, including fish, under his direct personal care. If a suitable authorized product is available, the veterinarian is expected to use it. However, if no authorized product is available, the veterinarian is expected to consider whether there is a product approved for use in another species whih may be suitable. If not he may have recourse to a product authorized for use in man, or he may prescribe an extemporaneous preparation. If no withdrawal period has been specified for the species concerned, the veterinarian must ensure that a withdrawal period of at least 500 degree x days is observed. All instances of extra label use must be recorded in writing, and the records held available for inspection by the competent authorities for a period of at least three years.

Conclusion

In conclusion, it appears that although a certain amount of progress has been made, there is still a certain amount of work which remains to be done towards the elaboration of appropriate and realistic regulatory standards for the control of veterinary medicines used in fish. In seeking to define appropriate standards, the Community will also take into consideration the work being undertaken in other relevant fora. For this reason, this OIE Conference is particularly timely.

Discussion

MCGILL Just one thing, if you could clarify for me. Do you intend to keep the same MRLs for fish as you have for meat?

HANKIN Unless there is a good reason not to, yes. Very often with MRLs, you are extrapolating, you often have the raw data put together for a mammalian product and then you have an incidental use of the product in fish, so the chances are, you are going the same MRL unless you have good reason not to, but it is always open to you to put in data to try to justify a different MRL.

FISH DRUGS IN GERMANY: NATIONAL REGULATIONS AND THE EEC

Eva-Maria Bernoth

Institute for Veterinary Medicine (Robert von Ostertag Institute),
Federal Health Office, P.O.Box 33 00 13,
W-1000 Berlin 33, Germany

Summary: *In Germany, only licensed drugs can be used in fish destined for human consumption. There are several ways by which authorisation to use a drug can be obtained:*

1 - Individual marketing authorisation of a new drug for use in fish, according to the Law on the Reform of Drug Legislation (1976),

2 - Re-authorisation of a drug which had been authorised for use in fish before the Reform,

3 - Standard marketing authorisation of a drug,

4 - Authorisation according to the Directives 81/851/EEC and 90/676/EEC for drugs already authorised in other EEC member states,

5 - Authorisation according to Directive 87/22/EEC concerning high-technology medicinal products.

6 - Furthermore, a drug which is licensed for other animals serving the production of food, may be used in fish.

The above alternatives will be explained and their particular requirements and importance for fish drugs will be discussed.

There are hardly any attempts by the pharmaceutical industry to obtain an individual marketing authorisation for a new drug. At present, several drugs, including antibiotics, are being re-authorised. Some chemicals, including malachite green and formaldehyde, are being discussed as candidates for a standard marketing authorisation. Applications for authorisation under Directive 81/851/EEC and No. 90/676/EEC or Directive 87/22/EEC have not so far been submitted. The use of drugs which are not licensed for fish but for other animals destined for human consumption is only possible in individual cases and is subject to several restrictions.

***Résumé*: Les médicaments à usage piscicole en Allemagne. La réglementation nationale et la CEE**

En Allemagne le traitement de poissons destinés à la consommation est placé sous le contrôle d'une loi de 1976 et ne peut être envisagé qu'avec des médicaments officiellement autorisés. Il existe 6 types d'autorisations.

1) Autorisations individuelles de commercialisation pour des produits nouveaux, qu'il s'agisse de substances actives ou de formules composées. Des allègements de procédures sont possibles lorsque le produit actif a déjà été autorisé.

2) Renouvellements d'autorisation pour des produits qui l'étaient déjà avant la promulgation de la nouvelle loi.

3) Autorisations standard de commercialisation, concernant des produits exempts de riques et librement distribués.

4) Autorisations délivrées selon la directive 81/851 de la CEE, pour des produits déjà autorisés dans d'autres pays membres.

5) Autorisations accordées selon la directive 87/22 de la CEE, pour des produits issus des nouvelles technologies.

6) Autorisations d'emploi des médicaments autorisés pour d'autres espèces que le poisson.

En pratique, très peu de dossiers sont présentés actuellement pour des produits nouveaux, et seule l'association sulfadiméthoxine-triméthoprime a été homologuée. Plusieurs formules, incluant des antibiotiques, attendent leur renouvellement d'autorisation (Tableau I). Certains produits, qui comme le formol et le vert malachite n'étaient pas classés comme médicaments (Tableau II) et devaient obtenir l'autorisation standard, butent sur la démonstration de leur innocuité. Aucune demande n'a encore été formulée au titre des directives CEE, et l'emploi ponctuel de médicaments autorisés pour d'autres espèces est soumis à restrictions.

Les mesures nationales devront être conciliées avec celles définies par la CEE dès janvier 1992. Toutefois le poids accordé aux problèmes d'impact sur l'environnement et la difficulté à établir les futures limites maximales de résidus risquent de peser très lourd sur le coût des études de dossiers.

Introduction

On January 1, 1978, a new Drug Law came into force in the Federal Republic of Germany (1). Its purpose is to guarantee quality, efficacy and safety of drugs for use in man and animals (§1 Drug Law). Drugs can be marketed only if they have been authorised. This is valid for finished drugs as well as for premix drugs (§4 (11) and §21 (1) Drug Law). The application for authorisation must be made by the manufacturer (§21 (3) Drug Law); the competent authority is the Federal Health Office (Bundesgesundheitsamt, BGA) in Berlin (§77 Drug Law).

The scope of this review is to present the various national approaches to legalise use of a drug in aquaculture, and to give an outlook to the EEC regulations.

Individual marketing authorisation of a new drug

In order to guarantee safety, comprehensive data must be submitted for drugs "containing substances or preparations the effects of which are not generally known in the field of medical science". These drugs are also subject to automatic prescription (§49 Drug Law). Among the documents which must be submitted, are the methods of quality control (control methods), the results of physical, chemical, biological or microbiological examinations and the methods used in their determination (analytical test), the results of the pharmacological and toxicological examinations (pharmacological-toxicological test), and the results of the clinical or other medical, dental or veterinarian tests (clinical trial) (§22 (1), (2) Drug Law).

In cases where the drug is intended for use in animals serving human consumption, data are required on withdrawal periods, residue testing and residue test procedures (§23 (1) Drug Law). In the case of premix drugs, the suitability of the premix drug for the manufacture of medicated feeding stuffs has to be proven (§23 (2) Drug Law).

Furthermore, there have to be expertises on the control methods, the results of the analytical test, the pharmacological-toxicological test and the clinical trial (§24 (1)1-3 Drug Law). Expertises on the residue test procedure have also to be presented (§24 (1)4 Drug Law).

A decision is made by BGA on the basis of the submitted documents. Additionally, the authority may utilise its own scientific results, summon experts or request expert opinions (§25 (5) Drug Law).

If a new drug is to be authorised, a marketing authorisation commission must be consulted (§25 (6) Drug Law). Members of these commissions are appointed by the Federal Minister (of Youth, Family Affairs, Women and Health), and the peculiarities of the drugs are considered when appointing the members.

If a new drug is for example just a new combination of drugs, the effects and side-effects of which are already known, scientific documents may be presented instead of the pharmacological-toxicological and clinical tests, which greatly facilitates application (§22 (3) Drug Law). This is also the case for the documents, proofs and test results on the residue test and the residue test procedure, as well as on the tests on the suitability of a premix drug (§23 (3) Drug Law). Expertises on these documents have to be presented, too (§24 (2) Drug Law).

According to §25 (7) Drug Law, BGA must have documents on "old" substances processed by certain commissions ("Aufbereitungskommission").

This processing refers to efficacy and safety of the drug according to §22 (3) and §23 (3) Drug Law. These commissions are established for specific fields of application, groups of substances etc. For instance, there is a special commission for veterinary drugs. Members are appointed as for new drugs (see above). Prior to final promulgation, drafts of the monographs are published for discussion among specialists. Concerning efficacy and safety, the monograph of a substance is the basis for a later decision on re-authorisation of a drug containing this substance (see below).

Re-authorization

The Law on the Reform of Drug Legislation from August 24, 1976 (2) also stipulates that drugs, which have been in use before the Reform, were considered as authorised until April 30, 1990. If the manufacturer wanted to market them beyond this date, an application for re-authorisation had to be made (Article 3 §7 (1), (2), (3) Law on the Reform of Drug Legislation). Some documents had to be submitted together along with the application form. The deadline for submission of for example the analytical test (§22 (2) No.1 Drug Law), the suitability of a premix drug (§23 (2) Drug Law), and the analytical expert opinion (§24 (1) Drug Law) are stipulated by BGA individually. If required, documents on bioavailability have also to be submitted (Article 3 §7 (4) Law on the Reform of Drug Legislation). Without this application for re-authorisation, the marketing authorisation of a drug has elapsed on April 30, 1990. Under certain conditions, the drug may still be marketed until December 31, 1992 (Article 3 §7 (3) Law on the Reform of Drug Legislation and §31 (4) Drug Law).

In the re-authorisation process, quality, efficacy and safety must be proven. The pharmaceutical manufacturer has to demonstrate the pharmaceutical quality, whereas efficacy and safety are validated mainly on the basis of the monographs (see above).

Standard marketing authorization

The Federal Minister can by ordinance exempt certain drugs from the obligation to obtain a marketing authorisation, "as far as a direct or indirect hazard to human or animal health is not to be feared, since it is evident that the requirements with regard to the necessary quality, efficacy and safety have been met". The exemption may be restricted to - for example - certain fields or ranges of application. Therefore, data are gathered on these drugs concerning for example use in fish, and monographs are published by BGA. Drafts of these monographs are promulgated in order to enable a public discussion among specialists. Finally, the Standard Marketing Authorisation Commission, consisting of representatives of human and animal medicine, pharmaceutical industry, pharmacologists, pharmacists, etc., is summoned by the Federal

Minister and the decision whether to exempt the drug from the obligation of authorisation is made (36 Drug Law). These drugs can then be obtained in pharmacies without a veterinary prescription.

Directives No. 81/851/EEC and No. 90/676/EEC

According to Article 17 (1) of the Directive No. 81/851/EEC (3), amended by the Directive 90/676/EEC (4), the manufacturer of a drug licensed in their own EEC member state has the following option: The manufacturer may apply to the competent authorities of at least two further member states for permission to place the drug on the market in those states. According to Article 17 (2), the manufacturer must inform the EEC Committee for Veterinary Medicinal Products about this application. Also, the home member state must be notified. When each of the member states has received the dossier, approval to the application is given within 120 days, or - alternatively - substantiated objections are made (Article 17 (3)). Further articles of this Directive stipulate the proceeding in cases of assessments differing among member states.

Directive No. 87/22/EEC

This Directive (5) approximates the member states' regulations concerning the placing on the market of high-technology medicinal products. The scope is to guarantee that these products, the development of which is extremely expensive, find a market large enough to justify investment and further research.

Use of a drug which is licensed for other animals

The Drug Law defines the situation where a veterinarian may prescribe, distribute or administer drugs to a certain animal species although they are not licensed for use in this species (§56a (2) Drug Law): This is only possible if the drug has a marketing authorisation for another animal species serving the production of food, if there is no drug authorised for marketing for the respective species, if, otherwise, the necessary medical care of the respective animal is seriously endangered, and if a direct or indirect risk to the health of man or animal need not be feared.

Discussion

The extent of the catalogue of data required for authorisation of new drugs is justifiable because of safety reasons. However, there are hardly any attempts by the pharmaceutical industry to apply for the authorisation of new drugs for use in fish, probably because of the high costs versus low profits expected from later sale. At present, only one drug - a trimethoprim-sulphadimethoxin combination - is licensed for use in fish in Germany.

Concerning re-authorisation of "old" drugs, there is a lack of published data on efficacy and safety of the substances concerned. Thus, the conclusion of a monograph could be that use of these substances cannot be permitted because - for example - its efficacy is not proven in fish. Table I shows a list of the substances, which were authorised for use in fish before the Reform of the Drug Law, and which may still be marketed today (April, 1991).

Table I
Antibacterial and antiparasitic active ingredients of drugs authorised for use in fish prior to the reform of the German Drug Law in 1976 and still marketable today (April 1991)

1 - Furazolidone
2 - Sulphadimidine
3 - Chloramphenicol (palmitate)
4- Chlortetracycline hydrochloride
5 - Dimetridazole
6 - Dibutyltin oxide
7 - Trimethoprim
8 - Paramomycinsulphate
9 - Metrifonate (Trichlorphon)
10 - Sulfadoxine

Several chemicals - some very common - have been used in aquaculture for decades without ever having been authorised. However, according to the Drug Law, they have to be regarded as drugs and thus require authorisation. These chemicals are candidates for standard marketing authorisation (Table II). The major problem is the exclusion of "direct or indirect hazard to human or animal health", which is the most important prerequisite for such an authorisation, as drugs authorised in such a manner may be obtained without prescription. The assessment of safety is especially difficult for malachite green.

Table II
Drugs proposed for a standard marketing authorisation in Germany

1 - $NaCl$
2 - CaO
3 - $Ca(OH)_2$
4 - Malachite Green
5 - Formaldehyde
6 - $KMnO_4$

Up to now, no use has been made of the possibility to extend a marketing authorisation among EEC member states according to the Directives No. 81/851/EEC and No. 90/676/EEC. This may partly be due to the national regulations on licensing fish drugs, differing substantially among EEC member states (6).

The requirements laid down in the Directive 87/22/EEC are not fulfilled by fish drugs.

Certainly, there are many cases where there is a lack of drugs authorised for use in fish, but where fish do need medical treatment. If veterinarians - in such cases - use a drug, which is licensed for example for pigs, it is the veterinarian's own responsibility to exclude any harm to the fish as well as to the human consumer of that fish. This implies the stipulation of a withdrawal period of the drug when given to fish. Of course, the veterinarian can only estimate such a time period. Thus, such a procedure implies many difficulties and possible consequences, not only to fish and to the consumer, but also to the veterinarian, if - for example - such a withdrawal period proved not to be sufficient.

Outlook

The EEC legislation will come into effect on January 1, 1993. By then, the drug legislation should have been assimilated among the member states. The Directive No. 81/851/EEC on the approximation of the laws of the member states relating to veterinary medicinal products has recently been amended by the Directive No. 90/676/EEC, which enters into force on January 1, 1992. For the first time, fish as an animal has explicitly been mentioned in these amendments. This means that in future Community legislation governing veterinary medicinal products fully applies to drugs for use in fish. Concerning authorisation, this has the following implications:

The EEC requirements concerning quality and efficacy (laid down in Directive No. 81/852/EEC, 7) do not differ substantially from the above mentioned national requirements on new drugs. However, from January 1, 1992 onward, the safety documentation implies safety for the environment (Article 1 #5 (7) Directive No. 90/676/EEC). Specific requirements for fish drugs will have to be stipulated by the Community. All drugs which are already on the market, must be made subject to these regulations within four years after January 1, 1992 (Article 2 (5) Directive No. 90/676/EEC).

Furthermore, from January 1, 1992 onward, no new active substance will be permitted to be used in food animals unless a maximum residue level (MRL) has been established by the Community. Until the end of 1996, the Community must establish MRLs for all existing compounds (Article 14 Regulation 2377/90/EEC, 8). The establishment of these data for fish is especially difficult because of the often lacking residue depletion studies.

The new Directive No. 90/676/EEC also clarifies the extra-label use of drugs in fish. The conditions under which such a usage is permitted are almost the same as in the national Drug Law, however, unlike the Drug Law, the EEC Directive says that, unless a withdrawal period has been specified, the prescribing veterinarian has to ensure that a withdrawal period of at least 500 degree days is observed. Also, he must keep records of any extra-label use instance for at least three years.

This EEC legislation will be binding on member states and override any national law. The ultimate goal of the realisation of the European Economic Community is to facilitate trade among the member states. However, it becomes clear that regulations on authorisation and use of veterinary medicinal products in fish will in the near future not be any easier. The extent of the testing required is considerable, but the fish segment of the animal health medicinals market is too small to justify heavy investments. This could deter many companies from attempting to license drugs for use in aquaculture to the detriment of the latter. It may therefore be essential to facilitate the authorisation of drugs for use in aquaculture. Ways may be found to facilitate the administrative procedure and lower the costs, for example by official sustaining of institutes which may be able and ready to specialise in conducting analytical, microbiological, pharmacological or clinical studies.

References

1). Drug Law (Arzneimittelgesetz) of 24.08.1976. *Federal Law Gazette* I p. 2448, in the version of the FOURTH LAW AMENDING THE Drug Law of 11.04.1990. *Federal Law Gazette* I p. 717.

2). LAW ON THE REFORM OF DRUG LEGISLATION of 24.08.1976. *Federal Law Gazette* I p. 2445, in the version of the LAW AMENDING THE LAW ON THE REFORM OF DRUG LEGISLATION of 22.12.1989. *Feeral Law Gazette* I p. 2462.

3). DIRECTIVE. 81/851/EEC of 28.9.1981 on the Approximation of the Laws of the Member States Relating to Veterinary Medicinal Products. *O. J.* No. L 317/1 of 6.11.1981.

4). DIRECTIVE 90/676/EEC of 13.12.1990 amending Directive No. 81/851/EEC. *O.J.* No. L 373/15 of 31.12.1990.

5). DIRECTIVE No. 87/22/EEC of 22.12.1986 Concerning High-Technology Medicinal Products. *O. J.* No. L 15/38 of 17.1.1987.

6). BERNOTH E.-M., 1990. Marketing authorisation of fish drugs - Current status and future intentions. *J. Vet. Med. B*, **37**, 401-406.

7). DIRECTIVE 81/852/EEC of 28.9.1981 Concerning the Analytical, Pharmaco-Toxicological and Clinical Standards and Protocols in Respect of the Testing of Veterinary Medicinal Products. *O. J.* No. L 317/16 of 6.11.1981, in the version of Directive No. 87/20/EEC of 22.12.1986 amending Directive No. 81/852/EEC. *O. J.* No L 15/34 of 17.1.1987.

8). REGULATION 2377/90/EEC of 26.6.1990 on the Creation of a Community Procedure for the Establishment of Maximum Residue Limits (MRLs) for Residues of Veterinary Medicines in Foodstuffs of Animal Origin. *O. J.* No. L 224/1 of 18.8.1990.

THE ROLE OF THE EXPERT IN THE EC'S MEDICINES LICENSING PROCEDURES

M.J. Chapman* & P.W. Scott**

* Zootechnology Ltd, P.O.Box 305, Cambridge, CB4 4WR, UK
** Zoo & Aquatic Veterinary Group,Keanter, Stoke Charity Rd, Kings Worthy, Winchester, SO23 7LS, UK

Summary : *The medicines' licensing authorities of the member states of the European Community (E.C) require submissions for marketing authorisations to be accompanied by reviews and critical expert reports by a pharmacist, a pharmaco-toxicologist, and a veterinary clinician. This still evolving system is reviewed, with especial reference to pharmaceuticals used in aquaculture. It is concluded that the system - if not misapplied to avoid difficult (political rather than scientific) issues, offers benefits to all parties (the licensing authority, the applicant, the customer and 'society-at-large'), and has much to commend it to other authorities outside Europe.*

Résumé : **Rôle de l'expert dans les procédures d'autorisation de mise sur le marché des médicaments de la CEE**

La sécurité du consommateur est une préoccupation primordiale en matière de traitements médicamenteux, et la présence d'éventuels résidus dans les aliments en constitue un point critique. L'aquaculture est particulièrement concernée par ce problème. La CEE s'est dotée de directives qui prévoient des rapports d'experts pour fonder la qualité, l'innocuité et l'efficacité des produits soumis à une procédure d'AMM, et en définir les conditions d'emploi. Ce recours à des experts indépendants s'inscrit dans une tradition d'exercice libéral solidement établie en Europe. Il présente des avantages pour toutes les parties et pourrait servir de modèle à d'autres régions du globe.

L'exigence d'un rapport écrit fournit l'occasion de faire le point sur le problème considéré et d'en préciser clairement les limites et les implications au travers de considérations critiques dont l'objectivité, garantie par l'indépendance de l'expert, n'est pourtant pas toujours facile à assumer. Si les règles et les modalités d'interprétation de l'expertise sont normalement explicitées par les textes, certains points sont sujets à caution. Dans l'attente de la définition des limites maximales de résidus auxquelles travaille la CEE, on se réfère encore à la notion un peu arbitraire de doses journalières acceptables. Récemment sont apparues des craintes liées à la contamination de l'environnement par les agents thérapeutiques. Leur importance ira sans doute croissant, mais les dispositions prévues souffrent d'un manque de références objectives.

L'application des directives de la CEE présente encore des variantes selon les pays membres, notamment dans le choix des experts. Tous les problèmes pendants ne sont pas résolus, et il faut bien convenir que le débat s'inspire parfois encore de considérations plus politiques que techniques.

Introduction

The safety of consumer products has been a matter of increasing concern over the last few decades. All medicines, whether to be used by 'consumers' (human and companion animal medicines) or by other 'users' (farm animal medicines) are subject to the same regulatory procedures in the European Community (E.C). And, ironically, whilst the 'safety of medicines' has been under strict control for many years, the possible incorporation of medicinal residues into the food chain still causes as much concern as any of the 'current food health scares'.

These concerns (ranging from the genuine to the alarmist) are currently acute in aquaculture. Important factors in the genesis of these concerns in aquaculture are:

- the perceived novelty of aquaculture,
- our still developing knowledge of the long term impact of the husbandry systems and of the techniques of medicine administration used in them;
- a dislike of technologies perceived either as 'novel', environmentally intrusive or unnatural by some campaigners
- the potentially intimate contact of any therapeutic used in aquaculture with an excellent solvent and carrier medium, be it sea or fresh water

Review

The European Communities' procedure for licensing any medicines, either for human use [1] or for veterinary use [2] requires that the dossier submitted in application for a marketing authorisation to the licensing authority is accompanied by reviews and expert reports on the data, dealing with each of the three main scientific sections of the application. These may loosely be termed quality, safety and efficacy, or more properly Article 5 of the E.C's Directive refers to :

physico-chemical, biological or microbiological tests, toxicological and pharmacological tests and clinical trials.

and continues:

Article.7
... the role of the experts shall be:

1. to carry out such work as falls within their particular discipline (analysis, pharmacology and similar experimental sciences, clinical trials) and to describe objectively the results obtained in both quantitative and qualitative terms;

2.

..., and in particular to state:

(a) in the case of analysts, whether the product conforms with the stated composition, providing any reasons for the control testing methods which the manufacturer is to use;

(b) in the case of pharmacologists and appropriately qualified specialists:

- the toxicity of the product and the pharmacological properties observed,

- whether, after administration of the veterinary medicinal product under normal conditions of use and observance of the recommended withdrawal period, foodstuffs obtained from the treated animals contain residues which might constitute a health hazard to the consumer;

(c) in the case of clinicians, whether they have found in animals treated with the product effects corresponding to the information furnished by the manufacturer pursuant to Article 5, whether the product is well tolerated, what dosage they recommend and what are the contra-indications and side-effects, if any;

... ...

There are advantages in such an independent review to the regulatory authority, to consumers (purchasers of the medicinal product, and 'consumers' in the wider sense of all in society); and it is argued not least advantages to the applicant. Despite its roots being very firmly in the European tradition of the 'liberal professions' and thus the 'independent expert', the system could be used, or adapted and used, by other authorities, and thus a detailed discussion of the system and its perceived strengths and weaknesses as currently implemented is presented.

Discussion

At the very least the expert report system allows a review of the data in an application dossier before presentation. The discipline of having to set out a synopsis of the case in writing is a strict one. Bingham [3], a judge of the English High Court, commented with regard to court judgements:

'I cannot, I hope, be the only person who has sat down to write a judgement, having formed the view that A must win, only to find in the course of composition that there were no sustainable grounds for that conclusion and that on any rational analysis B must succeed'.

The comparison with the judicial process is a justifiable one. The data has to be summed up by the expert, and whilst he will leave final judgement to the licensing authority, he will strongly indicate the merits, or otherwise, of the case for licensing in his approach.

Lord Donaldson [5] stated:

'Much of the art of giving a judgement lay in telling a story logically, coherently and accurately. That was something which required skill, but it was not a legal skill and it was not necessarily advanced by legal training.'

The expert system being limited to the three experts also allows a focus for queries outside an expert's field of expertise. In this system any query beyond one's own field can be directed to one of two other people. They should be able to either immediately explain the matter, or to do so after enquiry within their own subject area.

A team of three is convenient for this, at an interpersonal level, as well as being a convenient division of the scientific work to be reviewed. It is probable that separate Safety and (where relevant) Residue reports will be required in the near future. It seems likely though that as much of the data is common for these two reports, that they will be written by the same pharmacologist/toxicologist.

Experts may find 'poor' data in their work. This does not necessarily make an application inadequate, any inadequacy will often be in the current presentation of the information. An expert can do much to piece together what is a coherent whole from poorly planned, or poorly executed (sometimes both) experiments and trials. By highlighting what would anyway disturb the licensing authorities, the expert can divide the data into 'poor irrelevant' data, and data that is supportable on the evidence given, and use that smaller part in support of the application. It really is for the applicant (and his experts) to do this 'weeding;' and 'pruning' of the data for the authority. It is not the latter's job to construct a case out of a mass of data of varying quality. (Naturally all available data should be presented, and all should be commented on by the expert. It is the emphasis that the expert invites the authority to place on different parts that can ease everyone's job.)

Before the present system was introduced the authors always found it useful to introduce the data (to tell the story as Donaldson would have it). This was undoubtedly useful to the reader. It did though have the danger that it could be seen as leading, or even as attempting to prejudice, the reader. A need for such reviews, introductions, or whatever to be a specific individual's responsibility was obvious. As such they obviously ought to be signed by that individual, and he should justify his qualification for composing the review.

The present system goes beyond an introduction to the data, and requires a critical review. The expert is to use his own judgement on the data; to criticise the individual trials and laboratory studies; to criticise any conclusions that the applicant may seek to draw from those results; to make his own conclusions, where necessary; and, to form his own judgement, within his own area of expertise, of the product as a whole.

The expert is independent, even if employed by the applicant. This independence is vital, and is to be jealously guarded. Anyone signing these reviews should be a senior member of good professional standing, and as such should be able to withstand and redirect the pressures of perhaps over eager applicants. It would be disingenuous of us, as two independent consultants, without ties to any particular pharmaceutical company, not to see the position of the external expert as a better one than that of the in-house colleague.

An external expert can look on the data anew, often finding the trivial and the obvious overlooked during the necessities of getting other parts of the application completed (the dangers of losing the objective, due to the drive for early consensus is reviewed by Janis [7]). He is also not only (like his in-house colleague) independent, but should easily be seen to be manifestly and undoubtedly independent. The external expert is also likely to have a broader experience, from work with several applicants, and possibly to have a better knowledge of the licensing authority if much of his work is concentrated at this end stage of presenting application data, rather than across the spectrum of creating and helping to present data. In defence of the in house expert, it has to be said he will usually have been involved with a product since its conception. His knowledge will often be more intimate since he may have not only carried out the work described, but also have devised the protocols for that work.

New guidelines have been proposed by the European Communities for the presentation of both pharmacology/toxicology and clinical material in applications. These are still in draft stage at the time of preparing this paper. Those for the pharmacology/toxicology material would require separate consideration of safety, and of residues (for products in food producing animals). The basic philosophy remains that *'use of the product will not present unacceptable risks for the safety of:*

- *the target animal*
- *the environment, or,*
- *human beings administering the product to animals (users)'.*

With the additional proviso that for food-producing-animal medicines *'residues .. will not present any risk to consumers of foodstuffs of animal origin'.* (One must presume that the omission of the word 'unacceptable' from the second criterion is meaningful.)

The EC is proposing to issue MRLs (maximum residue levels) for all pharmacologically active compounds used in food-producing-animals by 1997. Currently where no figure exists an ADI (acceptable daily intake) has to be estimated and used in conjunction with arbitrary likely consumption figures (300g of muscle and skin in 'natural proportions' for fish), consideration of intake from other foods, and various 'safety factors', to achieve an MRL proposed by the applicant.

Consideration of safety to the target animal has the longest history, the safety of users [4], the problems for industrial processing of foods (e.g. antibiotic residues destroying fermentation cultures) and even of consumers (i.e. residues) are more recent, but environmental impact of medicines is - at least in terms of detailed analysis - the most recent.

Medicines have traditionally been used in small quantities (a U.K. enquiry mounted after worries about use in animal husbandry found that, at the time of the enquiry, 60% of total antibiotic usage was in man, or nearly four times as much on a mg/Kg BW basis [9]) and in situations where usage was deemed essential for preservation of life and/or welfare. Greater concern for the environment, the feeling (however justified or unjustified) that medicines are used to support bad and inhumane husbandry systems, an increasing knowledge of the actions and environmental fates of chemicals have all meant that questions have been posed.

Aquaculture is obviously at the acute edge of such concerns. Medication almost always means mass medication. The ecotoxicological implications of that are only too easy to conjure with, if not always known with any degree of certainty. The potential impact of antibacterial residues in fish farm effluent water entering river systems subsequently used for drinking water has only recently been raised. Factors of dilution- ie farm size and river flow, river recovery time - related to flow rate and distance, biodegradeability etc all enter the equation. In the same context, human pharmaceuticals entering sewage treatment systems which also dicharge to rivers will also come under scrutiny. The science of ecotoxicology is set to explore these areas. The persistance of compounds such as ivermectins in faecal pats and their persistance on the pasture has brought this subject very much to the fore. Indeed, when the same compounds are to be looked at with a view to their use outwith their Product Licence indications for treatment of salmon lice, some of the comments about the much more biodegradeable and to many more acceptable Dichlorvos (Aquagard,Nuvan, Ciba Geigy) [6], pale into insignificance.

The current (draft) guidelines state:

'When relevant, substances shall be tested for environmental toxicity and biodegredation. The extent and scope of the ecotoxicological documentation

may vary from product to product, but the following type of tests would frequently be required:

- *Biodegredation in soil,*
- *Biodegredation in aquatic environments,*
- *Inhibition of algal growth,*
- *Effects on terrestrial plants,*
- *(Acute) toxicity for earthworms,*
- *(Acute) toxicity for Daphnia,*
- *(Acute) toxicity for fish.*

Additional studies may be required as necessary.'

The lack (or perhaps the unattainability) of a well accepted, 'universally applicable model ecosystem must hinder such studies. ... .'

The guidelines for the clinical report require an initial overview supported by detailed critique of the chosen dose rate and method of administration and the effects that this could be expected by the clinician or farmer to achieve. Data on efficacy will usually be supported as appropriate by expected MIC's of appropriate organisms, determined in the laboratory and on the farm. Clinical trial data is presented in tabular form with comments on applicability to the application.

Where dose rates other than those on the application have been used these should be reported and commented upon for their value in palateability and tolerance studies as well as the obvious area of efficacy. Dose rates used in fish need always to be related to feed rates and palateability, since low feed rates enforce high inclusion rates which may further reduce feed intake with consequent loss of efficacy. A statistical analysis of the clinical data where possible is required, the clinical expert is then required to present comments gleaned from his own experience, published material or from clinical reports of any contraindications. The report then is drawn together with a conclusion highlighting the important areas presented in the body of the report.

Authorities produce guidance notes for experts. These though often differ in the manner and order of the questions asked rather than the overall standard, that is in substance rather than intent.

The implementation of the EC Directives varies between different member states. In the United Kingdom, for example, any suitably qualified person with the relevant experience can have his reports accepted. Though pharmacy/chemistry reports must be by a pharmacist (a restriction not in the original directive). In France, by contrast, the relevant ministers publish (as an *Arrete* in the *Journal Officiel*) a list of names of those that are accepted by the government as experts.

Such differing systems obviously pose problems where applications are made independently in more than one member state.

The responsibility of the licensing authority vis-a-vis the applicant is a complicated one and involves the expert in his role of collating and presenting a clear report. This almost came to judicial resolution in Britain in the 1980's with the 'Opren Case' where aggrieved users commenced legal proceedings both against the manufacturer (marketing authorisation holder) and against the licensing authority (the UK Department of Health). An out-of-court settlement meant that liability was neither determined or (if it existed) apportioned between the two possible defendants. The implications of this throw a heavy responsibility on the expert. Pending legal action in the UK will, if it goes to court, explore the responsibilities of the manufacturers of benzodiazepines to change their clinical guidance for the use of the products over the years from the initial launch. It may also explore the duty of the licensing authorities to keep licensed products under review. Due to the age of the products, the formal role of experts in the licensing procedure will most likely not be examined.

These matters are aired here. Perhaps rather loosely for a 'scientific paper', but it is not purely facetious to suggest that expert report writing may become the province of the retired, not because it is a nice sedentary occupation, but rather they are the only professionals 'old enough' to take the risk! That would be a sad day.

References

1). DIRECTIVE 75/319/EEC, *Official Journal of the European Communities*, no: L 147 of 9.6.75 pp: 13-23.

2). DIRECTIVE 81/851/EEC, *Official Journal of the European Communities*, no: L 317 of 6.11.81 pp: 1-15.

3). BINGHAM, Lord Justice. 1987 - Differences Between a Judgement and a Reasoned Award. in the series: *Judicial Techniques in Arbitration and Litigation*. London: Chartered Institute of Arbitrators.

4). CHAPMAN, M.J. 1990. - *Respiratory Penetration of Toxicants.* DIBT dissertation. London: Institute of Biology.

5). DONALDSON, J.F. 1981. - Westzucker GmbH v. Bunge AmbH. *Times Law Reports* 20.5.81

6). IPCS International Programme on Chemical Safety. 1989. *Environmental Health Criteria 79, Dichlorvos.* World Health Organisation, Geneva

7). JANIS, I.L. 1982. - *Groupthink.* Boston, USA: Houghton Mifflin.

8). LAROCHE, M.-J., FABIANI,P. & ROUSSELET,F. 1986. - *L'expertise toxicologique des medicaments.* Paris: Masson.

9). SWANN, M.M. 1969. - *Report of the Joint Committee on the use of Antibiotics in Animal Husbandry and Veterinary Medicine.* ['The Swann Report'] London: Her Majesty's Stationery Office: Cmnd.4190.

also

The Rules Governing Medicinal Products in the European Community.
Volume I. The rules governing medicinal products for human use in the European Community.
Volume V.Veterinary medicinal products.
Luxembourg: Office for Official Publications of the European Communities: 1989.

Glossary

ADI	Acceptable Daily Intake
MRL	Maximum Residue Level
NOEL	No Effect Level
SF	Safety Factor
MIC	Minimum Inhibitory Concentration

Marketing Authorisation (EC) = Product Licence (UK)
Composition (EC) = Formulation (UK)

SESSION 2

Toxicity and side effects of chemotherapeutants

INCIDENTS ET ACCIDENTS DES ANTIBIOTIQUES

J.P. Giroud

Laboratoire central de Pharmacologie clinique
G.H.Cochin, Pav. G. Roussy
27 r. du Faubourg Saint-Jacques, 75674 Paris Cedex 14, France

Résumé: *L'antibiothérapie se heurte chez l'homme à des contre-indications et à des effets indésirables dont les principales manifestations sont rappelées. Les béta-lactamines sont surtout responsables de phénomènes allergiques parfois graves. La toxicité des aminoglucosides s'exerce pincipalement au niveau de l'oreille interne et du rein, tandis que les phénicolés, avec surtout le chloramphenicol, sont responsables d'accidents hématologiques dont l'aplasie médullaire est une forme souvent fatale. Les tétracyclines entraînent des inconvénients variés, souvent fonction du terrain, liés à leur accumulation dans certains tissus: tractus digestif (intolérance, candidoses), dépôt dans les os et les phanères, désordres métaboliques, atteintes hépatiques ou vestibulaires (minocycline). Les macrolides enfin peuvent également donner lieu à des réactions d'hypersensibilité ou à des désordres hépatiques.*

Summary : **Incidents and accidents with antibiotics.**

A pharmacologist and clinician accustomed to dealing with a single human species reviewing the potential hazards of antimicrobial drugs in aquaculture is immediately struck by the range of species and problems encountered. For this reason, this review is confined to an examination of known hazards associated with beta-lactams, aminoglycosides, phenicols, tetracyclines and macrolides which are or may be in use in aquaculture.

The beta-lactam group consists of some 100 compounds, about 80 of which are used clinically. There are generally well tolerated and their main adverse effect is in the form of hypersensitivity reactions which can either occur quickly or after some delay both in people receiving therapy and in those handling the product. Beta-lactam hypersensitivity reactions include anaphylactic cardiovascular shock, dermal urticaria and oedema, asthmatic crises and, in the kidney, allergic nephropathies which are particularly associated with ampicillin.

Aminoglycosides, of which the most important is streptomycin have their main antibacterial effect on gram negative bacteria by interference with the bacterial ribosomes. These antibiotics, particularly kanamycin are strongly ototoxic, affecting both the vestibule and thus balance and also the cochlea leading to deafness. Such effects mainly arise after prolonged therapy, particularly when other ototoxic drugs are also being used. Nephrotoxicity may also be encountered, either as a simple proteinurea or as renal insufficiency. The onset of such effects can be insidious and difficult to recognise since they are often not accompanied by any clinical signs.

Phenicols as a group consist only of chloramphenicol and thiamphenicol. Their wide therapeutic range makes them very important, but they also produce some very undesirable side effects. Haematological damage can occur during treatment with both drugs and is normally reversible on cessation of the treatment, but with chloramphenicol if the occurrence of haematological damage is delayed, prognosis is poor, with more than 50% mortality associated with medullary aplasia. Nausea and vomiting and neurological and psychiatric effects have also been reported.

Tetracyclines comprise a group of about 12 antibiotics which are generally well tolerated, the principal reported adverse effects reported being associated with irritation of gut mucosa or gut flora changes with risk of selection of drug resistant bacterial strains. As with other wide spectrum antibiotics, tetracyclines can favour the development of candidiasis by eliminating competing intestinal microflora. Tetracycline therapy results in yellow staining of teeth and bone and rarely photosensitisation effects have been noted. Fatty degeneration of the liver accompanied in some cases by pyelonephritis.

Finally, the narrow spectrum macrolides which affect the 50S ribosomal subunit, can at high doses produce nausea, vomiting and diarrhoea. The most serious adverse effect is liver toxicity which can occur with erythromycin therapy.

Je tiens à remercier l'Office International des Epizooties et le Comité d'Organisation de la Conférence de m'avoir invité a participer à cette réunion sur le thème des thérapeutiques en aquaculture. En tant que pharmacologue clinicien j'ai vite été frappé par l'ampleur de la tâche que je venais d'accepter. En effet, mes connaissances dans ce domaine étant davantage celles d'un gourmet que d'un scientifique je ne pouvais imaginer que les problèmes de l'Ichtyopathologie puissent dépasser en complexité ceux que l'on rencontre chaque jour dans un hôpital sur une seule espèce, l'homme. Confronté ici avec une multitude d'espèces animales, et de plus avec des maladies qui sont tout à fait étrangères au médecin, je me suis référé à un certain nombre d'articles scientifiques spécialisés, et en particulier à une revue de C. Michel (1), avant de présenter cette synthèse de nos connaissances sur la toxicité des antibiotiques antibactériens les plus employés en aquaculure. L'exposé portera sur les béta-lactamines, les aminosides, les phénicolés, les tétracyclines, les macrolides et les quinolones (1,2,3,4).

Les béta-lactamines

Il existe une centaine d'antibiotiques appartenant à ce groupe, dont environ 80 sont utilisés en clinique. Ce sont des antibiotiques bactéricides qui agissent en bloquant la biosynthèse des protéoglycanes de paroi.

Ces antibiotiques sont généralement bien tolérés et le seul grand problème que l'on puisse rencontrer tient en fait à des réactions individuelles d'hypersensibilité. Celles-ci peuvent survenir, soit de façon immédiate, soit au contraire de manière plus ou moins retardée, et le phénomène de sensibilisation peut s'observer non seulement chez les sujets qui reçoivent ces antibiotiques mais également chez ceux qui sont amenés à les manipuler. La localisation de ces réactions peut conduire à des manifestations variées:

- Au niveau cardiovasculaire peuvent être observés des chocs anaphylactiques.
- Au niveau cutané peuvent survenir indifféremment un urticaire generalisé, un oedème de Quincke, des exanthèmes, souvent d'ailleurs maculopapulaires, plus rarement des erythèmes multiformes, voire un syndrome de Lyell ou de Stevens-Johnson.
- Au niveau pulmonaire, des crises d'asthme, associées à des infiltrats pulmonaires labiles et riches en éosinophiles ont été decrites.
- Au niveau rénal il a été rapporté des néphropathies allergiques, voire des néphrites interstitielles, en particulier avec l'ampicilline.
- La maladie sérique enfin, se caractérise par un état fébrile associé à un exanthème, à des douleurs articulaires, à une adénopathie avec leucopénie.

En dehors des accidents immunologiques, d'autres réactions, plus rares, peuvent apparaître après l'emploi des béta-lactamines. Certaines sont en relation avec des surdosages et se traduisent par l'apparition de vertiges, d'hallucinations, de contractions myocloniques voire de crises convulsives. D'autres manifestations, actuellement exceptionnelles, s'observent dans le cas particulier du traitement de la syphilis (reaction de Jarisch Herxheimer).

Les aminoglucosides

Ces produits, dont le chef de file est la streptomycine, sont des antibiotiques bactéricides qui agissent essentiellement sur les bacilles Gram-négatif aérobies, principalement au niveau du ribosome bactérien où ils déterminent une erreur dans la lecture du code génétique.

Ces antibiotiques, en particulier la kanamycine, sont très fortement ototoxiques. Leurs atteintes peuvent non seulement intéresser l'appareil vestibulaire, touchant alors l'équilibre, mais également la cochlée, entraînant un risque de surdité. Ces risques sont d'ailleurs accrus dans certaines conditions d'emploi, si on utilise des doses élevées ou si le traitement est prolongé au-delà de dix jours. Ils sont également majorés ou aggravés dans leurs manifestations par l'association d'autres médicaments eux-mêmes ototoxiques, ou si le sujet présente une insuffisance rénale.

La néphrotoxicité se traduit de façons diverses. Elle peut se limiter à une simple protéinurie, ou parfois se caractériser par une insuffisance rénale. Les atteintes rénales provoquées par les aminosides sont très insidieuses et souvent difficiles à reconnaître, car aucun signe clinique n'en trahit l'installation. Mais les études histologiques montrent qu'il existe une atteinte tubulaire avec oedème insterstitiel, sans infiltration cellulaire importante. Ces néphropathies semblent être favorisées chez les sujets âgés. Il est vraisemblable que la néphrotoxicité des aminosides est dose-dépendante, et que des facteurs comme l'insuffisance rénale préexistante, l'hypovolémie et les désordres acido-basiques sont de nature à en augmenter les effets. D'autres accidents, tels les phénomènes de curarisation, sont beaucoup moins fréquents et n'apparaissent que lorsque l'antibiotique est utilisé localement au niveau du péritoine.

Les phénicolés

Cette classe d'antibiotiques comprend le chloramphenicol et le thiamphenicol. Ils ont eu un grand interêt car ce sont des molécules à spectre large, dont l'activité de type bactériostatique peut dans certaines conditions devenir bactéricide.

Le chloramphenicol et le thiamphenicol, sont responsables de différents effets indésirables, et en particulier d'accidents hématologiques.

1. Accidents hématologiques

Il en existe deux types, l'un d'apparition précoce et réversible apres l'arrêt de la prise de l'antibiotique, et l'autre d'apparition retardée, de fréquence moindre, mais se manifestant par une pancytopénie, souvent mortelle.

Accidents hématologiques d'apparition précoce et reversibles

Ils se caractérisent par une cytopénie médullaire limitée souvent à une seule lignée cellulaire: érythroblastopénie, thrombopénie, granulopénie. Ces manifestations peuvent s'aggraver en cas de surdosage. Elles sont reversibles mais nécessitent l'arrêt du traitement. Le thiamphenicol aussi bien que le chloramphenicol en sont responsables.

Accidents d'apparition retardée.

Il survient essentiellement avec le chloramphenicol, plusieurs semaines voire plusieurs mois après l'arrêt de la prise de l'antibiotique. La toxicité médullaire se manifeste par une anémie, une leucopénie ou granulopénie, une thrombopénie. Le pronostic est d'autant plus sombre que les manifestations hématologiques apparaissent plus tardivement, et cette aplasie médullaire avec pancytopénie est mortelle dans plus de 50 p.100 des observations.

2. Réaction de Jarisch-Herxheimer

Lors du traitement de la typhoïde par le chloramphenicol à des doses d'emblée trop elevées, un etat de choc peut survenir, lié à la libération d'endotoxine.

3. Accidents digestifs

Dans moins 10 p.100 des cas lors d'ingestion de doses élevées ou prolongées de chloramphenicol ou de thiamphenicol, peuvent apparaître des nausées, des vomissements et des diarrhées. Des modifications de l'équilibre de la flore intestinale en sont responsables. Elles peuvent entraîner une hypovitaminose K, avec un allongement du temps de Quick, et plus rarement, favoriser les surinfections par des staphylocques dorés.

4. Accidents neurologiques et psychiatriques

Enfin, d'autres effets indésirables ont été exceptionnellement rapportés a la suite de posologies fortes et prolongées de chloramphenicol. On peut citer une névrite optique, des troubles de la vision des couleurs, des paresthésies digitales, unc surdité, et des troubles psychiques avec cephalées, délire, confusion et hallucinations.

Les tétracyclines

Ce sont des antibiotiques dont la structure de base est un noyau naphtacène-carboxamide sur lequel viennent se substituer diffèrents radicaux . On recense environ une douzaine d'espèces appartenant a cette classe. Ce sont des antibiotiques à large spectre, de type bactériostatique, qui inhibent les biosynthèses protéiques au stade de la traduction, en se fixant sur une protéine et en entravant la fixation ulterieure des ARN de transfert porteurs de leur acide aminé.

Généralement bien tolérées, les tétracyclines entraînent principalement des troubles digestifs liés à leur action irritante directe sur la muqueuse digestive ou à une modification des flores commensales (dysmicrobisme). Outre la sélection éventuelle de germes resistants, les tétracyclines, comme tous les antibiotiques à large spectre, peuvent également favoriser le développement des levures. La plupart des autres incidents et accidents observés sont en rapport avec les fortes concentrations retrouvées au niveau des différents organes (dents, foie, rein, os).

Tractus digestif

Muqueuses buccale et stomacale

Les tétracyclines peuvent provoquer des stomatites et des irritations de l'oropharynx, voire des ulcérations de l'oesophage. Des nausées, des vomissements, des brûlures épigastrique liés a l'effet irritant direct sur les muqueuses ont également été signalés.

Entérocolites

On décrit trois types d'entérocolites :

- Diarrhée sans signes généraux: cette diarrhée est en fait difficile a différencier des effets toxiques sur la muqueuse.

- Diarrhée associée à des signes généraux: on la rencontre dans des cas d'intoxication pouvant aller jusqu'au choc septique. Habituellement ces phénomènes se manifestent trois a cinq jours après le début du traitement.

- Intoxication sévère sans diarrhée: c'est le syndrome dit du "côlon toxique", l'intestin grêle pouvant lui-même présenter des signes de météorisme et un ileus paralytique. Ce tableau s'observe surtout après chirurgie intestinale.

Candidoses

Commes d'autres antibiotiques a large spectre, les tétracyclines sont susceptibles de favoriser le développement de *Candida albicans*. L'antibiothérapie en effet peut détruire la majeure partie de la flore digestive et créer un déséquilibre qui favorise la prolifération de ces levures naturellement résistantes. Le plus souvent digestives, les candidoses (muguet buccal, langue noire, candidoses anorectale et vulvovaginale) peuvent exceptionnellement conduire à des septicémies chez des sujets débilités.

Dents et os

Dents: une coloration anormale, jaune, brun-jaune ou brune, risque d'apparaître chez les enfants qui reçoivent des tétracyclines au cours de la minéralisation des dents de lait ou des dents définitives. Dans les cas graves, des défauts de l'email et des hypoplasies pourront favoriser le développement de caries.

Os: Chez des adultes traités par les tétracyclines, on a retrouvé des depôts d'antibiotique au niveau des zones de remodelage ou de réparation. Les os des enfants peuvent présenter d'importantes zones colorées en jaune, qui deviennent fluorescentes sous la lumière ultraviolete.

Peau et ongles

Les tétracyclines entraînent assez fréquemment des dermites phototoxiques revêtant l'aspect d'un coup de soleil plus ou moins intense accompagné eventuellement de phlyctènes ou d'éruptions bulleuses, ces réactions cutanées se limitant aux zones exposées. Plus rarement se développent de véritables dermites photo-allergiques caractérisées par une réaction urticariennes papuleuse ou eczémateuses. Elles ne surviennent alors qu'après un délai de photosensibiiisation, et leur topographie déborde les zones exposées a la lumière.

L'onycholyse, associée a la coloration anormale des ongles, peut atteindre aussi bien les ongles des mains que ceux des pieds.

Atteinte hépatique

Elle se traduit par des nausées, des vomissements, de la fièvre, un ictère, une acidose, une urémie avec hématémèse et méléna pouvant entraîner la mort. L'anatomopathologie précise qu'il s'agit d'une dégénérescence graisseuse du foie. Ce syndrome a tout d'abord été décrit chez des femmes enceintes et dans le proche post-partum. Certains malades présentaient simultanément une pyélonéphrite. Un dysfonctionnement renal et une dénutrition augmentent les probabilités de ces complications hépatiques. Elles sont heureusement très exceptionnelles et n'ont pas été décrites dans le cas des dérivés les plus récents.

Atteinte rénale

Lors du traitement par l'ensemble des tétracyclines à l'exception de la doxycycline, les malades sont particulièrement enclins a présenter une élevation de l'urée sanguine, sans augmentation de la creatinémie. Généralement les valeurs de l'urémie reviennent à la normale quelques jours après l'arrêt du traitement. Ce phénomène d'élévation transitoire de l'urémie semble du à un effet "anti-anabolique" des tétracyclines et peut être amplifié par l'association de diurétiques au traitement.

Atteinte vestibulaire

Fréquemment observés avec la minocycline, les troubles vestibulaires se traduisent par des manifestations d'ébrieté et des impressions de vertiges. Ils disparaissent spontanément à l'arrêt du traitement.

Myasthénie et bloc neuromusculaire

Une aggravation importante de l'état de faiblesse musculaire a été observée chez des malades myasthéniques recevant des tétracyclines par perfusion veineuse (oxytétracycline, rolitétracycline).

Les macrolides

Ce sont des antibiotiques à spectre étroit. Leur action s'exerce au niveau des sous-unités 50S du ribosome, en empêchant la translocation au cours des synthèses protéiques. Ce sont des antibiotiques bactériostatiques.

Les manifestations d'intolérance digestive sont fréquentes, surtout a fortes doses, mais rarement graves. Il s'agit essentiellement de nausées, voire de vomissements, de diarrhées et de douleurs abdominales. Elles se manifestent chez 5 a 30 p.100 des patients.

Les réactions d'hypersensibilité sont rares et se traduisent par des éruptions maculo-papuleuses, de l'urticaire, du prurit, de l'oedème vasculaire.

L'accident le plus sérieux est representé par des atteintes hépatiques observées surtout chez l'adulte. Une élévation des transaminases SGOT et SGPT survient dans 1 a 10 p.100 des cas après administration d'erythromycine sous forme de base ou de sels. Une hépatite cholestatique d'hypersensibilité peut également survenir avec toutes les formes d'erythromycine. Elle est souvent associée a une éruption cutanée et a une hyper-éosinophilie.

Les quinolones

Les quinolones, dont le chef de file est l'acide nalidixique, sont des antibiotiques actifs principalement sur les entérobactéries. Elles agissent en inhibant la biosynthèse de l'ADN bactérien par blocage de l'ADN-gyrase.

Les incidents et accidents de cette classe d'antibactériens se manifestent par des troubles digestifs à type de pesanteur gastrique, des nausées, des épigastralgies, et surviennent chez 8 % des sujets traités.

Les troubles cutanés, moins fréquents (5 %), sont de deux types:

. soit allergiques (érythème scarlatiniforme, morbiliforme, urticaire, purpura, oedème de Quincke).

. soit toxiques (dermites bulleuses apparaissant lors de l'exposition au soleil).

Les autres manifestations sont exceptionnelles:

. des atteintes neurologiques ont été décrites, se traduisant par des céphalées, vertiges, troubles visuels à type de vision colorée, de halo brillant. Elles sont plus particulièrement observées chez l'insuffisant rénal ou le sujet âgé.

. des atteintes hématologiques: neutropénie, thrombocytopénie, voire crises hémolytiques chez des sujets présentant un déficit en glucose-6-phosphate déshydrogénase.

Conclusion

Ce rapide tour d'horizon montre qu'il existe selon les produits une grande variété de mécanismes pouvant conduire à des phénomènes d'intolérances plus ou moins graves. Du point de vue de l'emploi de ces substances en aquaculture, le problème peut se poser à deux niveaux: celui des animaux traités, et celui des individus humains, professionnels ou consommateurs, appelés à entrer en contact avec les produits ayant servi aux traitements.

Les données publiées à ce jour sur la toxicité des antibiotiques pour les poissons sont loin d'être aussi riches qu'en clinique humaine, mais on ne peut manquer de retrouver certaines similitudes dans les incidents et accidents que l'on observe tant en médecine humaine qu'en ichtyopathologie. Celà tient évidemment aux propriétés des molécules et à leur tropisme marqué pour des tissus et des organes bien particuliers.

Le risque toxique représenté par les résidus de traitements dans les poissons de consommation doit être appréhendé en termes de doses très faibles. En dehors de l'aplasie médullaire causée par le chloramphenicol, pour laquelle l'effet-dose n'est pas établi, la plupart des accidents graves évoqués précédemment impliquent des situations thérapeutiques faisant appel à des doses élevées et prolongées. La probabilité de voir surgir de semblables accidents par ingestion de résidus est négligeable. Les risques de perturbation des flores intestinales ne semblent pas non plus de nature à se traduire par des désordres toxiques. En revanche il faut tenir compte des phénomènes allergiques, qui peuvent fort bien survenir sur des personnes déjà sensibilisées

De même le pisciculteur ou le praticien entreprenant de mélanger un médicament à l'eau ou à un aliment s'exposent à un risque d'inhalation ou de contact cutané avec le produit de base. Là encore ce sont les accidents allergiques qui paraissent les plus à craindre, et il sera avisé de s'entourer de précautions et de prévoir le port de vêtements de protection.

Références

1). MICHEL, C., 1986. Intérêt pratique, dangers potentiels et règles d'emploi des thérapeutiques antibactériennes chez les poissons. *Rev. sci. tech. Off. int. Epiz.*, **5**, 635-657.

2). PHILIPPON, A., VILDE, J.L. & GIROUD, J.P. 1988. Antibactériens et antiviraux. In, *Pharmacologie clinique, bases de la thérapeutique*, J.P. Giroud *et al.*eds, Expansion scientifique française, pp. 1295-1484.

3). MERCK, 1989. The Merck Index, 11th Edition, Merck & Co, Rahway, New Jersey.

4). REYNOLDS, J.P., 1989. Martindale, the Extra Pharmacopoeia. 29th Edition, The Pharmaceutical Press, London, 1896pp.

Discussion

HILL Thank you Dr Giroud. That was quite a catalogue of adverse effects in people under treatment with antibiotics and antimicrobials and of course also some examples of similar effects which have been seen in fish under treatment. What is not clear in my mind is what risk, if any, is there of such adverse effects occurring in people treating animals? What about the farmer using antibiotic in feed, particularly in mixing the feedstuffs? Is there a risk to that person?

GIROUD I think the main side effect which could occur with the farmer using these antibiotics, is an hypersensitivity reaction due to contact with the antibiotics. Otherwise I don't think that breathing the substance would result in any other type of toxicity. So only an hypersensitivity reaction may be expected to occur This is well known risk even when you are working in a drug company where you are preparing antibiotics. We do know that people can be sensitized by antibiotics and this sometimes results in some very acute symptoms.

HILL I wonder if anybody is aware of any actual cases of adverse effects in fish farmers. If anybody is, could they give examples?

CHAPMAN To confirm that we don't know of any cases, but that we did spend quite a long time doing literature reviews and asking round for examples. The only ones do seem to be hypersensitivity ones, predominantly in the pharmaceutical industry, rather than in users and we couldn't find any reports of hypersensitivities on farms or inded any other reactions at all.

WITHNELL We did have one case of hypersensitivity in one of our mills, a case of hypersensitivity to oxytetracycline.

HILL On the face of it, it would seem to be a very small risk or we would know of more cases than we are observing. So this particular aspect of the 'problems of chemotherapy in aquaculture' would seem to be a fairly small risk, but nonetheless with the long list of medical disorders we heard about, it is something people should be aware of and keep an eye open for adverse effects in farmers who are handling antibiotics if they are mixing feed or handling medicated feed.

FATE AND EFFECTS OF ANTIBACTERIAL AGENTS IN AQUATIC ENVIRONMENTS

B.T. Lunestad

Department of Microbiology and Plant Physiology
Jahnebakken 5, N - 5007 Bergen, Norway

Summary: *The consumption of antibacterial agents in aquaculture has reached a level calling for concern. The major part of the drugs end up in the environment, either directly due to exessive feeding and reduced appetite of the cultured organism, or indirectly after having passed through the organism under treatment. A careful examination of the fate and possible effects of such substances in the environment is required. Some of the drugs entering the environment will be taken up by wild fish, shellfish and crustaceans, resulting in concentrations that exceed those acceptable in food for human consumption. In this presentation the antibacterial agents oxytetracycline (OT), furazolidone (FU) and oxolinic acid (OXA) will be covered most extensively since they are among the most frequently used in the treatment of cultured organisms.*

Résumé: **Devenir et effets des agents antimicrobiens dans le milieu aquatique.**

La consommation de médicaments par l'aquaculture ne cesse d'augmenter. En 1989, 19 350 kgs d'antibiotiques ont été employés dans les élevages norvègiens, où depuis 1987 l'acide oxolinique a supplanté l'oxytétracycline et la furazolidone. La distribution d'un aliment médicamenteux s'accompagne d'inévitables pertes, résultant d'une solubilisation partielle du produit actif avant ingestion et du manque d'appétit des animaux malades. Les déchets s'accumulant sous les cages, peuvent être directement ingérés par des espèces sauvages, tandis que les principes solubilisés risquent d'affecter la végétation, les organismes filtrants (moules, tuniciers), et surtout sur les populations microbiennes. L'oxytétracycline notamment, est encore massivement utilisée (5000 kgs en 1989) malgré une absorbtion intestinale très médiocre en milieu marin.

Pour préciser les risques résultant des traitements, des mesures ont été exécutées par dosages microbiologiques et par chromatographie liquide à haute performance sur des échantillons de faune sauvage (poissons, crabes, moules), collectés au voisinage de cages d'élevage recevant de l'acide oxolinique. A la fin du traitement les concentrations musculaires des poissons s'élevaient jusqu'à 10000 ppb (moyenne 3800 ppb), avant de disparaître en 12 jours.

La stabilité des produits dans l'eau de mer a également été appréciée. L'oxytétracycline n'est pas dégradée par les microorganismes et apparaît très stable à l'obscurité. En conditions d'éclairage sa demi-vie de 30 heures à pH=8, contre 30 jours en eau douce à pH=7, indique que la photodécomposition joue probablement un rôle majeur dans l'épuration du milieu marin. Les effets les plus connus de l'oxytétracycline sont l'apparition de résistances bactériennes transmissibles et l'induction de malformations au stade larvaire de certains poissons. L'acide oxolinique est toujours très stable, même à la lumière, et bien qu'il précipite, conserve son activité antimicrobienne au moins 2 mois en eau de mer. Il n'induit que des résistances chromosomiques, mais des effets toxiques ont été décrits chez les mammifères et en lignées cellulaires. La furazolidone en revanche est très vite dégradée (demi-vie de 10 jours à la lumière). On lui connait des effets carcinogènes et mutagènes sur mammifères, et elle peut être toxique pour certains crustacés.

Enfin, on s'est intéressé à la formation de complexes entre l'oxytétracycline et les cations divalents. Mg^{++} est plus avide que Ca^{++}. Près de 95% de l'antibiotique peuvent ainsi se trouver complexés aux doses thérapeutiques usuelles. Le transport à travers les membranes cellulaires s'en trouve affecté, y-compris dans le tractus digestif, car les poissons marins boivent beaucoup afin de compenser l'hypertonie ambiante. En milieu marin l'absorbtion de l'oxytétracycline est réduite de 70% par rapport à l'eau douce. On sait par ailleurs que l'activité de l'antibiotique complexé est très amoindrie.

Ces exemples montrent que bien des études devraient être menées avant d'envisager le traitement des poissons. On recourt trop facilement à la thérapeutique actuellement, et d'autres méthodes de contrôle des maladies, comme la vaccination, devraient être développées d'urgence.

Introduction

Precise statistics on the total use of medical substances in aquaculture are available in Norway. A monopolized import and distribution system combined with strict veterinarian control of the prescription of antibacterial agents ensures reliable figures for the use of these substances. In 1989 the total amount of antbacterial agents used in Norwegian aquaculture was 19350 kg, measured as active components. Oxytetracycline (OT) and furazolidone (FZ) were for many years the agents of first choice. Since 1987 the quinolones, especially oxolinic acid (OXA), have found a broader use, at the expense of OT and FZ. In 1989 OXA represented 65 % of all antimicrobial agents prescribed in the treatment of fish in Norwegian aquaculture (Statistics provided by The Norwegian Medicinal Depot, Oslo).

Diseased fish are commonly treated using food-pellets containing the desired antibacterial agent. When the pellets are thrown into the water, fragments containing the antibacterial agent may fall off, or drug may be released from the pellet in a dissolved form (13). Particlebound medicine could be taken up by filter-feeders such as appendicularians or mussels. Depending on the substance, dissolved drugs may be chemically or microbiologically degraded or diluted in the watermasses. Some dissolved drugs can also be absorbed by organisms directly from the water (20). Provided the organism has a satisfactory appetite some of the medicine given will reach the organism under treatment. However an effective treatment is often obstructed due to the reduced appetite of the diseased organism. Thus considerable amounts of antibacterial agents will fall through the net-pens and may subsequently be eaten by pelagic or demersal organisms.

When investigating possible effects of antibacterial agents in the environment the persistence of the substances in question should be taken into consideration. If otherwise equal, the effects of persistent drugs could be expected to be most pronounced. Residues of antibacterial agents may have several environmental impacts. One being toxic effects on plants or animals caused by the substance itself or its degradation products (1, 2, 6, 9, 10, 11, 18, 33). However most antibacterial agents are characterized by their selective toxicity, thus being most active against bacteria.

Oxytetracycline (OT) has for many years been the agent of first choice. Even if a reduction in the use of OT has been observed during the last years, considerable amounts are still prescribed. In 1989 the amount of OT used in Norwegian aquaculture was 5000 kg, measured as active component. The standard dose recommended for treatment of fish is five to ten times higher than doses commonly used in medical practice, indicating that OT is poorly absorbed by the fish. This is especially true for fish held in seawater, where the intestinal uptake is substantially reduced as compared to fish in fresh water. To compensate for the poor intestinal uptake higher amounts of OT are used during medication, giving an increased environmental load of this agent.

1. Materials and methods

Oxolinic acid (OXA) in wild fauna

As a model for the distribution of drugs in the environment the dispersion of OXA in the wild fauna has been studied on organisms harvested in the vicinity of two aquaculture facilities during and after a treatment with this drug. The occurrence of OXA in the gut content of fish was examined using a standard microbiological method applying an OXA-sensitive organism (*E. coli* B^6. The residues of OXA in wild fish (plasma, muscle and liver), crab (haemolymph) and blue mussels (homogenate of soft parts) was quantified by a chemical method (HPLC) (unpubl. res.). The detection limit for OXA applying the HPLC method was 5 ppb (5 $\mu g\ kg^{-1}$).

Stability of OT, OXA and FZ

The stability of aqueous solutions of OT, FZ or OXA has been investigated under laboratory conditions on samples kept in the dark or illuminated using fluorescent tubes. Chemical measurements of reduction in the concentrations of drugs were performed spectrophotometrically or by HPLC. The reduction in antibacterial activity was examined applying a microbiological method with a sensitive bacterium.

OT in seawater

Complexes with magnesium and calcium are formed when oxytetracycline (OT) is dissolved in seawater. The complex constants for OT-hydrochloride with magnesium and calcium were determined spectrophotometrically on the basis of the absorption characteristics for free and complexbound OT (17).

Bacterial sensitivity to OT complexes was determined both as minimum inhibitory concentration (MIC) and by the disc-diffusion test as described by Lunestad and Goksoeyr (1990) (17).

2. Results and discussion

Oxolinic acid (OXA) in wild fauna

In this investigation we examined a total of 250 fishes, 50 crabs and 30 samples of blue mussels during and after medication with OXA. Residues of OXA were detected in coalfish, mackerel, cod, whiting, ling, pollack, ballan wrasse, escaped salmon, crab and blue mussels. At the day of termination of medication the average concentration of OXA in the muscle of wild fish was 3800 ppb. This exceeds by far the drug concentrations tolerated in food for human consumption. The highest concentration found was 10 000 ppb (muscle of coalfish). Even in fish caught 400 m away from the cages, residues could be detected. Twelve days after termination of the medication, detectable concentration of OXA could not be found in any species examined (detection limit 5 ppb). This example clearly shows that during medication, the wild fauna is exposed to considerable amounts of medicine.

If wild fish are caught in the vicinity of an aquaculture facility during medication, drugs could find their way to the dinner table. OXA is rather thermostable and no reduction in the antibacterial activity could be observed when aqueous solutions of the drug was boiled for 15 minutes to simulate conditions resembling normal treatment of fish.

Fate and effects of OT, OXA and FZ

Even though OT is produced by microorganisms, this substance is surprisingly resistant to degradation by microbial enzymes (16, 27). OT added to marine aquaculture sediments during medication is washed out rather than degraded in the sediment (15, 23, 25). Aqueous solutions of OT kept dark at 15°C for a period of two months seems to be very stable with only a small reduction in concentration and antibacterial activity. When OT-solutions are illuminated the situation is dramatically changed. A half-life of 30 days could be observed when a fresh water solution (pH = 7) of OT at an initial concentration of 0,5 $mg.ml^{-1}$ was kept illuminated under a 20 W, 1200 lumen Philips, PLC 20 Electronic fluorescent tube at 15°C. OT dissolved in seawater (pH = 8) and kept illuminated as described above had a half-life of only 30 hours. Thus, provided a sufficient light intensity, the main degradation mechanism for OT in the environment, especially seawater, seems to be photodecomposition. The effect of light on OT solutions has previously been described by Samuelsen (1989) and Oka *et al.* (1989) (21, 23).

Aquatic environments contain a large number of bacteria that could be affected by antibacterial agents. The sensitivity to drugs differ between various bacterial biotypes giving the possibility for a change in a bacterial community under the selective force of an antibacterial agent. Such changes may also include an increase in the number of resistant bacteria (5, 14, 25). In most situations where antibacterial agents are in common use bacterial resistance has become a more or less serious problem (7). Several authors report resistance problems in connection with fish pathogens (3, 4, 12, 26, 29). The importance of horizontal gene transfer in natural environments has been emphasized during the last years (8, 28, 31). Special concern has been paid to the possibility for transfer of resistance genes to bacteria pathogenic to fish or humans (14, 19). The following broad statement can be made: the presence of antibacterial agents in the aquatic environment will enhance the multiplication of resistant bacteria and increase the possibility for transfer of genes carrying resistance.

OT inhibits bacterial growth by interfering with procaryote protein synthesis (11). R-plasmid located resistance towards this drug has frequently been reported (7, 11, 16). Some reports state that the level of tetracycline-resistance in bacteria isolated from fish has increased as a result of the use of this agents in aquaculture (3, 4, 26, 29, 30).

The ability of OT to induce deformities in fish larvae has been studied. Halibut larvae hatched and kept in natural seawater containing OT in a concentration of 100 ppm, had a ten times higher incidence of deformities of the notocord and yolk sack as compared to a control group belonging to the same cohort (unpubl.res.) This effect is probably in part caused by the ability of OT to bind irreversibly to bone structures.

Preliminary examinations of solutions of OXA in seawater indicate that this drug is stable over a period of at least two months even when illuminated. A white precipitate was formed in the solution, but this precipitate still possess antibacterial activity and resembles OXA when examined chemically.

In higher animals OXA is associated with incidences of CNS toxicity (9, 11). Furthermore, OXA is suspected to interfere with gene transcription in mammalian cells (9). Single step mutations leading to high-level resistance towards OXA in bacteria have been reported to occur with frequencies in the range of 10^{-6} to 10^{-8} (33). On the other hand there has been no description of transferable or plasmid-mediated resistance towards OXA (33).

In contrast to OT and OXA, FZ is readily degraded when present in sediments following degradation pathways previously described (24). However when kept dark in natural seawater the concentration was only slightly reduced over a period of 300 hours. When illuminated the substance was readily degraded giving a half-life of 10 days. In addition to the antibacterial effect of FZ this substance excerts an effect on several protozoans (11), thus also affecting bacterial predators. The toxicity of FZ has been reviewed by several authors (1, 2, 6). FZ is reported to have several adverse effects on higher animals, including mutagenicity and carcinogenicity, making the use of this drug questionable (1, 2, 6, 11). Furthermore toxicity tests have been performed on various aquatic organisms (18). This investigation revealed a significant toxicity of the compound on *Culex pipiens* and *Daphnia magna*, while *Artemia salina* proved to be less sensitive.

OT in seawater

The complex constant for a 1:1 complex between Mg^{2+} or Ca^{2+} and OT at pH 8 were found to be 2.9 * 10^2 M^{-1} for MgOT and 1.9 * 10^2 M^{-1} for CaOT, indicating that Mg^{2+} is the most avid ligand. Seawater with a salinity of 35 p.1000 contains 54 mM Mg^{2+} and 10 mM Ca^{2+} (29). When OT is dissolved in seawater in concentrations ranging from 0,5 to 100 $\mu g.ml^{-1}$, as much as 95 % of the drug will be bound to Mg^{2+} and Ca^{2+}.

To pass a biological lipid membrane a substance must be in an undissociated state i.e. without molecular charge. When OT is complex bound to cations a positive charge is obtained reducing this agents ability to pass the membranes of cells lining the intestine of fish or the membranes surrounding bacteria. Thus the formation of complexes has two important implications: 1) The antibacterial activity of OT is strongly reduced and 2) OT is poorly absorbed from the intestine in fish held in seawater.

When marine fish are medicated OT will unavoidably come into contact with Mg^{2+} and Ca^{2+}. This occurs when the OT-coated food pellet is in the seawater or in the gastrointestinal tract. Marine teleosts are hypoosmotic and must drink seawater to compensate for loss of water. Rainbow trout (*Salmo gairdneri*) kept in seawater have been reported to drink approximately 130 $ml.kg^{-1}.day^{-1}$, while rainbow trout kept in fresh water drink little or not at all (32) Drinking rates were independent of gut food content but were significantly higher in feeding fish (32). Thus, under marine conditions the gut fluid of fish can be regarded as slightly modified seawater. Since OT in the intestine must be taken up from the liquid phase, the presence of Mg^{2+} and Ca^{2+} will affect the absorption. This has lately been confirmed by Rogstad and Ellingsen (in prep). In this experiment two groups of rainbow trout from the same cohort where medicated, using a standard OT regime, while kept in either fresh water or seawater. On the average, the tissue level of OT in the seawater group was approximately 30 % of what could be found in the fresh water group.

In all cases the presence of Mg^{2+}/Ca^{2+}, whether provided as seawater or as laboratory chemicals, led to a profound increase in the ability of the test bacteria to tolerate OT. The effect of Mg^{2+} was greater than the effect of Ca^{2+}.

The example of oxytetracycline clearly visualizes the need for a careful examination of drugs to be used in the treatment of fish. In addition to investigations of absorption, distribution, elimination and effects of the substances in the fish, an evaluation of the half-life and possible effects on the environment should also be carried out. Additionally the methods of drug production should be considered. The amounts of hazardous by-products during production is relevant information when evaluating the total environmental burden caused by any substance.

Concluding remarks

The amount of antibacterial substances used in aquaculture is much too high. Although fragmentary, our present knowledge calls for a prudent use of these drugs. Disease problems can not be solved by increased use of chemicals. Efforts should be made on improving the general health condition of cultured organisms, including the development of effective vaccines.

References

1) ALI B.H., 1983. Some pharmacological and toxicological properties of furazolidone. *Vet. Res. Com.*, **6**, 1-11.

2). ALI B.H., 1989. Minireview: Pharmacology and toxicity of furazolidone in man and animals: some recent research. *Gen. Pharmac.*, **20**, 557-563.

3) OKI T. & KITAO T., 1981. Drug resistance and transferable R plasmids in *Edwardsiella tarda* from fish culture ponds. *Fish Pathol.*, **15**, 277-281.

4). AOKI T., KANAZAWA T. & KITAO T., 1985). Epidemiological surveillance of drug resistant *Vibrio anguillarum* strains. *Fish Pathol.*, **20**, 199-208.

5). BAYA A.M., BRAYTON P.R., BROWN V.L., GRIMES D.J., RUSSEK-COHEN E. & COLWELL R.R., 1986). Coincident plasmids and antimicrobial resistance in marine bacteria isolated from polluted and unpolluted atlantic ocean samples. *Appl. Environ. Microbiol.*, **51**, 1285-1292.

6). BLIJLEVEN W.G.H., KORTSELIUS M.J.H. & KRAMERS P.G.N., 1977. Mutagenicity testing of H-193, AF-2 and furazolidone in *Drosophila melanogaster*. *Mutat. Res.*, **56**, 95-100.

7). BRYAN L.E., 1984. *Antimicrobial drug resistance*. Academic Press Inc.

8). COUGHTER J.P. & STEWART G.J., 1989. Mini Review; Genetic exchange in the environment. *Antonie van Leeuwenhoek*, **55**, 15-22.

9). CRUMPLIN G.C., 1988. Aspects of chemistry in the development of the 4-Quinolone antibacterial agents. *Rev. Inf. Dis.*, **10**, 2-9.

10). FASULO M.P., BASSI M. & DALL'OLIO G., 1981. Effects of erythromycin on *Euglena gracilis* as a model for testing the toxicity of antibacterial drugs. *Arzneim. Forsch.* **31** (II), 1869-1872.

11). GOODMAN L.S. & GILMAN A.G., 1985. *The pharmacological basis of therapeutics*. Seventh edition 1985, Macmillan Publishing Company, New York.

12). HASTINGS T.S. & McKAY A., 1987. Resistance of *Aeromonas salmonicida* to oxolinic acid. *Aquaculture*, **61**, 165-171.

13). HIGUERA-CIAPARA I., BROWN J.H. & JAUNCEY K., 1990. Leaching of oxytetracycline from pelleted shrimp feeds. Abstract from the scientific symposia: *Bacterial diseases of fish*, 26-29th June 1990, University of Stirling, Scotland, UK.

14). HUSEVÅG B., LUNESTAD B.T., JOHANNESSEN P.J., ENGER, O.E & SAMUELSEN O.B., 1991. Simultaneous occurrence of *Vibrio salmonicida* and antibiotic resistant bacteria in sediments at abandoned aquaculture sites. *J. Fish. Dis.*, in press.

15). JACOBSEN P. & BERGLIND L., 1988. Persistence of oxy-tetracycline in sediments from fish farms. *Aquaculture*, **70**, 365-370.

16). LEVY S.B., 1988. Tetracycline resistance determinants are widespread. *ASM News*, **54**, 418-421.

17). LUNESTAD B.T. & GOKSØYR J., 1990. Reduction in the antibacterial effect of oxytetracycline in sea water by complex formation with magnesium and calcium. *Dis. Aquat. Org.*, **9**, 67-72.

18). MACRI A., STAZI V. & DELUPIS D.D., 1988. Acute toxicity of furazolidone on *Artemia salina*, *Daphnia magna*, and *Culex pipiens molestus* larvae. *Ecotoxicol. Environ. Safety*, **16**, 90-94.

19). NAKAJIMA T., SUZUKI M., HARADA K., INOUE M. & MITSUHASHI S., 1983. Transmission of R plasmids in *Vibrio anguillarum* to *Vibrio cholerae*. *Microbiol. Immunol.*, **27**, 195-198.

20). NUSBAUM K.E. & SHOTTS Jr. E.B., 1981. Absorption of selected antimicrobic drugs from water by channel catfish, *Ictalurus punctatus*. *Can. J. Fish. Aquat. Sci.*, **38**, 993-996.

21). OKA H., IKAI Y., KAWAMURA N., YAMADA M., HARADA K., ITO S. & SUZUKI M., 1989. Photodecomposition products of tetracycline in aqueous solution. *J. Agric. Food Chem.*, **37**, 226-231.

22). POTTS W.T.W. & PARRY G., 1964. *Osmotic and ionic regulation in animals*. Pergamon Press, Oxford.

23). SAMUELSEN O.B., 1989. - Degradation of oxytetracycline in seawater at two different temperatures and light intensities, and the persistence of oxytetracycline in the sediment from a fish farm. *Aquaculture*, **83**, 7-16.

24). SAMUELSEN O.B., SOLHEIM E. & LUNESTAD B.T., 1991. The fate and microbiological effects of furazolidone in a marine aquaculture sediment. *Sci. Tot. Environ.*, in press..

25). SAMUELSEN O.B., TORSVIK V. & ERVIK A., 1990. Long-range changes in oxytetracycline concentration and bacterial resistance towards oxytetracycline in a fish farm sediment after a medication. *Sci. Tot. Environ.*, in press.

26). SCHLOTFELDT H.J, NEUMANN W., FUHRMANN H., PFORTMUELLER K. & BOEHM H., 1985. Remarks on an increasing resistance of fishpathogenic and facultative-fishpathogenic bacteria in Lower Saxony (FRG). *Fish Pathol.*, **20**, 85-91.

27). SPEER B.S. & SALYERS A.A., 1989. Novel aerobic tetracycline resistance gene that chemically modifies tetracycline. *J. Bacteriol.*, **171**, 148-153.

28). STEWART G.J. & SINIGALLIANO C.D., 1990. Detection of horizontal gene transfer by natural transformation in native and introduced species of bacteria in marine and synthetic sediments. *Appl. Environ. Microbiol.*, **65**, 1818-1824

29). TAKASHIMA N., AOKI T. & KITAO T., 1985. Epidemiological surveillance of drug-resistant strains of *Pasteurella piscicida. Fish Pathol.*, **20**, 209-217.

30). TORANZO A.E., COMBARRO P., LEMOS M.L. & BARJA J.L., 1984. Plasmid coding for transferable drug resistance in bacteria isolated from cultured rainbow trout. *Appl. Environ. Microbiol.*, **48**, 872-877.

31). TREVORS J.T., BARKAY T. & BOURQUIN A.W., 1987. Gene transfer among bacteria in soil and aquatic environments: a review. *Can. J. Microbiol.*, **33**, 191-198.

32). USHER M.L., TALBOT. C & EDDY F.B., 1988. Drinking in atlantic salmon smolts transferred to seawater and the relationship between drinking and feeding. *Aquaculture*, **73**, 237-246.

33). WOLFSON J.S. & HOOPER D.C., 1989. *Quinolone antimicrobial agents*. American Society for microbiology, Washington, DC.

Discussion

The discussion for this paper is presented after the following complementary paper by Samuelsen.

THE FATE OF ANTIBIOTICS/CHEMOTHERAPEUTICS IN MARINE AQUACULTURE SEDIMENTS

O. B. Samuelsen

Department of Pharmacology and Toxicology, University of Bergen, Armauer Hansens Hus, N-5021 Bergen, Norway.

Summary: *The fate of three antibiotics/chemotherapeutics (oxytetracycline, oxolinic acid and furazolidone) in marine aquaculture sediments has been examined in aquaria and in situ (oxytetracycline). Furazolidone was metabolised by bacteria in the sediment, the main metabolite being 3-(4-cyano-2-oxobutylideneamino)-2-oxazolidone, a well known degradation product of furazolidone in mammals, fish and cultures of* Escherichia coli. *The half-life of furazolidone in the sediment at 4°C was calculated to be 18 hours. The bulk of oxytetracycline and oxolinic acid disappeared during the first weeks, but they persisted in the sediment in lower concentrations for months after the enrichment. Oxolinic acid and oxytetracycline are probably not metabolised by the sediment bacteria and a solvation and diffusion process is the most likely mechanism whereby they escape from the sediment.*

Résumé: **Devenir des antibiotiques et des agents thérapeutiques dans les sédiments d'aquaculture marine.**

Les problèmes pathologiques de l'aquaculture marine intensive ne sont pas toujours résolus par la vaccination, et il est alors indispensable de traiter. On incorpore généralement le médicament à la ration alimentaire, mais une partie de celle-ci peut n'être pas consommée, et certains composés, mal absorbés, sont pour partie rejetés dans les fécès. Ils risquent alors d'être ingérés par des espèces sauvages, de se déposer au fond, ou d'induire des modifications dans la composition des flores microbiennes naturelles. Pour évaluer ces risques on a entrepris des études expérimentales et des mesures in situ *sur le devenir de l'oxytétracycline, de l'acide oxolinique et de la furazolidone. Les expériences étaient menées à 4°C dans des réservoirs d'1 m^3, garnis d'une couche de sable surmontée de sédiments marins où l'on incorporait l'antibiotique à 400 mg/l. Les prélèvements étaient exécutés par carottage. Dans des fermes de saumons peu exposées aux courants, des séries de cages ont été traitées par l'oxytétracycline à 75 mg/kg pendant 10 jours. La 1ère cage, peuplée d'animaux vaccinés, servait de témoin non traité. La 2ème recevait un rappel de 5 jours après 51 jours. La 3ème était débarrassée de ses poissons au 23ème jour. Les prélèvements, effectués sur les fonds à intervalles réguliers, étaient congelés jusqu'à analyse. Toutes les mesures ont été exécutées en HPLC par référence à des courbes préalablement étalonnées pour chaque antibiotique.*

Les courbes standards s'avèrent linéaires pour les 3 antibiotiques. La persistance dans les différents systèmes de la furazolidone, et de son métabolite principal, la 3-(4-cyano-2-oxobutylideneamino)-2-oxazolidone, ainsi que celles de l'oxytetracycline et de l'acide oxolinique, sont reportées dans les Fig.1 à 3. Les demi-vies sont indiquées dans le Tableau I. La dégradation microbienne de la furazolidone en moins de 24 heures constitue une certaine garantie. Par contre l'oxytétracycline et l'acide oxolinique persistent bien plus longtemps, et ne s'éliminent probablement que par dilution. L'oxytétracycline s'inactive plus rapidement en surface, et sa persistance est augmentée si on la recouvre de 4 cm de sédiment (cas des cages 1 et 2 du tableau I). On note également que la demi-vie est plus longue en conditions naturelles que dans l'expérience contrôlée. Si la diffusion des antibiotiques est rapide dans les 1ères semaines, de faibles concentrations résiduelles peuvent persister pendant des mois. Il est vraisemblable qu'en dehors des phénomènes de sédimentation, des mécanismes physico-chimiques contribuent compliquer la situation dans l'environnement naturel.

Introduction

Problems with diseases are frequently encountered in modern intensive salmon farming. Although vaccination of fish has reduced the occurrence of some diseases, e.g "cold water vibriose" due to *Vibrio salmonicida* chemotherapy against bacterial infections must frequently be employed. The antibiotic/chemotherapeutic agents are administered in the form of medicated food pellets. Since fish suffering from bacterial diseases usually show reduced appetite, much of the drug intended for therapy passes through the cages and into the environment. Oxytetracycline, a widely used antibiotic in fish farming, has shown to be very poorly absorbed from the intestinal tract of the fish (3), probably due to complexation with Ca_2^{++} and Mg_2^{++} ions (7). Accordingly, much of the oxytetracycline passes through the fish and enters the environment in an unchanged, active form via faeces. The drug-containing faeces from the farmed fish and the medicated food are either eaten by the wild fauna or reach the bottom deposit under the farm (4, 13). Both oxytetracycline and oxolinic acid are known to induce resistance to antibiotics/chemotherapeutics in the marine microflora including bacteria pathogenic to fish (2,5,14,16,17). For bacteria resistant to oxytetracycline, the resistance determinant is often found in R-plasmids which can be transferred to other bacteria by conjugation (17). The sediment under a fish farm contains large amounts of bacteria, including bacteria pathogenic to fish. If drug is present in the sediment for a prolonged period following medication, the chance of bacteria developing resistance to the drug may increase. Information on the occurrence, persistence and metabolism of drugs in marine aquaculture sediments was therefore deemed necessary. Experiments, both *in situ* and under controlled aquaria conditions were therefore carried out in order to obtain this information. In this article, I will discuss the fate of the three drugs, oxolinic acid, oxytetracycline and furazolidone in marine aquaculture sediments, based on newer findings and results published in three papers, (10,12,13).

1. Materials and methods

Experimental design

To study the persistence and metabolism of furazolidone, oxytetracycline and oxolinic acid in a marine aquaculture sediment eight circular aquaria, (90 cm i.d. x 100 cm height) were each supplied with a 10 cm thick layer of shell sand at the bottom and a 10 cm thick sediment layer from a salmon farm on top. The sediment was, prior to addition to the aquaria, homogenised with one of the three drugs to give a concentration of 400 µg per ml. Two portions of the sediment were homogenized without drugs and served as controls. A sea water supply of approximately 10 l per min per aquarium created a stable current. The average temperature of the seawater was 4°C. Sediment samples were collected from the aquaria using a polypropylene coring tube (1.5 cm i.d., 20 cm length). The tube was forced into the sediment and a rubber stopper was placed on the top of the tube to prevent the sample from falling out on withdrawal from the sediment. Before withdrawal, a larger polypropylene tube (2.5 cm i.d., 10 cm length) was forced into the sediment on the outside of the coring tube to prevent the sediment from collapsing when the coring tube was withdrawn. The lower end of the coring tube was sealed with a rubber stopper immediately after withdrawal from the sediment. The larger tube remained in the sediment.

A salmon farm outside Bergen on the West coast of Norway, with a weak water-flow for waste transport and therefore a rather large sedimentation under the cages, was treated with oxytetracycline for 10 days (23 March - April 2). The dosage was 75 mg oxytetracycline per kg body weight per day. The sediments under three cages were selected for sampling. The cages contained approximately 1500 kg fish (cage 1 and 2) and 1150 kg (cage 3). The salmon in cage 1 took part in a vaccination program against "vibriose" (*Vibrio anguillarum*) and "cold water vibriose" (*Vibrio salmonicida*) directed by the Institute of Marine Research (Bergen, Norway) and were therefore not given the antibiotic. At day 23 the fish in cage 3 were removed. Therefore, no feeding took place and further sedimentation of excess food pellets and feces under this cage was minimized. At day 51 another 5 days of medication with oxytetracycline took place in cage 2, using the same dosage as described previously. Sediment samples, six cores from under each of the three selected cages, were taken by divers. One meter long white sticks were placed in the sediment as markers and ensured that the sampling was taking place within the same area of the sediment each time. The sediment samples were frozen and kept at -30° C until analysed.

Chromatography

For the high-performance liquid chromatographic analysis (HPLC), a Spectra-Physics SP-8800 ternary HPLC-pump connected to a Spectra-Physics SP-8780 X R Autosampler was used. The UV-detector was a Spectra-Physics model SP-8480 XR Scanning detector, and the integrator a model SP-4270, also from Spectra-Physics. The analytical column, 100 x 5 mm, was packed with ODS-Hypersil 3 µm particle size packing material for HPLC (Shandon Southern Products, Cheshire, UK), and the guard column with ODS-Hypersil 40 µm pellicular material. For the quantitative measurement of furazolidone and oxytetracycline in sediment, the mobile phase used was acetonitrile : water (14 : 86, v/v) containing 1x 10^{-3} M Na_2-EDTA, 0.1 M potassium nitrate and nitric acid giving pH 3.2 (6). The detection was made at 365 nm, and the flow-rate was 1.5 ml per min. For the qualitative and quantitative analysis of any metabolites of furazolidone in both the polar and non-polar fractions of the enriched interstitial water, a linear 22 min. gradient of 2 - 60 % acetonitrile in water : 1.0 M acetate buffer pH 5 : acetonitrile (900 : 75 : 25, v/v) at a flow rate of 1 ml per min. was used. Detection was made at 268 nm (18).

For the quantitative measurement of oxolinic acid in the sediment, the mobile phase used contained A = 0.025 N oxalic acid pH 3.2 : acetonitrile : methanol : tetrahydrofuran (80 : 2.5 : 15 : 2.5 v/v) and B = oxalic acid pH 3.2 : acetonitrile : methanol : tetrahydrofuran (50 : 20 : 25 : 5 v/v). The solutions were filtered through a 0.2 µm Millipore filter. The elution profile was as follows: 0-5min, 0-100% B linear gradient; 5-10min isocratic 100% B; 10.1-15min isocratic 100% A. The flow rate was 1 ml per min. Detection was made at 280 nm (11).

Preliminary HPLC-experiments were carried out using sediment samples (1 ml) to which small amounts of one of the three drugs had been added. No interfering material was seen in the area of the chromatograms in which the drugs appeared. A reference curve was prepared using sediment homogenate (1 g) to which drugs in amounts ranging from 10 to 400 µg were added. By using a standard method with respect to sample preparation and injected sample volume, the peak areas from the chromatograms from the sediment samples could be directly compared with those from the standard curves to give the value of the amount of drugs present in the sediment samples.

Sample preparation

Since furazolidone is light sensitive all samples containing furazolidone were processed in subdued incandescent light. The coring tube was gently heated 1 min. with a current of warm air before the frozen sediment was pushed out with a piston. The sediment was homogenized by vigorous shaking for 5 min. To determine the concentration of drug in the sediment, the homogenate (1g) was transferred to a polypropylene tube (8 ml) with a screwcap and extracted with two 5 ml aliquots of 0.1 M acetic acid in water (furazolidone), McIlvaine buffer pH 4.0 containing 0.1M EDTA (oxytetracycline) or 0.1M NaOH (oxolinic acid). After centrifugation at 4200 g for 10 min. and 4° C, (Sorvall RC-5B), 1 ml of the combined supernatants was transferred to a 1.5 ml polypropylene micro centrifuge test tube, centrifuged at approximately 15000 g for 2 min. (Beckman Microfuge B). The resultant supernatant (20 µl) was used for HPLC analysis.The non-polar metabolites in the sediment enriched with furazolidone were extracted using the method of Vroomen *et al.*, (20) with minor modifications. The homogenate (5 ml) was centrifuged at 7000 g for 10 min. and the supernatant was extracted twice with equal volumes of ice-cold ethyl acetate. The ethyl acetate was evaporated to dryness (*in vacuo*) and the residue dissolved in 100 µl of water : acetonitrile (80 : 20), centrifuged (2 min. at 15000 g) and the supernatant (20 µl) used for HPLC. Following extraction with ethyl acetate, the supernatant containing possible polar metabolites was evaporated to dryness (*in vacuo*), dissolved in 500 µl distilled water, centrifuged for 5 min at 15000 g and analysed by HPLC under the same conditions as described for the non-polar metabolites.

The qualitative mass-spectral data of the furazolidone metabolite was obtained using a Hewlett Packard 1090 HPLC-system connected to a Vestec-Thermospray LC/MS-system.

2. Results and discussion

The standard curves had a high degree of linearity for the three drugs (R= 0.97-1.0).The main metabolite of furazolidone in the sediment was identified using LC/MS and shown to be 3-(4-cyano-2-oxobutylideneamino)-2-oxazolidone. The persistence of oxytetracycline, oxolinic acid, furazolidone and 3-(4-cyano-2-oxobutylideneamino)-2-oxazolidone is shown in Figs. 1-3. The half-lives of the drugs in the aquarium sediment and under a salmon farm are shown in Table I.

Furazolidone remained in the sediment for a very short time ($t_{1/2}$ = 18 h) and was actively metabolized to 3-(4-cyano-2-oxobutylideneamino)-2-oxazolidone by the microorganisms in the sediment, probably following the metabolic pathway described for *E. coli* (1) and for other organisms (1, 9, 15, 19). According to Vroomen *et al.*, (18) this metabolite exhibits no mutagenicity and according to our findings no antibacterial activity (12). Based on peak areas measured at 268 nm, the chromatograms of the ethylacetate extracts (non-polar fraction) showed that 3-(4-cyano-2-oxobutylideneamino)-2-oxazolidone accounted for at least 80% of the metabolites in this fraction. Numerous unidentified metabolites were detected in the water-phase (polar fraction).

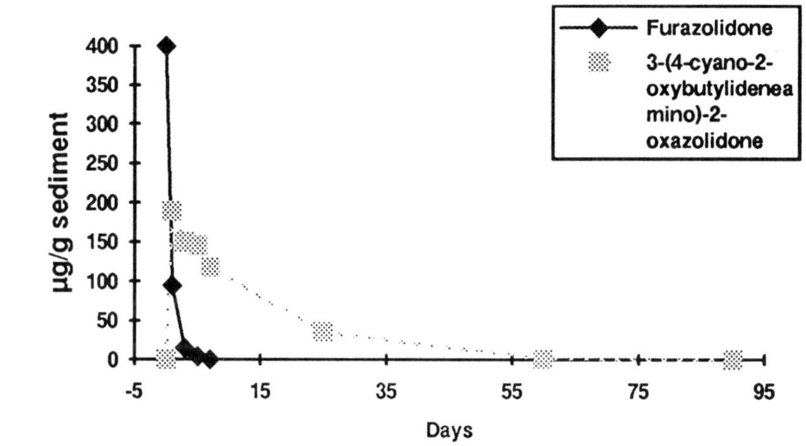

Fig 1 Metabolism of furazolidone and the persistence of the main metabolite 3-(4-cyano-2-oxobutylideneamino)-2-oxazolidone in marine aquaculture sediment.

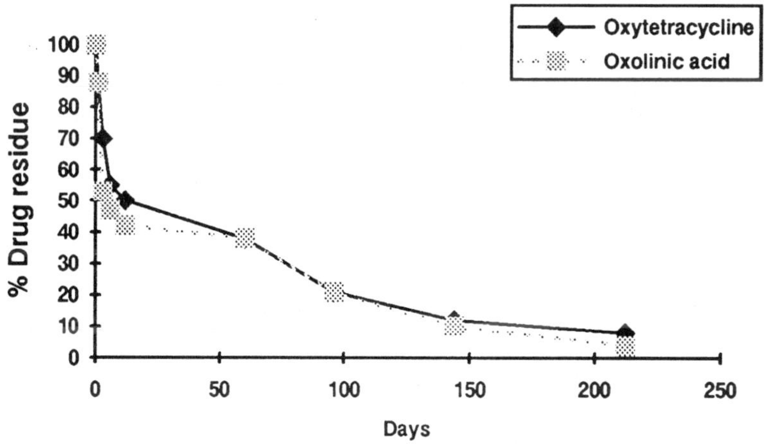

Fig 2 Persistence of oxytetracycline and oxolinic acid in a marine aquaculture sediment.

Table I. Halflives of oxytetracycline, oxolinic acid and furazolidone in the sediment from the aquaria and of oxytetracycline under three cages at a salmon farm.

Drug	**Half-lives(salmon farm)** Cage 1	Cage 2	Cage 3	**Half-lives (aquaria)**
Furazolidone				18h
Oxolinic acid				48d
Oxytetracycline	125 d	142 d	89d	55 d

h = hours, d = days

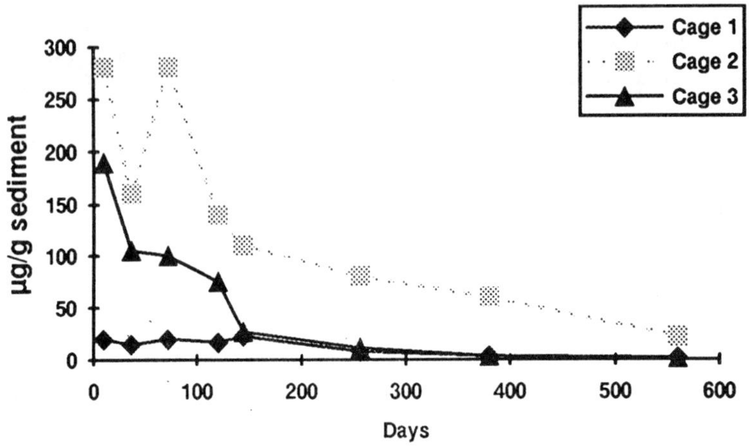

Fig 3. Persistence of oxytetracycline in the sediment under three cages at a salmon farm.

Figs. 2 and 3 and Table I show that oxytetracycline and oxolinic acid persist in the sediment for a much longer period than furazolidone. No reports concerning possible microbiological degradation processes of oxytetracycline or oxolinic acid in marine sediments have yet been published. It is therefore likely that since oxytetracycline and oxolinic acid are soluble in sea water, a solvation and diffusion process is the most likely mechanism whereby they escape from the sediment. Samuelsen (10) showed, in experiments using aquaria, that the persistence of oxytetracycline in a fish farm sediment was highly dependant on the sedimentation after a medication. Based on measurements for 224 days and with the oxytetracycline containing layer (crushed medicated pellets) placed on top of the 8 cm thick sediment, $t_{1/2}$ was found to be 72 days. When the oxytetracycline-containing layer was immediately covered with an additional 4 cm of sediment, the calculated $t_{1/2}$ was nearly doubled (135 days). Examination of the vertical distribution of oxytetracycline in the sediment from the salmon farm showed that the oxytetracycline first disappeared in the upper layer of the sediment. A gradually increasing cover on the drug-containing layer with additional faeces and food residues will inevitably result in a reduced contact with the overlying sea water and decrease the solvation and diffusion process. At the salmon farm the further sedimentation after the medication was larger in cage 1 and 2 than in cage 3 due to removal of the fish in cage 3. Therefore, considering the different extent of sedimentation in the cages, Table I shows that $t_{1/2}$ calculated for cage 1, 2 and 3 (125, 142 and 89 days respectively) are close to the values calculated in the aquaria experiment using crushed medicated pellets (135 and 72 days). Mixing pure compound into the sediment as was done in the last aquarium experiment, reduced the persistence of the drugs in the sediment considerably (Table I). However, the results show that oxytetracycline and oxolinic acid probably follow the same pathway (solvation and diffusion) to escape from the sediment.

Both Figs. 2 and 3 and other experiments (10) show that the bulk of oxytetracycline and oxolinic acid disappears from the sediment during the first few weeks, but that residual concentrations persist in the sediment for months after an enrichment or a medication. However, decreased diffusion due to sedimentation may not be the only reason for these drugs to persist for such an extended period in the sediment. Crystallisation (8), complexation (7) and binding to particles may prevent the drug molecules from dissolving in the interstitial water.

Acknowledgments: The excelent assistance of the academic and technical staf at the Department of Pharmacology and Toxicology at University of Bergen is highly appreciated. I am grateful for the financial support of the Norwegian Research Council for Fisheries Science.

References

1). ABRAHAM, R.T., KNAPP, J.E., MINNIGH, M.B., WONG, L. K., ZEMATIS, M. A. & ALVIN, J. D., 1984. Reductive metabolism of furazolidone by *Escherichia coli* and rat liver *in vitro. Drug. Metab. Dispos.* **12**, 732-741.

2). AUSTIN, B., 1985. Antibiotic pollution from fish farms: effect on aquatic microflora. *Microbiol. Sci.*, **2**, 113-117.

3). CRAVEDI, J.-P., CHOUBERT, G. & DELOUS, G., 1987. Digestability of chloramphenicol, oxolinic acid and oxytetracycline in rainbow trout and influence of these antibiotics on lipid digestability. *Aquaculture*, **60**, 133-141.

4). JACOBSEN, P. & BERGLIND, L., 1988. Persistence of oxytetracycline in sediments from fish farms. *Aquaculture*, **70**, 365-370.

5). JONES, J.G., GARDENER, S., SIMON, B.M. & PICKUP, R.W., 1986. Factors affecting the measurement of antibiotic resistance in bacteria isolated from lake water. *J. Appl. Bacteriol.*, **60**, 455-462.

6). KNOX, J.H. & JURAND, J., 1979. Mechanism of reversed-phase separation of tetracyclines by high-performance liquid chromatography. *J. Chromatogr.* **186**, 776-782.

7). LUNESTAD, B.T. & GOKSØYR, J., 1990. Reduction in the antibacterial effect of oxytetracycline in sea water by complex formation with magnesium and calcium. *Dis. Aquat. Org.*, **9**, 67-72.

8). MERCK, 1983. *The Merck Index*, 10th edition, 1983. Merck & Co. Inc., Rahway, N.J., USA.

9). NAKABEPPU, H. & TATSUMI, K., 1984). Metabolism of Furazolidone in eels. *Chem.Pharm.Bull.*, 32, 4193-4196.

10). SAMUELSEN, O.B., 1989. Degradation of oxytetracycline in seawater at two different temperatures and light intensities and the persistence of oxytetracycline in the sediment from a fish farm. *Aquaculture*, **83**, 7-16.

11). SAMUELSEN, O. B., 1990. Simple and rapid method for the determination of flumequine and oxolinic acid in salmon (*Salmo salar*) plasma by high performance liquid chromatography and fluorescence detection. *J. Chromatogr.* **530**, 452-457.

12). SAMUELSEN, O. B., LUNESTAD, B. T. & SOLHEIM, E. (1991) - The fate and microbiological effects of furazolidone in a marine aquaculture sediment. *Sci. Tot. Environ.*, (in press).

13). SAMUELSEN, O. B., TORSVIK, V. L. & ERVIK, A., 1991. Long-range changes in oxytetracycline concentration and bacterial resistance towards oxytetracycline in a fish farm sediment after a medication. *Sci. Tot. Environ.*, in press.

14). SCHLOTFELDT, H.J., NEUMANN, W., FUHRMAN, H., PORTMUELLER, K. & BOEHM, H., 1985. Remarks on increasing resistance of fish pathogenic and facultative fish pathogenic bacteria in lower Saxony (FRG). *Fish Pathol.*, **20**, 85-91.

15). TATSUMI, K., OU, T., YAMADA, H., YOSHIMURA, H., KOGA, H. & HORIUCHI, T., 1978. Isolation and identification of the metabolite of N-(5-nitro-2-furfurylidene)-3-amino-2-oxazolidone (furazolidone). *J. Pharm. Dyn.*, **1**, 256-261.

16). TORANZO, A. E., COMBARRO, P., LEMOS, M.L. & BARJA, J. L., 1984. Plasmid coding for transferable drug resistance in bacteria isolated from cultured rainbow trout. *Appl. Environ. Microbiol.*, **48**, 872-877.

17). TORSVIK, V.L., SOERHEIM, R. & GOKSØYR, J., 1988. Antibiotic resistance of bacteria from fish farm sediments. *ICES-report*, C.M. 1988/**F:10**.

18). VROOMEN, L.H.M., BERGHMANS, M.C.J., HEKMAN, P., HOOGENBOOM, L.A.P. & KUIPER, H.A., 1987. The elimination of furazolidone and its open-chain cyano-derivative from adult swine. *Xenobiotica*, **17**, 1427-1435.

19). VROOMEN, L.H.M., VAN OMMEN, B. & VAN BLADEREN, P.J., 1987. Quantitative studies of the metabolism of furazolidone by rat liver microsomes. *Toxic. in vivo*, **2**, 97-104.

20). VROOMEN, L.H.M., BERGHMANS, M.C.J., GROTEN, J.P., KOEMAN, J.H. & VAN BLADEREN, P. T., 1988. Reversible interaction of a reactive intermediate derived from furazolidone with glutathione and protein. *Toxicol. Appl. Pharmacol.*, **95**, 53-60.

Discussion

(Combined discussion with the previous complementary paper by Lunestad)

HILL We can see from the two papers just presented why there is growing concern amongst environmental protection agencies at the use of antibiotics in aquaculture. Although that was specifically a Norwegian situation, with marine cages and 35 tonnes of antibiotic used in a year, nonetheless it does have implications and it does point to potential problems of environmental adverse effects of chemotherapy in aquaculture in other areas.

LEWIN Have you done any work with oxolinic acid on the chelating effects of magnesium, because oxolinic acid is also chelated by magnesium and it has exactly the same effects in reduction of absorption from the gut in humans and increase in MIC?

LUNESTAD We have not so far been doing any experiments with oxolinic acid, just oxytetracycline so far.

RICHARDS The licensed treatment dose rate for oxytetracycline and oxolinic acid in both fresh and salt water is the same in the UK. Following the findings of Dr Lunestad and the comment just made, is there a difference made between the dose rates used in fresh and salt water in Norway?

LUNESTAD As far as I know, it is the same.

WITHNELL I was wondering, have any studies been carried out on sediments and fauna close to major sewage outfalls?

SAMUELSEN No I don't think so, not to my knowledge.

WITHNELL I was wondering whether you might find high levels of drugs used in human medicine in those sediments and fauna?

THOMEN There is a girl at the University of Copenhagen who is studying this question.

HILL But no information available yet?

THOMEN I think she has published some information already.

LEWIN Essentially you to tend to find, where there is very wide indiscriminate use of antibiotics a high degree of resistant organisms in and around sewage. In developed countries in the western world, the antibiotic consumption is relatively low and you don't actually seem to find that sort of problem, but you do in certain parts of the world.

FIJAN I would like to know what is the fate of these antibiotics in animal faeces in the terrestrial situation.

MIDTVEDT I can give you a general comment. None of the quinolones are broken down in faeces in any animals or human beings. You do have just a very few bacteria (as was mentioned by Lunestad) which are capable of breaking down the tetracyclines. You may have some of these bacteria in humans and animal species but they are not that active.

ANON Just for information from Dr Lunestad. He presented some figures on the concentration of the drug in fish in crabs and mussels. Were these wild animals?

LUNESTAD Yes, all of them were wild fish.

ANON Near a farm?

LUNESTAD Just at the farm site, underneath the farm. We have also examined for oxolinic acid in wild fauna up to 400 metres away from the farm and found rather high concentrations of the drug.

LYDIA BROWN Really in Norway's defence, although 37 tonnes of antibiotic seems a lot, they do appear to be the only country in the world that actually have strict control of the amount of drug that is used and they do seem pretty much know every kilogram that is used on fish farms. I think that we should point that out, because some other countries in the world that have aquaculture, quite frankly don't have much real idea of how much antibiotic is used. My question is to Dr Lunestad. You quoted, in respect of the use of oxolinic acid, where it was detected in wild fish, and you said the levels that you found exceeded the drug concentrations tolerated in food for human consumption. Who has set this limit and what is the limit? I was unaware that JECFA had made any MRLs for any quinolone?

LUNESTAD The limits employed differ in different countries as far as I know, but the concentrations we found were 1000 times more than should be found in food for human consumption. As far as I know the limit in Norway is 10ppb or something. I don't know if it is official, but that's approximately what is used. We found 10ppm and even more in fish.

DEAR Has anybody ever tried measuring these antibiotics in the people who are actually administering it to the fish? We spend a lot of time looking in the mussels and the in the fish and, with respect to M. Giroud's paper, should we not be spending some time looking at the risk to the user?

LUNESTAD That is a good idea.

HILL So long as it is non lethal sampling it should be OK!

ECOLOGICAL EFFECTS OF ANTIBIOTICS AND CHEMOTHERAPEUTANTS FROM FISH FARMING

P. Kupka Hansen*, B.T. Lunestad*, & O.B. Samuelsen**

* Department of Microbiology and Plant Physiology, University of Bergen, Jahnebakken 5, N-5007 Bergen, Norway
** Department of Pharmacology and Toxicology, University of Bergen, Armauer Hansens Hus, N-5021 Bergen, Norway

Summary: *The effects of four antibacterial agents (oxytetracycline (OTC), oxolinic acid (OXA), flumequine (FLU) and furazolidone (FUR)) on fish farm sediment were investigated in a mesocosm experiment. The residence time of the compounds in the sediment, the total bacterial numbers, the bacterial resistance towards the antibacterial agents and the anaerobic bacterial activity was measured.*

Résumé: **Conséquences écologiques des traitementts antibiotiques employés en aquaculture.**

L'emploi croissant des antimicrobiens en aquaculture entraîne l'accumulation dans les sédiments de produits actifs ou dégradés, qui résultent de gaspillages ou du rejet fécal des poissons. Le devenir de 4 antibiotiques (oxytétracycline, acide oxolinique, fluméquine et furazolidone) a été suivi expérimentalement dans des réservoirs de 200 l alimentés en continu par de l'eau à 6-9°C, et garnis d'une couche de 10 cm de sédiments . Des échantillons de sédiments étaient régulièrement analysés en chromatographie HPLC, la flore microbienne étant estimée quantitativement, et qualitativement pour sa résistance aux antibiotiques. La réduction des sulfates était également étudiée.

Bien que 50% de l'oxytétracycline et de l'acide oxolinique contaminant les sédiments disparaissent en 15 jours après traitement, il en persiste 5 à 10% au bout de 6 mois. La furazolidone par contre n'est plus détectable après 4 jour. La flore sédimentaire, détruite à 50% par les traitements, retrouve son niveau initial après 11 semaines. On observe une nette augmentation des résistances à l'oxytétracycline et à l'acide oxolinique, la dernière présentant un caractère croisé avec l'oxytétracycline et la fluméquine. L'activité de réduction des sulfates, due aux bactéries anaérobies, est également affectée et se restaure totalement en 70 jours. Ces études soulignent que la durée de détection de l'oxytétracycline et de l'acide oxolinique dans les sédiments se prolonge bien au-delà de leurs effets immédiatement perceptibles sur la flore microbienne ambiante.

Introduction

The enhanced use of antibiotics/ chemotherapeutics in aquaculture has increased the need for knowledge on the effects of these substances in the aquatic environment. Norway has, as the only country so far, precise statistics on the use of antibacterial agents in aquaculture (3). Drugs provided in fish feed are introduced to the environment either through excess feed or through fecal pellets. These particles sink to the sea bottom, and accumulation of organic waste and antibacterial agents can occur under fish farms (5)

To study the effects of antibacterial agents on marine sediments under controlled conditions an experimental system, consisting of mesocosms, was constructed.

Materials and methods

The mesocosms were six 200 l circular polyethylene tanks (height: 50 cm, diameter: 70 cm) containing 10 cm of sediment from under a fish farm and 115 l of water. Two tanks were used as controls, but in the rest of the tanks the fish farm sediments were added a dose of either oxytetracycline (400 ppm), furazolidone (400 ppm), oxolinic acid (100 ppm) or flumequine (100 ppm) which was mixed into the sediment. Constant seawater flowthrough controlled the temperature (6-9°C from July to April) and ensured water exchange (320 l per hour per tank). Sediment was sampled in plexiglass tubes (diameter: 3 cm, height: 25 cm).

Residue analysis of antibiotics/ chemotherapeutics was measured by High Performance Liquid Chromatography (HPLC).

Total bacterial numbers were measured applying DAPI-staining. Bacterial resistance was measured by plating sediment samples on media containing the various antibacterial agents.

Sulfate reduction rates were measured by incubating the sediment with a radioactive sulfate isotope and then measuring the hydrogen sulfide produced (2). Sulfate reduction was measured through the entire sediment column in 2 cm aliquots.

Results

During the first two weeks after medication, fifty percent of the oxytetracycline and the oxolinic acid in the sediments disappeared. This was probably due to washing out, degradation of these compounds in the sediment has not yet been reported. During the next six months most of the remaining residue vanished, leaving five to ten percent of the initial concentration (see O.B. Samuelsen Fig.2 this vol.).

Furazolidone was rapidly degraded and only remained in the sediment in an active form for about four days (see O.B. Samuelsen Fig. 1 this vol.).

The flumequine samples have not yet been analysed.

Table I shows total bacterial numbers in all the sediments. Two days after medication the total bacterial numbers in the treated sediments were reduced by up to fifty percent relative to the control. During the next eleven weeks the number of bacteria slowly increased towards the initial values.

Table I. Total bacterial numbers in all sediments as percent of the control at different times after medication. Numbers in brackets refer to total bacterial numbers in ml/g wet sediment.

	DAY 2	DAY 7	DAY 24	DAY 80
CONTROL	($3.0x10^{10}$)	($3.3x10^{10}$)	($4.3x10^{10}$)	($3.2x10^{10}$)
CONTROL	100 %	100 %	100 %	100 %
OXA	63 %	65 %	77 %	84 %
FLU	67 %	71 %	93 %	84 %

Table II shows the bacterial resistance towards the investigated antibacterial agents. The bacteria in the control sediment showed only very low levels of resistance. Tests on the oxytetracycline-treated sediment showed elevated levels of bacterial resistance to oxytetracycline, relative to the control, shortly after medication. The oxolinic acid treated sediment, when tested for bacterial resistance towards oxolinic acid, already showed a high resistance level two days after treatment. After eighty days an elevated level of bacterial resistance could still be observed in all medicated sediments. Additional tests of multiple resistance to all the investigated antibacterial agents were performed on all the sediments. Bacteria exposed to oxolinic acid developed cross resistance towards flumequine and oxytetracycline (not shown).

Table III shows the sulfate reduction rates in all the investigated sediments. Bacterial activity was measured as sulfate reduction, since this is one of the most important degradation pathways of organic matter in these sediments.

Table II. Bacterial resistance towards the investigated antibacterial agents in percent of plate counts at different times after medication.

	DAY 2	DAY 7	DAY 24	DAY 80
CONTROL	0.05 %	0.07 %	0.21 %	0.13 %
OTC	2.8 %	13.5 %	-	8.7 %
OXA	14.2 %	8.5 %	20.3 %	2.7 %
FLU	0.2 %	0.4 %	-	4.0 %

Seven days after medication, there was no activity in the treated sediments. Twenty-nine days later, sulfate reduction was present in all sediments, although the rates were lower in the treated sediments than in the control. After seventy days the sulfate reduction rates in the treated sediments were indistinguishable from the the control.

Table III. Sulfate reduction rates (mM SO4/m2/day) in all sediments at different times after medication

	DAY 7	DAY 29	DAY 70
CONTROL	14	312	309
OTC	0	55	412
OXA	0	18	541
FLU	0	18	295

Discussion

By introducing antibacterial agents to the sediment a considerable reduction in the total number of bacteria occurs. The results of the sulfate reduction measurements suggest that the number of sulfate reducing bacteria is reduced. The inactivation of this group of anaerobes is complete seven days after medication. At this point half of the initial amounts of oxytetracycline and oxolinic acid have been flushed out of the sediment.

The culturable fraction of the aerobic bacteria in the sediment establish an elevated level of resistance towards the antibacterial agents and keep it elevated as the concentration of the compounds slowly decreases. The sulfate reducing bacteria may similarly develop resistance since they regain activity after four weeks when there still is about forty-five percent of the oxytetracycline and oxolinic acid left in the sediment. However, the resistance of anaerobic bacteria was not measured.

Cross resistance between oxyetracycline and oxolinic acid has previously been described for fish pathogenic bacteria by Barnes *et al.* (1) and for sediment bacteria by Nygaard *et al.* (4). This is in accordance with our results.

Conclusions

Oxytetracycline and oxolinic acid are not degraded in the sediment and can be detected up to seven months after medication. The addition of antibacterial agents to sediments gives a dramatic initial reduction in the total number of bacteria present. An elevated number of bacteria resistant towards oxytetracycline and oxolinic acid in the sediment could be detected for at least eleven weeks.

Bacterial activity, measured as sulfate reduction, was strongly inhibited by medication but returned gradually to control levels within ten weeks.

References

1). BARNES A.C., LEWIN, C.S., HASTINGS,T.S. & AMYES, S.G.B. 1990. *FEMS Microbiology Letters*, **72**, 337-340.

2). FOSSING H. & JOERGENSEN, B.B. 1989. Measurement of bacterial sulfate reduction in sediments: Evaluation of a single-step chromium reduction method. *Biogeochemistry*, **8**, 205-222.

3). GRAVE K., ENGELSTAD, M., SOERLI, N.E. & HÅSTEIN, T. 1990. Utilization of antibacterial drugs in salmonid farming in Norway during 1980 - 1988. *Aquaculture*, **86**, 347-358.

4). NYGÅRD K., B.T. LUNESTAD, H. HEKTOEN, J.A. BERGE and V. HORMAZABAL. *Aquaculture* (submitted).

5). SAMUELSEN O., V. TORSVIK, P. KUPKA-HANSEN, K. PITTMAN and A. ERVIK, 1988. Organic waste and antibiotics from aquaculture. ICES C.M. 1988/F:14.

CONTROL OF THE USE OF THE INSECTICIDE DICHLORVOS IN ATLANTIC SALMON FARMING

J.G. McHenery, W.R. Turrell & A.L.S. Munro

The Scottish Office Agriculture and Fisheries Department
Marine Laboratory, PO Box 101, Aberdeen AB9 8DB, UK

Summary: *Unrestricted release of the insecticide dichlorvos (DDVP) into the coastal marine environment after it has been used to treat lice infestation of farmed Atlantic salmon may pose an environmental hazard. Techniques for the derivation of safe levels are described and how these are used to set Environmental Quality Standards (EQS). Two EQS levels are currently proposed, an average annual value of 40 ng/l which will regulate total annual usage and a 24 hr value of 600 ng/l which will regulate daily usage. The appli-cation of these EQS levels to specific sites is dependent on local hydrographic factors which are described. Final application of the EQS levels is dependent on theoretical modelling studies.*

Résumé: **Surveillance de l'emploi du dichlorvos dans les élevages de saumon atlantique.**

L'élevage en cages du saumon atlantique en Ecosse souffre des attaques de 2 copépodes, **Lepeophtheirus salmonis** *et* **Caligus elongatus**, *qui causent une mortalité et déprécient les poissons par leurs blessures. On les contrôle par l'Aquagard, qui contient 50 p.100 de dichlorvos, un insecticide organochloré récemment autorisé sur prescription vétérinaire. Cependant les effluents relargués (le traitement d'une cage représente 500 à 1000 m³ d'eau à 1 ppm de dichlorvos) ont vite suscité des craintes pour l'environnement. En accord avec le Control of Pollution Act de 1974, il a fallu s'attacher à rassembler de nouvelles données.*

La demie-vie du dichlorvos est de 4 à 7 jours en eau de mer. Ses dérivés ne sont pas actifs. L'étude dynamique de la dilution du produit devait donc permettre d'établir des seuils de concentration tolérables pour l'environnement. Aussi a-t'on essayé de modéliser les forces physiques en action dans les lochs écossais où se pratique l'élevage en cages. La dilution s'opère dans des masses d'eau soumises à l'influence des marées et des courants, mais il faut également tenir compte des apports d'eaux douces fluviales, qui stagnent en surface en créant des gradients de mélanges horizontaux et verticaux plus ou moins importants. La dispersion est affectée par les courants, les tourbillons, et les forces de cisaillement résultant de la rencontre de courants distincts.

La toxicité du dichlorvos constitue surtout une menace pour les organismes planctoniques, telles les larves de crevettes et de homards. On peut l'évaluer par les mortalités occasionnées, mais aussi sur des échantillons, par mesure de l'activité cholinestérasique. Des espèces-sentinelles (larves de homards, moules) ont été spécialement soumises à surveillance. Jusqu'à présent, on n'a pas détecté de différence entre les communautés d'invertébrés provenant de lacs soumis ou non à des rejets.

Pour prévenir tout risque à court ou long terme, des standards de qualité de l'environnement, fondés sur la sensibilité au produit des larves de homard, sont désormais imposés aux élevages: L'emploi du dichlorvos ne doit pas entraîner de rejet supérieur à la dose moyenne de 600 ng/l en 24 heures (1/5 de la concentration affectant l'acetylcholinestérase en 24 heures), et de 40 ng/l en 1 an (1/5 de la concentration tolérée pendant 23 jours). Ces normes ne peuvent guider l'administration et la périodicité des traitements que dans des situations où les phénomènes de dilutions sont conformes au modèle décrit. Des études complémentaires seront requises pour les adapter spécifiquement à chaque site d'élevage.

Introduction

In the sea cage rearing stage of Atlantic salmon farming two species of louse (*Lepeophtheirus salmonis* and *Caligus elongatus*) infest the fish. Unless infestation is controlled the lice cause such serious skin damage that fish may die or be unmarketable. Treatment of this condition in the UK is achieved by the use of a proprietary veterinary medicine, trade name Aquagard (Ciba-Geigy), containing 50% of the insecticide dichlorvos (DDVP) as its active ingredient, to be used as a bath formulation. Consequent upon a single cage treatment volumes of approximately 500-1,000 m^3 of sea water containing 1 ppm DDVP are released. As many farm sites may contain 50 or more cages the quantity of DDVP released to the environment over a short period of time may be significant. This paper considers information on the impact of DDVP on biological components of the marine ecosystem and how this information may be used to determine acceptable levels for release of DDVP.

Dichlorvos

DDVP is an organophosphorous compound (2,2 dichloroethenyl dimethylphosphate) with insecticidally active properties at low concentrations caused by the inhibition of the enzyme acetylcholinesterase (AChE). It has been widely used since the 1960s as an insecticide for crop protection, in animal husbandry for deinfesting animal houses, in human hygiene eg as an ingredient in fly sprays and as an ectoparasiticide for cattle, dogs and cats.

Dichlorvos as a Fish Medicine

Recently reclassified under UK law as a medicine for treatment of lice on fish Aquagard had to be licensed by the appropriate Ministers on the recommendation of the Veterinary Products Committee. Aquagard was licensed under that category of medicine which requires that it be prescribed only by the veterinarian responsible for the care of the fish. To give this recommendation the Committee required significant new data to be collected about the effects of DDVP in the marine environment and to be convinced that the agencies responsible for the protection of the environment had appropriate powers and knowledge to carry out their duties. Control of the release of fish medicines to the environment is by powers under the Control of Pollution Act 1974. Discharges from farms are controlled by consents issued by the relevant Pollution Control Authority. It is recognised that the use of dichlorvos is potentially dangerous to marine life. The basis of the control of its use depends on the assessment of data on its toxicity and behaviour, and based on this, the setting of Environmental Quality Standards (EQS) to protect marine life.

Dichlorvos in the Sea Water Environment

DDVP has a half life of 4-7 days in sea water (11). As the degradation products of DDVP are not considered to pose a toxic hazard they are not considered further. The steps in achieving safe environmental concentrations will be dilution and control of input. It is therefore important to establish how quickly and what factors are important in achieving dilution to a safe level. To achieve safe concentrations of DDVP dilution is obtained through naturally occurring hydrographic dispersive forces. Therefore a dual study of the hydrographic forces controlling dispersion of the large volumes released from fish cages and what should be considered a safe level of DDVP were initiated with the aim of defining Environmental Quality Standards (EQS).

Hydrographic Forces

The physical forces which disperse the released DDVP are complex being three-dimensional and time varying. To rationalise the diversity of this problem simple models of the first order physical processes occurring in Scottish sea lochs, the most common area for cage sites, were used to estimate the dilution of dichlorvos following simulated release from sea cages (13). Sea lochs are fjordic inlets characterised by a deep inner basin, often separated from the sea by a shallow sill. The principal physical forces which occur within a sea loch are advection and dispersion.

Advection: This refers to the gross movement of the volume or patch of water released from the treated cage. This movement can be brought about by tidal and residual currents. Tidal currents are those movements within the loch that occur twice each day and are driven by the rise and fall of the coastal ocean at the loch's open entrance. Residual motions are considered to be those currents with a period longer than that of the tide. They can arise from such factors as the tide itself, wind forcing and from the results of fresh water runoff. Wind may drive surface currents along the loch by direct application of frictional stress. These wind driven currents will be greatest at the surface.

Of fundamental importance within a fjord-like system such as a sea loch is the continuous vertical circulation driven by the addition, often at the head of a loch, of significant amounts of fresh water from land drainage. This fresh water, being more buoyant than the coastal water remains at the surface since in the deep slow moving parts of the loch there is insufficient energy supplied by the tide to mix this water through the water column. The depth of this surface layer where the DDVP is released may be determined by the amount of tidal energy available, the strength of the wind and the magnitude of the runoff. The horizontal gradients caused by the addition of fresh water may force vertical circulation, with flow towards the sea in the upper surface layer. The salinity of the upper layer increases towards the sea as saline water is entrained into it from below. Thus, a deep counter-current may be formed flowing towards the head of the loch to replace the entrained fluid.

Dispersion: This refers to the spread or dilution of a volume or patch of water following release while it is being advected around the loch by the currents acting on it. Molecular diffusion is negligible on this scale of activity. Of importance is the effect of turbulent eddies. These exist on many different scales. In the horizontal the largest size will be determined by the width of the loch, with a full spectrum of sizes down to smallest whorls and vortices of microturbulence. Their energy is predominantly derived from the tide. Eddies much smaller than the patch have little effect on its dilution, while eddies much larger advect the patch around the loch. Eddies of a similar size to the patch will stretch parts of the patch out into clearer water, while moving water containing less DDVP into the patch. Thus the concentration in the patch decreases while the size of the patch increases. As the size increases, the larger eddies may then act upon the patch and increase the rate at which it spreads. The size and nature of tidal eddies will alter along a loch with protruding headlands, and shallow sills.

Turbulence also exists in the vertical dimension. Here the maximum size of the eddy is often set by the depth of the upper layer as it is difficult for eddies to overturn the stable density gradients that exist at the pycnocline (the boundary between waters of differing temperature and/or salinity). Vertical mixing may derive energy from the wind and hence will vary with time, and will alter with total water depth, becoming more intense in the shallows.

In addition to dilution by turbulent diffusion shear dispersion is also important. This force results from the interaction between turbulent diffusion itself and the vertical and horizontal changes and shears of the tidal and residual currents as described. Shear dispersion results in the stretching out of a three-dimensional patch along the axis of the current shear. If it was generated by the wind the patch would elongate along the direction of the wind. An account of the modelling of these forces and measures of dilution has been reported (13, 14, 15).

Measures of DDVP Toxicity

Because of dispersion of DDVP in the water column planktonic, rather than benthic, organisms are deemed at greatest risk. Determination of mortality and LC_{50} values for examples of a resident planktonic crustacean (lobster larvae, *Homarus gammarus* L.), mollusc (*Mytilus edulis* L.), fish (herring larvae, *Clupea harengus* L.) and other invertebrates showed that the crustacean was the most sensitive (3, 8, 9) with a 552 hr LC_{50} of 1.25 µg/l (7). The sensitivity of crustaceans to dichlorvos is well established with the 96 hour LC_{50} of 4 µg/l for the sand shrimp, *Crangon septemspinosa*, (Eisler, 1970) being the lowest reported for a marine animal for this exposure period. The ability of organophosphates to inhibit AChE activity provides another more sensitive indicator of DDVP exposure than mortality effects, and has been utilised in a number of studies (1, 6).

Where a potentially dangerous compound is already in use ecological surveys provide the ultimate bioassay (12). Based on invertebrate community surveys in Scottish sea lochs without fish farms and with farms using DDVP (10) reported no obvious differences.

Another alternative means of assessing potential impact is the use of sentinal species which are selected because of their known sensitivity. This method does not require the extended exposure required for ecological surveys and may be used during field trials on prospective medicines. This approach has been reported by two groups in relation to fish farming. In studies in which larvae of the American lobster, *Homarus americanus,* were deployed in the vicinity of farms using DDVP in Canada, no significant mortalities were recorded (2). In studies conducted in Scotland larvae of the common lobster and mature mussels were deployed in the vicinity of farms administering DDVP. Significant mortalities of the lobster larvae were observed outwith the treatment cages at one farm (5). In the same studies AChE levels were depressed in sentinal groups of larval lobsters and mussels indicating that DDVP exposure had been significant at both farms (5). The results of the AChE studies were in broad agreement with the chemical determinations conducted at the same sites (5).

Conclusions

The periodic use of dichlorvos by fish farmers in Scottish sea lochs is recognised in the proposal to set two Environmental Quality Standards (EQS): an annual average value of 40 ng/l, and a 24 hour value of 600 ng/l. These values were determined to protect indigenous life from adverse effects of both the potential dangers of long term exposure to low concentrations, and the periodic inputs of relatively high concentrations found following release on completion of treatments at farms. The first standard was determined by applying a division factor of 100 to the 96 hour LC_{50} value for the sand shrimp, 4 µg/l. Whilst this species is not found in British waters the values were set on the basis that any derived values must relate to the most sensitive indicators known to protect the environment, the application factor of 100 providing a further margin of safety. The 40 ng/l standard was also derived by applying a factor of five to the maximum acceptable toxicant concentration (MATC) for effect on lobster larvae AChE levels over 23 days exposure (7). This value will control the quantities of DDVP used per year at a given site. The second EQS value is derived by applying a factor of five to the MATC for effect on lobster larvae AChE levels over 24 hours exposure (8). This EQS value, 600 ng/l, will regulate the amount of DDVP which can be administered each day per site. The EQS values related to all areas outwith a defined mixing zone around the site of application in which it is recognised that the EQS values may be exceeded. The application of these values to specific sites will depend on measures of certain local parameters, their use in modelling studies backed by field measures of concentrations and dispersion (14, 13, 15).

References

1). COPPAGE, D.L. & MATTHEWS, E., 1974. Short-term effect of organophosphate pesticides on cholinesterases of estuarine fishes and pink shrimp. *Bull. Environ. Contam. Toxicol.*, **11**, 483-488.

2). CUSACK, R. & JOHNSTONE, G., 1989. A study of dichlorvos, a therapeutic agent for sea lice. ERDA (Nova Scotia) Report No. 14, 56pp.

3). EGIDIUS, E. & MOSTER, B., 1987. Effect of neguvon and nuvan treatment on crabs (*Cancer parugus*, *C. moenas*) lobster (*Homarus gammarus*) and the blue mussel (*Mytilus edulis*). *Aquaculture*, **60**, 165-168.

4). EISLER, R., 1970. Acute toxicities of organochlorine and organophosphorous pesticides to estuarine fish. Bureau of Sport Fisheries and Wildlife Technical Paper No 46, Washington DC, 12pp.

5). McHENERY, J.G., 1990. Effects of dichlorvos treatment at fish farms upon lobster larvae and mussels maintained in cages. *Scot. Fish. Work. Pap.*, No. 10/90.

6). McHENERY J.G. & FRANCIS, C., 1990. Effects of repeated short exposures to dichlorvos on lobster larvae. *Scot. Fish. Work. Pap.*, No. 9/90.

7). McHENERY, J.G. & FRANCIS, C., 1990. Toxicity of dichlorvos to stage 4 *Homarus gammarus* larvae. *Scot. Fish. Work. Pap.*, No. 8/90.

8). McHENERY, J.G., SAWARD, D. & SEATON, D.D., 1990. Toxicity of dichlorvos to larvae of the common lobster (*Homarus gammarus* L.) and herring (*Clupea harengus* L.). *Scot. Fish. Work. Pap.*, No. 6/90.

9). McHENERY, J.G., FRANCIS, C., MATTHEWS, A., MURISON, D. & ROBERTSON, M., (1990. Comparative toxicity of dichlorvos to marine invertebrates. *Scot. Fish. Work. Pap.*, No. 7/90.

10). MURISON, D.J., MOORE, D.C., DAVIES, J.M. & GAMBLE, J.C., 1990. Survey of invertebrate communities in the vicinity of salmon farm cages in Scottish west coast sea lochs. *Scot. Fish. Work. Pap.*, No. 11/90.

11). SAMUELSEN, O.B., 1987. Aeration rate, pH and temperature effects on the degradation of trichlorfon to DDVP and the half lives of trichlorfon and DDVP in sea water. *Aquaculture*, **66**, 373-380.

12). SPRAGUE, J.B., 1971. Measurement of pollutant toxicity to fish-III. Sublethal effect and "safe" concentrations. *Water Research*, **5**, 245-266.

13). TURRELL, W.R., 1990. Simulation of advection and diffusion of released treatments in Scottish sea lochs. *Scot. Fish. Work. Pap.*, No. 16/90.

14. TURRELL, W. & MUNRO, A.L.S., 1988. A theoretical study of the dispersal of soluble and infectious wastes from farmed salmon net cages in a hypothetical Scottish sea loch. ICES CM 1988/F:36.

15). WELLS, D.E., ROBSON, J.N. & FINLAYSON, D.M., 1990. Fate of dichlorvos (DDVP) in sea water following treatment for salmon louse, *Lepeophtheirus salmonis*, infestation in Scottish fish farms. *Scot. Fish. Work. Pap.*, No.13/90.

Discussion

HILL Thank you Dr Munro. I think that is a very good example of the type of sophisticated environmental modelling work that is increasingly going to be called for, not just with dichlorvos of course, but with any chemicals that are being used in the aquatic environment for treatment of disease etc. As the authorities start seeing this type of approach, I am sure that they will ask for this type of investigation to be done before any large scale usage is permitted, such as, for example, with Atlantic salmon in marine cages.

HORSBURG According to my information the annual production of salmon in Scotland was about 45,000 tonnes in 1990 and 160,000 tonnes in Norway. But 15,000 kilograms of dichlorvos were used in Scotland, while only 3,416 kilograms were used against sea lice in Norway. I was wondering, knowing the that structure of marine fish farming is not very different in Scotland and Norway, about this very big difference. First of all are the figures unreliable for Scotland, and secondly do you have any explanation for this difference?

MUNRO Our surveys suggest that two out of three farms are using dichlorvos. I think you have got to recognise that sea lice are a bigger problem in Scotland. Although you say the structure of the industry is the same, it doesn't necessarily mean that the ecological situation is the same. Also, as I pointed out to you, we have quite large numbers of sites in some sea lochs. Those sea lochs have runs of wild Atlantic salmon which are a continuous seed supply of *Lepeophtheirus*. It does seem that, once you introduce farmed stocks, and if you don't operate a system of rotational stocking (a fallow period in which the levels of juvenile lice or lice eggs have time to die off in the system) so that you have a continuous supply of fish that the lice can multiply on, then the problem becomes that much worse. This is a situation which has applied in many of our farm sites. The large numbers of fish in these areas of restricted water exchange cause the numbers of lice populations to build up and therefore the need to use dichlorvos for control has been very great indeed. Our farmers have very much taken the situation to heart and it has been obvious from stocking patterns in 1989 and 1990 that the industry is attempting to manage the numbers of stocks and the numbers of fallow periods in such a way to minimise the numbers of lice. My own experience of one particular sea loch where there were at one time three commercial sites. We had our own experimental cage, largely to get supplies of sea lice. The three commercial sites have been fallow over the last year and we have had great difficulty in getting any numbers of lice over the last year from the cages that we have there. This from populations of salmon from which in the past we have been able to get good supplies of lice, simply because of the removal of the commercial sites. At least that is what we deduce. So, I think that our local conditions are a little different from Norway. The structure of the industry is the same, but the management practices have been a little different in our situation.

HILL We will move on now to look at another side of chemotherapy and that is the adverse effects on the fish themselves. We heard a little earlier about possible toxic effects on fish, now we have a paper on what is a more subtle effect, that is, a detrimental effect on the immune system.

THE MANAGEMENT OF ENVIRONMENTAL RISK: A CASE STUDY BASED ON THE USE OF DICHLORVOS TO CONTROL SEA LICE INFESTATIONS ON ATLANTIC SALMON

J.S. Buchanan

Scottish Salmon Growers Association, Drummond House Scott Street, Perth, Scotland, UK

Summary: *The measures which have evolved to manage the risks to the ecology of Scottish sea lochs that might arise from the use of dichlorvos to control sea lice infestations on salmon are reviewed. The work of the Department of Agriculture and Fisheries for Scotland (DAFS) in assessing hazard and in setting margins of safety is discussed and the role of pressure groups in bringing about legislative changes is illustrated. This case study shows the complicated nature of environmental risk management when technical, political and economic considerations combine.*

Résumé**: Gestion des risques pour l'environnement: - Cas du dichlorvos, employé dans la lutte contre les copépodes parasites du saumon atlantique.**

Le développement des élevages de saumon atlantique au cours des 20 dernières années s'est accompagné de pathologies que les résistances aux traitements et les restrictions à l'emploi des médicaments ne permettent pas de contrer efficacement. L'exemple du dichlorvos illustre bien les types d'obstacles auxquels se heurte l'adoption de toute nouvelle thérapeutique. Son emploi contre Lepeophtheirus et Caligus fut proposé au milieu des années 70. Son statut de pesticide exigeait alors d'en garantir l'efficacité et l'innocuité pour le consommateur et l'environnement. Reclassé dans les médicament en 1985, et régi par les dispositions du Medicine Act, il dut être soumis à de nouvelles procédures d'autorisation et n'obtint sa licence qu'en 1989. A ce moment, la prescription médicale s'en trouvait limitée par l'application de la directive CEE de 1981, et de vives campagnes écologistes parvenaient à le faire classer dans la liste rouge du Ministère de l'Environnement, subordonnant son emploi au contrôle des Conseils d'Epuration des Rivières. Son autorisation n'en a pas moins suscité des réactions et ne vaut que jusqu'en 1992, malgré l'effort des producteurs et des industriels pour financer des études d'impact ou organiser des programmes d'information et de formation sur la manière de traiter. L'inscription sur la liste rouge constitue un sérieux handicap, car il n'est pas besoin de démontrer le caractère nocif des produits concernés pour leur appliquer des restrictions. C'est aux utilisateurs qu'il appartiendra de prouver l'innocuité du dichlorvos ou de promouvoir des programmes de recherche destinés à lui trouver des substituts moins contestés.

Introduction

Over the past twenty years the production of farmed Atlantic salmon has increased almost exponentially reaching a total of 33,000 tonnes in Scotland last year. This phenomenal expansion has resulted in the industry becoming a victim of its own success. Pests and diseases are becoming increasingly difficult to control because of restrictions on the use of medicaments to overcome them and the development of resistance to existing medicines. Nonetheless, the industry is being targeted by numerous pressure groups on the grounds that it is insufficiently regulated. The present paper illustrates this dilemma by reviewing the case of dichlorvos (an organophosphorous pesticide) used by the industry as an ectoparasiticide for at least the last fifteen years (1,2).

Environmental activists have campaigned for precautionary safeguards to protect the ecology of Scottish lochs and have long called for an immediate ban on dichlorvos (3,4,5). In certain cases these demands have now been modified following the introduction of protective measures designed to minimise the risk (6,7).

This paper reviews the sequence of events to date in this conflict between (a) legislators divided between a fear of underestimating toxicity and a state of ultra-caution which might lead to unnecessary restrictions on a fledgling industry (b) licensing authorities whose remit does not allow considerations of economic need (c) regulatory authorities whose state of knowledge and resources have not kept pace with the expansion of the industry and (d) the forceful and skilfully applied political pressures from environmental activists. The measures that have evolved to manage risks to the environment are viewed against this background.

Environmental risk assessment

The assessment of risk requires a value judgement regarding the probability of harm to the aquatic environment resulting from known or potential concentrations of a substance in that environment. The skill of the assessor lies in setting margins of safety necessary to protect marine life having due regard to the results from toxicity testing, the type of pollutant and the conditions of its use (8).

The management of risk

Once the degree of exposure to a hazard has been quantified and a judgement has been made regarding the acceptability of the risk then it is important to put in place measures necessary for the management of that risk in order to minimize the likelihood of harm to the environment.

Assessing the hazards of non-bioaccumulatory, non- persistent chemicals depends upon the difference between "exposure" concentrations and the "no-effects" concentration (NOEC). In practice these are usually as taken 1/100 of the 96 hour LC_{50} or 1/10 of the NOEC (8).

Dichlorvos as an ectoparasiticide

Epizootics of sea lice have continuously plagued salmon in sea pens. The species causing most damage is *Lepeophtheirus salmonis* although *Caligus elongatus* can be troublesome. In the mid 70's a control method was developed by Rae and others (1) involving the use of dichlorvos as a 1 ppm bath for 1 hour and was completely effective. Dichlorvos was available as NOGOS 50EC^R (Ciba Geigy) this was later superceded by NUVAN 500EC^R a 50/50 mix of di-chlorvos and dibutylphthalate.

This commonly used pesticide was then under a limited clearance from the Pesticides Branch of the Ministry of Agriculture Fisheries and Food (MAFF) UK Pesticides Safety Precautions Scheme (UKPSPS). It was necessary to satisfy an independent body of experts on environmental and consumer safety as well as on the quality and efficacy of the product in order to gain this licence. Only one company (Marine Harvest International) was actually licenced to treat sea lice although the use of dichlorvos as NUVAN became widespread throughout the rapidly expanding salmon farming industry.

Dichlorvos reclassified as a medicine

In 1985, the MAFF UK PSPS scheme was superceded by the Food and Environment Protection Act (FEPA). Efforts to continue using dichlorvos as a pesticide were thwarted by the realisation that by statutory definition dichlorvos (NUVAN) had to be treated as a medicine when added to water to control ectoparasites on fish. It now came under the Medicines Act of 1968 and 1971 and was required to be licensed as a veterinary medicine. A product licence application was lodged by Ciba Geigy with the MAFF Veterinary Products Committee (VPC). The VPC is empowered to consider aspects of safety to the treated animal, the person administering the product as well as the consumer of the food from the treated animals and the environment generally, in addition to quality and efficacy; the question of need is specifically excluded. (MAFF pers. comm.) From 1st January 1988 dichlorvos as NUVAN was only available to treat fish under written veterinary direction. However, it was not until June 1989 that it became licenced as AQUAGARD SLT ¨.

Restrictions on the "right to prescribe"

Under Section 9(2) of the Medicines Act, veterinarians were allowed to prescribe any medicine or animals under their care, whether licensed or unlicensed, providing that it was specially prepared to order. Following the promulgation of the proposals for the amendment of the EC Directive 81/851 regarding Veterinary Medicines, a much stricter interpretation of the law was issued to UK veterinarians on the grounds that prescribed use of a number of compounds was proceeding with no official assessment of their environmental impact, the health and safety of operatives and of possible residues reaching consumers.

The influence of pressure groups

What appeared to be an exploitation of a loophole in the regulations in the case of dichlorvos caused a number of conservation groups to spring to the defence of the environment (3,4,5). The strength of their feeling was evident to the media, politicians and regulatory bodies. Intense lobbying resulted in the inclusion of dichlorvos on the department of the Environment's so called "Red List" in a paper entitled "Inputs of Dangerous Substances to Water; proposals for a unified system of control". When challenged, the Department of the Environment justified the inclusion of dichlorvos (a non-persistent, non-bioaccumulatory compound of low input to the environment (17)) on the grounds that it was toxic to marine life forms below the current limit of detection (then 1 µg/l) and that the effects of repeated short exposures on the survivability of marine larval life forms was not known. Subsequent studies (9,10,11,12,13) have shown the existence of a toxicity threshold although repeated doses at low levels resulted in enzyme inhibition and significant mortalities of larval lobsters.

The legislative framework to manage risk

The Department of the Environment's proposals for Environmental Quality, Objectives and Standards (EQOs and EQSs) have been accepted by Government and EQSs are in the process of being applied for receiving waters into which "Red List" substances are being discharged. An EQS is being discussed for dichlorvos which is likely to be 1/10 of the NOEC to lobster larvae (11) which will be the maximum permissible level and will be monitored by the River Purification Boards (RPBs). Since nanogram amounts cannot be measured, a system of mathematical modelling is being developed by DAFS. This will be based on the hydrographic characteristics, the carrying capacity and interactions within Scottish sea lochs in order to define water quality standards and to help to assess acceptable limits of environmental degradation.

The legal framework governing point source discharges is covered by discharge consents issued by those authorities responsible for protecting the purity of the aquatic environment under the Control of Pollution Act 1974. In Scotland these statutory bodies are the RPBs and Islands Councils. Until such time as an EQS is established discharge consents issued to fish farmers have been contained in an EQO stating that "No significant concentrations of toxic matter should be present at any time which would be detrimental to fish life or invertebrate life on which they depend". The RPBs require 48 hours prior notice before AQUAGARD SLT is used so that they can check that the recommended working concentrations are adhered to. Farmers are obliged to keep records of all medications and inspectors are entitled to have access to the farms at any time.

A voluntary quality assurance scheme run by Food from Britain with its own independent inspectorate has been adopted by the majority of farmers under the Scottish Quality Salmon scheme. Under this scheme random samples are taken for residue analysis by MAFF. The current limit of detection of dichlorvos in fish flesh is 1 ppb.

Risk management by training of operatives

The Scottish Salmon Growers Association (SSGA) initiated a training scheme for operatives using Aquagard SLT to be run by the Agricultural Training Board. Consultation with the manufacturer and with statutory and voluntary conservation organisations were held and a one day training scheme has been attended by over 500 operatives to date. Great stress has been laid on the safc and effective use of the chemical in accordance with the comprehensive instructions given on the label. A training video was produced to augment the scheme.

Licencing of dichlorvos as Aquagard SLT[R]

Because dichlorvos was on the "Red List" the licencing of NUVAN 500EC as AQUAGARD SLT in June 89 provoked a storm of protest from the pressure groups who accused the licensing authority of being a tool of Government. This and other concerns provoked politicians into setting up a Select Committee to look at fish farming. In 1990 this report by the House of Commons Agriculture Committee on Fish Farming in the UK recommended that DAFS should coordinate the completion of research to assess the environmental impacts of salmon farming. The Committee noted that "the banning of dichlorvos in the treatment of sea lice would have a disastrous effect on the salmon farming and would only be warranted in extreme circumstances; but the Government should bring forward proposals for eliminating its use in the medium to longer term".

The Government response to the special report was published in December 1990 (14). In Section 34, the Government noted that the current licence (expiring June 1992) was granted subject to some modifications to the label being made and to a further environmental impact report being provided by the manufacturers during 1991. The Parliamentary Secretary emphasized that studies to develop an alternative strategy to control sea lice in farmed salmon would continue. At the present time Government and industry are sponsoring research into chemical substitutes for dichlorvos, into vaccines, and into biological controls such as for sea lice such as the "cleaner" species, wrasse.

In Section 35 the Government stated that the presumption is that dichlorvos-based treatments should be replaced by suitable alternatives in order that their commitment made at the North Sea Conference to reduce by 50% the inputs of substances on the "Red List" to the marine environment can be met by 1995.

The Scottish Fisheries working papers on the effects of dichlorvos in the marine environment

In June 1990 at the same time as the Minister announced the extension of the licence for dichlorvos AQUAGARD SLT until 1992, a series of DAFS working papers were placed in the House of Commons library. Each bore the injunction "in confidence not to be quoted without the permission of DAFS Marine Laboratory and the SSGA". (9,10,11,12,13).

These studies set out to answer a number of questions set by the Veterinary Products Committee in 1989 (15) and were paid for by the SSGA and the manufacturer, Ciba Geigy. Copies of these papers can be supplied to interested parties by writing to the author at the above address or to the DAFS Marine Laboratory, Aberdeen, AB9 8DB.

A recent study by Cusak and Johnson (16) in Canada is in broad agreement with the conclusions of the DAFS Marine Laboratory findings.

Conclusion

The salmon farming industry must find alternatives to dichlorvos to combat the sea louse menace, not because the chemical has been proved to be harmful to marine life but because of UK Government's stated commitment to a precautionary approach to marine pollution and the fact that it is on the "Red List". In effect the precautionary approach requires that action be taken to control inputs of certain substances even before a causal link has been established by absolutely clear scientific evidence (18). The onus is therefore on the industry to prove that a substance is safe rather than on environmentalists to prove that it is damaging. This requirement to prove a negative will be a serious brake to progress in the development of alternatives.

Acknowledgements: The author wishes to express his gratitude to Mr Gordon Rae, Ciba Geigy Agrochemicals, the Marine Conservation Society and the Department of Agriculture and Fisheries for Scotland for their advice in the preparation of this manuscript.

References

1). RAE G.H., 1979. On the trail of sea lice. *Fish Farmer*, **6**, 22-25.

2). SAWARD D., SEATON D.D. and MCLEOD B.A., 1982. Summary of marine toxicity tests undertaken in connection with the use of dichlorvos for the control of salmon lice. *DAFS Marine Laboratory*, Aberdeen.

3). ROSS A. and HORSMAN P.V., 1988. The use of NUVAN 500EC in the salmon farming industry. *Marine Conservation Society*, Ross-on-Wye, England.

4). SCOTTISH WILDLIFE AND COUNTRYSIDE LINK, 1988. Marine fish farming in Scotland - A discussion paper. 70 pp.

5). THE NATURE CONSERVANCY COUNCIL, 1989. Fish farming and the safeguard of the natural marine environment of Scotland. *NCC Headquarters*, Edinburgh, 136pp.

6). SCOTTISH WILDLIFE AND COUNTRYSIDE LINK , 1990. Marine salmon farming in Scotland - a Review. 27 pp.

7). MARINE CONSERVATION SOCIETY, 1990. "A Future for the Coast?" Proposals for a UK Coastal Zone Management Plan. *Marine Conservation Society*, Ross on Wye, England.

8). LLOYD R. , 1988. Hazard assessment of toxicity tests - Results on marine species. *Chemistry and Industry*, **1**, 12-15.

9). McHENERY J.G., SAWARD D. and SEATON D.D., 1990. Toxicity of Dichlorvos to larvae of the common lobster (*Homarus gammarus* L) and

Herring (*Clupea harengus* L)). *Scottish Fisheries Working Paper 6/90*, DAFS Marine Laboratory Aberdeen.

10). McHENERY J.G., FRANCIS C., MATTHEWS A., MURISON and ROBERTSON M., 1990. Comparative toxicity of Dichlorvos to marine invertebrates. *Scottish Fisheries Working Paper 7/90.*

11). McHENERY J.G. and FRANCIS C., 1990. The toxicity of Dichlorvos to stage 4 *Homarus gammarus* larvae. *Scottish Fisheries Working Paper 8/90.*

12). McHENERY J.G. and FRANCIS C., 1990. Effects of repeated short exposures to Dichlorvos on lobster larvae. *Scottish Fisheries Working Paper 9/90.*

13). McHENERY J.G., 1990. Effects of Dichlorvos treatment at fish farms upon lobster larvae and mussels maintained in cages. *Scottish Fisheries Working Paper 10/90.*

14). GT BRITAIN, PARLIAMENT, 1990. Response by the Government and the Crown Estate to the Fourth Report from the Agriculture Committee, Session 1989-90 "Fish Farming in the UK". (HC141), HMSO, 16 pp.

15). BUCHANAN J.S., 1989. The threats and benefits to fish farming - the case of the Atlantic salmon. In *Animal health in a safer environment, parasiticides in the environment, Proceedings of the Association of Veterinarians Symposium*, RCVS, 32 Belgrave Square, London pp 14-32.

16). CUSAK, R. and JOHNSON G, 1990. A study of dichlorvos (NUVAN; 2,2 dichloroethenyl dimethylphosphate), a therapeutic agent for the treatment of salmonids infected with sea lice (*Lepeophtheirus salmonis*). *Aquaculture*, **90**, 101-112.

17). IPCS International Programme on Chemical Safety, 1989., Environmental Health Criteria, 79, DICHLORVOS. *World Health Organisation*, Geneva 160pp.

18) DEPARTMENT OF THE ENVIRONMENT, 1988. Second International Conference on the Protection of the North sea, London 1987, Ministerial Declaration, 25pp.

TRAITEMENT PAR LE DICHLORVOS DES COPEPODOSES DE LA TRUITE ARC-EN-CIEL ELEVEE EN MER : MODALITES DE TRAITEMENT ADAPTEES AUX CONDITIONS ENVIRONNEMENTALES FRANCAISES

J.L. Messager & F. Esnault

Centre National d'Etudes Vétérinaires et Alimentaires
Laboratoire de Pathologie des Animaux Aquatiques (CNEVA-LPAA)
B.P. 70 - 29280 Plouzané, France.

Résumé: *En France, l'incidence pathologique et économique des copépodoses s'est nettement accrue au cours des dernières années, en salmoniculture marine. Des possibilités de traitement par bain de dichlorvos, adaptées aux conditions d'élevage et d'environnement de la région, ont été recherchées.*

- Des essais d'efficacité contre Lepeophtheirus salmonis *et de tolérance pour la truite arc-en-ciel élevée en eau salée ont permis de fixer les modalités d'application d'un traitement de courte durée (15 ppm de dichlorvos pour une exposition de 1 min.) et d'un traitement en bain prolongé (1 ppm pendant 30 min.). Les indices thérapeutiques de ces deux modes de traitement varient alors de 5 à 10 (ou même davantage) selon la température du milieu et l'état de santé des poissons.*

- Les essais cliniques en station expérimentale ont démontré l'efficacité et l'innocuité des deux modes de traitement, alors que les conditions de traitement étaient à priori défavorables : très forte infestation, fortes températures et mauvais état général des poissons.

*- La toxicité aiguë du produit vis-à-vis d'un crustacé de l'environnement, la crevette rose (*Palaemon serratus*), a été étudiée. Cette espèce s'est révélée particulièrement sensible : les concentrations létales 50 p.100 (LC_{50}) 24 h, 48 et 96 h observées sont respectivement 36, 15 et 6 µg/l.*

- Une solution traitante à 15 ppm de dichlorvos, après traitement à la lessive de soude puis réajustement du pH, s'est révélée inoffensive pour la truite arc-en-ciel mais aussi pour la crevette.

Le traitement en bain court, mis au point dans cette étude, est particulièrement adapté aux conditions d'élevage en mer des salmonidés en France. La qualité essentielle de ce mode de traitement, comparativement aux modes en bains prolongés, est de réduire considérablement la quantité de dichlorvos à utiliser et de permettre en outre la dénaturation des solutions traitantes avant leur rejet dans l'environnement. Ce mode de traitement peut être aisément mécanisé et être appliqué à l'occasion d'autres opérations de manipulation du poisson (tri, transport, etc...). Des essais ponctuels de traitement réalisés sur le terrain ont enfin pleinement convaincu les aquaculteurs de l'intérêt d'utiliser ce mode de traitement, dès les premières infestations significatives qui apparaissent au printemps.

Summary : **Treatment with dichlorvos of copepod infections in rainbow trout reared in seawater : modes of treatment adapted to French environmental conditions.**

In France the frequency and economic costs of copepod infections has clearly built up in recent years in marine salmoniculture. This study reports an investigation into the possibilities of adapting the use of dichlorvos bath treatments under different rearing and environmental conditions.

Efficacy trials against Lepeophtheirus salmonis *and tolerance in rainbow trout reared in sea water have allowed us to determine the limits of a short duration treatment (15ppm dichlorvos for an exposure of 1 min.) and long duration treatment (1ppm for 30 min). The therapeutic indices of these two modes of treatment vary from 5 to 10 (or better) dependant on the temperature and the state of health of the animals being treated.*

Clinical trials at an experimental site have demonstarted the safety and efficacy of the two modes of treatment although the conditions for treatment were a priori unfavourable, with high infestation levels, high temperatures and animals in poor condition.

Acute toxicity for a crustacean from the local environment, the pink shrimp, Palaemon serratus *was investigated. This species was found to be particularly susceptible to dichlorvos : 24h, 48h and 96h LC_{50} being respectively 36, 15 and 6µg/l.*

A 15ppm dichlorvos treatment solution, after treatment with sodium hydroxide and readjustment of pH was found harmless both for rainbow trout and pink shrimp.

The short bath treatment, a major point of this study has been found to be particularly well adapted to French salmon culture conditions. The particular advantage of this mode of treatment compared to prolonged bath treatments is to considerably reduce the amount of dichlorvos used and to permit the denaturisation of treatment solutions before disposal to the environment. This mode of treament is susceptible to mechanisation and may be added to other fish handling proceedures (grading, transport etc). Trials carried out in the field have convinced fish farmers of the value of this method of treatment from the first significant infections which ordinarily occur in the spring.

Introduction

Au cours des dernières années, les copépodoses ont eu une incidence pathologique et économique croissante sur la salmoniculture marine française, à tel point qu'aujourd'hui leur traitement s'avère être indispensable.

Le traitement des copépodoses par bains de dichlorvos (1 ppm environ, pendant quelques dizaines de min.) est aujourd'hui largement utilisé dans les élevages de saumons de Norvège, d'Ecosse et d'Irlande notamment. Cependant, du fait de certaines particularités d'ordre climatique, environnemental, écologique, épidémiologique ou liées à la pratique de l'élevage marin des salmonidés en France, quelques adaptations peuvent ou doivent être apportées au protocole classique de traitement.

En effet, la salmoniculture marine française concerne essentiellement la truite arc-en-ciel. Elle est pratiquée en de petites exploitations, situées pour la plupart dans les rias de Bretagne, qui abritent d'autres exploitations aquacoles (ostréiculture, conchyliculture...) ou qui constituent des zones de pêche côtière. Les infestations par *Lepeophtheirus salmonis* ou, à un moindre degré, par *Caligus elongatus*, ne sont importantes qu'au printemps et en été, dans les sites en eau légèrement déssalée (27 à 33 °/₀₀ de salinité) implantés dans les rias, et atteignent aussi bien la truite arc-en-ciel (*Oncorhynchus mykiss*) ou la truite fario (*Salmo trutta*) que le saumon Atlantique (*Salmo salar*). Au moment des plus fortes infestations, la température est élevée, jusqu'à 20° C au mois de juillet. Le "syndrome estival des salmonidés" décrit par Aldrin *et al.* (1) se surajoute alors à la parasitose et peut d'autre part sensibiliser les poissons aux effets du dichlorvos. Enfin, le produit thérapeutique couramment utilisé NUVAN R, CIBA-GEIGY inc.) n'est pas disponible sur le marché français.

Par conséquent, des modalités de traitement par le dichlorvos adaptées à ce contexte environnemental (température et salinité élevées, impact écologique) et pathologique (syndrome estival) ont été recherchées : a) un mode de traitement par bain prolongé (30 min.), selon des modalités classiquement utilisées, plus spécialement destiné aux exploitations en barges situées en haute mer, b) un mode de traitement par bain de courte durée (1 min.) qui permettrait une récupération ou une dénaturation des solutions traitantes après usage. Ceci serait destiné aux exploitations de cages situées en rias.

I. Etudes préliminaires d'efficacité et de tolérance

Matériel et méthodes

Le produit de traitement utilisé est une microémulsion contenant 35 % de dichlorvos (w/w) proposée par les Laboratoires SANOFI.

Pour les essais d'efficacité du produit de traitement contre *Lepeophtheirus salmonis*, des truites arc-en-ciel parasitées sont pêchées dans un site d'élevage infesté, et aussitôt plongées (2 par récipient) dans des récipients de 50 l contenant diverses concentrations de dichlorvos. Après 1 min. (bain de courte durée) ou 30 min. d'exposition (bain prolongé), elles sont placées séparément dans des récipients contenant de l'eau de mer normale. Une heure environ après le début des traitements, la mortalité des parasites est notée.

Pour les essais de tolérance de la truite arc-en-ciel vis-à-vis du produit de traitement, des truites non parasitées, élevées au laboratoire, sont utilisées. Lors des essais "bains de courte durée", les poissons à jeun depuis 48 h sont pêchés dans un grand bassin (7 m^3), placés pendant 1 min. dans des récipients contenant 50 l de solutions de dichlorvos à diverses concentrations puis placés dans des bacs d'expérience contenant 200 l d'eau de mer normale. Pour les essais "bains prolongés", les solutions de dichlorvos sont déversées dans les bacs d'expérience de 200 l d'eau de mer qui contiennent les truites. Après 30 min. d'exposition, celles-ci sont transférées dans des bacs semblables contenant 200 l d'eau de mer normale. La mortalité des truites est toujours observée 24 h après le début des traitements.

D'autres précisions qui concernent les températures, les salinités, le nombre d'animaux d'expérience, les concentrations de dichlorvos utilisées, sont présentées dans les tableaux de résultats (1 à 2).

Résultats

Essai d'efficacité : selon les résultats présentés dans le tableau 1a, un bain de 1 min. à une concentration de 10 ppm de dichlorvos suffit à tuer la quasi-totalité des parasites portés par les truites arc-en-ciel. Pour un bain prolongé de 30 min., une concentration de 0,5 ppm a suffi à obtenir ce même résultat (Tableau 1b).

Tableau I - Mortalité des parasites (*Lepeophtheirus salmonis*) de truites arc-en-ciel exposées à diverses concentrations de dichlorvos

a) pendant 1 minute :

Concentration (ppm dichlorvos)	Nombre des parasites encore vivants sur la truite n° 1	Nombre des parasites encore vivants sur la truite n° 2	Conditions expérimentales
40	1	0	Température : 16-17° C
30	1	0	Salinité : 32 ppt
	0	0	3 réplicats à 0 ; 10 et 15 ppm
15	1	3	Observation du nombre de
	1	3	parasites survivants :
			30 à 45 min. après remise
	0	0	en eau de mer normale
10	3	4	
	1	3	
5	7	25	
	26	63	
0	65	54	
	50	42	

b) pendant 30 minutes :

Concentration (ppm dichlorvos)	Parasites encore vivants sur les truites n° 1	Parasites encore vivants sur les truites n° 2	Parasites décrochés des truites n° 1 et n° 2	Conditions expérimentales
	1	0	70	T° : 16-17°C
1	0	1	83	S : 25 ppt
	0	0	26	3 réplicats par concentration
	1	0	65	Observation du
0,5	0	1	34	nbre de parasi-
	2	0	72	tes survivants : 30 à 45 min.
	47	44	2	après remise en
0	14	16	0	eau de mer
	45	35	0	normale

Essai de tolérance : la concentration maximale sans effet sur la mortalité des truites élevées en eau de mer est 150 ppm (dichlorvos) à 15° C, pour une exposition de 1 min. (Tableau 2a). Les différences de mortalité notées lors du traitement à 300 ppm pendant 1 min., à 11 et 15° C, laissent penser que la tolérance de la truite arc-en-ciel vis-à-vis du dichlorvos est augmentée quand la température d'élevage diminue.

Lorsque la température est élevée (18 à 20° C), la concentration maximale sans effet sur la mortalité de la truite est 5 ppm (Tableau 2b), pour une exposition de 30 min.).

II. Efficacité et innocuité des traitements expérimentaux

Bain de courte durée : 15 ppm de dichlorvos/1 min.
Bain prolongé : 1 ppm de dichlorvos/30 min.

Matériel et méthodes

Les deux modes de traitement sont éprouvés dans un site naturel d'élevage de truites arc-en-ciel de 950 g environ, fortement infestées par *L. salmonis*. Trois répétitions, pour chacun des traitements, ont été réalisées en 3 jours consécutifs. Lors de chaque répétition, 20 truites parasitées sont pêchées au hasard dans une cage d'élevage (environ 500 m^3) et plongées pendant 1 min. (bain de courte durée) ou 30 min. (bain prolongé) dans des bacs (200 l) contenant les solutions de dichlorvos à 0 ppm (témoins) et 15 ppm pour l'épreuve du bain de courte durée, à 0 ppm (témoins) et 1 ppm pour l'épreuve du bain prolongé. Une bonne oxygénation (7 ppm O_2) est maintenue dans les

Tableau II - Mortalité des truites arc-en-ciel après exposition à diverses concentrations de dichlorvos

a) pendant 1 minute :

Concentration (ppm dichlorvos)	Mortalité des truites (Effectives)	Temperature °C	Conditions expérimentales
350	18/20	15	Salinité : 23-27 ppt Truites arc-en-ciel de 400 g environ 3 réplicats à 150 ppm 24 h après remise en eau de mer normale
300	0/6 15/20	11 15	
250	4/20	15	
200	1/20 0/20	15	
150	0/20 0/20	15	

b) pendant 30 minutes :

Concentration (ppm dichlorvos)	Mortalité des truites (Effectifs)	Conditions expérimentales
10	15/21	Température : 18-20° C Salinité : 22-28 ppt Truites arc-en-ciel de 500 g environ Observation de la mortalité 24h après remise en eau de mer normale
5	0/21 1/20	
0	0/20 0/20	

bacs de traitement. Puis les poissons sont placés en observation dans le milieu d'élevage, dans une cage (500 m^3) séparée en 4 parties par des filets. 24 h après les traitements, le nombre de poissons morts et le nombre de parasites portés par les poissons de chaque lot, témoins et traités, sont notés.

D'autres précisions (température, salinité, pH...) sont données dans les tableaux 3 et 4.

Résultats

Efficacité des deux modes de traitement : 24 h après traitement, le nombre de parasites vivants portés par les poissons varie de 0 à 4 chez les traités, de 16 à plus de 100 chez les témoins. La majorité des poissons porte de 0 à 1 parasite chez les traités, d'une soixantaine à plus d'une centaine chez les témoins, selon les répétitions.

Tableau III - Efficacité et innocuité du traitement expérimental par bain de courte durée : 15 ppm d dichlorvos pendant 1 mn

Réplicat	Concentration (ppm de dichlorvos)	Mortalité des truites	Nb median de parasites survivants/poisson	Conditions experimentales
1	0 15	4/20 0/19	> 100 1	Temp. : 16-17°C Salinité : 29 ppt pH : 8.0 Observations : 24 h après remise en eau normale
2	0 15	1/18 1/20	> 100 1	Temp. : 17-17,5° C Salinité : 31ppt pH : 7.8 Observations : 24 h après remise en eau norm.
3	0 15	1/21 1/22	59 0	Temp. :18-19° C Salinité : 31 ppt pH : 7,95 Observations : 24 h après remise en eau norm.

Tableau IV - Efficacité et innocuité du traitement expérimental par bain prolongé : 5 ppm de dichlorvos pendant 30 min.

Réplicat	Concentration (ppm de dichlorvos)	Mortalité des truites	Nb median de parasites survivants/poisson	Conditions experimentales
1	0 1	1/20 1/19	> 100 1	Temp. : 16-17°C Salinité : 29ppt pH : 8,0 Observations 24 h après remise en eau normale
2	0 1	1/20 0/21	91 1	Temp. : 16-17.5° C Salinité :ppt pH : 7.8 Observations 24 h après remise en eau normale
3	0 1	1/22 2/17	62 0	Temp. :18-19° C Salinité : 31 ppt pH : 7.95 Observations : 24 h après remse en eau norm.e

Innocuité des deux modes de traitement : la mortalité enregistrée 24 h après traitement par bain de courte durée s'élève à 2 poissons parmi les 61 traités, et à 6 parmi les 59 témoins, pour l'ensemble des 3 répétitions. De même, elle s'élève à 3 poissons parmi les 62 témoins et à 3 poissons parmi les 57 traités par bain prolongé.

III. Essai de dénaturation d'une solution traitante à 15 ppm de dichlorvos

Matériel et méthodes

A 180 l d'une solution de dichlorvos (15 ppm) dans de l'eau de mer (35 °/oo - 15° C) est ajoutée de la lessive de soude afin d'augmenter le pH à 10. Dix huit heures plus tard, le pH est ramené à 8.3 avec de l'acide chlorhydrique puis 10 truites arc-en-ciel élevées en mer, d'environ 500 g, et 10 crevettes roses (*Palaemon serratus*) sont placées dans la solution. Après 4 h 30 min. de maintien en conditions statiques, avec une bonne oxygénation du milieu, un renouvellement en eau de mer (100 l/h) est établi.

Résultats

Aucune mortalité, ni parmi les poissons, ni parmi les crevettes, n'a été relevée au cours des 10 jours qui ont suivi leur exposition à la solution traitante dénaturée.

IV. Premiers essais de traitement par bain de courte durée en stuctures d'élevage

Matériel et méthodes

Ces essais ont été réalisés au mois de juillet, à température élevée, sur des populations de truites très infestées par *L. salmonis* et affectées par le syndrome estival des salmonidés.

Essai 1 : traitement des truites arc-en-ciel : 680 truites de 950 g environ, infestées par *L. salmonis* (> 100 parasites/poisson), sont pêchées 10 par 10 dans une cage d'élevage, plongées pendant 1 min. dans un bac contenant 200 l d'une solution de dichlorvos (15 ppm) dans de l'eau de mer, puis placées dans une cage d'observation divisée en deux parties : la première partie accueille les 350 premières truites traitées, les 300 truites suivantes étant placées dans la seconde partie. L'ensemble de l'opération a duré 1 h 50 min. La température est élevée (19.5° C), la salinité est de 32 °/oo et le pH = 8,0. Le taux d'oxygène est maintenu entre 6.5 et 8.3 ppm O_2. La même solution traitante a été utilisée durant toute l'opération.

Essai 2 : traitement de truites fario (*Salmo trutta*) : 2,000 truites de 500 g environ, amaigries, infestées par *L. salmonis* (> 50 copépodes/poisson), porteuses de lésions ulcératives sur la tête et derrière la nageoire dorsale sont pêchées, 30 par 30, dans une cage d'élevage (température = 18.3° C; salinité = 34 °/oo ; pH = 8.08), plongées pendant 1 min. dans 210 l d'une solution de traitement (15 ppm de dichlorvos) bien oxygénée, puis remises dans le milieu d'élevage dans une cage voisine. La même solution traitante a été utilisée durant toute l'opération qui a duré 50 min.

Résultats

Le traitement par bain de courte durée s'est montré efficace lors des deux essais réalisés en structure d'élevage. En effet, seuls quelques copépodes (< 5) étaient encore visibles sur quelques poissons, 24 h après traitement. Lors de l'essai 1, une bonne conservation des propriétés thérapeutiques de la solution au cours de la deuxième heure de traitement a pu être vérifiée. L'essai 2 a montré qu'une tonne de poissons pouvait être efficacement traitée avec seulement 3.15 g de dichlorvos.

Cent cinquante six truites sur les 680 traitées (essai 1) sont mortes dans les jours qui ont suivi le traitement. Mais une telle mortalité préexistait, attribuée essentiellement au syndrome estival, et ne permet pas d'apprécier l'innocuité du traitement. Au cours des 8 jours qui ont suivi le traitement 2, la mortalité enregistrée est du même ordre que celle notée avant traitement (< 5 %) pour une même période.

Discussion - conclusion

A notre connaissance, peu de données de la littérature, depuis celles fournies par Rae (2) puis par Wootten *et al.* (4), se sont attachées à préciser les limites d'efficacité et d'innocuité du traitement des copépodoses par le dichlorvos, notamment lorsque les conditions d'application du traitement diffèrent de celles rencontrées en Norvège et en Ecosse.

Dans le présent travail, des concentrations de l'ordre de 10 et 0.5 ppm de dichlorvos ont été déterminées comme étant les concentrations minimales efficaces contre *L. salmonis* pour des durées de traitement de 1 et 30 ppm, respectivement, dans les conditions utilisées, de salinité et de température élevées.

Les résultats montrent par ailleurs qu'une min. d'exposition à 150 ppm de dichlorvos n'entraîne pas de mortalité chez la truite arc-en-ciel élevé en eau salée (23-27 °/oo) à 15° C, et laissent penser qu'elle puisse supporter des concentrations supérieures à des températures plus basses. A 18-20° C, pour une exposition de 30 min., une concentration maximale de 5 ppm a été trouvée.

Compte-tenu de ces données, les modalités d'un traitement par bain de courte durée, 1 min. d'exposition à 15 ppm, et d'un traitement par bain prolongé, 30 min. à 1 ppm, ont été fixées. Les indices de sécurité thérapeutique sont ainsi de 5 pour un bain prolongé effectué à 17-20° C, de 10 pour un bain de courte durée effectué à 15° C et seraient probablement augmentés à des températures inférieures.

Ces modalités de traitement éprouvées expérimentalement sur des truites parasitées ont donné des résultats satisfaisants tant sur le plan de leur efficacité que sur celui de leur innocuité. La présence de quelques parasites, 24 h après traitement, peut s'expliquer par une réinfestation des poissons qui sont restés dans le milieu naturel infesté, durant le délai d'observation. D'autre part, la mortalité enregistrée, tant chez les témoins que chez les traités, est sensiblement du même ordre que celle relevée quotidiennement, à cette époque de l'année, pour l'ensemble de l'élevage duquel ont été extraites les populations expérimentales. Ces essais réalisés à température élevée, avec des truites fragilisées, sensibilisées, pourraient même plaider en faveur d'une grande innocuité des deux modes de traitement étudiés.

Ainsi, deux possibilités de traitement contre les copépodoses pourront être rapidement proposées aux salmoniculteurs français, selon la localisation et les structures de leurs élevages. Le bain de courte durée semble le plus adapté aux exploitations françaises, situées dans des zones sensibles à toute forme de pollution, car il permet de réduire considérablement la quantité de dichlorvos à utiliser. De plus, les solutions traitantes peuvent être dénaturées avant leur rejet dans l'environnement. La faible toxicité de ces rejets de traitement est matérialisée par l'absence d'observation de mortalité parmi les crevettes roses (*Palaemon serratus*) exposées pendant plus de 4 h 30 min. au cours de ces essais. La crevette rose est en effet très sensible aux effets du dichlorvos. Au laboratoire, une concentration de 0,32 ppm a tué 50 % de la population de crevettes roses en 4 h 35 min. et les concentrations létales 50 p.100 (LC_{50}) 24 h, 48 h, 96 h déterminées pour cette espèce étaient de 36, 15, 6 µg/l, respectivement, proches des valeurs établies (5) pour *Crangon septemspinosa* (LC_{50} 48 h = 12 µg/l ; LC_{50} 96 h = 4 µg/l) ou (3) pour la larve de *Crangon crangon* (LC_{50} 96 h = 4,4 µg/l).

L'inconvénient majeur du traitement par bain de courte durée est de nécessiter de longues et lourdes manipulations de poissons. Mais il peut être aisément mécanisé et être appliqué à l'occasion d'autres opérations de manipulation, de tris par exemple. Dans ce cas, il devient le mode de traitement de choix, notamment dans le contexte français de petites unités de production, dans lesquelles un traitement ne devrait s'imposer qu'une à deux fois par an, au printemps de préférence.

Références

1). ALDRIN J.F., BARDOU L.G., BAUDIN LAURENCIN F., GUILLAUME J., MESSAGER J.L., MEVEL M., STEPHAN G. & TIXERANT G., 1984. Quelques conséquences physiopathologiques de l'élévation de la température et de la salinité chez la truite arc-en-ciel, *Salmo gairdneri* R. Communications présentées à la *réunion de la Société d'Ichtyopathologie Fondamentale et Appliquée (SIFA)*, Toulouse, Octobre 1984.

2). RAE G.H., 1979. On the trail of the sea lice. *Fish Farmer*, **2**, 22-23 et 25.

3). THAIN J.E., MATTHIESSEN P. & BIFIELD S., 1990. The toxicity of dichlorvos to some marine organisms. Résumé du *ICES Statutory Meeting*, 1990.

4). WOOTEN R., SMITH J.W. & NEEDHAM E.A., 1982. Aspects of the biology of the parasitic copepods *Lepeophtheirus salmonis* and *Caligus elongatus* on farmed salmonids, and their treatment. *Proceedings of the Royal Society of Edinburgh*, **81B**, 185-197.

5). WORTHING C.R. & WALKER S.B., 1983. *The pesticides Manual (a World Compendium)*, 7th edition, The British Crop Protection Council.

INHIBITION OF ACETYLCHOLINESTERASE IN RAINBOW TROUT FOLLOWING DICHLORVOS TREATMENT AT DIFFERENT DISSOLVED OXYGEN LEVELS

T. Høy, T. E. Horsberg & R. Wichstroem

The Norwegian College of Veterinary Medicine, Department of Pharmacology and Toxicology, P.O.box 8146 Dep.,N-0033 Oslo 1, Norway

Summary: *Two groups of rainbow trout weighing 135 ± 40 g were subjected to a normal dichlorvos treatment (1.136 ppm for 60 min) at a water temperature of 9 °C. During treatment, the water oxygen levels were 15 mg/l (TH) and 3 mg/l (TL), respectively. Brain samples were taken at certain intervals up to 21 days after administration and analyzed for AChE activity. Immediately after treatment, the TH group showed significantly higher AChE activity in brain tissue than the TL group. The average degree of AChE inhibition was 45 % (TH) and 58 % (TL). Average TH values were above TL values for more than 14 d. However, both groups reached normal values at about the same time.*

Résumé: **Inhibition de l'acétylcholinestérase chez la truite arc-en-ciel, après traitement au dichlorvos dans différentes conditions d'oxygènation.**

Inhibiteurs de l'acétylcholinestérase, les organophosphorés sont souvent administrés par bain pour combattre les copépodes marins responsables d'ectoparasitoses. Une oxygénation active est généralement conseillée lors des traitements, car sous faible pression d'oxygène l'augmentation des mouvements respiratoires et du rythme cardiaque tendent à accélérer l'absorbtion du produit. Des truites, élevées à 9°C et traitées par le dichlorvos dans des conditions d'oxygènation variables (3 à 15 mg/l), ont été sacrifiées. Les taux d'acétylcholinestérase étaient alors mesurés par spectrophotométrie. Ces taux n'ont pas varié chez les animaux témoins soumis aux diffèrentes conditions d'oxygénation. Chez les poissons traités ils étaient abaissés de 45% sous forte oxygènation, de 58% sous faible oxygènation. Ces effets physiologiques s'estompaient lentement, en 3 semaines, alors que le dichlorvos disparaît des tissus en 3 à 6 jours. Ces taux d'inhibition élevés et persistants (des accidents aigüs ont été observés au taux d'inhibition de 80%) justifient d'autant plus l'usage d'un apport d'oxygène au moment des traitements que la résistance des parasites au stade Chalimus impose souvent de répéter l'opération après un délai de 15 jours

Introduction

The usual method of treatment against infestations with sea lice is to bathe the fish in a solution of one or other of the organophosphates dichlorvos or trichlorfon. These act by inhibition of the enzyme acetylcholinesterase (AChE). Oxygenation of the organophosphate solution during treatment is recommended (2, 8).

During short term exposure to low water oxygen levels, the ventilation volume of fish is increased, as is the cardiac blood output (7). If fish are exposed to a chemical at the same time, the substance concerned may be absorbed more rapidly due to the increased water flow over the gills and the elevated blood flow. As organophosphates act by inhibition of AChE, a higher degree of enzyme inhibition in the fish would be expected when treatment is carried out at low water oxygen levels compared to high.

The dual purpose of this study was to determine the degree and duration of AChE inhibition in rainbow trout after normal dichlorvos treatment, and to investigate the actual significance of the water oxygen level during exposure to the drug.

Materials and methods

Healthy rainbow trout weighing 135 ± 40 g (N=90) were kept in fibreglass tanks supplied with through-flowing sea water pumped from a depth of 20 m. After an adaptation period of 2 weeks, 30 fish, intended to serve as controls, were marked by removal of the adipose fin. After a further fortnight of adaption, the study was initiated.

At the day of treatment, the temperature of the inflow water was 9 °C, salinity was 30‰ and oxygen level was 6.2 mg/l. These parameters remained constant throughout the study.

Treatment was performed in a fibre-glass tank containing 440 l of static sea water. The dissolved oxygen level was continuously measured and kept constant. It was reduced as described by Swift & Lloyd (10) by introducing nitrogen into the tank via perforated rubber tubes, and increased by introducing oxygen through the same system. During treatment, the oxygen level was kept at 3 mg/l in the low oxygen level groups, and 15 mg/l in the high oxygen level groups. Each group was exposed for exactly 60 min. The water in the treatment tank was changed between each group. The design of the study is presented in Table I.

Samples of 5 fish from the TH and TL groups were taken immediately after treatment and at 2, 4, 7, 14 and 21 days after treatment. Corresponding samples were taken from the control groups immediately after treatment, and after 7 and 21 days.

The sampled fish were anaesthetized in a 50 ppm benzocaine solution and weighed. They were killed by decapitation. The heads were rinsed and frozen in 0.1 M phosphate buffer, pH 8. The samples were kept frozen at - 70 °C until analysis was performed.

Table I: Design of the experiment. Each group (rainbow trout) was exposed for exactly 60 minutes in a 440 l exposure tank.

Group	Numberof fish	Wateroxygen level mg/l	Dichlorvos mg/l
Controls low (CL)	15	3	0
Controls high (CH)	15	15	0
Treated low (TL)	30	3	1.136
Treated high (TH)	30	15	1.136

AChE levels in brain were determined by the spectrophotometric method of Ellman et al. (3). The enzyme activity was related to the amount of protein in the sample, as measured by the Biorad method of Bradford (1) using lyophilized bovine serum albumin (1.21 mg/ml) (Sigma) as standard.

The Wilcoxon test of rank sums was used to determine whether the differences between groups were statistically significant.

Results

The two control groups, CH and CL, did not differ in any of the measured parameters, nor did the AChE activity vary significantly between different sampling times. This showed that neither the nitrogen gas nor the low oxygen level had any effect on the AChE activity per se. The two control groups were therefore pooled into one control group to gain a better basis for the statistical evaluation of the results.

The results of the analyses performed on brain tissue are presented in Fig. 1. Immediately after treatment, the AChE activity was reduced by 45 % in the TH group and by 58 % in the TL group, the difference between the two groups being significant ($p < 0.05$), as was the difference between the controls and both the TH and the TL group ($p < 0.01$).

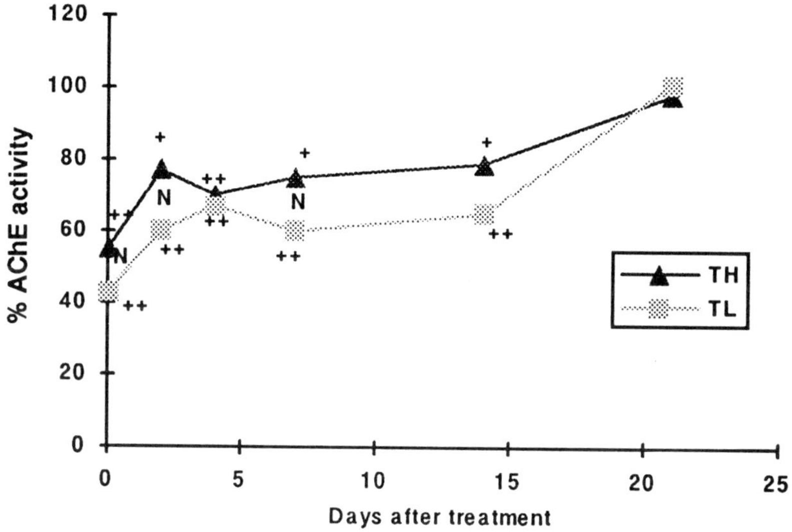

Figure 1: **AChE activity in brain tissue of rainbow trouts after exposure to 1.136 ppm dichlorvos in water with an O_2 concentration of 3 mg/l (TL) and 15 mg/l (TH). Each point represents the average of 5 fish, and is expressed as the percentage of the average AChE activity in 30 controls.**
(+ = significant difference from controls (p<0.05)
++=highly significant difference from controls (p<0.01
N= significant differnce between treated groups (p<0.05))

The average enzyme activity in brain tissue gradually increased during the next two weeks, but remained significantly lower than the activity in the controls. After three weeks, enzyme activity was approx. normal.

Discussion

In a residue study of dichlorvos in salmon, Horsberg et al. (5) were unable to detect residues in muscular tissue for more than 3 d after treatment. Traces of dichlorvos were found in liver tissue for up to 6 d after treatment. The enzyme AChE is, however, inhibited for a longer period than residues can be detected in tissues. The reason for this is the irreversible inhibition of AChE due to phosphorylation of the esteratic site of the enzyme. Normal AChE activity is reestablished by hydrolysis of the bond and new synthesis in connection with normal turnover of the enzyme (4).

Intoxication incidents with high mortality rates (5, 9) revealed AChE inhibition of 80 % or more in fish that died in connection with delousing.

AChE inhibition rates of 45 % (TH) and 58 % (TL) in brain tissue of treated groups immediately after exposure to a dichlorvos concentration of 1.136 ppm are surprisingly high. The inhibition in the TL group approaches the toxic level of 80 % inhibition.

This study showed that the average AChE inhibition is significantly greater in trout exposed to dichlorvos at low water oxygen levels compared to high. It indicated that mortalities after insufficient aeration of the treatment solution might not only be due to hypoxia, but also to toxic reactions. This confirms the great importance of maintaining a high oxygen concentration in the treatment solution during exposure if mortalities in connection with delousing are to be avoided.

The study revealed that the water oxygen level per se has no influence on AChE activity. It showed that the duration of AChE inhibition in rainbow trout after normal dichlorvos treatment is 2 - 3 weeks in brain. In connection with salmon lice infestations, repeated treatment is often required due to the pesticide only being effective against post *Chalimus* stages of the parasites (11). Repeat treatments at an interval of 2 weeks are sometimes necessary. As the AChE inhibition may persist for a longer period, a regimen of repeat treatments may involve a higher risk of mortalities, especially when water oxygenation is insufficient.

References

1). BRADFORD M.M., 1976. A rapid and sensitive method for the quantification of microgram quantities of protein utilizing the principle of protein-dye binding. *Analyt. Biochem.*, **72**, 248 - 254.

2). BRANDAL P.O. & E. EGIDIUS 1979. Treatment of salmon lice (*Lepeophtheirus salmonis* Kroeyer, 1838) with Neguvon^R - description of method and equipment. *Aquaculture*, **18**, 183 - 188.

3). ELLMAN, G.L., K.D. COURTNEY, V. ANDERS & R.M. FEATHER- STONE, 1961. A new and rapid colorimetric determination of acetylcholinesterase activity. *Biochem. Pharmacol.*, **7**, 88 - 95.

4). ETO, M., 1974. *Organophosphorus pesticides: Organic and biological chemistry*. CRC press, Cleveland, Ohio, pp 123 - 144.

5). HORSBERG, T.E., T. HØY & I. NAFSTAD, 1989. Organophosphate poisoning of Atlantic salmon in connection with treatment against salmon lice. *Acta Vet. Scand.*, **30**, 385 - 390.

6). HORSBERG, T.E., T. HØY & O. RINGSTAD, 1990. Residues of dichlorvos in Atlantic salmon (*Salmo salar*) after delousing. *J. Agric. Food Chem.*, **38**, 1403 - 1406.

7). HUGHES, G.M., 1973. Respiratory response to hypoxia in fish. *Am. Zool.*, **13**, 475 - 489.

8). RAE, G.H., 1979. On the trail of the sea lice. *Fish Farmer*, **2**, 22-23, 25.

9). SALTE, R., C. SYVERTSEN, M. KJØNNØY & F. FONNUM, 1987. Fatal acetylcholinesterase inhibition in salmonids subjected to a routine organophosphate treatment. *Aquaculture*, **61**, 173 - 179.

10). SWIFT, D.J. & R. LLOYD, 1974. Changes in urine flow and haematocrit value of rainbow trout *Salmo gairdneri* (Richardson) exposed to hypoxia. *J. Fish Biol.*, **6**, 379 - 387.

11). WOOTTEN, R., 1985. Experience of sea lice infestations in Scottish salmon farms. *International Council for the Exploration of the Sea*. Mariculture Committee. CM 1985/F:7/Ref.M. (mimeo).

TRIALS ON THE EFFICACY OF AZAMETHIPHOS AND ITS SAFETY TO SALMON FOR THE CONTROL OF SEA LICE

M. Roth & R.H. Richards

Institute of Aquaculture, University of Stirling
Stirling, Scotland, FK9 4LA

Summary

Toxicity trials were conducted, in vivo, on the salmon louse Lepeophtheirus salmonis *and its salmonid host,* Salmo salar, *to the organophosphorus compound azamethiphos. Data is presented on: the efficacy of azamethiphos to remove sea lice from salmon; toxicity to salmon and sublethal effect on brain acetylcholinesterase activity. In one hour bath treatments, azamethiphos was effective at removing adult lice at concentrations of 0.01 ppm. When given one hour bath exposures, salmon tolerated doses of 0.5 ppm (1 treatment) and 0.3 ppm (repeated three times at weekly intervals). Reduction in brain acetylcholinesterase activity was not cumulative with repeated exposure. Mortalities occurred when brain acetylcholinesterase activity was reduced by 74%. Due to its relatively high toxicity to sea lice and high tolerance by salmon, it is suggested that azamethiphos would be a promising candidate as an alternative chemotherapeutic agent for the control of sea lice infestations.*

Résumé **Evaluation de l'efficacité et de l'innocuité de l'azamethiphos dans le traitement de la lépeophtheirose des saumons.**

Les copépodes Lepeophtheirus salmonis, *et* Caligus elongatus, *constituent une menace permanente pour les élevages marins de saumon atlantique. Le contrôle en repose actuellement sur l'administration en bain de 2 produits organochlorés, le trichlorfon et le dichlorvos. Des essais ont été entrepris sur des saumons élevés en eau de mer à 10°C, afin d'évaluer l'efficacité et la toxicité de l'azamethiphos. La mesure de l'activité acétylcholinestérasique était effectuée dans le cerveau d'animaux sacrifiés, la répétition de traitements hebdomadaires permettant d'éprouver la tolérance à long terme.*
Un contact d'une heure avec des doses de 0,01 à 1 ppm suffit à entraîner la disparition des parasites adultes et des post-larves, mais les larves Chalimus *se montrent plus résistantes. L'activité acétylcholinestérasique du cerveau est réduite de 36% à la dose de 0,1 ppm, des accidents pouvant survenir à 1 ppm. Cependant les saumons tolèrent bien une exposition unique à 0,5 ppm ou 3 expositions répétées à 0,3 ppm. Les effets sur l'acétylcholinestérase ne paraissent pas cumulatifs et le retour à la normale s'effectue en 8 jours. Ces*

résultats attestent la grande sensibilité des parasites, une certaine marge de sécurité existant entre la dose active et la dose toxique. L'azamethiphos est considéré comme un produit d'avenir dans le contrôle de la lepeophteirose

Introduction

The salmon louse, *Lepeophtheirus salmonis*, and a related copepod *Caligus elongatus*, is currently one of the most serious pests to salmon farmers in Scotland and Norway (2, 4, 8). The effective control and management of sea lice infestations on salmon farms is currently dependent on AQUAGARD® (Ciba Geigy) and NEGUVON® (Bayer). Since the number of compounds available to salmon growers is limited to these two compounds, a project was initiated to evaluate the feasibility of alternative compounds for the treatment of sea lice. The active ingredients in NEGUVON® and AQUAGARD® are trichlorfon and dichlorvos (respectively) and are used in bath treatments (1 hour) at 300 ppm and 1ppm (respectively) (2, 4, 6, 8). Both contain organophosphorus compounds which act by inhibiting cholinesterase activity in cholinergic nervous systems (1).

This report summarizes results from preliminary trials on the efficacy of azamethiphos to remove sea lice from, and its toxicity to, salmon. Azamethiphos is a broad spectrum organophosphorus insecticide used against nuisance flies and other arthropods.

Materials and Methods

Azamethiphos (S-(6-chloro-oxazolo[4,5-b]pyridin-2(3H)-on-3-yl-methyl)-0,0,di methyl-phosphorothioate) was obtained from Ciba Geigy Agrochemicals as a 50% W/W wettable powder. Salmon of an average weight of 1.5 kg were obtained from local salmon growers. Fish were held in fibreglass tanks with flow through sea water with a temperature of 10°C ±1°C and a salinity of 32 ppt. Fish were exposed to various concentrations of azamethiphos for a period of one hour during which flows to the tanks were shut off and the tanks aerated. After treatment, flows to the tanks were reinstated and the tanks flushed for a period of thirty minutes. All trials were conducted between May and November, 1989.

Acute toxicity of azamethiphos to sea lice *in vivo*

Two experiments were conducted to evaluate the efficacy of azamethiphos to remove sea lice from salmon at various concentrations. Salmon were obtained from Golden Sea Produce's Loch Creran Site (Argyll). In the first trial, trial A, groups of five salmon were exposed to concentrations of 1.0 and 0.1 ppm azamethiphos. Fish were sacrificed twenty four hours after treatment and the number/stage of lice recorded. In the second experiment, trial B, groups of ten infected salmon were exposed to concentrations of 0.1 and 0.01 ppm azamethiphos (9 control fish). Fish were sacrificed two hours after treatment and the number/stage of lice recorded.

Acute toxicity of azamethiphos to salmon

In combination with trial B (above) two trials were conducted with azamethiphos at 1.0 and 3.0 ppm (5 fish/group including controls). The same protocols were used as given above (for trial B). All the fish were sampled for brain acetylcholinesterase activity (see below). For analysis both control groups were combined.

Time tolerance trial

Groups of 30 salmon, obtained from Otterferry Salmon Ltd, Otterferry, Tighnabruaich, Argyll, were given three weekly treatments at 0.1, 0.3 and 0.5 ppm azamethiphos. In addition, one group of ten salmon was given a single dose of 1.0 ppm. Throughout the experimental period fish were hand fed to satiation three times daily. At the end of each treatment ten fish were removed and assayed for brain AChE activity (see below). Fish treated in the 1.0 ppm group were only treated once and subsequently assayed for brain AChE activity. Ten control fish were used as a reference for normal brain AChE activity levels.

Brain AChE assay

Fish sampled for brain AChE assays were sacrificed with a blow to the cranium. Brains were excised and frozen in dry ice prior to transport to the laboratory where they were stored at -20°C. Brain tissues were held in storage for no longer than one week and were thawed prior to the assay. Tissues were weighed (wet)and then homogenized (50% W/V) in 50 mM tris buffer, pH 7.4 (Sigma). The preparation was then centrifuged for ten minutes at 5000 rpm. AChE activity was determined spectrophotometrically (3).

The assay contained: 0.43 mM 5,5' -dithiobis-(2-nitrobenzoic acid (3-carboxy-4-nitrophenyl disulfided (DTNB) (Sigma), 36 mM tris buffer, pH 7.4 (Sigma) and 2.6 mM acetylthiocholine iodide (Sigma). Changes in absorbency were recorded at 0 and 3 min. Specific enzyme activities were calculated, number of μmoles of substrate converted per min. per gram., using molar absorptivity coefficient of 13.6 for the product (5-thio-2-nitrobenzoic acid)at 410 nm.

Using the mean value obtained from the control group as 100% activity, the relative activities of the treated fish were calculated as a percentage ([treated fish/controls] x 100)

Results

Acute toxicity of azamethiphos to sea lice *in vivo*

After treatments, adult and preadult stages of lice were removed at all concentrations (Table I). However, chalimus (larval) stages were still observed on treated fish 24 hours after treatment at 1.0 and 0.1 ppm. All figures given are for lice which were found alive. None of the fish died during any of the treatments.

Table I Mean No. sea-lice fish (std.dev.) following 1 hour bath/treatments with azamethiphos

Trial A (5 fish/group)			
	Controls	0.1 ppm	1.0 ppm
AS	42.2 (35.3)	0	0
LS	34.0 (31.8)	10.4 (14.0)	1.6 (1.9)
Trial B (10 fish/group)			
	Controls	0.1 ppm	0.01 ppm
AS	17.4 (12.2)	0	0
LS	0.5 (1.1)	0	0

Where AS = Adult & preadult stages (male & female)
LS = Larval stages (chalimus)

Acute toxicity of azamethiphos to salmon

Results for the acute toxicity of azamethiphos to salmon are summarized in Figure 1. At concentrations of 0.01 and 0.1 ppm no mortalities were observed with brain acetylcholinesterase activity reduced by 8 and 36% respectively (from control values). At 1.0 ppm brain AChE activity was reduced by 80% with 3 (of 5) fish dying. At 3.0 ppm brain AChE activity was reduced by 92% and all fish in the group died within 10 minutes of the beginning of the treatment.

Fig 1 Brain AChE activity and mortality in Atlantic salmon following 1 hour bath treatment with azamethiphos

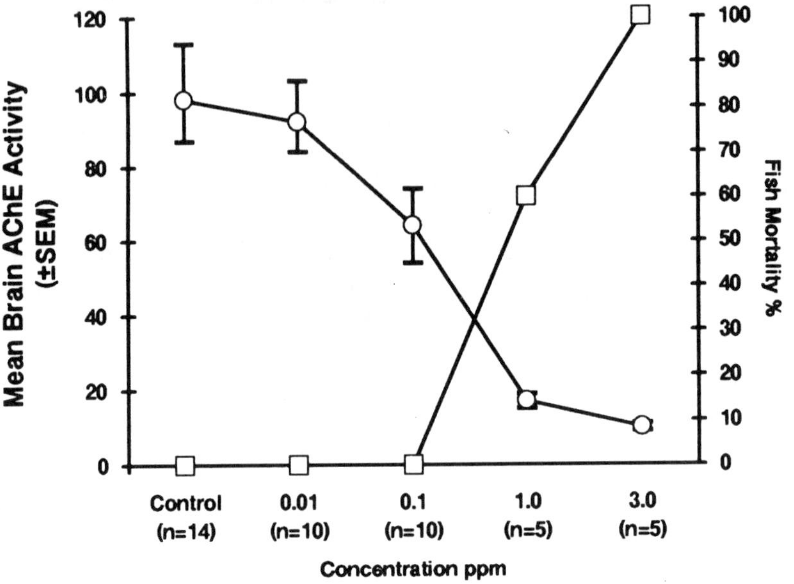

Fig 2 Brain AChE activity in Atlantic salmon following repeated weekly treatments with azamethiphos

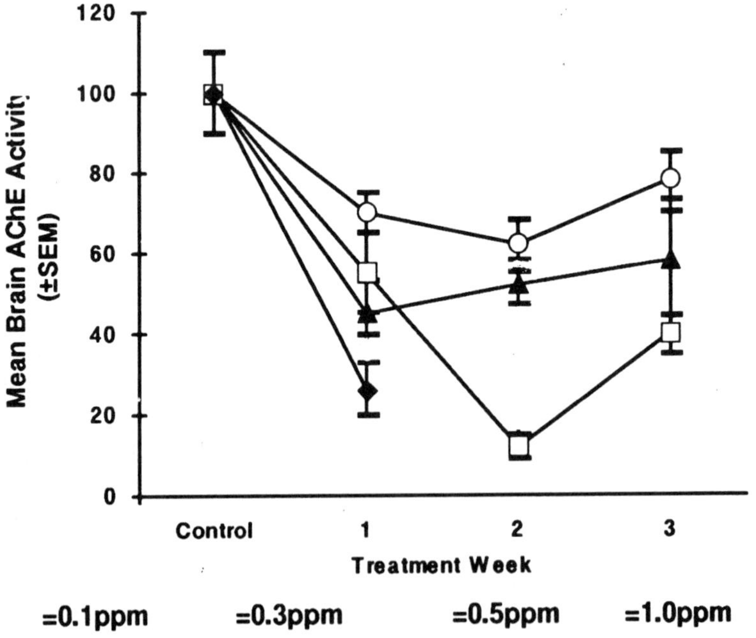

Time tolerance trial

Results for the time tolerance trial are summarized in Figure 2. No mortalities were recorded during or after any of the treatments at .1 and .3 ppm. At .5 ppm mortalities were observed only after the second treatment; two fish before the end of the treatment, another fish within 24 hours and another after 48 hours. At 1.0 ppm brain AChE activity was reduced by 74 % with two fish dying before the end of the 1 hour treatment period.

No appreciable loss in appetite was noticed by technicians feeding the fish. At concentrations of .3 ppm and higher many fish became lethargic during treatment and required a few hours to recover as indicated by their swimming behaviour.

Discussion

The present data found that azamethiphos was effective, in the laboratory, at a concentration of 0.01 ppm. Toxicity to salmon was observed at much higher doses giving the compound a wide safety margin. Unlike other organophosphates (7), azamethiphos toxicity was not cumulative.

The data also suggests that fish exposed to azamethiphos were recovering within a week. Even at dose rates 10 and 30 times that required to kill sea lice, fish appeared to recover significantly by the third treatment. Recovery from AChE inhibition is related to dose and exposure time (7) and it is possible that the relatively short exposures to azamethiphos resulted in quick recovery periods; however, this is unlikely given the relatively high dose rates. Another factor which may have contributed to the tolerance of the salmon to azamethiphos is their size as size is inversely related to anti cholinesterase toxicity (5)

Similar to both dichlorvos and trichlorfon, azamethiphos did not appear to kill larval chalimus stages. Although some reduction in numbers of chalimus intensity was seen, it is difficult to speculate on the compounds efficacy as prevalence and intensity of chalimus infection was highly variable.

Although the data presented herein is preliminary; the high toxicity to sea lice combined with a good tolerance by salmon indicate that azamethiphos would make a suitable candidate as an alternative anti louse agent. However, additional trials will be required to establish dose regimes for field conditions.

Acknowledgments : The project was carried out for and funded by the Scottish Salmon Growers Association. The authors wish to thank Dr. Christina Sommerville, Institute of Aquaculture, University of Stirling, for use of laboratory facilities and valuable advice. We also wish to acknowledge Dr. Shona Whyte for technical assistance both in the laboratory and at the study sites.

References

1). BAILLE, A. C. 1985. The biochemical mode of action of Insecticides. *In* N. F. James [ed.] *Recent advances in the chemistry of insect control.* The Royal Society of Chemistry, London.

2). BRANDAL, P. O. & EGIDIUS, E. 1979. Treatment of salmon lice (*Lepeophtheirus salmonis*, Krøyer, 1838) with Neguvon - description of method and equipment. *Aquaculture* 18: 183 - 188.

3). ELLMAN, G. L., COURTNEY, K. D., ANDRES, V., JR, & FEATHERSTONE, R. M. 1961. A new and rapid colorimetric determination of acetylcholinesterase activity. *Biochem Parmacol.* 7: 88 - 95.

4). RAE, G. H. 1979. On the trail of the sea lice. *Fish Fmr* 2, 22-23, 25.

5). RATH, S. & MISRA, B. N. 1979. Relative toxicity of dichlorvos (DDVP) to *Tilapia mossambica* Peters of 3 different age groups. *Exp. Gerontol.* 14: 307 - 309.

6). REYES, P. & BRAVO, S.S. 1983. Nota sobre una copepodosis en salmones de cultivo. *Invest. Mar Valparaiso* 11: 55 - 57.

7). WEISS, C. M. 1961. Physiological effect of organic phosphorus insecticides on several species of fish. *Trans. Am. Fish. Soc.* 90: 143 - 152.

8). WOOTEN, R., SMITH, J. W., & NEEDHAM, E. A. 1982. Aspects of the biology of the parasitic copepods *Lepeophtheirus salmonis* and *Caligus elongatus* on farmed salmonids, and their treatment. *Proc. R. Soc. Edinburgh* 81B, 185 - 197.

IMMUNOMODULATING EFFECTS OF ANTIBIOTICS

M.H.T. van der Heijden*, W.B. van Muiswinkel, J.L. Grondel** & J.H. Boon***

* Department of Fish Culture and Fisheries
** Department of Experimental Animal Morphology and Cell Biology,
Wageningen Agricultural University, P.O. Box 338,
6700 AH Wageningen, The Netherlands

Summary: *Studies have been performed on the interaction between antibiotics and the immune system in mammals, birds and fish. Some antimicrobial agents have been shown to interfere with the immunological defence mechanisms. Impairment of the immune system may have serious implications for the results of therapy, especially when recovery needs a relatively long period. Only a limited number of studies is available concerning the effects of antibiotics upon basic immunological processes such as antigen processing, proliferation and maturation of antigen sensitive lymphocytes and the production of immunoregulatory factors in fish. To estimate the immunomodulating effect of antibiotics, different tests for the determination of the activity of immunological defence mechanisms must be optimized and standardized.*

***Résumé*: Effets immunomodulateurs des antibiotiques.**

Les antibiotiques sont très employés, parfois à titre préventif, contre les infections bactériennes qui menacent les élevages de poissons. Parmi les effets secondaires ou indésirables de ces traitements (résidus, résistances, impact sur l'environnement), certains peuvent affecter les poissons eux-mêmes. On ne connaît rien d'analogue à l'anaphylaxie, mais des interactions avec le système immunitaire ont été prouvées. La plupart des mécanismes de défense des homéothermes se retrouvent, en moins complexe, chez les poissons: réactions non spécifiques très développées, classe unique d'immunoglobulines, proches des IgM mais présentant les fonctions des IgG de mammifères, existence probable de cellules T et d'un complexe majeur d'histocompatibilité.

Bien que la référence aux homéothermes n'ait guère de sens, les connaissances pharmacodynamiques relatives au poisson ne permettent pas encore d'optimiser les traitements. La diversité des espèces et des conditions de milieux induisent dans la diffusion tissulaire et le temps d'élimination des produit des différences qui ne tiennent pas seulement à la température (voir le cas de la fluméquine chez l'anguille, Fig.1).

Les cellules lymphoïdes, à division rapide, sont très sensibles à l'action des agents xénobiotiques. Cependant l'action immunomodulatrice des antibiotiques, immunosuppresseur de l'oxytétracycline a été montré sur les composantes cellulaires et humorales de la réponse spécifique, chez la carpe, puis chez la truite. En revanche la phagocytose paraissait augmentée. Le Romet (sulfamide + ormétoprime) paraît sans action, et pour d'autres substances l'effet observé semble dépendre de la dose (Fig.2). En fait, les méthodes classiques pour étudier ces phénomènes (mesure des anticorps, stimulation mitogénique) se heurtent à la variabilité des facteurs pouvant influencer la réponse. Pourtant les résultats doivent inciter à la prudence, conduire à prohiber l'emploi préventif des antibiotiques, et à préférer les molécules autorisées et si possible dotées de propriétés immunostimulantes. Des travaux plus poussés pourraient étayer de nouvelles stratégies de traitements.

Introduction

The increase in fish culture production is a result of an increasing number of fish farms as well as a further intensification of the applied culture techniques. Higher densities of fish stocks are accompanied by an increased prevalence of infectious diseases as a result of a high infection pressure and decreased resistance of the fish. Economical losses related to diseases are caused by reduced growth and an increased morbidity and mortality of the animals. To minimize these losses medicines are used. Primary and secondary bacterial infections are treated with antibiotics, curatively as well prophylactically (3).

Recently a growing interest in the possible hazards due to the intensive use of antibiotics in aquaculture can be observed. Bernoth (6) paid attention to the risks concerning public health, i.e. residues of drugs in fish destined for human consumption, development of resistance in human pathogenic bacteria and the direct toxic effects on humans from handling the drugs. Furthermore, treatment of fish may cause pollution of the environment. This results in a negative selective influence on the growth of the normal bacterial fauna revealing in an increase of drugs resistant pathogens (4).

Although antibiotics have proven to be of remarkable value for the treatment and prevention of bacterial infections (5, 26), they often exhibit undesired side effects. Among these, gastro-intestinal disturbances, superinfections with (i) viruses, (ii) resistant bacteria, (iii) fungi (iv) parasites and damage to liver and kidneys have been well described in clinical reports (11). Moreover, it is reasonable that anaphylactic reactions following antibiotic administration as have been reported in man (8) also occur in fish.

Especially rapidly dividing cells like intestinal epithelium and activated lymphocytes seem to be prone to toxic effects of antibiotics. Studies have been performed on the interaction between antibiotics and the immune system in mammals, birds and fish. Some antimicrobial agents have been shown to interfere with the immunological defence mechanisms. This interference may result in immuno-stimulation or immuno-suppression. The former case perfectly fits in the purpose of an antibacterial treatment, whereas immunosuppression is greatly undesirable. This is obvious, because the defence mechanisms have to execute the final elimination of the pathogens.

The scope of the present paper is to point out the possible immunomodulating effects of the still generally accepted use of antimicrobial agents for the maintenance of fish health.

The immune system of teleost fish

Like other vertebrates fish possess an extensive defence system, which enables the individual to survive and maintain its integrity in a hostile environment. The protective mechanisms are directed against foreign matter recognized as non-self, including pathogens and malignant cells, and comprise a number of 'innate' or non- specific and 'adaptive' or specific reactions. The immune system of fish is less complex than that of higher vertebrates (9), but it is well adapted to the aquatic environment by natural selection during more than 300 million years. During this time the defence mechanisms of fish also diverged and one must be aware of the danger of assuming conformity between different species.

The non-specific immune response differs from the specific immune response by the absence of an enhanced reaction after repeated infection (no memory formation) and the non-specific character. Mechanisms of non-specific defence comprise the relatively stable physical and chemical barriers, e.g. epithelia and their secretions, several protective (glyco) proteins (e.g. lectins, lytic enzymes, complement and interferon), natural killer cells, phagocytosis by macrophages and granulocytes without lymphoid interaction (19).

The specific immune response, characterized by specificity and memory formation, can be subdivided in cellular and humoral immunity. The latter involves the processing and presentation of the antigen by macrophages to antigen sensitive B-lymphocytes which subsequently proliferate and differentiate under influence of T-helper-cells into antibody secreting plasma cells (19). The specific antibody binds to the antigen. Subsequent elimination may take place by way of several mechanisms (25). Teleost fish possess only one class of antibodies, comparable to mammalian IgM, which functionally mimic several mammalian Ig classes (9).

Cell mediated immunity is attributed to the activity of T-lymphocytes. These cells can act as regulatory (T-helper and -suppressor) or effector cells (T-killer) (19). The regulation and control of T-lymphocyte activation can be ascribed to a regulatory system analogous or homologous to the mammalian major histocompatibility complex as has been postulated by Stet and Egberts (27).

The immune response is a rather complex interaction between distinct leucocyte populations and immunoregulatory factors (cytokines) produced by the interacting cells. Therefore major disturbances in the immunological response may result from a subtle modification at a crucial point. On the other hand resistance against for example an infection can be achieved by means of several routes and inactivation or impairment of one route will not always lead to a recognizable pathology.

Pharmacokinetics of antibiotics

To date only limited data on the pharmacokinetics of antimicrobial drugs in fish are available. The growing concern about possible harmful effects of antibiotic treatment probably accounts for an increasing attention and number of publications on this subject, mainly dealing with salmonids. However, the momentary available data are difficult to compare due to a variety of drugs, routes of administration, dose rates, fish species and water temperatures used.

Due to the still insufficient scientific basis for the use of antibiotics in aquaculture doses, treatment times and formulations are often the same as those used in treatment of homeothermic farm animals. Furthermore, pharmacokinetic data which describe drug behaviour in fish can differ markedly from endothermic animals a mammals and birds (14, 21). Most prominent is the usually extended elimination half-life of drugs in fish compared to homeothermic animals (7, 17, 18, 23, 28). This may be explained by the differences in physiological and morphological properties of those species including ambient temperature and corresponding metabolic rate and membrane lipid composition, potentialities for biotransformation and excretion of xenobiotics, other tissue composition and blood perfusion and a single versus a double circulatory system in fish and homeothermic animals respectively.

Clear differences in pharmacokinetics of drugs also occur between fish species. The constraints of the aquatic environment enforced many morphological and physiological similarities between fish species but the wide variety of habitats that have been occupied in time account for diversity within this group. Differences in ambient temperature seem to be an obvious explanation for pharmacokinetic disparities. However, from recent results it may be concluded that pharmacokinetic differences are not merely due to the ambient temperature. This becomes clear when fishes with a different environmental adaptation are compared at the same temperature. Groups of five fish of three species, European eel (*Anguilla anguilla*), African catfish (*Clarias gariepinus*) and carp (*Cyprinus carpio*) received the same dose of 18 mg flumequine/kg body weight. Each species was kept at 24°C in an aquarium with aerated tap water. Blood samples were taken at previously fixed time intervals and were analyzed by means of high performance liquid chromatography. The results (Fig. 1) are in agreement with preliminary results on plasma disposition of flumequine in eel (7) and show a significant difference in the elimination rate of the antibiotic between carp and catfish on one hand and eel on the other.

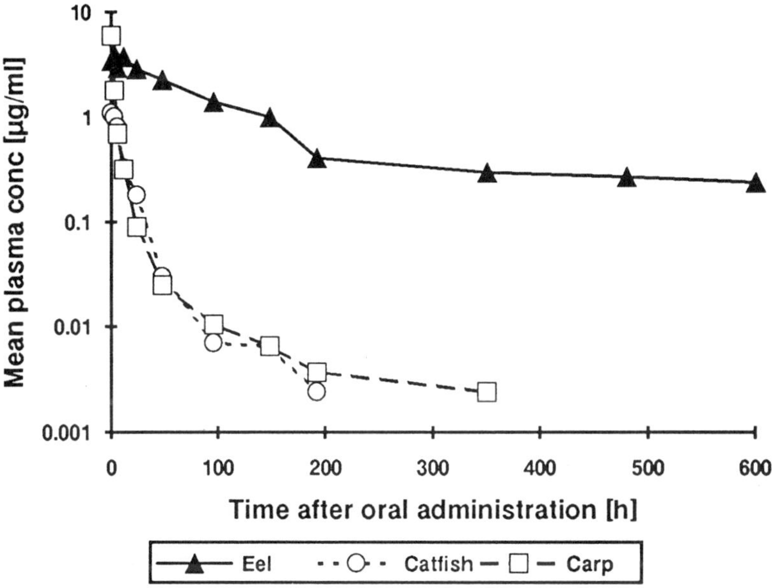

Figure 1: Mean plasma concentrations of flumequine in eel (n=5), catfish (n=5) and carp (n=5) following a single oral administration of 18 mg/kg. Fish were kept at 24 °C.

Moreover, antibiotic distribution across the body is heterogeneous and depends on for instance the composition and blood perfusion of the tissues/organs and the biochemical characteristics of the drug. Consequently, a drug may accumulate in specific tissues/organs of a certain fish species. Accumulation of antibiotics in lymphoid organs may enhance the probability of immunomodulating effects.

Immunomodulation

Immunomodulation is a consequence of a change in number and/or function of the cells involved in the immune response. Thus, each step of the non-specific or specific defence mechanisms may be affected by an immunomodulating agent. Selective interference of xenobiotics with cells of the immune system, particularly lymphocytes, may be explained by an enhanced vulnerability of these cells as a consequence of a high division rate besides slight abilities for detoxification. Lymphocytes are morphologically characterized by a large nucleus and a narrow rim of cytoplasm with few organelles. Up to now little is known about other than immunological functions of lymphocytes, but a drug metabolizing system is presumably hardly present.

Many reports have described either positive, negative or no effect at all of antimicrobial agents on the defence system (14). However, the picture is based on effects of different antibiotics studied in many animal species by means of a variety of immunological tests. Only a limited number of studies is available concerning the effects upon specific immunological processes such as antigen processing, proliferation and maturation of lymphocytes and the production of immunoregulatory factors in fish.

Immunomodulating characteristics of oxytetracycline (OTC) were investigated in common carp (*Cyprinus carpio*) kept under experimental conditions (12, 13, 16, 22, 23). Oral administration of OTC had no influence upon the median survival time (MST) of allografts, but injections with OTC significantly prolonged the MST (22). Grondel and Boesten (12) found that the in vitro mitogenic response of leukocytes of several organs to T-cell mitogens showed an OTC dependent decrease indicating an impairment of the cellular immunological functions by OTC. These results are in agreement with those obtained by Grondel et al (17), who observed a low bioavailability of OTC after oral treatment of carp.

A suppressive effect of OTC on the humoral immune response in carp was clearly demonstrated by Rijkers *et al* (22, 23) at the level of antigen binding and antibody producing cells. On the peak day of the response the numbers of antibody producing cells were reduced up to 95% during a primary response, whereas secondary responses were not significantly inhibited. A differential sensitivity may be observed after antibiotic treatment started at different moments during the development of the immune response. Moreover, an increased number of granulocytes was observed in the spleen of OTC-treated animals. It was speculated that in those cases where specific defence mechanisms are blocked, the phagocytic defence system becomes more active.

Results from pharmacokinetic research (17) showed that OTC was accumulated and retained in scales, bone tissue and pronephros, one of the main lymphoid organs in carp (24). It was suggested that the negative influence of OTC on the carp immune responses could be explained by the specific affinity of OTC for lymphoid tissues.

Suppression of the antibody production after oral application of OTC has also been demonstrated in rainbow trout, whereas a potentiated sulfonamide

(Ro5-0037) did not affect the antibody production in this species when tested by the same regimens (1).

Grondel *et al.* (13) tested various doses of several antibiotics for their possible interference with the mitogenic response of isolated pronephric leukocytes in an *in vitro* system from carp (Fig. 2). From the results it can be concluded that at lower concentrations nearly all the chemotherapeutants stimulates the immune response while using higher doses some are inhibitors Furthermore, in this respect it is worthwile to mention that physical or physiological stress, which may be a consequence of antimicrobial drug treatment (e.g. handling of fish or changes in water quality), may cause changes indirectly by inducing higher concentrations of plasma cortisol. High levels of cortisol may affect lymphocyte and other cellular functions of the immune system (1).

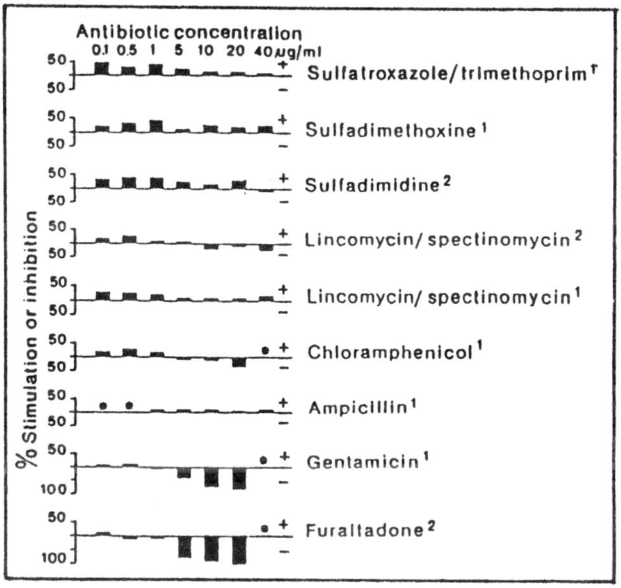

Figure 2 Several antibiotics were tested for their possible interference with DNA synthesis. PHA-activated pronephric leucocytes were cultured for 4 days in the absence or presence of varying concentrations of different antibiotics. The results of triplicate cultures are expressed in percentage stimulation or inhibition compared with mitogen-treated control cultures. 1 injectable, 2) powder and *) not tested (13).

Immunological indicator assays

At present, many *in vitro* and *in vivo* immunological assays are available and provide us with sensitive tools for monitoring drug effects. The immune indicator assays can be divided into three broad categories (2): non-specific assays of general characteristics that indicate the health of fish without the use of antigenic stimulation, indicators that can be used with or without antigenic stimulation and indicators that usually include immunization of the fish and measurement of the specific immune response.

Table 1 Immunological indicator assays that can be used to determine immunomodulation (slightly modified after Anderson (2)).

Assay	Time in immune kinetics	Sensitivity	Specific immune impairment	Statistical variation
Non-specific assays				
Haematocrit (blood cell level)	Any	Low	None	Wide
Leukocrit (white blood cell levels)	Any	Low	None	Wide
Cell differentials	Any	Low	None	Wide
Nonspecific or specific assays used with in vivo or in vitro stimulation				
Phagocytic index	Early	Good	Phagocytosis	Wide
Pinocytosis	Early	Good	Phagocytosis	Wide
Chemoluminescence (oxidative burst)	Early	Excellent	Phagocytosis	Medium
Blastogenisis (T and B cell mitogenisis)	Mid	Fair	Lymphocytes	Wide
Blastogenisis (radioimmunoassay)	Mid	Good	Lymphocytes	Medium
Specific immune response assays				
Melanomacrophage centre antigen accumulation	Mid	Good	Haemopoeisis	Medium
Passive haemolytic plaque assay	Late	High	Antibody production	Narrow
Agglutination (bacterial particle)	Late	Good	Antibody production	Medium
Passive haemaglutination	Late	Excellent	Antibody production	Narrow
Preciptin (gel)	Late	Fair	Antibody production	Narrow
Counterimmunoelectro-phoresis	Late	Good	Antibody production	Narrow
ELISA (enzyme linked immunosorbent assay)	Late	Excellent	Antibody production	Narrow
Neutralisation	Late	Good	Antibody production	Narrow
Fluorescent antibody	Late	Excellent	Antibody production	Narrow
Challenge with pathogen (LD50'S; rel % survival)	Mid	Excellent	Protection against disease	Narrow

The production of humoral and circulatory antibodies is one of the most often used determinants of immunomodulation. Measurement of the presence and amount of humoral antibodies include some of the simplest and fastest immunoassays with the added advantage of a narrow statistical variation. Moreover, the amount of antibody in the serum is often linked to the degree of protection against a specific disease. However, for a complete picture of the immune status indicators of the cellular and non-specific response are needed as well.

In order to study the activity of the specific cellular immune system the allogeneic scale transplantation technique (22) and the mitogenic response to T-cell mitogens, e.g. phytohaemaglutinin (PHA) and concanavalin A (Con A), can be used. These tests represent an *in vivo* and an *in vitro* indicator assay respectively with a high coefficient of variation. Whether this is due to suboptimal technical conditions or genetic variation in the fish is yet unknown.

Measuring or evaluating the effects of antibiotics on the immune system by immunological indicators is difficult because of the large number of variables that can influence the immune response. Water temperature, nutritional status, species, sex, reproductive and developmental stages, and physiological condition influence different physiological functions and, therefore, render repetitive tests difficult to compare. However, it is essential to standardize and optimize the immunological indicator assays.

Conclusions

Many arguments can be put forward against the use of antibiotics in aquaculture. These include the occurrence of superinfections as a result of immunosuppression. To minimize the risks antibiotic prophylaxis should be avoided, unless knowledge about the identity of the invading pathogenic micro-organisms and their sensitivity to antimicrobial agents is present. Preference should be given to licenced drugs with rapid and complete absorption, distribution throughout the proper tissues without accumulation, short elimination times, low toxicity and little or no development of resistant bacteria. Moreover, the drug should be easy to administer.

Inasmuch as fish farming is still a rapidly expanding industry, more chemicals and drugs will be used in attempts to increase production. Drugs which negatively affect the immune system will enhance the probability of disease outbreaks and mortality. In this respect attention should be paid to the promising immuno-stimulating effects of some antibiotics, caused by interference with the bacterial physiology and/or stimulation of the host immune system. The combined action of immune system and drug will increase the defensive potential and consequently support an effective medical management in aquaculture.

Up till now only few data are available about the possible mechanisms of the immunomodulating effects of xenobiotics. Pharmacokinetic investigations, including tissue distribution, combined with immunotoxicological research, extended over a wide range of species and drugs is necessary to correlate immunomodulating effects and drug concentrations in various tissues/organs at several times after administration of a drug. Along these lines it will be possible to get more insight into the action of antibiotics and other drugs upon immunomodulation. Until then cautious use of antimicrobial drugs is recommended.

References

1). ANDERSON, P.D., VAN MUISWINKEL, W.B. & F ROBERSON, B.S., 1984. Effects of chemically induced immune modulation on infectious diseases of fish. *Progress in Clinical and Biological Research,* **161**, 187-211.

2). ANDERSON, P.D., 1990. Immunological indicators: Effects of environmental stress on immune protection and disease outbreaks. In: *Biological ind icators of stress in fish.* Adams, S.M. ed, American Fisheries Society Symposium 8., Bethesda, Maryland, 38-50.

3). AUSTIN, B., 1984. The control of bacterial fish diseases by antimicrobial compounds. In: *Antimicrobials in agriculture*. Woodbine, M. ed., Butterworths, London, 255-268.

4). AUSTIN, B., 1985. Antibiotic pollution from fish farms: effects on aquatic microflora. *Microbiol. sciences*, **2**, 113-117.

5). BAUER, O.N., MUSSELIUS, V.A. & STRELKOV, Yu.-A., 1969. *Diseases of pond fishes*. Izdatel'stvo Kolos, Moskva, 220 pp.

6). BERNOTH, E.-M., 1991. Possible hazards due to fish drugs. *Bull. Eur. Ass. Fish Pathol.*, 11, (1), 17-21.

7). BOON, J.H., NOUWS, J.F.M., VAN DER HEIDJEN M.H.T., BOOMS, G.H.R. & DEGEN, M., 1991. Disposition of Flumequine in plasma of European eel (*Anguilla anguilla*) after a single intramuscular injection. *Aquaculture*, in press.

8). CLAASSEN, E., 1991. Immunotoxicologie; nieuwe ontwikkelingen en perspectieven. *Ned. Tijdschr. Geneeskd.*, **135**, 20, 879-885.

9). ELLIS, A.E., 1982. Differences between the immune mechanisms of fish and higher vertebrates. In: *Microbial diseases of fish*, R.J. Roberts, ed., Academic Press, London, 1-29.

10). ELLIS, A.E., 1991. Tissue residues of chemotherapeutants in fish. *Bull. Eur. Ass. Fish Pathol.*, 11 (1), 22-29.

11). GARROD, L.P., LAMBERT, H.P. & O'GRADY, F., 1973. *Antibiotics and chemotherapy*. V. Livingston, London, 149-166.

12). GRONDEL, J.L. & BO_STEN, H.J.A.M., 1982. The influence of antibiotics on the immune system. I. Inhibition of the mitogenic leucocyte response in vitro by oxytetracycline. *Dev. Comp. Immunol.*, **6**, Suppl. 2, 211-216.

13). GRONDEL, J.L., GLOUDEMANS, A.G.M. & VAN MUIS-WINKEL, W.B. , 1985. The influence of antibiotics on the immune system. II. Modulation of fish leucocyte responses in culture. *Vet. Immunol. Immunopathol.*, **9**, 251-260.

14). GRONDEL, J.L. & VAN MUISWINKEL, W.B., 1986. Immunological defence mechanisms as a target for antibiotics. In: *Comparative veterinary pharmacology, toxicology and therapy.* Van Miert ed., 3rd EAVPT Congress, Ghent, Belgium. MPT Press Ltd., Lancaster, 263- 282.

15). GRONDEL, J.L., NOUWS, J.F.M. & HAENEN, O.L.M., 1986. Fish and antibiotics. Pharmacokinetics of sulphadimidine in carp (*Cyprinus carpio*). *Veterinary Immunology Immunopathology*, **12**, 281-286.

16). GRONDEL, J.L., NOUWS, J.F.M. & VAN MUISWINKEL, W.B., 1987a. The influence of antibiotics on the immune system: immuno-pharmacokinetic investigations on the primary anti-SRBC response in carp, *Cyprinus carpio* L., after oxytetracycline injection. *J. Fish Diseases*, **10**, 35-43.

17). GRONDEL, J.L., NOUWS, J.F.M., DE JONG, M., SCHUTTE, A.R. & DRIESSENS, F., 1987b. Pharmacokinetics and tissue distribution of oxytetracycline in carp (*Cyprinus carpio*) following different routes of administration. *J. Fish Diseases*, **10**, 153-163.

18). GRONDEL, J.L., NOUWS, J.F.M., SCHUTTE, A.R. & DRIESSENS, F., 1989. Comparative pharmacokinetics of oxytetracycline in rainbow trout (*Salmo gairdneri*) and African catfish (*Clarias gariepinus*). *J. Vet. Pharmacol. Therap.*, **12**, 157-162.

19). LAMERS, C.H.J., 1985. The reaction of the immune system of fish to vaccination. PhD thesis, Agricultural University, Wageningen, The Netherlands. 256 pp.

20). NOUWS, J.F.M. & VREE, T.B., 1983. Effect of injection site on the bioavailabilty of an oxytetracycline formulation in ruminant calves. *The Veterinary Quarterly*, **5**, 165-175.

21). NOUWS, J.F.M., VREE, T.B., BREUKINK, H.J., VAN MIERT, A.S.J.P.A.M. & GRONDEL, J.L., 1986. Pharmacokinetics, hydroxylation and acetylation of sulphadimidine in man, ruminants, horses, pigs and fish. In: *Comparative Veterinary Pharmacology, Toxicology and Therapy*, Van Miert ed., 3rd EAVPT Congress, Ghent, Belgium. MPT Press Ltd., Lancaster, 301-318.

22). RIJKERS, G.T., TEUNISSEN, A.G., VAN OOSTEROM, R. & VAN MUISWINKEL, W.B., 1980. The immune system of cyprinid fish. The immunosuppressive effect of the antibiotic oxytetracycline in carp (*Cyprinus carpio* L.). *Aquaculture*, **19**, 177-189.

23). RIJKERS, G.T., VAN OOSTEROM, R. & VAN MUISWINKEL, W.B., 1981a. The immune system of cyprinid fish. Oxytetracycline and the regulation of the humoral immunity in carp (*Cyprinus carpio*). *Vet. Immunol. Immunopathol.*, **2**, 281-290.

24). RIJKERS, G.T., 1981b. Introduction to fish immunology. *Comp. and Dev. Immunol.*, **5**, 527-534.

25). ROITT, I., BROSTOFF, J. & MALE, D., 1985. *Immunology*. Gower Medical Publishing, London / New York.

26). SCHAPERCLAUS, W., 1967. Erfolgereiche Bekämfung der infectiösen Bauchwassersucht des Karpfens mit antibiotischen Mitteln in 11 Jahren, *Z. Binnenfish. DDR*, 14, 64-66.

27). STET, R.J.M. & EGBERTS, E., 1991. The histocompatibility system in teleostean fishes: from multiple histocompatibility loci to a major histocompatibility complex. *Fish & Shellfish Immunology*, **1**, 1-16.

28). ZIV, G., SOBACK, S., BOR, A. & KURTZ, B., 1986. Clinical pharmacokinetics for flumequine in calves. *J. Vet. Pharmacol. Therap.*, **9**, 171-182.

EFFECT OF ORGANOPHOSPHORUS INSECTICIDES USED AGAINST ECTOPARASITES ON IMMUNE RESPONSE OF CARP (*Cyprinus carpio*)

M. Dunier* & A. K. Siwicki **

*Laboratoire d'Ecotoxicologie INRA-Ecole Nationale Vétérinaire de Lyon, 69280 Marcy l'Etoile, France
**Inland Fisheries Institute in Olsztyn, Departments in Zabieniec, 05-500 Piaseczno, Poland

Summary : *Trichlorfon is one of the most widely used organophosphorus insecticides for the control of ectoparasites in finfish in Eastern European countries. Dichlorvos is its degredation product in water. Carp were bathed in trichlorfon at medium (10 and 20ppm) and high concentrations (20,000ppm). No statistically significant effect was observed on antibody production against* Yersinia ruckeri *vaccine.*

Contamination of organs by trichlorphon was estimated by acetylcholinesterase activity. At 20,000ppm only liver and brain were highly contaminated, spleen and lymphocytes were not. At 20ppm, no residue was found in organs. The effect of 20,000ppm and 10,000ppm trichlorfon on the nonspecific immune response of carp experimentally infected by Pseudomonas alcaligenes *and* Aeromonas punctata *was also examined.*

Leucocyte number, phagocytic ability of neutrophils, percentage NBT-positive PMN cells, phagocytic index, lysozyme level in serum, and ceruloplasmin activity in plasma were examined on Days 2, 4, 6, 10, 14, 22 and 26 after carp exposure. After intoxication a leucopaenia was observed as a decrease in the phagocytic ability of neutrophils and in the phagocytic index. Lysozyme level in serum was also decreased. The percentage of NBT-positive PMN cells decreased whilst the ceruloplasm activity in plasma increased in intoxicated fish.

A 30ppm bath in dichlorvos did not affect antibody production against Yersinia ruckeri *even though spleen and kidney were obviously contaminated from results of acetylcholinesterase activity analysis. Some blood parameters of non-specific immunity (ceruloplasmin, lysozyme, haemagglutinins) were affected.*

Lymphoid cells isolated from the pronephros were incubated in vitro *with trichlorfon and with dichlorvos. Dose-dependent supressive effects were observed for lymphocyte proliferation and for myeloid cells respiratory burst activities.*

Résumé : **Effets sur la réponse immunitaire de la carpe (*Cyprinus carpio*), des insecticides organophosphorés employés a titre antiparasitaire**

Largement employé en Europe de l'Est, le trichlorfon est un insecticide organochloré liposoluble, qui inhibe la cholinesterase et provoque l'accumulation tissulaire d'acetylcholine. Il est rapidement dégradé en dichlorvos, également actif et davantage utilisé en eau de mer. La toxicité de ces poisons du système nerveux est particulièrement marquée chez les invertébrés mais également reconnue chez les vertébrés, ce qui impose des précautions dans leur emploi. Une bonne connaissance de la cinétique d'absorbtion et de la toxicité n'implique pourtant pas qu'il en aille de même pour les aspects physiologiques. Comme des effets sur le système immunitaire ont été rapportés chez les mammifères, des investigations de même type ont été entreprises sur poissons.

Effets sur la production d'anticorps: Des carpes étaient baignées dans le trichlorfon à 20 et 20 000 ppm. Dans le 2ème cas, la mesure de l'activité cholinestérasique a permis de déceler des résidus dans le cerveau et le foie, mais ni dans la rate, ni dans les lymphocytes. La réponse humorale après injection de Yersinia ruckeri *n'était pas affectée. Dans le cas du dichlorvos à 30 ppm, des traces étaient retrouvées dans le cerveau, le foie, la rate et le rein, la réponse humorale demeurant également inchangée.*

Effets sur la réponse non spécifique: à 20 000 ppm, le trichlorfon provoquait une leucopénie, abaissait le taux de lysozyme et la fonction phagocytaire, la réponse à une infection bactérienne apparaissant sensiblement déprimée. Cette cytotoxicité pour les cellules circulantes et cette sensibilisation aux infections bactériennes sont en revanche peu évidentes avec le dichlorvos à 30 ppm.

Effets in vitro *sur les cellules myeloides et lymphoides: l'activité fonctionnelle de lymphocytes et de macrophages extraits de la rate et du rein a été étudiée en présence de doses croissantes de trichlorfon (1 mg à 20 g/l) ou de dichlorvos (1 à 50 mg/l). La stimulation lymphocytaire par la PHA est faiblement abaissée par le trichlorfon <10 mg, plus fortement au delà, totalement abolie à 250 mg/l. La phagocytose, partiellement affectée dès la dose de 0,25 g, est inhibée à 2,5 g/l. Le dichlorvos inhibe également l'activité lymphocytaire (partiellement dès 3 mg, totalement à 24,5 g/l) et la phagocytose (partiellement à 12,2 mg, totalement à 49 mg/l).*

In vivo, *seul le trichlorfon présentait un effet perceptible sur les défenses non spécifiques.* In vitro *son action est plus marquée, de même que celle du dichlorvos, qui s'exerce à doses bien plus faibles. Cette dernière constatation est inquiétante quand on songe à l'importante dégradation du trichlorfon en dichlorvos. Même si les effets immunodépresseurs de ces produits sont difficiles à démontrer* in vivo *les éleveurs ne devraient pas perdre de vue les risques encourus par les poissons à l'occasion des traitements.*

Introduction

Organophosphorus insecticides (OP's) are some of the most widely used substances in agriculture and, in fisheries practice, are used for the control of populations of planktonic invertebrates and for treatment and prevention of parasite infestation. Previous studies into the effects of organophosphorus compounds on fish health have concentrated on aspects of toxicity for fish, the kinetics of elimination from the of fish, and determination withdrawal periods (16, 17).

The effects of organophosphorus insecticides on different biochemical and physiological parameters of carp have been studied by a nuber of workers (5, 9, 11). Such organophosphorus compounds are known to produce a specific inhibition of acetylcholinesterase, which in some cases is accompanied by the inhibition of neuro target esterase (10, 16). Dichlorvos and trichlorphon are the organophosphates most frequently used in the aquaculture industry to eliminate ectoparasites from finfish. Trichlorfon is also sometimes employed on crops and fields as an insecticide under the name Neguvon. Its inhibition of acetylcholinesterase activity is not limited to target arthropods, but occurs also in fish (8) and even low concentrations of have been found to be lethal for fish (9). For control of fish ectoparasites, it has been recommended either that this compound is spread on ponds, or that fish are immersed for 1-20 min. in concentration between 2.2 and 25 g/l. In practice, a wide range of doses (0.25 to 25,000 mg/l) and exposure times (1 minute to 4 days) are used in the treatment of different parasitic diseases caused by the monogenetic Trematodes (*Gyrodactylus* and *Dactylogyrus*), *Hirudinea*, and by parasitic Copepods (9).

Dichlorvos (DDVP) belongs to the same family of organophosphorus compounds and is the degradation product of trichlorfon (8). According to Samuelsen (12) in sea water, 90 % of trichlorfon is transformed to dichlorvos within 4 hours. Both compounds are liposoluble insecticides, inhibitors of cholinesterase and lead to the accumulation of acetylcholine. Dichlorvos, commonly used as an anti-parasitic agent on marine fish, depresses mouse spleen weight after oral delivery (6) and depresses the humoral immune response of rabbits (2). In this paper, we discuss the results of an investigation into the effects of the organophosphorus insecticides trichlorfon and dichlorvos on immune response of carp (*Cyprinus carpio*).

Effect of trichlorfon and dichlorvos exposure on antibody production in relation to residue level in organs.

Two baths of trichlorfon were applied, the first at 20 ppm for 1 hr, the second at 20,000 ppm for 15 min. Levels of organophosphate in tissues (brain, liver, spleen and lymphocytes) were analyzed by acetylcholinesterase activity. At 20 ppm, no residue was found in any tissue. At 20,000 ppm only brain and liver were highly contaminated. This high-level exposure had no effect on lymphocytes and spleen, and thus no effect on lymphoid organs (4). Also no statistically significant effect was observed on antibody production against *Yersinia ruckeri* vaccine. Individual kinetics of anti-*Yersinia ruckeri* antibodies in sera were compared by statistical analysis. Analysis of variance confirmed the lack of statistically significant differences between control and OP exposed fish. Under these experimental conditions, trichlorfon had no immunodepressive effect on humoral response either at 20 ppm or with the short 20,000 ppm bath (4). After a 30 ppm bath in dichlorvos, tissue contamination levels on day 3 were statistically significantly different from control fish. The activity of acetylcholinesterase was decreased in brain, liver and spleen. After 8 days, the enzyme was still significantly decreased in kidney (7). The 30 ppm bath in dichlorvos did not affect antibody production against *Yersinia ruckeri* vaccine. Analysis of variance showed no statistically significant difference between intoxicated and control fish. Dichlorvos therefore had no immunodepressive effect on the humoral immune response (7).

Effect of trichlorfon and dichlorvos on nonspecific immune response.

On Day 3 after exposure to 20,000 ppm trichlorfon, the total leucocyte count, phagocytic ability of neutrophils and lysozyme level in serum were significantly decreased whilst ceruloplasmin activity in plasma increased. On Day 56 after exposure, the total leucocyte number and lysozyme level in serum was still diminished, compared with those of the control fish (unexposed), but the phagocytic ability of neutrophils and ceruloplasmin activity in plasma were equivalent (14). The effect of 10,000 ppm of trichlorfon on nonspecific immune response of carp experimentally infected by *Pseudomonas alcaligenes* and *Aeromonas punctata* was also examined (14). Leucocyte count, phagocytic ability of neutrophils, percentage NBT-positive PMN cells, phagocytic index, lysozyme level in serum, and ceruloplasmin activity in plasma were examined (in exposed and in control fish) on Days 2, 4, 6, 8, 10, 14, 18, 22 and 26 after the injection of bacteria. In OP intoxicated fish, immune response was significantly reduced after injection of live bacteria. From the time of experimental infection, the leucocyte count, phagocytic ability of neutrophils, percentage NBT-positive PMN cells, phagocytic index and lysozyme level in serum were significantly diminished in intoxicated fish, compared with those of the control fish. There was a significantly increase in the ceruloplasmin activity in plasma, compared with that of control fish. In this study we investigated the cytotoxic effect of trichlorfon on the immunocompetent cells in blood. The immunosuppressive effect was found to correlate with the disease condition of the carp. On Day 6 after experimental infection clinical disease appeared, and from Days 18 to 26, fish mortalities occurred. No mortality was observed in the control fish and the ulceration tended to cicatrize (14).

The effect of dichlorvos on the nonspecific immune response was also examined (7). Fish were bathed for 30 min in a 30 mg/l solution of dichlorvos. Leucocyte level in blood, phagocytic ability of neutrophils (NBT index), ceruloplasmin activity in plasma, lysozyme level and natural antibodies (haemaglutinins, haemolisins) in serum were measured at Day 0, 7, 23, 35 and 53. The ceruloplasmine level was between 140 and 190 mg/l with no statistical difference between the exposed and the control groups. Lysozyme was found to be between 0.4 and 0.5 ug/ml; only a slight decrease was observed on Day 7 in exposed fish. Haemaglutinin titer varied between 2 and 2.6 (log2) and seemed to be slightly reduced by intoxication from Day 7 to 35. The levels of haemolysins varied over the same range and seemed to be affected only from Day 23 to 35. Immune cells in blood were also examined. From the NBT index (0.36 to 0.4), the phagocytic ability of neutrophils was not affected but a slight leucopenia (16%) was observed on day 23. After dichlorvos intoxication, a lower suppression effect on the nonspecific immune responses was observed, compared with those produced by trichlorfon.

In vitro effect of trichlorfon and dichlorvos on lymphoid and myeloid cells

For *in vitro* studies, trichlorfon concentrations from 1 mg/l to 20 g/l and dichlorvos from 1 mg/l to 50 mg/l were used. The pronephros and spleen of carp were removed and single cell suspensions produced by teasing the tissues in medium through a nylon mesh. The subpopulation of lymphocytes and macrophages were isolated using technique of Bayne (1) and Siwicki and

Cossarini-Dunier (15). Lymphocyte proliferation and macrophage activity were measured by the methods of Cossarini-Dunier (3) and Siwicki and Cossarini-Dunier (15). Trichlorfon at concentrations of 250 mg/l and above was totally suppressive on proliferation of spleen and pronephros lymphocytes with PHA-P. From 1.0 to 100 mg/l, suppression was partial to weak up to 10 mg/l and strong from 12.5 to 100 mg/l. Similarly, dichlorvos at 24.5 mg/l or higher concentrations totally suppressed lymphocyte proliferation with PHA-P. It was partially suppressive at 12.2 mg/l and had no effect below 3 mg/l (7).

Trichlorfon was completely suppressive to spleen and kidney macrophage phagocytosis at high concentrations (20 to 2.5 g/l), whilst there was a dose-dependent suppression between 1.25 and 0.25 g/l, and no effect at lower concentrations. Similarly, dichlorvos concentrations of 49 mg/l and higher were totally suppressive on phagocytosis of macrophages, 32.6 to 12.2 mg/l were partially suppressive and no significant effect was observed at 10.9 mg/l and below (7).

Discussion

In a previous study, we observed no effects of trichlorfon and dichlorvos at different concentration on the humoral immune response of carp, but certain parameters of nonspecific immunity were decreased. This *in vivo* depression of nonspecific cellular and humoral immunity was observed with concentrations of 20,000 mg/l and 10,000 mg/l of trichlorfon. This affects the non-specific cells of the fish immune system, which is the first line of defence in fish (18, 13). In further investigations of the effects of organophosphorus insecticides, we have studied the effects of the degradation product of trichlorfon in water: dichlorvos, using the same parameters. We obtained similar results, but the suppressive effect on nonspecific cellular and humoral immunity was less obvious for dichlorvos than for trichlorfon, even though lymphoid organs still contained detectable levels of dichlorvos a week after exposure to 30 mg/l.

In vitro studies of the effect of organophosphorus insecticides used against ectoparasites on lymphocyte and macrophage activities showed a strong depressive effect for high concentrations of trichlorfon. It must be pointed out that, in practice, concentrations of 10,000 mg/l and below are used in intensive fish farming (9, 14). For dichlorvos, the range of depressive effect is much lower: 10 mg/l to 20 mg/l which is alarming because trichlorfon is quickly transformed into dichlorvos in water (8) and thus trichlorphon could be much more active in that way.

Conclusion

In this paper, we have described a series of studies into the effects on carp nonspecific and specific immune response of the two organophosphorus insecticides commonly used used against fish ectoparasites. A strong *in vitro* immunosuppressive effect on two major functions of cellular immunity was demonstrated clearly : phagocytosis of macrophages and proliferation of lymphocytes in a range of concentrations including those used in the treatment of fish ectoparasites. *In vitro* effects of trichlorfon on nonspecific cellular and humoral immunity were also clearly shown, whilst dichlorvos seemed to produce much less effect and neither had an effect on specific humoral immunity. Ichthyopathologists and fishfarmers, especially in eastern Europe, should be aware of the hazards of chronic exposure to organophosphates to fish defense mechanisms against environment pathogens.

References

1). BAYNE C.J. (1985) - Pronephric leucocyte of *Cyprinus carpio*: Isolation, separation and characterization. *Vet. Immunol. and Immunopathology*, **12**, 141.

2). CASALE G.P., COHEN S.D. & DICAPUA R.A. (1983) - The effects of organophosphate induced stimulation on the antibody response to sheep erythrocytes in inbred mice. *Toxicol. Appl. Pharmacol.*, **68**, 198.

3). COSSARINI-DUNIER M. (1987). Effects of pesticides atrazine and lindane and of manganese ions on cellular immunity of carps, *Cyprinus carpio.*, *J. Fish Biol*,. **31**, 67.

4). COSSARINI-DUNIER M., DEMAEL A. & SIWICKI A.K. (1990) - *In vivo* effect of an organophosphorus insecticide: Trichlorfon on immune response of carp (*Cyprinus carpio*). I.Effect of contamination on antibody production in relation with residue levels in organs. *Ecotox. Env. Safety*, **19**, 93-99.

5). DEMAEL A., DUNIER M. & SIWICKI A.K. (1990) - Some effects of dichlorvos on carp metabolism. *Comp. Biochem. Physiol.*, **93**, 237-240.

6). DESI I., VARGA L. & FARKAS I. (1978) - Studies on the immunosuppressive effect of organochlorine and organophosphoric pesticides in subacute experiments. *J. Hyg. Epidemiol. Immunol.*, **22**, 115.

7). DUNIER M., SIWICKI A.K. & DEMAEL A. (1991) - Effect of organophosphorus insecticides: Trichlorfon and dichlorvos on immune response of carp (*Cyprinus carpio*). III. *In vitro* effect on lymphocyte proliferation and phagocytosis and *in vivo* effect on humoral response. *Ecotox. Env. Safety* (in press)

8). EGIDIUS E. & MOSTER B. (1987) - Effect of Neguvon and Nuvan treatment on crabs (*Cancer pagurus, C. maenas*), lobster (*Homarus gammarus*) and blue mussel (*Mytilus edulis*). *Aquaculture*, **60**, 165.

9). JENEY Z. & JENEY G. (1986) - Studies on the effect of Trichlorfon on different biochemical and physiological parameters of common carp (*Cyprinus carpio* L.). *Aquacultura Hungarica*, **5**, 79.

10). REPETTO G., SANZ P. & REPETTO M. (1988) - *In vivo* and *in vitro* effect of trichlorfon on esterases of the red crayfish *Procambarus clarkii. Bull. Contam. Toxicol.*, **41**, .

11). SAKAGUCHI H. (1972) - On the effects of agricultural chemicals upon fish. I. Changes of chemical components in serum and liver of carp exposed to organophosphate compounds. *Bull. Japan Soc. Sci. Fish.*, **38**, 555.

12). SAMUELSEN O.B. (1987) - Degradation of Trichlorfon to Dichlorvos in Seawater. *Aquaculture*, **60**, 161-164.

13). SIWICKI A.K. & STUDNICKA M. (1987) - The phagocytic ability of neutrophils and serum lysozyme activity in experimentally infected carp, *Cyprinus carpio* L.. *J. Fish Biol.* **31**, 57.

14). SIWICKI A.K., COSSARINI-DUNIER M., STUDNICKA M. & DEMAEL A. (1990) - *In vivo* effect of an organophosphorus insecticide: Trichlorfon on immune response of carp (*Cyprinus carpio*). II. Effect of high doses of Trichlorfon on nonspecific immune response. *Ecotox. Env. Safety*, **19**, 99-11

15). SIWICKI A.K. & COSSARINI-DUNIER M. (1990) - Effect of levamisole on the lymphocyte and macrophage activity in carp (*Cyprinus carpio*). *Ann. Rech. Vet.*, **21**, 95-100.

16). STUDNICKA M. (1970) - Investigation on the remanants of the phospho-organic compound Foschlor in tissues of fishes subjected to the treatment with this preparation. *Pol. Arch. Weterynaryjne*, **13**, 107-121.

17). STUDNICKA M. & SOPINSKA A. (1983) - Toxicity of technical Foschlor for fish. *Rocznik Nauk Rol.*, **2**, 111-116.

18). STUDNICKA M., SIWICKI A.K. & RYKA B. (1986). Lysozyme level in carp (*Cyprinus carpio*). *Bamidgeh* **38**, 22.

Discussion

LE GOUVELLO Do you have the same study made on copper sulphate which is widely used on fish?

DUNIER Copper sulphate? No, only copperoxychoride. We are working on the effect of pesticides and heavy metals on fish immune systems. We have worked six years on carp, now we are working on trout. So we checked the effect of copper, magnesium, manganese, lead and mercury on trout and carp immune response. Copper has, as I have shown here on the same kind of test, a dose dependant immunomodulating effect, which could be stimulation at one concentration, inhibition at another concentration and without effect at a third.

THOMEN Have you any idea what effect such chemicals will have on the ability to immunise the fish with a vaccine?

DUNIER The first example I gave was a commercial vaccine available in France named Yersivax against *Yersinia ruckeri*. It is a model. I did not do a trial with virulent bacteria. The response here was no effect. When I tested atrazine, lindane and organophosphorus and when I followed the immune response in sera through antibody production, I never got a decrease. Then what we will do next is to check the effect of those pesticides on the early stages of antibody production when B cells are producing antibodies and not later when these antibodies are released in circulation. That could make a difference. Most of the time we see an effect when we test in vitro those pesticides or heavy metals on phagocytic activity, on proliferation of lymphocyte and so on. Sometimes it is difficult to contaminate fish. When you have a positive effect, a decreasing effect in vitro, then you check in vivo. The route could be oral. With lindane and atrazine, we gave in feed pellets every day some lindane or atrazine and fed the fish for one month. Lindane for example is bioaccumulating and we monitored the level in lymphoid organs. Trichlophon was applied as a bath, because we want to simulate what is done practically in ponds. Trichlorphon or dichlorvos is used on ponds or they bath the fish as Dr Munro showed previously. People have tried to inject, but I think that that is not so biological, if you inject the product directly. If we want to mimic real contamination, we have no other choice to contaminate fish, we are limited in the practical way to contaminate fish. Each time we checked in lymphoid organs and in the liver what was the residue level of the pollutant.

STRUB Dr Dunier, est-ce que vous pouvez spéculer sur l'importance pratique de vos résultats relativement à la résistance aux infections? Est-ce que vous avez fait des challenge tests? Est-ce que vous pensez qu'on pourrait éviter cette imunosuppression par les moyens que Viviane Verlhac avait indiqués?

DUNIER Pour le moment nous n'avons pas fait de challenge ni travaillé avec des bactéries et des virus, pour des raisons pratiques: on ne veut pas introduire d'agents pathogènes dans nos circuits puisque mes collègues travaillent sur la détoxification par les cytochromes P450 au niveau du foie. Le travail en pathologie ne peut donc pas se faire dans notre laboratoire et doit plutôt être envisagé en collaboration avec nos collègues de Jouy-en-Josas, qui sont équipés pour celà. C'est prévu dans le futur, il est évident que l'efficacité de ce que nous montrons dans nos modèles non pathogènes devra être vérifiée sur des modèles pathogènes viraux ou bactériens. Donc, il est prouvé que nous avons un effet immunodépresseur sur certaines fonctions immunitaires. Maintenant, comment restituer une réponse? Le Dr Siwicki pourrait vous en parler, car il est spécialiste de l'immunité non spécifique et a beaucoup travaillé en Pologne sur ce problème. Il a essayé de restituer une réponse immunitaire déprimée avec par exemple le lévamisole, les vitamines et divers autres produits, et a obtenu des succès au cours d'infections expérimentales. Je pense qu'il serait très intéressant que vous en discutiez avec lui.

SESSION 3

Problems of resistance

LES PROBLEMES D'ANTIBIORESISTANCE CHEZ LES ESPECES ANIMALES AUTRES QUE LES POISSONS

E. Chaslus-Dancla

Institut National de la Recherche Agronomique,
Centre de Recherches de Tours,
37380 Nouzilly, France

Résumé: *L'utilisation d'antibiotiques conduit à la sélection de bactéries résistantes. Le délai d'isolement, le mécanisme et le support génétique de ces résistance dépendent de nombreux facteurs, parmi lesquels la nature de la molécule et son mode d'utilisation. Nous présentons un exemple d'émergence et de diffusion d'un nouveau mécanisme de résistance aux aminosides apramycine (Apr) et gentamicine (Gm), quelques années après leur introduction en médecine vétérinaire en 1980 en France. En milieu vétérinaire, les premières souches résistantes AprGm ont été isolées au début de 1984, en élevages bovins. En 4 ans, on en a isolé de prélèvements pathologiques de plus de 200 élevages répartis dans les principales zones de production bovine en France. Les bactéries résistantes étaient principalement des* Escherichia coli *et des* Salmonella typhimurium. *Dans tous les cas, la résistance était due à la production d'une acétylase AAC(3)IV, et dans la majorité des cas, véhiculée par des plasmides. Ce mécanisme de résistance a été ultérieurement détecté en milieu hospitalier en Belgique. Une comparaison génétique a été effectuée entre des plasmides d'origine bovine ou hospitalière. Une forte homologie génétique était détectée entre ces plasmides, qui illustre les échanges possibles entre les deux communautés.*

Summary: **Experience of problems with farmed animals other than fish**
Any use of antibiotics leads to the selection of resistant bacterial strains. Time delay before detection, mechanism and genetic supports of this resistance depend on many factors including the nature of the drug and the way it is used. We present an example of emergence and spread of a new mechanism of resistance to the aminoglycosides apramycin (Apr) and gentamicin (Gm) some years after their introduction in bovine therapy in France in 1980. In veterinary medicine, the first AprGm resistant strains were isolated early in 1984 in bovine herds. Within four years, such resistant bacteria were detected in pathological samples from more than 200 herds widely distributed in France. The resistant strains were mainly Escherichia coli *and* Salmonella typhimurium. *In all cases, this resistance was due to the production of an aminoglycoside-acetyltransferase AAC(3)IV, and was in most cases plasmid-mediated. This mechanism of resistance was further encountered in hospital strains in Belgium, and a genetic comparison was initiated between plasmids of bovine and human origins from both Belgium and France. A high genetic*

homology was detected between these plasmids which underline the probable exchanges between the two communities.

Introduction

Toute utilisation d'un nouvel antibiotique conduit à la sélection de souches bactériennes résistant à cet antibiotique. Ce schéma simplifié a été observé aussi bien en milieu hospitalier qu'en milieu vétérinaire. Les délais d'apparition ou d'isolement de bactéries résistantes dépendent principalement de la nature de la molécule antibiotique et de son mode d'utilisation (1).

Dès le premier usage de l'antibiotique débute une phase muette, précoce qui prend fin lors de la mise en évidence de la présence de mécanismes de résistance parmi les bactéries pathogènes. Cette phase précoce est caractérisée par la mise en contact de l'antibiotique avec un réservoir bactérien. Les chances de sélection de bactéries résistantes varient selon la quantité d'antibiotique utilisée, mais aussi selon la taille et la diversité du réservoir bactérien concerné. Lors de cette première phase, l'antibiotique est principalement en présence de bactéries non pathogènes, et c'est parmi celles-ci que s'opèrent les premières sélections et échanges génétiques dont certains seulement joueront un rôle ultérieurement. L'acquisition d'un mécanisme efficace par des bactéries pathogènes amorce la deuxième phase, illustrée par la diffusion détectable de ce mécanisme de résistance que peuvent alors mettre en évidence les laboratoires de diagnostic. L'importance de cette seconde phase est liée à de nombreux facteurs parmi lesquels la stabilité du ou des gènes de résistance, la nature du support génétique (plasmides, chromosome) et l'intensité de la pression de sélection.

Ainsi, lorsqu'apparait un mécanisme de résistance à un nouvel antibiotique parmi des bactéries pathogènes,on peut prédire que ce mécanisme est déjà bien implanté parmi les bactéries commensales, et que les agents pathogènes auront dorénavant ces vastes réservoirs potentiels d'échange et d'acquisition de nouveaux gènes de résistance aux antibiotiques.

Les travaux que nous rapportons constituent l'un des rares exemples ou, après mise sur le marché thérapeutique d'un nouvel antibiotique, ait pu être prévue et suivie en médecine vétérinaire l'apparition d'un mécanisme de résistance à cet antibiotique.

I. Conditions générales d'étude

Nous nous sommes intéressés à 2 aminosides, la gentamicine et l'apramycine lors de leur mise sur le marché thérapeutique vétérinaire au début des années 80. Si la gentamicine a une utilisation hospitalière, en revanche, l'apramycine est à usage exclusivement vétérinaire. Ces antibiotiques sont principalement utilisés en élevages bovins.

Les données bibliographiques hospitalières permettaient de recenser différents mécanismes d'inactivation enzymatique de la gentamicine (adénylases, phosphorylases, acétylases), ces enzymes étant fréquemment rencontrés (2). En revanche, un seul enzyme rarement signalé, était décrit, capable de modifier et d'inactiver l'apramycine : l'acétylase 3 de type IV[AAC(3)IV], inactivant également la gentamicine.

Dès cette époque, nous pouvions bénéficier de la mise en place par le Laboratoire de Pathologie Bovine (CNEVA, Lyon, Dr. J.L. Martel) d'un Réseau National de Surveillance de l'Antibiorésistance de bactéries pathogènes isolées en élevage bovin, basé dans les laboratoires départementaux.

Jusqu'en 1984, aucune bactérie résistante à la gentamicine (Gm) et/ou à l'apramycine (Apr) n'était signalée dans le Réseau de Surveillance. Les premières bactéries résistantes AprGm ont été isolées en Février 1984.

Notre étude a été réalisée en 3 étapes :

- Surveillance de l'émergence et de la diffusion de la résistance AprGm en milieu vétérinaire en France 1984-1988.

- Détection de la résistance AprGm en milieu hospitalier 1986-1989.

- Comparaison génétique de plasmides AprGm d'origine vétérinaire et hospitalière 1988-1990.

Pour chacune de ces 3 études nous avons adopté une même démarche (schéma I).

En quelques mois, plus de 30 élevages ont été recensés dans lesquels des bactéries résistantes AprGm étaient isolées de prélèvements pathologiques.

Ces germes étaient en majorité des *Escherichia coli* mais également des *Salmonella typhimurium*. Une étude a été réalisée dans 2 élevages indépendants parmi les premiers recensés ou des *Salmonella typhimurium* AprGm étaient responsables d'épisodes pathologiques néonatals (5, Table I). Cette étude a montré que l'ensemble des souches résistaient à de nombreux antibiotiques, et qu'un seul mécanisme de résistance AprGm était présent : la production de l'acétyltransférase 3 de type IV [AAC (3) IV], encore non décrit en Europe.

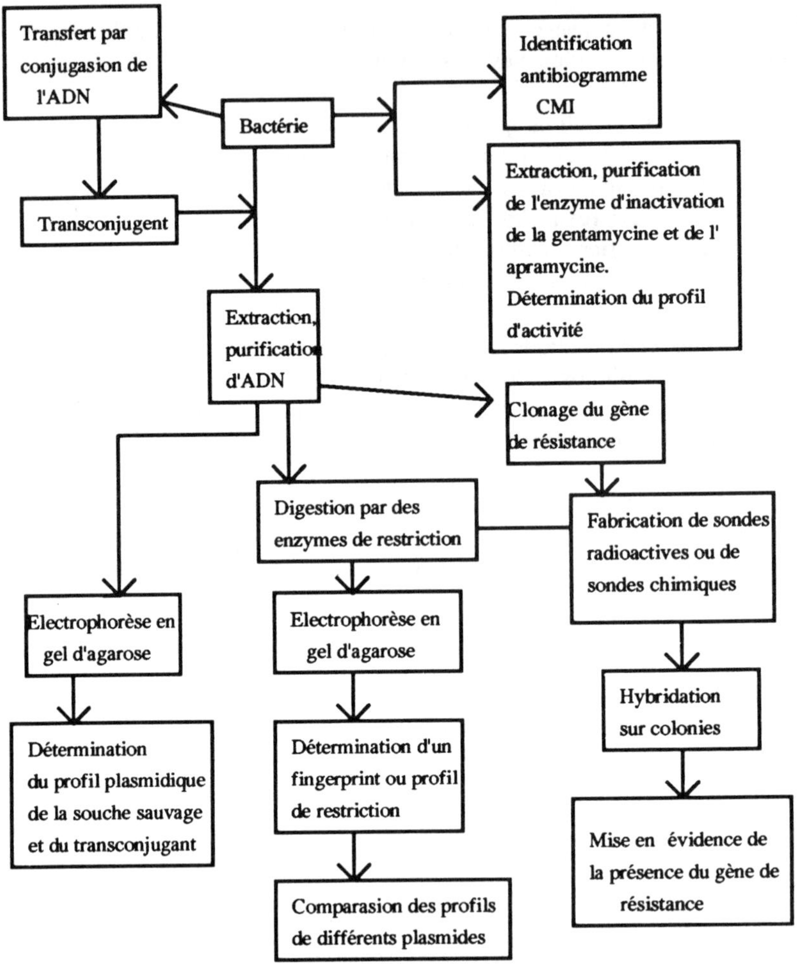

Schéma I Stratégie d'analyse des bactéries résistantes aux antibiotiques (Refs 3,4)

II. Emergence et diffusion de la Résistance AprGm en milieu vétérinaire en France.

Dans la totalité des souches étudiées les gènes codant pour cet enzyme étaient présents sur des plasmides tranférables. Enfin il a été mis en évidence un réservoir de plasmides AprGm véhiculés par des *E. coli* de la flore digestive dominante d'animaux sains de chacun des 2 élevages. Ces *E.coli* avaient été isolés de prélèvements effectués un ou deux mois après l'arrêt des traitements donc postérieurement à l'épisode pathologique. Une forte homologie génétique ayant été démontrée entre des plasmides de *E. coli* issus des 2 élevages, l'hypothèse d'une circulation précoce et rapide des plasmides par des voies parfois non suspectées a pu être avancée.

Table I. Strains from animal isolates : origin and properties of the strains.

Strain	Species, origin (a)	Resistance pattern(b)	Resistance transferred
BN8700	S.typhimurium herd 1	Ap Apr Cm Gm Km Sm Tc Tp Nal	Apr Gm Sm
BN5110	E.coli herd 1	Ap Apr Cm Gm Km Sm Tc Tp Nal	Apr Gm
BN5120	E.coli herd 1	Ap Apr Cm Gm Km Sm Tc Tp Nal	Apr Gm
BN5210	E.coli herd 1	Ap Apr Cm Gm Sm Tc Tp	Apr Gm
BN523O	E.coli herd 1	Ap Apr Cm Gm Km Sm Tc Tp Nal	Apr Gm
BN532O	E.coli herd 1	Ap Apr Cm Gm Km Sm Tc Tp Nal	Apr Gm
BN5400	E.coli herd 1	Apr Cm Gm Sm Tc Tp	Apr Gm Tp
BN8900	S.typhimurium herd 2	Ap Apr Cm Gm Km Sm Tc Tp	Apr Gm Sm
BN4100	E.coli herd 2	Ap Apr Cm Gm Km Sm Tc Tp	Apr Gm
BN4230	E.coli herd 2	Ap Apr Cm Gm Sm Tc Tp Nal	Apr Gm
BN4310	E.coli herd 2	Ap Apr Cm Gm Sm Tp Nal	Apr Gm
BN4740	E.coli herd 2	Ap Apr Cm Gm Km Sm Tc Tp	Apr Gm

(a) *S.typhimurium* were isolated from septicemic animals. *E.coli* were from healthy animals, sampled one or two months after the pathological episode.

(b) Ap, Ampicillin; Apr, apramycin; Cm, Chloramphenicol; Gm, gentamicin; Km, kanamycin; Sm, streptomycin; Tc, tetracycline; Tp, trimethoprim; Nal, nalidixic acid.

Cette étude qui a été ultérieurement élargie aux souches d'autres élevages a confirmé la présence d'un unique mécanisme de résistance AprGm à support plasmidique. Le bilan effectué en cours d'année 1988 a révélé la présence de telles souches AprGm associées à des épisodes pathologiques dans plus de 200 élevages bovins, et quelques élevages porcins, répartis dans les grandes régions de production bovine. Il illustre la diffusion rapide de ce nouveau mécanisme de résistance.

III. Détection de bactéries AprGm en milieu hospitalier

Alors que l'on assistait à la diffusion de ce nouveau mécanisme de résistance en milieu vétérinaire en Europe (6,7), aucune souche semblable n'était isolée de prélèvements hospitaliers. La raison pouvait être la difficulté de détection de ce phénotype de résistance par l'antibiogramme, les disques chargés en apramycine n'étant pas utilisés à l'hôpital.

Table II. Strains from hospital isolates : Origin and properties of the strains.

Strain	Hospital	Resistance pattern(a)	Resistance transferred
-E.coli BN8910	Brugmann, Brussels	Ap Apr Gm Km Sm Tc Tp	Apr Gm Sm Tp
S.typhimurium BN8911	A.Z.V.U.B.Brussels	Apr Gm Sm Tc	(b)
E.coli BN8912	Brugmann, Brussels	Apr Gm Sm Tc Tp	Apr Gm Sm Tp
E.coli BN8913	Brugmann, Brussels	Apr Gm Sm Tp	Apr Gm Sm Tp
E.coli BN8914	Brugmann, Brussels	Ap Apr Cm Gm Km Sm Tc Tp	(b)
E.coli BN8915	Brugmann, Brussels	Ap Apr Cm Gm Km Sm Tc Tp	Ap Apr Gm Sm
E.coli BN8916	Brugmann, Brussels	Ap Apr Gm Sm Tc Tp	Ap Apr Gm Sm Tp
E.coli BN8917	Brugmann, Brussels	Ap Apr Cm Gm Km Sm Tc Tp	Ap Apr Gm Km Sm Tc
E.coli BN8918	Brugmann, Brussels	Ap Apr Cm Gm Km Sm Tp	Ap Apr Gm Sm
K.pneumoniae BN8919	A.Z.V.U.B.Brussels	Ap Apr Cm Gm Sm Tc Tp	(b)
E.coli BN8920	St Jan, Brugge	Ap Apr Gm Sm Tc Tp	Apr Gm Sm Tc
E.coli BN8921	Brugmann, Brussels	Apr Gm Sm	Apr Gm Sm
E.coli BN8922	Brugmann, Brussels	Ap Apr Cm Gm Km Sm Tc Tp	Ap Apr Gm Tc
E.coli BN8923	Brugmann, Brussels	Ap Apr Cm Gm Sm Tc Tp	Apr Gm Sm Tc Tp
E.coli BN8924	Bordet, Brussels	Ap Apr Gm Km Sm Tc Tp	Ap Apr Gm Tp
E.coli BN8925	A.Z.V.U.B.Brussels	Ap Apr Gm Km Sm Tc Tp	Apr Gm Km Tp
E.coli BN8926	St Rafaël, Leuven	Ap Apr Gm Km Sm Tc	Apr Gm

(a) cf table 1

(b) Apr Gm resistance not transferred - transfer of the other characters not tested.

Le Docteur Y. Glupczinsky (Hôpital Universitaire Brugmann, Bruxelles) a entrepris dès 1985 une étude sur la résistance aux aminosides dans son hôpital et dans des hôpitaux proches, en relation avec l'utilisation des aminosides en première intention, et a recherché la présence de l'ensemble des mécanismes de résistance à cette famille d'antibiotiques décrits jusqu'alors. Des souches AprGm ont été isolées de différents services ou hôpitaux, entre 1986 et 1988, et nous ont été envoyées (Table II). Nous avons étudié 17 souches (15 *E. coli*, 1 *S. typhimurium* et 1 *Klebsiella pneumoniae*) qui toutes hébergeaient plusieurs plasmides (8). Par hybridation à une sonde spécifique, le gène codant pour l'acétylase AAC(3)IV a pu être mis en évidence dans chacune des souches et dans la majorité d'entre elles sur des plasmides tranférables. Les plasmides de 12 souches étaient de taille identique. Certains d'entre eux étaient génétiquement proches et provenaient de souches isolées dans le même service, ou dans des services différents illustrant les possibilités de diffusion de plasmides à l'intérieur d'un hôpital et entre différent hôpitaux.

Mais la moitié de ces plasmides ne présentaient entre eux aucune homologie génétique. Ceci suggère la présence, dans la communauté hospitalière et plus largement humaine, de plusieurs réplicons génétiquement différents véhiculant la résistance AprGm, et met l'accent sur l'importance de la diffusion d'un tel mécanisme, méconnu du monde hospitalier malgré son incidence sur des antibiotiques utilisés à l'hôpital tels que la gentamicine et la nétilmicine.

Une étude multicentrique de la prévalence de la résistance aux aminosides, effectuée en 1989 dans 10 hôpitaux français a permis l'isolement de souches AprGm dans 3 d'entre eux (9).

IV. Comparaison génétique de plasmides AprGm d'origine hospitalière ou vétérinaire

Une comparaison des plasmides issus des bactéries de ces 2 communautés et isolées dans des zones géographiques proches a été entreprise sur 9 souches d'origine vétérinaire belge, une souche vétérinaire française et sur les souches hospitalières de l'étude précédente (10, Table III).

Dans l'ensemble des souches vétérinaires le gène codant pour l'AAC(3)IV était présent sur un plasmide tranférable. Le plasmide de la souche BN8915, d'origine hospitalière, présentait une forte homologie génètique avec ceux, de même phénotype de résistance et hébergés par les souches d'origine vétérinaire BN8932 et BN8934. De même, une forte parenté génètique était démontrée entre le plasmide de la souche BN8913 et les plasmides des souches BN8935, BN8936, et BN8931. Malgrè la faible taille de notre échantillon, nous avons pu mettre en évidence une forte parenté génétique entre des plasmides véhiculés par des bactéries d'origine animale et humaine. Ces résultats illustrent la possibilité des échanges génétiques entre ces deux communautés, sans que l'on soit toutefois en mesure d'en quantifier l'importance.

Table III. Comparaison of plasmids from animal or hospital isolates: origin and properties of the strains.

Year of isolation	Strain	Source	Resistance pattern (a)	Resistance transferred
1986	E.coli BN8936	France, calf	Ap Apr Cm Gm Km Sm Tc Tp	Apr Gm Tp
1988	E.coli BN8935	Belgium, calf	Ap Apr Cm Gm Km Sm Tc Tp	Apr Gm Tp
1988	E.coli BN8934	Belgium, calf	Ap Apr Cm Gm Sm Tc	Ap Apr Gm Sm
1988	E.coli BN8933	Belgium, calf	Ap Apr Gm Km Sm Tc Tp	Apr Gm Tp
1988	E.coli BN8932	Belgium, calf	Ap Apr Cm Gm Km Sm Tc Tp	Ap Apr Gm Sm
1988	E.coli BN8931	Belgium, calf	Ap Apr Cm Gm Sm Tc	Apr Gm Tp
1988	E.coli BN8930	Belgium, calf	Ap Apr Cm Gm Sm Tc	Apr Gm Tp
1987	E.coli BN8929	Belgium, pig	Ap Apr Gm Sm Tc Tp	Apr Gm Sm
1985	S.typhimurium BN8928	Belgium, calf	Ap Apr Cm Gm Km Sm Tc Tp	Ap Apr Gm Sm
1985	S.typhimurium BN8927	Belgium, calf	Ap Apr Cm Gm Km Sm Tc Tp	Ap Apr Gm Sm
1986	E.coli BN8915	Belgium,hospital patient (b)	Ap Apr Cm Gm Km Sm Tc Tp	Ap Apr Gm Sm
1987	E.coli BN8913	Belgium,hospital patient (b)	Apr Gm Sm Tp	Apr Gm Sm Tp

(a) cf table I

V. Discussion

En un peu moins de 10 ans, un mécanisme de résistance qui n'était alors décrit que chez quelques rares souches aux Etats-Unis, a diffusé très rapidement et largement en milieu vétérinaire, en France mais également dans d'autres pays d'Europe, et, bien qu'il y soit encore peu fréquemment rencontré, il est dorénavant présent parmi les souches isolées à l'hôpital.

Plusieurs facteurs ont pu favoriser cette rapide évolution :

- la structure de l'élevage bovin qui permet le transport et le brassage d'un nombre important de très jeunes animaux d'origines différentes, dans des conditions favorables au stress, à la fatigue, et aux contaminations bactériennes.

- l'utilisation d'antibiotiques

En milieu vétérinaire, les 2 molécules ont été utilisées, avec une prépondérance pour la gentamicine. Les enquêtes effectuées en élevage n'ont pas permis de dégager clairement l'importance du rôle joué par l'une ou l'autre des 2 molécules dans la sélection des bactéries résistantes.

En milieu hospitalier, l'utilisation de la netilmicine, de la tobramycine ou de la gentamicine a pu concourir à la sélection de souches productrices d'AAC(3)IV.

Dans la majorité des souches, la résistance AprGm était présente parmi d'autres mécanismes de résistance à des antibiotiques de familles différentes, associés ou non sur le même plasmide. Dans ces conditions, des phénomènes de co-sélection ont pu s'exercer.

- la nature du support génétique de la résistance

Dans la grande majorité des souches étudiées, les gènes codant pour l'AAC(3)IV étaient localisés sur des plasmides conjugatifs. Ce support génétique hautement transférable est sans doute l'un des éléments essentiels de cette dissémination rapide.

Une étude comparative des gènes codant pour l'acétylase AAC(3)IV réalisée sur des plasmides d'origine humaine ou vétérinaire a montré que dans tous les cas la résistance AprGm était associée à la résistance à l'hygromycine (substance utilisée en milieu vétérinaire mais pas comme antibiotique) et également à une séquence d'insertion (IS). Cette association évoque une structure de transposon (11).

- <u>les échanges entre les deux communautés vétérinaire et hospitalière</u>

Bien que portant sur de faibles échantillons, nos études comparatives ont pu mettre en évidence de fortes homologies génétiques entre des plasmides circulant dans l'une ou l'autre des deux communautés. Les échanges entre les deux communautés peuvent être de natures différentes (bactéries, plasmides, transposons). De tels échanges ont été abondamment illustrés dans la littérature (12,13,14). Mais aucune structure de surveillance et d'étude ne permet actuellement de quantifier l'importance de ces échanges.

Ainsi, l'utilisation en milieu vétérinaire de nouveaux antibiotiques, la gentamicine et l'apramycine, a conduit, en quelques années, à l'émergence et à la large diffusion d'un nouveau mécanisme de résistance à ces antibiotiques, responsable d'échecs thérapeutiques. Cette diffusion a dépassé la seule communauté vétérinaire puisque ce mécanisme est actuellement mis en évidence dans quelques souches hospitalières. Plusieurs facteurs ont contribué favorablement à cette rapide diffusion. Un tel exemple plaide pour une utilisation plus prudente des antibiotiques, et pour un renforcement des systèmes de surveillance de l'évolution de l'antibiorésistance, associant microbiologistes vétérinaires et hospitaliers.

Références

1). ACAR J.F., 1989. Début et réussite d'un mécanisme de résistance. In *Mises au point en microbiologie clinique. Usage des antibiotiques et résistance bactérienne.* pp.10-12.

2). COURVALIN P. & CARLIER C., 1981. Resistance towards aminoglycosides resistance in France. *J. Antimicrob. Chemother.*, **8** (suppl A.), 57-69.

3). MANIATIS T., FRITSCH E.F. & SAMBROOK J., 1982. *Molecular cloning : a laboratory manual.* Cold Spring Harbor Laboratory, Cold Spring Harbor, N.Y.

4). TAKAHASHI S. & NAGANO Y., 1984. Rapid procedure for isolation of plasmid DNA and application to epidemiological analysis. *J.Clin. Microbiol.*, **20**, 608-613.

5). CHASLUS-DANCLA E., MARTEL J.L., CARLIER C., LAFONT J.P. & COURVALIN P., 1986. Emergence of aminoglycoside 3-N-Acetyltransferase IV in *Escherichia coli* and *Salmonella typhimurium* isolated from animals in France. *Antimicrob. Agents Chemother.*, **29**, 239-243.

6). WRAY C., HEDGES R.W., SHANNON K.P. & BRADLEY D.E., 1986. Apramycin and gentamicin resistance in *Escherichia coli* and Salmonellas isolated from farm animals. *J.Hyg. Camb.* **97**, 445-456.

7). THRELFALL E.J., ROWE B., FERGUSSON J.L. & WARD L.R., 1986. Characterization of plasmids conferring resistance to gentamicin and apramycin in strains of *Salmonella typhimurium* phage type 204c isolated in Britain. *J. Hyg. Camb.* **97**, 419-426.

8). CHASLUS-DANCLA E., GLUPCZYNSKI Y., GERBAUD G., LAGORCE M., LAFONT J.P. & COURVALIN P., 1989. Detection of apramycin resistant Enterobacteriaceae in hospital isolates. *FEMS Microbiol. Lett.*, **61**, 261-266.

9). COLLATZ E., BISMUTH R., CHANAL C., CROIZE J., DRUGEON H., FLANDROIS J.P., GOLDSTEIN F.W., HARE R., LAMBERT T., MENARD Y., MICHEL-BRIAND Y., MILLER G., MINOZZI C., THABAUT A., WEBER M. & COURVALIN P., 1989. Distribution des enzymes modificatrices des aminosides chez les Enterobactéries résistantes : premiers résultats d'une étude multicentrique réalisée dans dix hôpitaux français. Réunion interdisciplinaire de Chimiothérapie antiinfectieuse, Décembre 1989, Paris, p. 107.

10). CHASLUS-DANCLA E., POHL P., MEURISSE M., MARIN M. & LAFONT J.P., 1991. High genetic homology between plasmids of human and animal origins conferring resistance to the aminoglycosides gentamicin and apramycin. *Antimicrob. Agents Chemother.* (in press).

11). SALAUZE D., OTAL I.,GOMEZ-LUS R. & DAVIES J., 1990. Amino-glycoside acetyltransferase 3-IV(*aacC4*) and hygromycin B 4-I phosphotransferase (*hph B*) in bacteria isolated from human and animal sources. *Antimicrob. Agents Chemother.*, **34**, 1915-1920.

12). CAMPBELL G. & MEE B.J., 1987. Mapping of trimethoprim resistance genes from epidemiologically related plasmids. *Antimicrob. Agents Chemother.*, **31**, 1440-1441.

13). TOWNER K.J., WISE P.J. & LEWIS M.J., 1986. Molecular relationship between trimethoprim R plasmids obtained from human and animal sources. *J. Appl. Bacteriol.*, **61**, 535-540.

14). HOLMBERG S.D., WELLS J.G., COHEN M.L., 1984. Animal-to-man transmission of antimicrobial-resistant *Salmonella* : investigations of U.S. outbreaks, 1971-1983. *Science*, **225**, 833-5.

Discussion

MICHEL Thankyou Dr Chaslus-Dancla, I think it is very interesting to see the direct demonstration of the relations between human and veterinary problems.

GROOTHUIS Just to check up, because it was rather difficult for me to follow it in French. If I understood well, you did DNA sequencing of the resistance plasmids and concluded more or less that there is no transfer of resistance from animals to humans. Is that correct?

MICHEL Could you repeat the beginning of the question please?

GROOTHUIS Yes, you typed the plasmids by DNA restriction patterns?

CHASLUS DANCLA Yes

GROOTHUIS I had the impression that you concluded that types of plasmid from the veterinary field are not transferred to the human population. Is that correct?

CHASLUS DANCLA Non, nous n'avons pas du tout tiré ce type de conclusion! Nous avons en milieu vétérinaire un certain nombre de plasmides qui ont diffusé. Et là, je veux limiter les conclusions: nous avons pu voir en milieu hospitalier des plasmides tout à fait similaires, mais dans nos systèmes de surveillance on ne peut pas aller plus loin. Nous avons en milieu hospitalier des plasmides génétiquement très proches de ceux qui ont diffusé en milieu vétérinaire.

GROOTHUIS Merci, compris.

MUNRO Can you tell us how long these antibiotics have been circulating in human and veterinary medicine....for how many years?

CHASLUS-DANCLA En médecine vétérinaire ils circulent toujours. Ils ont été introduits au début des années quatre-vingts, et sont actuellement utilisés. Ils le sont un peu moins, parce qu'on a essayé d'en freiner l'utilisation non raisonnée au vu de ces travaux. En milieu hospitalier, la gentamycine est un produit utilisé depuis le milieu des années soixante, la méthylmycine est actuellement bien utilisée, la gentamycine moins.

PRESENT AND FUTURE PROBLEMS CONCERNING THE DEVELOPMENT OF RESISTANCE IN AQUACULTURE

T. Aoki

Department of Biological Resources,
Faculty of Agriculture, Miyazaki University, Miyazaki 889-21, Japan

Summary: *The widespread use of chemotherapeutic agents has promoted the emergence of drug-resistant fish-pathogens in fish farms in Japan. Multiple drug-resistant strains of fish pathogens;* Aeromonas hydrophila, A. salmonicida, Edwardsiella tarda, Pasteurella piscicida, *non-hemolytic* Streptococcus *sp.,* Vibrio anguillarum, *and* Vibrio *sp., have appeared at high frequencies in fish farms. Almost all strains carried with a transferable R plasmid. Additionally, strains of gram-negative fish-pathogens encoded with chromosomal resistance to pyridonecarboxylic acids and nitrofuran and have also increased. The properties of the R plasmids from fish-pathogens differ from those of R plasmids from human and domestic animal pathogens. The properties and DNA structures of the R plasmids also differ depending on the species of the fish-pathogens. However, R plasmids with the same resistance markers in similar DNA structures were found in some* A. hydrophila *and* E. tarda, *and in some* A. hydrophila *and* A. salmonicida.

Chemotherapy will be increasingly important to treat bacterial infections in cultured fry as well as adult fish from year to year. The incidence of drug resistant infections will occur frequently in fish farms. Unfortunately, designing strategies for controlling the spread of resistance to chemotherapeutics will be difficult. Drug-resistant infections will become a big problem throughout the world.

Résumé : **Problèmes passés, présents et futurs liés au développement des résistances en aquaculture.**

Les antibiotiques sont largement utilisés pour combattre les infections bactériennes majeures des poissons, causées par Aeromonas salmonicida, Renibacterium salmoninarum, Edwardsiella tarda, Vibrio anguillarum, Pasteurella piscicida *et* Streptococcus *spp. Très tôt pourtant des résistances multiples sont apparues, intéressant l'ampicilline (AP), le chloramphenicol (CP), la colistine (CL), la kanamycine (KM), la lincomycine (LIM), les macrolides (MLs), les nitrofuranes (NF), l'acide pyridone-carboxylique (PA), la streptomycine (SM), les sulfamides (SA), les tétracyclines (TC) et le triméthoprime (TMP). Diverses combinaisons de résistances existent selon les bactéries, mais sauf de rares cas de déterminisme chromosomique (nitrofuranes et acide pyridone-carboxylique) presque toutes mettent en jeu des facteurs R portés par des plasmides.*

L'aptitude de ces derniers à se transmettre par conjugaison et à s'intégrer au génome bactérien explique leur rapide dissémination. L'épidémiologie des facteurs R, très étudiée au Japon, montre que dans toutes les espèces ils peuvent se classer en groupes de grande homogénéité génétique.

Les premières résistances transmissibles ont été décrites en 1971 chez A.hydrophila, *d'abord sur l'anguille, puis sur d'autres espèces. Les marqueurs les plus fréquents combinent les associations SA+TC ou CP+SA+SM et appartiennent respectivement aux groupes d'incompatibilité Inc A-C ou U. Les souches qui les hébergent, très répandues dans l'environnement et la flore digestive des poissons, se comportent en pathogènes opportunistes. La combinaison SA+TC est également classique chez* A.salmonicida, *chez qui des plasmides du groupe Inc U codant les resistances à CP+SA+SM ont été identifiés plus tard. Récemment, avec l'introduction de nouvelles molécules, les résistances chromosomiques à NF et PA se sont multipliées.*

Décrites en 1972 les résistances de E.tarda *sont répandues chez les anguilles du Japon et de Taïwan. Elles combinent SA+TC et CP+SA+TC et appartiennent toutes au groupe A-C. Des souches de* P.piscicida *multirésistantes à 6 ou 7 antibiotiques, AP+CP+KM+NF+PA+SA+(TC) ont été découvertes chez la sériole en 1980. Très difficiles à combattre elles ont permis de définir un nouveau groupe Inc.*

Les élevages de Plecoglossus altivelis *(ayu), très sensibles à la vibriose, ont permis d'observer une évolution complexe des résistances de* V.anguillarum. *Détectée en 1973 l'association CP+SA+SM+TC s'est restreinte dès 1974 à CP+SA+TC, avant d'être concurrencée par des résistance chromosomiques à NF, PA et TMP. Après une accalmie, des résistances transmissibles à AP et TMP sont apparues vers 1980, ainsi qu'une résistance non transférable à CL. Toutes ces variations reflètaient fidèlement les modifications des habitudes thérapeutiques. Trois groupes Inc ont pu être individualisés chez* V.anguillarum.

Depuis 1976 on connaît chez les streptocoques non hémolytiques de la sériole des résistances intermédiaires ou élevées à CP, LIM, MLs ou TC. Les premières ne sont pas transférables, mais des plasmides associant MLs+LIM+TC ou MLs+LIM+CP supportent les plus élevées.

On ne peut guère envisager de se passer des antibiotiques et d'éviter la sélection de nouvelles résistances dans le futur. Toutefois des progrès dans la prévention hygiénique et l'immunoprophylaxie pourraient contribuer à abaisser la pression sélective imputable aux traitements.

Introduction

The most significant bacterial diseases are furunculosis caused by *Aeromonas salmonicida*, bacterial kidney disease caused by *Renibacterium salmoninarum*, edwardsiellosis caused by *Edwardsiella tarda*, vibriosis by *Vibrio anquillarum*, pseudotuberculosis by *Pasteurella piscicida*, and streptococcal infection by non-hemolytic *Streptococcus* sp. in Japan (23).

Accordingly, chemotherapeutic agents (amoxicillin, ampicillin, colistin, florfenicol, lincomycin, macrolide antibiotics, novobiocin, pyridonecarboxylic acids, sodium nifurstyrenate, sulfonamides, the sulfa monomethoxine: ormetoprim complex, tetracycline derivatives, and thiamphenicol) have been used for the treatment of bacterial infections in cultured fish (3). Following the wide use of these agents in therapy, the multiple drug resistant strains of fish-pathogens have increased in fish farms (3). The drug-resistant strains are encoded with resistance to ampicillin (AP), chloramphenicol (CP), colistin (CL), kanamycin (KM), lincomycin (LIM), macrolide antibiotics (MLs), nitrofuran derivatives (NF), pyridonecarboxylic acids (PA), streptomycin (SM), sulfonamides (SA), tetracycline (TC), and/or trimethoprim (TMP). Drug resistance markers depend on each fish-pathogen. There is a distinction between plasmid-mediated and chromosomally-mediated resistance mechanisms. Transferable R plasmids have been detected in almost all drug-resistant strains of fish-pathogens. Only chromosomal resistance to NF and PA has been found so far.

Two important features of R plasmids which have contributed to their rapid evolution and spread are their conjugative ability and the presence of transposons. The origin of the R plasmids from fish pathogens is very interesting. The R plasmids detected from fish pathogens have been compared with their properties and DNA structures.

In this paper, epidemiological studies of the drug resistance of fish-pathogens carrying R plasmids and the properties of R plasmids from fish-pathogens are discussed.

Hemorrhagic septicemia caused by *Aeromonas hydrophila*

Drug resistant strains of *A. hydrophila* carrying transferable R plasmids were isolated in 1971 in cultured eel (1). The resistant strains carrying R plasmid were isolated at high frequencies in cultured amago (*Oncorhynchus masou macrostomus*), ayu (*Plecoglossus altivelis*), carp, and eel. (1, 10). Almost all strains had chromosomally mediated resistance to AP (24). Moreover, the detection of a transferable R plasmid encoded with resistance to AP from *A. hydrophila* was reported (19). The transferable R plasmids were detected from these multiple drug-resistant strains. The typical marker detected from *A. hydrophila* had resistance to SA and TC, and CP SA and SM (1). Drug resistant strains of *A. hydrophila* carrying R plasmids were found in France, Ireland, Taiwan and the U.S.A (20, 21, 25). Recently, these strains were also isolated from cultured snakehead fish (*Ophicephalus striatus*) in Thailand(16).

The DNA of transferable R plasmids classified into incompatibility (Inc) group A-C, which had been detected in different areas, had resistance to SA and TC and showed a high homology to each other (1). R plasmids classified into Inc group U with resistance to CP SA and SM showed homology within a specific species and with R plasmids from *A. salmonicida*(1).

The resistant strains of *A. hydrophila* were widely distributed in the intestinal tracts of cultured eel and carp, as well as in the water of these fish ponds (2, 17). In general, *A. hydrophila* is an opportunistic pathogen that can cause extensive mortality among freshwater fish. Recently, this infection with either sensitive or drug-resistant strain, has not been observed in freshwater farms in Japan.

Furunculosis caused by *Aeromonas salmonicida*

The drug-resistant infection with *A. salmonicida* occurred first in 1957 in salmonids farms in the U.S.A. A transferable R plasmid with resistance to SA and TC was detected in *A. salmonicida* which had been isolated during the same period (7). On the other hand, the incidence of drug resistant strains of *A. salmonicida* has increased in amago farms in parallel with the development of cultured amago in Japan (8). Almost all strains harbour transferable R plasmid with resistance to CP, SA, and SM, all of which are classified into Inc group U (18). There was high homology within the DNAs of the R plasmids from *A. salmonicida* from various sources. In particular, drug-resistant strains of *A. salmonicida* carrying R plasmid with the same DNA structure are distributed widely in amago farms (12). Non-transferable R plasmids encoded with resistance to TC were also isolated in *A. salmonicida*(14). The drug resistant strains of *A. salmonicida* carrying R plasmids were isolated in France, Ireland, and U. K. as well (20). Recently, the drug resistance marker of *A. salmonicida* has changed to resistance to NF and PA, and these strains have increased in fish farms, because pyridonecarboxylic acids have been used frequently in Japan (11).

Edwardsiellosis (Parocolon disease) caused by *Edwardsiella tarda*

The drug resistant strains of *E. tarda* carrying R plasmid were isolated first in 1972 (5). By this time, drug-resistant infections with *E. tarda* have occurred frequently in eel farms(13). This infection was also observed in cultured eel in Taiwan (21).

The detected R plasmid were classified into Inc group A-C and encoded with resistance to SA, TC, and to CP, SA and TC (4). The DNA structure of these R plasmids showed high homology within a specific species. The DNA structure of R plasmids with resistance to SA and TC from *E. tarda* was similar to the same resistance marker of the R plasmids from *A. hydrophila*. These results show that drug-resistant strains of *E. tarda* carrying R plasmid with the same DNA structure have prevailed in eel ponds (4).

Pseudotuberculosis caused by *Pasteurella piscicida*

Drug-resistant infection with *P. piscicida* occurred suddenly in yellowtail (*Seriola quinqueradiata*) farms in 1980. The following year the infection spread to all districts of Japan (26). The isolates were resistant to combinations of AP, CP, KM, NF, PA, SA and/or TC. It has been very difficult to combat this infection with commercial chemotherapeutics. Transferable R plasmids were classified into a new Inc. group and encoded with combinations of CP, KM, SA and/or TC. There was high homology of these R plasmids with different resistance markers. Consequently, the multiple drug-resistant strains of *P. piscicida*, carrying an R plasmid with the same DNA structure, were distributed in yellowtail farms in various areas (26).

Vibriosis caused by *Vibrio anquillarum*

An epidemic spread of drug-resistant strains of *V. anquillarum* has occurred in ayu farms in various districts of Japan since 1973 (6). These strains show resistance to various combinations of nine drugs: AP, CL, CP, NF, PA, SA, SM, TC and TMP. The detected R plasmids in 1973 showed resistance to CP, SA, SM and TC. Transferable R plasmids detected from 1974 to 1977 still carried markers for resistance to CP, SA and TC. The number of NF and/or PA resistant strains has increased, and the number of CP, SA and TC resistant strains has correspondingly decreased since 1976. A TMP resistant strain appeared first in 1976 and could not be transferred to *Escherichia coli*. The incidence of drug-resistant infections decreased in ayu farms in 1978 and 1979. However, drug-resistant strains increased again in 1980. Transferable R plasmids encoded with resistance to AP and TMP were detected for the first time in *V. anquillarum* in that year. Drug resistance to CL was also first seen in *V. anquillarum* in 1980; however, CL resistance could not transfer to *E. coli*. The kinds of chemotherapeutics used correlated with the appearance of the corresponding drug-resistant marker of *V. anquillarum* in cultured ayu (9).

The R plasmids from *V. anguillarum* were classified into three groups by their DNA structures (22). Two of these R plasmids groups were detected before 1976, and the other one has been detected since 1980 in ayu farms.

Streptococcal infection caused by *Streptococcus* sp.

Drug-resistant strains of non-hemolytic *Steptococcus* sp. appeared in 1986 in yellowtail farms (15). These resistant strains were classified into intermediate- and high-level resistance to CP, LIM, macrolide antibiotics(MLs), and

TC. Intermediate-level resistant strains were not transferred to recipient strains of *S. faecalis* and non-hemolytic *Streptococcus* sp. On the other hand, high-level resistance to MLs, LIM and TC, and to MLs, LIM and CP was transferable to recipient strains. Transferable R plasmids were detected first in the drug-resistant strains of non-hemolytic *Streptococcus* sp. isolated in cultured yellowtail. These multiple drug resistant infections with *Streptococcus* sp. have been observed in yellowtail farms (15).

The Future

New chemotherapeutics will be developed for treatment of drug-resistant fish-pathogens. As mentioned above, fish pathogens will soon acquire resistance to the new drugs as well. It is very difficult to discover new chemotherapeutics to which fish pathogens will not be able to acquire drug resistance. Drug-resistant infections are assumed to prevail in fish farms. I admit I have no idea how to prevent the increase and spread of drug-resistant strains of fish pathogens in fish farms. If chemotherapeutics were not applied to bacterial infections in fish farms, however, drug-resistant strains would decrease and would not appear in fish farms. In the control of bacterial infections, it is very important to maintain natural favorable conditions in fish ponds and to follow a policy of prevention with vaccines.

References

1). AKASHI A. & AOKI T., 1986. Characterization of transferable R plasmids from *Aeromonas hydrophila. Bull. Japan. Soc. Sci. Fish.*, **52**, 649-655.

2). AOKI T., 1974. Studies of drug-resistant bacteria isolated from water of carp-ponds and intestinal tracts of carp. *Bull. Japan. Soc. Sci. Fish.*, **40**, 247-254 (in Japanese).

3). AOKI T., 1991. Chemotherapy and drug resistance in fish farms in Japan. In, *Diseases in Asian Aquaculture*. Fish Health Section, Asian Fisheries Society, Manila, Philippines (in press).

4). AOKI T., AKASHI A. & SAKAGUCHI T., 1986. Phylogenetic relationships of transferable R plasmids from *Edwardsiella tarda. Bull. Japan. Soc. Sci. Fish.*, **52**, 1173-1179.

5). AOKI T., ARAI T. & EGUSA S., 1977. Detection of R plasmids in naturally occurring fish-pathogenic bacteria, *Edwardsiella tarda. Microbiol. Immunol.*, **21**, 77-83.

6). AOKI T., EGUSA S. & ARAI T., 1974. Detection of R factors in naturally occurring *Vibrio anquillarum* strains. *Antimicrob. Agents Chemother,* **6**, 534-538.

7). AOKI T., EGUSA S., KIMURA T. & WATANABE T., 1971. Detection of R factors in naturally occurring *Aeromonas salmonicida* strains. *Appl. Microbiol.*, **22**, 716-717.

8). AOKI T., EGUSA S., YADA C. & WATANABE T., 1972. Studies of drug resistance and R factors in bacteria from pond-cultured salmonids. I. Amago (*Oncorhynchus rhodurus macrostomus*) and Yamame (*Oncorhynchus masou ishikawae*). *Japan. J. Microbiol.*, **16**, 233-238.

9). AOKI T., KANAZAWA T. & KITAO T., 1985. Epidemiological surveillance of drug resistant *Vibrio anguillarum* strains. *Fish Pathol.*, **20**, 199-208.

10). AOKI T., KITAO T. & ARAI T., 1977. R plasmids in fish pathogens. In, *Plasmids - Medical and Theoretical Aspects*, Mitshuhashi S., Rosival L. & Krcmery V. eds, Avicenum-Czechoslovak Medical Press, Prague, pp.39-45.

11). AOKI T., KITAO T., IEMURA N., MITOMA Y. & NOMURA T., 1983. The susceptibility of *Aeromonas salmonicida* strains isolated in cultured and wild salmonids to various chemotherapeutics. *Bull. Japan. Soc. Sci. Fish.*, **49**, 17-22.

12). AOKI T., MITOMA Y. & CROSA J. H., 1986. The characterization of a conjugative R-plasmid isolated from *Aeromonas salmonicida. Plasmid*, **16**, 213-218.

13). AOKI T., SAKAGUCHI T. & KITAO T., 1987. Multiple drug-resistant plasmids from *Edwardsiella tarda* in eel culture ponds. *Nippon Suisan Gakkaishi*, **53**, 1821-1825.

14). AOKI T. & TAKAHASHI A., 1986. Tetracycline resistant gene of a non-transferable R plasmid from fish-pathogenic bacteria *Aeromonas salmonicida. Bull. Japan. Soc. Sci. Fish.*, **52**, 1913-1917.

15). AOKI T., TAKAMI K. & KITAO T., 1990. Drug resistance in a non-hemolytic *Streptococcus* sp. isolated from cultured yellowtail *Seriola quinqueradiata. Dis. aquat. Org.*, **8**, 171-177.

16). AOKI T., UMEDA T., TAKAMI K., KITAO T., SAITANU K., CHONGTHALEONG A. & PUNYARATABANDHU P., 1990. Drug-resistant *Aeromonas hydrophila* in Thailand. In, *The Second Asian Fisheries Forum*,

Hirano R. & Hanyu I. eds, Asian Fisheries Society, Manila, Philippines, pp. 693-696.

17). AOKI T & WATANABE T., 1973. Studies of drug resistant bacteria isolated from eel-pond water and intestinal tracts of the eel (*Anguilla japonica* and *Anquilla anquilla*). *Bull. Japan. Soc. Sci. Fish.* **39**, 121-130 (in Japanese).

18). BRADELY D. E., AOKI T., KITAO T., ARAI T. & TSCHAPE H., 1982. Specification of characteristic for the classification of plasmids in incompatibility group U. *Plasmid*, **8**, 89-93.

19). HAYASHI F. & INOUE M., 1988. Oxacillin hydrolyzing penicillinase produced by *Aeromonas hydrophila. Nippon Suisan Gakkaishi*, **54**, 2227.

20). HEDGES R. W., SMITH P. & BRAZIL G., 1985. Resistance plasmids of Aeromonads. *J. Gen. Microbiol.*, **131**, 2091-2095.

21). KOU G. H. & CHUNG H. Y., 1980. Drug resistant R^+ bacteria in eel-cultured pond. *Reports on Fish Diseases Research* (III), CAPD Fisheries Series No 3, pp. 1-8 (in Chinese)

22). MITOMA Y., AOKI T., & CROSA J. H., 1984. Phylogenetic relationships among *Vibrio anquillarum* plasmids. *Plasmid*, **12**, 143-148.

23). MUROGA K.,1990. Bacterial infections of cultured fishes in Japan. In, The Second Asian Fisheries Forum, Hirano R. & Hanyu I. eds, The Asian Fisheries Society, Manila, Philippines, pp. 963-966.

24). SAWAI T., MORIOKA K., OGAWA M., & YAMAGISHI S., 1976. Inducible oxacillin-hydrolyzing penicillinase in *Aeromonas hydrophila* isolated from fish. *Antimicrob. Agents Chemother.*, **10**, 191-195.

25). SHOTTS E. B., VANDERWORK V. L. & CAMPBELL L. M., 1985. Occurrence of R factors associated with *Aeromonas hydrophila* isolates from aquarium fish and waters. *J. Fish. Res. Board Can.*, **33**, 736-740.26).

26). TAKASHIMA N., AOKI T. & KITAO T., 1985. Epidemiological surveillance of drug-resistant strains of *Pasteurella piscicida. Fish Path.*, **20**, 209-217.

Discussion

STRUB Dr Aoki, how much antibiotics are being used in Japan in a year for fish farms and what do you recommend for fish farmers in the future?

AOKI The first question, I don't know correctly the value. [Consults Japanese colleagues] Sorry we don't know the value and the second question is very difficult and I have no ideas, because it is a difficult question. [to DeKinkelin]...Do you have any ideas?

DE KINKELIN Difficult question!

FIJAN Have there been any experimental models to simulate R plasmid development within a farm like situation or within an enclosed environment?

AOKI I mentioned my opinion is that plasmid mediated resistance will decrease rapidly if a chemotherapeutant is not used. Chromosomal resistance does not decrease. I surveyed for two years, but this drug is subject to chromosomal mutational resistance.

LAFONT You have shown that in many bacterial species the same plasmid is present in different strains from different farms. Is this due to commercial exchange of animals between farms or do the pathogens travel by themselves from farm to farm?

AOKI May be cultured fish with resistant strains are moved by fish culturists and drug resistance with R plasmid is widely distributed in various areas.

LAFONT It is due to commercial exchange of animals?

AOKI And also fish culturists use chemotherapeutants widely, chemotherapeutic pressure is important for increasing drug resistance.

LAFONT To maintain the resistance of the strain?

AOKI Yes.

JOSEPH Could you explain how your administration can establish such a list of withdrawal periods and dosage regimen for all the antibiotics that you mentioned in your presentation? How is it possible?

AOKI This is difficult. Unfortunately Mr Okamoto of the Japanese Fishery Agency is not here and I cannot answer for him.

RESISTANCE TO ANTIBIOTICS USED IN THE TREATMENT OF FRESHWATER FISH DURING A TEN YEAR PERIOD (1979-1988) IN SWITZERLAND

W. Meier, M. Schmitt & T. Wahli

Fish Disease Laboratory, University of Berne, Länggasstr. 122, CH-3012 Berne, Switzerland

Summary: *Between 1979 and 1988, 2389 bacterial examinations were carried out. A resistance test was performed in 272 cases and included the following antibiotics: Chlortetracycline (A); Chloramphenicol (C); Erythromycin (E); Furoxone (FX); Gentamycin (GM); Oxolinic acid (OA); Sulfadiazin (SD); Sulfonamide-Trimethoprim (ST) and Oxytetracycline (TE).*

The analysis of all bacteria species revealed that the lowest degree of resistance was detected by using OA (19%) and GM (13%). By contrast, SD showed the highest degree (75%). The values for all other antibiotics were between 40-50%.

The analysis of the occurrence of a multiple resistance phenomenon revealed that 118 strains were resistant to less than 33.5%, 73 strains were resistant to 33.5%-66.5%, and 70 strains to more than 66.5% of all antibiotics tested. Altogether, 11 strains were resistant against all tested antibiotics.

The development of single and multiple resistance phenomena against selected bacteria was compared between two 5-year periods (1984-1988 vs 1979-1983). Aeromonas salmonicida *showed a distinct increase of single (18%) and multiple resistance in the period 1984-1988, whereas in contrast, both, the degree of single and multiple resistance of* Aeromonas *sp., most often found in ornamental fish (88%), remained constant or was even lower.* Pseudomonas *sp., most frequently diagnosed in ornamental fish (79%) demonstrated the highest increase of single (39%) and multiple resistance.*

Bacteria detected in non salmonid fish from farms showed a lower degree of resistance to antibiotics (14%) in the period 1984-1988. Inversely, the degree of single and multiple resistance of bacteria of farmed rainbow (11%) and brown trout (22%) clearly increased. The highest level of resistance was found in ornamental fish.

Résumé: **Antibiorésistances observées dans le traitement des poissons d'eau douce en Suisse, pendant 10 ans (1979-1988).**

Au cours des années 1979-1988, 272 antibiogrammes ont été effectués à partir de 2389 analyses bactériologiques, en incluant les antibiotiques suivants: chlortétracycline (A), oxytétracycline (TE), chloramphenicol (C), érythromycine (E), furoxone (FX), gentamycine (GM), acide oxolinique (OA), sulfadiazine (SD), et association sulfamide-triméthoprime (ST).

Toutes bactéries confondues, les antibiotiques les moins concernés par les résistances sont OA (19 %) et GM (13 %) les plus grandes fréquences de résistances s'observant avec SD (75 %), et les autres antibiotiques se situant dans la moyenne (40 à 50 %). L'analyse des résistances multiples montre que dans 118 cas les bactéries résistaient à moins d'1/3 des antibiotiques étudiés. Dans 73 cas la résistance intéressait 1/3 à 2/3 des molécules, et dans 70 cas elle était supérieure à 2/3. Fait plus inquiétant, 11 des souches isolées résistaient à tous les antibiotiques éprouvés.

Le développement de résistances simples ou multiples en fonction des groupes bactériens (Tableaux I et II) a été comparé entre deux périodes de 5 ans (1979 à 1983 et 1984 à 1988). Aeromonas salmonicida *présente une nette augmentation des résistances simples et multiples, dans la période 1984-1988. Au contraire, les fréquences de résistance des autres Aeromonadacées (*Aeromonas *sp.), très souvent isolées de poissons d'ornement (88 %), restent stables, ou régressent.* Pseudomonas *sp., souvent associé aussi aux poissons d'ornement (79 %), révèle la progression la plus sensible des résistances simples (39 %) et multiples.*

Les bactéries isolées de poissons provenant d'élevages autres que salmonicoles présentaient des pourcentages d'antibiorésistances assez faibles (14 %) dans la période 1984-1988. Inversement, les taux de résistances simples et multiples des bactéries isolées de truite fario (22 %) ou arc-en-ciel (11 %) issues de piscicultures n'ont cessé d'augmenter. On remarque néanmoins que les taux de résistances les plus élevés ont été observés chez les bactéries des poissons d'ornement.

Introduction

Together with the expansion of fish culturing in Switzerland, also bacterial infections in fish have increased.

The diagnostic work in the fish disease laboratory also includes proposals for efficient therapeutic measures. In order to have an overview of the bacterial resistance to therapeutically administered antibiotics, all resistance tests performed between 1979 and 1988 were analysed with regard to the presence of a single and multiple resistance phenomenon. Furthermore to investigate the development of the resistance of selected bacteria against various antibiotics, a 5 year period (1984 - 1988) was compared to a previous period of 5 years (1979 - 1983).

Material and Methods

Most of the cases (44 %) from a total of 4272 cases examined in our laboratory (1979-1988), originated from private fish farms, 32 % from aquaria, the others either from rivers and lakes (15 %) or from governmental fish farms (10 %) (Fig. 1). Trout, either rainbow trout (36 %) or brown trout (23 %) were dominating whereas all other species together represented 41 % (Fig. 2).

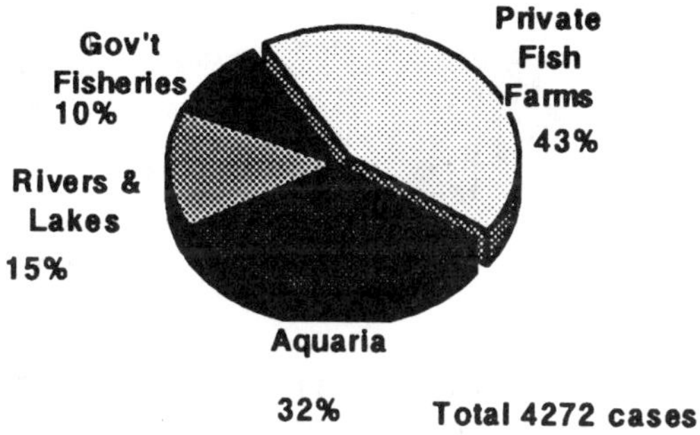

Fig 1 Origin of fish investigated at the fish diseases laboratory 1979-1988

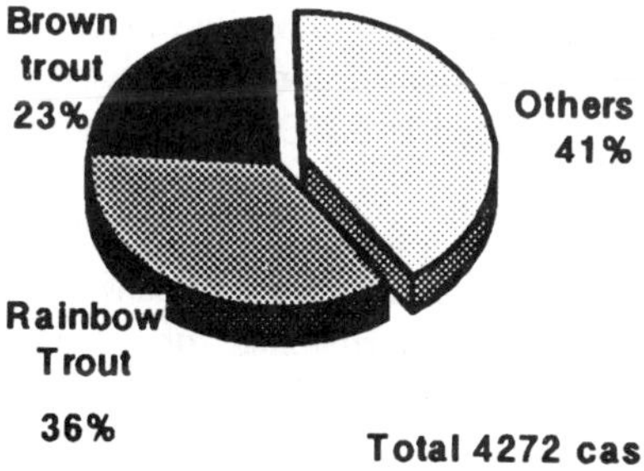

Fig 2 Species of fish investigated at the fish diseases laboratory 1979-1988

Two thousand three hundred and eighty nine bacterial examinations were carried out, either in culture or by means of specific stainings such as Gram or Ziehl-Neelsen (Fig. 3). For the isolation of bacteria, samples from liver, spleen and kidney were taken routinely with a sterile loop and spread on a culture-plate. The following agars were used: Columbia blood with 5 % sheep-blood (4), Bromothymol blue-lactose with 5 % saccharose (5). The identification of bacteria was performed using a slide agglutination test (specific rabbit antisera kindly donated by Dr. P.E.V. Jorgensen, DK) and/or standard tube tests (8).

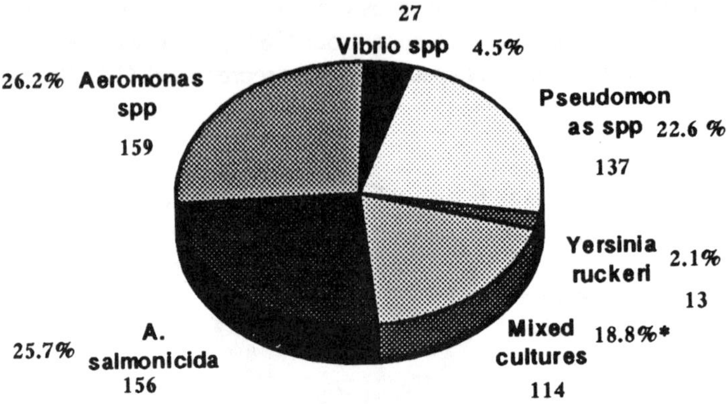

Fig 3 Total of cultured bacteria at the fish diseases laboratory 1979-1988
(* Mixed cultures = growth of two or more species from the same case)

Two hundred and seventy two resistance tests using the following antibiotics were performed: Chlortetracycline (A), Chloramphenicol (C), Erythromycin (E), Furoxone (FX), Gentamycin (GM), Oxolinic acid (OA), Sulfonamide (SD), Sulfonamide - Trimethoprim (ST), Oxytetracycline (TE). These tests were carried out on Mueller-Hinton agar (1), using commercial sensi disks (5). The dishes were incubated at 24°C and the results were read after 24 - 48 hours.

The following groups of bacteria were classified and analysed: all bacteria, *Aeromonas salmonicida*, other *Aeromonas* spp. , *Pseudomonas* sp..

Results

I Analysis of resistance according to bacteria and origin (way of production) of fish

- <u>Single resistance to various antibiotics</u>

Overall, 13 % and 19 % of the bacteria were resistant to GM and OA, resp., whereas 75 % of all cultures were resistant to SD (Fig. 4).

Only 5 of 114 cases with the diagnosis *Aeromonas salmonicida* did not originate from fish farms. The degree of resistance of *A. salmonicida* differed only slightly from the average of all tests. The lowest resistance was noticed for GM and OA and the highest for SD (Fig. 5).

Most of the 51 detected other aeromonads originated from aquaria. Their degrees of resistance were higher than those of *A. salmonicida* (Fig. 6). These strains were nearly as resistant to A (61 %) and E (68 %) as to SD (74 %).

Fig 4 Common Bacterial Resistance

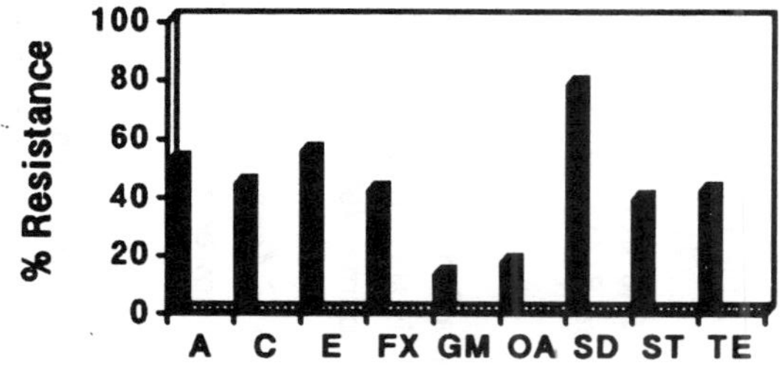

Fig 5 Aeromonas salmonicida

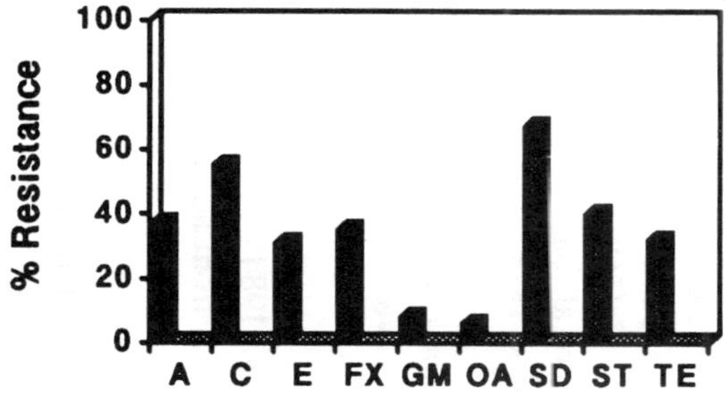

Fig 6 Other Aeromonads

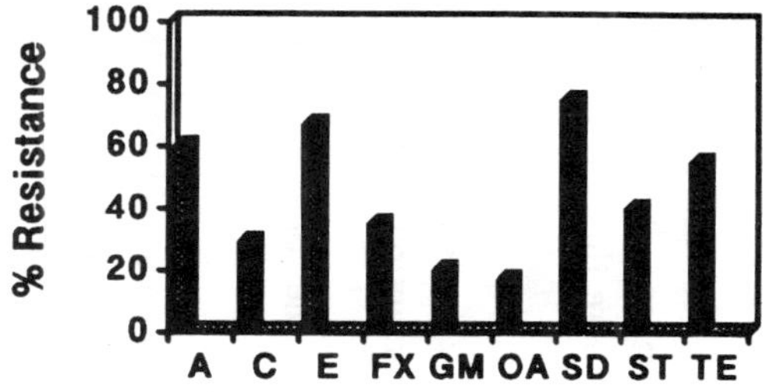

Fig 7 Pseudomonas sp.

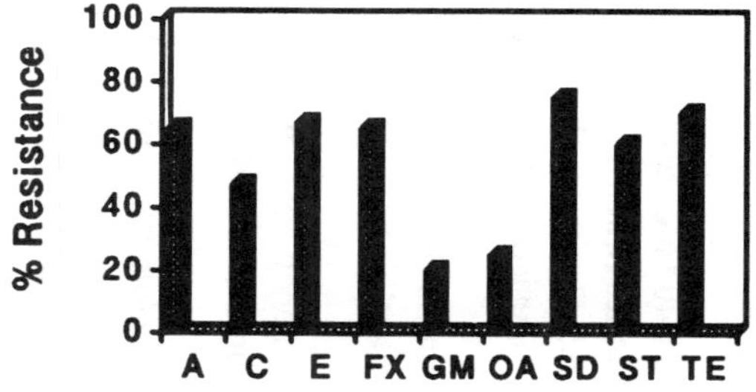

Figs 4 to 7 : Number of bacterial cultures resistant to selected antibiotics in % (1979-1988)

The highest degree of resistance was found in the 43 cases of *Pseudomonas* sp., which again mostly originated from aquaria. Of these, 19 % were resistant to GM and 26 % to OA, whereas 78 % were resistant to SD, 67 % to A, 70 % to E, 68 % to FX, 57 % to ST and 70 % to TE (Fig. 7).

- Multiple resistance to various antibiotics

More than 40 % of all bacterial cultures were resistant to less than 33.5 % of the antibiotics when tested for multiple resistance. 27 % were resistant to 33.5 % - 66.5 %, 26 % showed resistance to more than 66.5 % and 4 % of cultures were resistant to all antibiotics.

More than 50 % of all strains of *A. salmonicida* were resistant to less than 33.5% of all antibiotics, but in 3 cases not a single antibiotic was effective.

The frequency of multiple resistance of the other aeromonads was essentially similar to that of all other bacteria.

In the *Pseudomonas*-group, the high number of cases (44 %) with a multiple resistance to more than 66.5 % of all antibiotics is considerable. A 100 %-resistance was found in three cases.

II Development of resistance comparing between two 5-year periods (1984-1988 vs 1979-1983)

- Single resistance to various antibiotics

Generally, an increase of resistance was detectable (Fig. 8-11). The highest levels were found for SD, (41%), followed by TE (24%), FX (21%), potentiated ST (21%), C (16%), A (13%) and GM (8%). The development of a single resistance of all bacteria tested is shown in Figure 8. OA and E were not used in the period 1979 - 1983.

Aeromonas salmonicida - 93% of cases with a furunculosis originated from farms - showed a distinct increase of single resistance (Fig. 9). The minimal difference of increase was 3% for GM and the maximal 46% for SD.

Between 1984 - 1988, the degree of resistance of *Aeromonas* sp., most often found in fish from aquaria and ponds (88%), was lower except for SD and TE (Fig. 10).

The highest increase of resistance with differences from 24% for GM to 62% for ST was found in *Pseudomonas* sp., 80% of them were isolated from fish of aquaria and ponds (Fig. 11).

Fig 8 Total of all bacteria tested

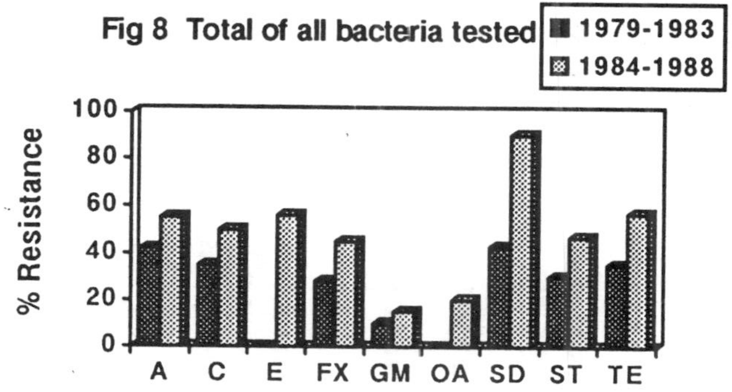

Fig 10 Other Aeromonas spp

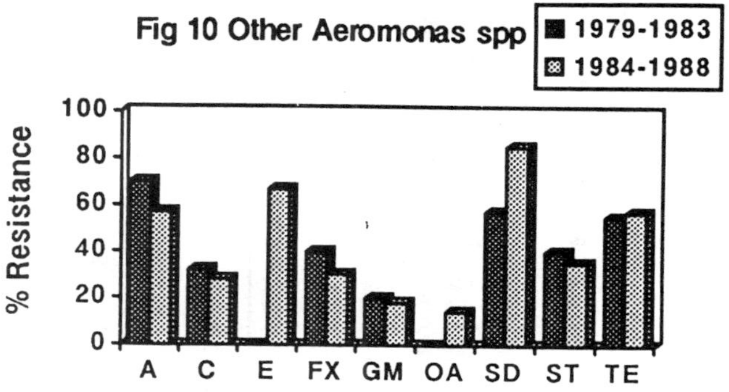

Fig 9 Aeromonas salmonicida

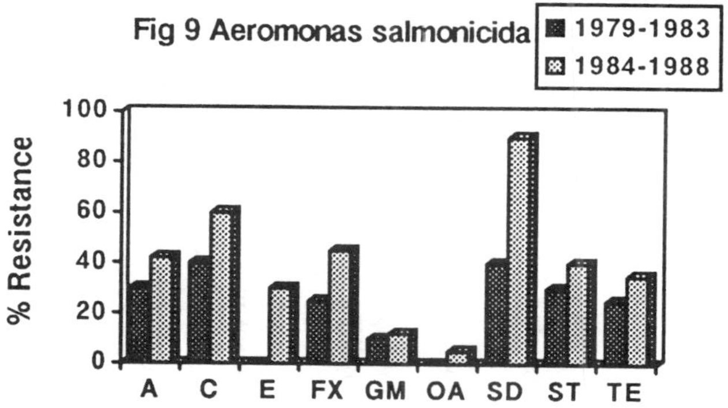

Fig 11 Pseudomonas spp

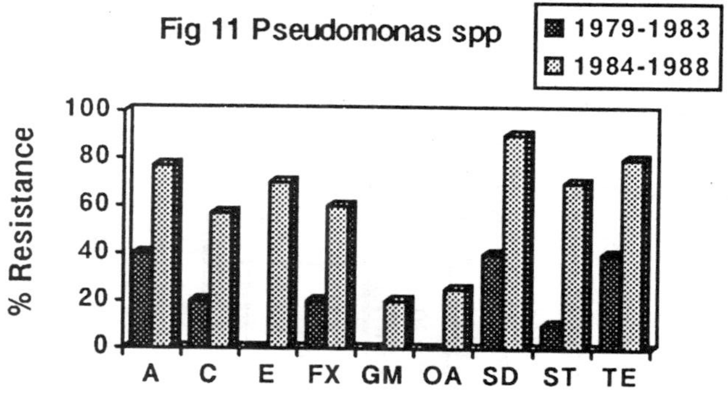

Figs 8 to 11 Resistance (%) of different strains to selected antibiotics : Comparison of two 5 year periods

- Multiple resistance to various antibiotics

In the period 1984 - 1988 more bacterial strains proved to be resistant to a higher number of antibiotics and only in *Aeromonas* sp., a decrease was noticed (Tab. 1).

Table I: Development of multiple resistance: changes of multiple resistance (%) between the period 1979 - 1983 and the period 1984 - 1988.

Resistance to antibiotics (in %)	Total	*Aeromonas salmonicida*	*Aeromonas* sp	*Pseudomonas* sp
< 33.5	- 18	- 15	+ 9	- 52
33.5-66.5	- 2	+ 7	- 3	- 6
> 66.5	+ 20	+ 8	- 6	+ 58
100	+ 4	+ 5	- 6	+ 9

- Single resistance in relation to fish species and origin

The comparison of the origin and fish species revealed that the bacteria isolated from farmed rainbow trout were more resistant to all substances except GM in the period 1984 - 1988 (Fig. 12).

Similar results were achieved with bacteria from farmed brown trout, but the increase of resistance was clearly higher (Fig. 13).

In farmed non-salmonids, a decrease of the average degree of resistance of 13% was noticed although in these cases the resistance to SD was higher in the period 1984 - 1988 (Fig. 14).

The highest increase (ranging from 14% for GM to 41% for SD) was seen in bacteria isolated from fish from aquaria and ponds (Fig. 15).

- Multiple resistance in relation to fish species and origin

The multiple resistance in relation to the origin and fish species developed in a rather different way (Table II).

A decrease of the number of strains resistant to a few and an increase of those resistant to a high number of antibiotics was noticed in fish from aquaria and ponds. A similar, but less pronounced tendency was seen for brown trout, whereas in rainbow trout only an increase of strains resistant to more than 66.5% of the antibiotics was observed. In contrast, strains isolated from non-salmonid fish clearly showed less multiple resistance compared to the period 1979 - 1983.

Fig 12 Farmed rainbow trout

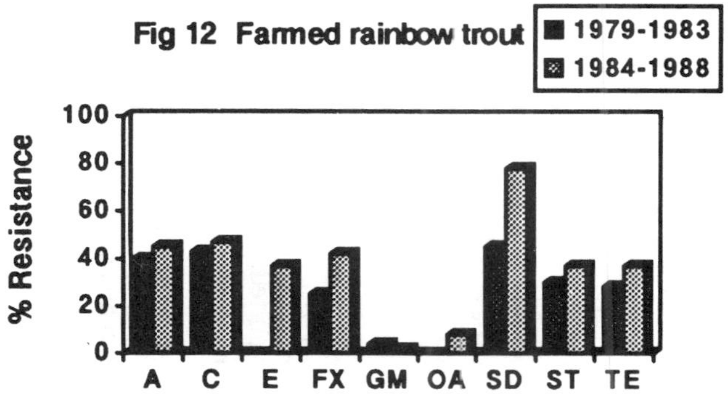

Fig 14 Farmed non-salmonids

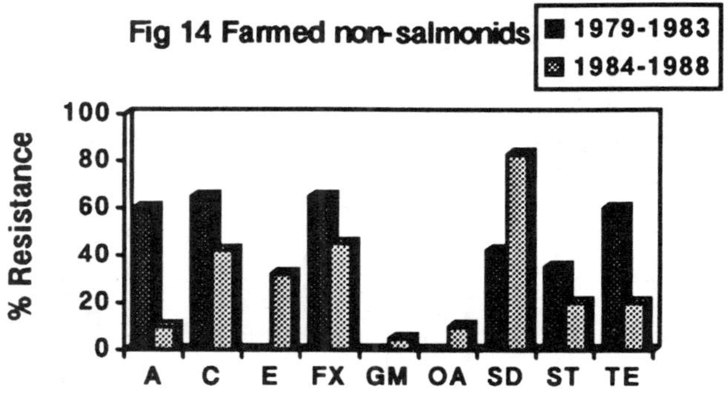

Fig 13 Farmed brown trout

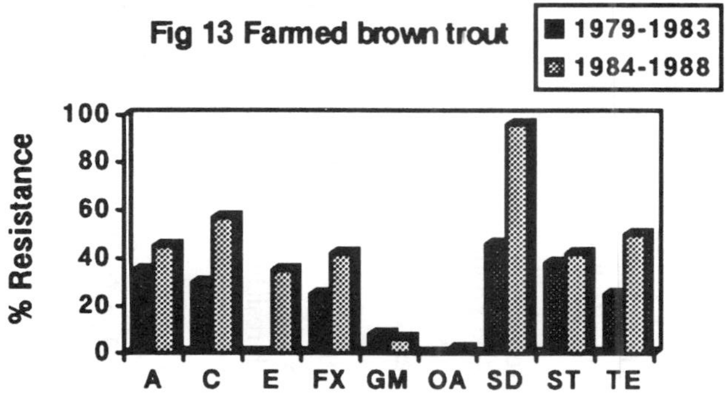

Fig 15 Fish from aquaria/ponds

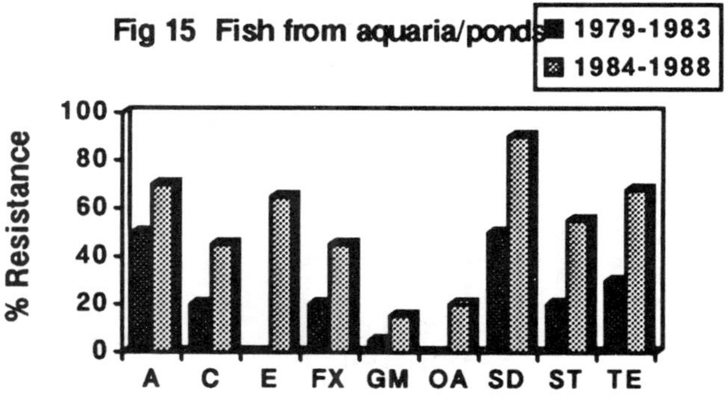

Figs 12 - 15 Resistance of bacteria in relation to fish species : Comparison of two 5 year periods

Table II: Development of multiple resistance in relation to fish species: Change of multiple resistance (%) between the period 1979 - 1983 and the period 1984 - 1988.

Resistant to antibiotics (in %)	Rainbow trout	Brown trout	Non salmonid fish	Fish from aquaria and ponds
< 33.5	- 3	- 22	+ 29	- 29
33.5-66.5	- 3	+ 17	- 25	- 2
> 66.5	6	+ 5	- 4	+ 31
100	0	+ 2	0	+ 8

Discussion

The results of this investigation clearly demonstrated a high proportion of single and multiple resistance of bacteria to commonly used antibiotics in Switzerland. In 3 cases of furunculosis none of 9 widely used broad spectrum antibiotics was effective. This is alarming when the damaging effect of this disease is considered (7, 2).

In general, the degree of single and multiple resistance was higher in the period 1984 - 1988. This essentially coincides with the results of studies from other countries (9, 3, 6, 10).

However, our findings reveal some exceptions: *Aeromonas* sp. were considered - in contrast to all other strains - as less resistant in the second period, although most isolates like *Pseudomonas* sp. originated from fish from aquaria and ponds. The reasons still remain to be explored.

The difference of resistance between rainbow trout and brown trout could be fish species-independent. In our country, an increase of brown trout raised in hobby-farms was noticed in the last few years and was often combined with a rather careless use of antibiotics.

The distinct increase of bacterial resistance in fish originating from aquaria or ponds is not surprising. A lot of ornamental fish proved to be already infected with resistant strains of bacteria when imported. This situation became worse by the common and unprofessional application of chemotherapeutics.

In conclusion it can be stated that the use of antibiotics in fish must be very restrictive and performed under professional supervision only.

References

1). BAUER, A.W., KIRBY, W.M.M., SHERRIS, J.C. and TURCK, M., 1966. Antibiotic susceptibility testing by a standardized single-disk method. *Amer. J. Clin. Path.*, **45**, 493-496.

2). BULLOCK, G.L., CIPRIANO, R.C. and SNIESZKO, S.F., 1983. Furunculosis and other diseases caused by *Aeromonas salmonicida*. *U.S. Fish and Wildlife Service., Fish Disease Leaflet 66*, 29 pp.

3). CHESCHIA, G., GIORGETTI, G., BERTOLINI, G., FONTEBASSO, S., 1987. The *in vitro* sensitivity of *Yersinia ruckeri* to specific antibiotics. *Journal of Fish Disease*, **10**, 65-67.

4). ELLNRER P.D., STOESSEL, C.I., DRAKEFORD, E. and VASI, F., 1958. A new culture medium for medical bacteriology. *Amer. J. Clin. Path.*, **29**, 181-183.

5). FEY, H., 1978. *Kompendium der allgemeinen medizinischen Bakteriologie*. Verlag Paul Parey, Berlin, Hamburg.

6). HERBST, J., SCHLOTFELDT, H.-J., ALVARADO, V., STANISLAWSKI D., BÖHM, K.H., 1989. Gegenwärtige Resistenzlage fischpathogener Bakterien gegenüber einigen antibiotischen Wirkstoffen im Einsatz bei Nutz- und Zierfischen in Niedersachsen. *Fischer und Teichwirt* **2**, 42-47.

7). McCARTHY, D.H., 1975. Fish furunculosis. *Journal of the Institute of Fisheries Management*, **6**, 13-18.

8). MEIER, W., 1986. Enteric redmouth disease: outbreak in rainbow trout in Switzerland. *Dis. aquat. Org*. **2**, 81-82.

9). SCHLOTFELDT, H.-J., FUHRMANN, H., PFORTMÜLLER, K., BÖHM, K.H., SALTZMANN H.A., 1984. Anmerkungen zur Frage der (zunehmenden) Resistenz fischpathogener - und fakultativ - fischpathogener Bakterien in Niedersachsen gegenüber den für Fische zugelassenen Arzneimittel - Vermischungs - Wirkstoffen, mit besonderem Hinweis auf Lincospectin und Oxolinsäure. *Tierärztl. Umschau*, **39**, 974-988..

10). TSOUMAS, A., ALDERMAN, D.J., RODGERS, C.J., 1989. *Aeromonas salmonicida*: Development of resistance to 4-quinolone antibiotics. *Journal of Fish Disease*, **12**, 493-507.

Disussion

MICHEL I was interested in that your paper outlined the importance of ornamental fish in the increase of resistance. I would like some additional information. Is the ornamental fish market an important one in Switzerland?

MEIER Yes, we have some figures and think that people are spending more money for ornamental fish than for dogs and cats, so we are sure that it is important, especially since marine and ornamental fish are imported from all over the world, especially from the far east.

MICHEL. It seems like the USA a few years ago. Are there any restrictions or controls on the selling of treatments for pet fish? In France it is not the case.

MEIER No, there is no control for ornamental fish, that is a problem. It is so difficult to control, so they let them be. It is nearly impossible to control all importation because most are coming in small quantities from different countries.

MICHEL Certainly I think its a problem and we don't pay enough attention to this aspect of pet fish.

SCHLOTFELDT This increase of resistance in ornamental fish is not surprising. In the N part of Germany there are several big rearers and traders of ornamentals and we have been there several times when we have examined the water of the bags that the fish come in from Singapore and Hong Kong. Tetracyclines are used at enormous concentrations in the transport water. We have cases of ornamentals in which the bacteria isolated are resistant against all the compounds officially used for fish in Germany. There is a clear danger of spread, because these rearers not only maintain warm water ornamentals, but outside in ponds they have a lot of "cold water" ornamentals, shubunkin, koi etc, so I think that spread will be automatic since it is the same facility.

LAFONT You have reported a frequency of gentamicin resistance of about 10% in 1983, increasing to about 20% in 1988 and, as you have mentioned, gentamicin is not used in therapy of fish.

MEIER No, it is not used in farmed fish, but it is used in ornamentals.

LAFONT Does it mean that all the strains of gentamicin resistant *Aeromonas* you had were isolated from ornamentals.

MEIER You did see correctly, we have found some in farmed fish. Gentamicin is not used in our country, but more than two thirds of our fish are imported so I cannot say what is happening in other countries, but I am nearly sure it will not be used for farmed fish because of the cost, so there must be some transmission, but I don't know how it is happening.

VARIATION IN ANTIBIOTIC RESISTANCE PATTERNS OF *AEROMONAS SALMONICIDA* ISOLATED FROM ATLANTIC SALMON *SALMO SALAR* L IN SCOTLAND.

R. H. Richards, V. Inglis, G. N. Frerichs & S. D. Millar

Institute of Aquaculture, University of Stirling, Stirling, FK9 4LA, Scotland

Summary: *Antibiotic resistance patterns of 444 isolates of* **Aeromonas salmonicida** *from 354 outbreaks of furunculosis in Scotland between 1988 and 1991 were investigated.*

Although a slight increase in overall resistance appeared to occur during the period, a remarkable stability in multiple resistance was present with the notable exception in 1990-91 of an increase in multiple resistance to 5 of 6 available licenced compounds. The level of resistance to amoxycillin, the most recently licenced and utilised antibiotic, was very low and occurred in the absence of known exposure to the compound. The number of antibiograms associated with a single outbreak remained relatively constant during the period, 23-29% of outbreaks being associated with two or more antibiogram variants. The possible route of spread of resistance is discussed together with the need for constant monitoring of antibiograms as an aid to disease control.

***Résumé*. Variations des résistances aux antibiotiques chez les souches d'**Aeromonas salmonicida **isolées du saumon d'Ecosse.**

La furonculose est la bactériose économiquement la plus grave pour les élevages de saumons écossais. Le développement des fermes intensives, l'apparition de résistances aux antibiotiques autorisés et l'efficacité limitée des vaccins en rendent le contrôle difficile. L'évolution des résistances peut être appréhendée à travers les données de l'Institut d'Aquaculture de Stirling: de 1988 à 1991, 354 flambées de maladie dans 46 exploitations ont permis d'isoler 424 souches dont le spectre de sensibilité a été étudié (tableau I).

Le tableau II montre que les pourcentages de résistances observés contre les différents antibiotiques ont peu varié, sauf pour l'oxytétracycline, et que les résistances combinées sont fréquentes. On note en 1990-91 une poussée des résistances à l'acide oxolinique et à la tétracycline, ainsi que l'apparition de résistances multiples pouvant intéresser 5 moléculess (tableau III). De rares résistances à l'amoxicilline sont apparues indépendemment de son emploi. Enfin le tableau IV indique que plusieurs types de résistances combinées peuvent être observés sur un même site.

L'interprétation doit évidemment tenir compte des aléas d'échantillonnage liés à l'activité de diagnostic, mais la progression des multirésistances est d'autant plus inquiétante que peu de formules sont disponibles. L'amoxicilline, récemment autorisée, est encore peu employée, l'association sulfamides-triméthoprime pose des problèmes d'inappétence, et l'inefficacité croissante de l'acide oxolinique et de l'oxytétracycline conduit certains éleveurs à préférer l'abattage suivi d'une période de jachère aux aléas d'un traitement coûteux. L'association fréquente de souches bactériennes différentes semble désigner le transfert des smolts en mer comme mode principal de propagation de l'infection, de même que la rareté des types nouveaux ne plaide pas pour une survie durable de l'agent dans le milieu marin. Il reste que davantage de produits autorisés et des vaccins plus efficaces seraient bienvenus pour limiter l'impact majeur de la furonculose sur les élevages.

Introduction

Furunculosis continues to be the most significant bacterial disease affecting the Scottish salmon farming industry, producing substantial financial loss through mortality and the cost of treatment.

With recent expansion of the industry, control of the disease has become more problematical, not least because of recent development of multiply resistant strains of *A salmonicida* to antibiotics currently licenced for fish use, (6,7). Though vaccines are currently used as an aid to furunculosis control, their use has not led so far to a significant reduction in the necessary use of antibiotics in order to reduce mortality from the disease. The use of such antibiotics has been reviewed (2). In this study, the records of the diagnostic service of the Institute of Aquaculture, University of Stirling have been utilised to quantify the development of resistance.

Materials and Methods

Field isolates were obtained during the period April 1988 to January 1991 from 354 sequential outbreaks of furunculosis at 46 salmon or smolt rearing units distributed over 36 geographically separate seawater or freshwater sites. The majority of sites were located on the west coast of mainland Scotland and the Hebridean islands with a few inland and one on the Shetland Islands and are referred to in a previous paper (7).

The frequency of submission of samples varied between sites and years but almost all sites were represented on a number of occasions. Repeat specimens, sent within two weeks from the same site, giving bacteria with identical sensitivity patterns, were included in the analysis only once. The location and frequency of sampling was in response to requests to the diagnostic service and includes variation in stock origin and ownership. Figures therefore are seen as demonstrating trends.

Four hundred and twenty four *A salmonicida* isolates were tested for sensitivity to the range of antibiotics and at the antibiotic levels given in Table I according to the disk diffusion method (1). Bacterial lawns were prepared from colony suspensions spread on Iso-sensitest agar (Oxoid), zones of inhibition read after incubation for 24 hours at 2_C and sensitivity assessed according to the scale in Table 1. In this analysis only the first two categories (+++ and ++) were considered sensitive, since antibiotic treatment would not be recommended if there was a lesser degree of sensitivity.

Table I Bacterial inhibition zone diameter and level of antibiotic sensitivity

Antibiotic disk	Dose mg/ml	Relationship of minimum size (mm) to sensitivity			
Oxytetracycline	30	28	19	12	0
Oxolinic acid	2	25	17	12	0
Co-trimoxazole	25	29	20	11	0
Sulphafurazole	100	29	20	11	0
Nitrofurantoin	100	26	17	11	0
Furazolidone	50	26	17	11	0
Amoxycillin	10	29	23	14	0
Level of sensitivity		+ + +	+ +	+	0

Results

The sample consisted of 444 isolates. The percentage of isolates resistant to a range of antibiotics between 1988-1991 is shown in Table II. Though fluctuation in the figures occurs from year to year, a slight increase in resistance appears to be developing over the time period concerned, with the exception of resistance to oxytetracycline, which was already high (53% of isolates) in 1988-1989 and produced similar values (50%) by 1990-91.

The presence of multiple antibiotic resistance shows remarkable stability over the three years examined, with the exception of a particular increase in 1990-1991 in resistance to all 5 antibiotics used in the original screening procedure.

The patterns of multiple resistance were examined in Table III. Perhaps of greatest note is the increase in 1990-91 in the number of strains resistant a) only to oxolinic acid b) to all five original antibiotics and c) to a lesser extent to both oxolinic acid and oxytetracycline.

Table II Resistance of *A. salmonicida* to antibiotics between 1988-1991

Resistance to:	Resistant strains % in		
	1988-89	1989-90	1990-9 1
None	18	19	12
Oxytetracycline	55	55	50
Oxolinic acid	41	31	54
Co-trimoxazole	11	10	15
Nitrofurantoin	26	26	37
Furazolidone	26	28	34
Amoxycillin		0(n = 31)	2
Number of strains tested	122	144	178

In 1988-89 there were 14 distinct antibiotic resistance patterns (antibiograms) among 122 isolates, 18 in 1989-90 among 144 and 19 in 1990-91 among 1128. Twelve such patterns were common in all years. Resistance to the nitrofurans was usually shared (in 108 out of 124 cases of nitrofuran resistance).

Four new antibiograms were also found in 1990-91 and of four strains found for the first time to be resistant to amoxycillin in 1990-91, three were also resistant to the two nitrofurans and one was resistant to only oxytetracycline and oxolinic acid in addition to amoxycillin.

Isolates with different antibiograms were obtained together from the same site but from separate sampling points, e.g. cages within a site. On other occasions, they were isolated from the same sampling point and on a very few occasions from the same fish. Distribution of antibiograms to individual site or outbreak samples is shown in Table IV. Isolates with a single antibiogram were obtained from 74% of outbreaks in 1988-89, 77% in 1989-90 and 71% in 1990-91, the remaining outbreaks being associated with two or more sensitivity patterns.

Discussion

Resistance in this study was determined by a fixed cut-off point on a semi-quantitative scale, observing quality control procedures in disk testing (5). A good reproducibility was obtained using different technical staff as indicated by previous authors (8) although some slight variation may have occurred near the specified cut-off points. The use of the disk diffusion method as a means of judging appropriate "field" treatments over the three years in question has met with excellent clinical success.

Table III Antibiotic resistance patterns of *A. salmonicida* isolates.

Resistance to	Isolates in: 1988-89	1989-90	1990-91
0	22	28	22
1	30	43	24
2	11	7	30
4	0	1	3
5	0	1	1
1,2	19	13	25
1,3	2	3	1
1,5	1	2	1
2,3	1	2	1
2,4	0	1	0
2,5	0	1	0
4,5	12	17	19
1,2,3	4	6	8
1,2,4	0	0	1
1,3,4	1	0	0
1,4,5	0	2	9
2,4,5	9	7	10
3,4,5	0	0	1
1,2,3,4	0	0	2
1,2,4,5	5	6	6
1,3,4,5	4	2	0
1,3,4,5	0	0	2
1,2,3,4,5	1	2	12
Number of strains	122	144	178

Key

1 = oxytetracycline
2 = oxolinic acid
3 = co-trimoxazole
4 = nitrofurantoin
5 = furazolidone

The nature of the sampling programme used, relying on the isolation of bacteria from clinically affected fish provided by clients of the Institute's diagnostic service, demands considerable caution in the interpretation of results. Such results should perhaps be reviewed qualitatively rather than quantitatively as vigilance and submission of samples varied considerably from site to site and multiple submissions were received from some sites in the absence of serious clinical disease outbreaks. By contrast, of course, all outbreaks eventually requiring antibiotic therapy were included in these results.

Table IV Antibiotic sensitivity patterns (antibiograms) of A. salmonicida isolates from outbreaks of furunculosis

Number of antibiograms per outbreak	Number of outbreaks %		
	1988-89	1989-90	1990 91
1	74%	77%	71%
2	16%	17%	20%
3	7%	6%	6%
4	3%	0%	2%
5	0%	0%	1%
Total outbreaks	89	140	125

Of greatest concern is the increase in multiply resistant strains noted in 1990-91. The figure has changed from 19% resistant to 3 or more antibiotics in 1988-89 to 17% in 1989-90 to 29% in 1990-91. The number of strains resistant to all of the original five antibiotics in the same period changed from <1 to 1 to 7%. Perhaps of particular significance is the fact that this increase has taken place despite attempts at fallowing sites with multiply resistant strains of *A. salmonicida*.

Essential to the survival of the industry during the latter part of the period of study was the introduction of amoxycillin. Disk sensitivity to amoxycillin was introduced into this study during the latter months of 1989-90 so that the last 31 strains were tested and all found to be sensitive to amoxycillin. In 1990-91, all 178 strains were tested, and of these only 4 strains were found to be resistant to amoxycillin.

Furthermore, all 4 strains were isolates from sites and fish in which amoxycillin had never been used and probably represented naturally present resistance. Of those sites where amoxycillin has been used in recent months, no resistance has yet been observed though this early result should be treated with great caution - the same was true of other antibiotics when originally used.

Table II clearly demonstrates considerable resistance to oxytetracycline and oxolinic acid, the two most frequently used antibiotics, a considerable rise occurring in 199091 in oxolinic acid resistance compared to the previous year. Resistance to potentiated sulphonamides remained low but problems of palatability with the currently licenced preparations reduces the success of these compounds in practice. The nitrofurans are also clearly still usable but are affected by palatability problems and concern over their longer term future as antimicrobials - there are currently no licenced forms available as specific fish medicines but other formulations can be used under the dispensation allowing the use of licenced antimicrobials from other species when there is no available licenced alternative for the species concerned.

When sensitivity to a number of antimicrobials is present, the choice of treatment depends very much on efficacy of uptake (esp. absence of palatability problems) as well as price so that when considering development of new antimicrobials a significantly increased price per treatment will result in restricted usage except in cases of multiple resistance. Slaughter of affected stock may also be decided upon either a) to harvest the stock and avoid treatment cost and withdrawal period problems or b) to allow fallowing to take place, despite the complete loss of that stock, in an attempt to remove a mixture of strains or a particularly resistant strain from the environment.

Table IV demonstrates the number of antibiograms obtained in single outbreaks of disease. This is dependent on several factors including the number of fish examined per outbreak, the percentage of that number which yield isolates and the variation in isolates obtained from a single fish. When obtaining pure cultures from single colonies on secondary isolation, one colony only is usually sub-cultured from a plate, but to obtain a rapid idea of field sensitivity, disks are also placed onto parallel primary cultures. Any variation in apparent sensitivity is checked by further subculturing so that the results obtained probably do represent the variation present. Multiple antibiograms per outbreak are most commonly obtained from sites purchasing multiple smolt stocks from a variety of producers, indicating the probable role of infected smolts in spreading the disease. This contrasts with the reported situation in Japan (3) where drug resistance appeared to be directly related to enhanced usage.

The use of antibiotics is however strictly controlled by the veterinary profession in Scotland ensuring a correct dosage rate for an appropriate period, unlike the uncontrolled drug usage situation in Japan. Many sites purchasing from the same smolt source every year, though suffering endemic furunculosis, have had the same single antibiogram both prior to (unpublished observation) and during the 3 year period, despite frequent treatment with a single antibiotic drug during the period.

The situation is less clear cut when sites are situated in close proximity to one another, when wild or escapee fish, or simply flow of water and suspended solids from one site to another, appear to introduce new strains of the bacteria. Fortunately, as will be seen from Table III, this problem does not seem to have become dramatically worse over the period in question. The variations obtained do underline the need for extensive and continual routine monitoring of sites, so that an emergency treatment choice can be made prior to detailed bacteriology results being available .

Attempts have been made to utilise antibiograms in the study of furunculosis epidemiology in the absence of multiple serotypes of this organism. It is generally considered, however, that such variation is not sufficiently specific or stable for this use except in the demonstration of generalised trends (9, 4).

The lack of increase in numbers of antibiogram in the face of antibiotic therapy may well indicate that, though new resistant variants are likely to be created during treatment of an outbreak, such strains may not have particularly good survival characters.

This study demonstrates the need for continued bacterial monitoring in order to improve control of furunculosis, the urgent need for a wider range of licenced antibacterial agents for fish and suggests the significant advantage that could be created with really efficient furunculosis vaccines. Furunculosis, with sea-lice infection, continues to be the major problem facing the salmon farming industry in Scotland .

Acknowledgements We are grateful to the members of the Diagnostic Service, Institute of Aquaculture, University of Stirling whose work provided the basis for this study.

The work was supported by the Scottish Salmon Growers Association.

References

1). ACAR, J.F. & GOLDSTEIN, F.W. 1986. Disk susceptibility Test. In: *Antibiotics in Laboratory Medicine*, 2and ed. (ed. by W. Lorain), pp. 27-60. Williams and Wilkins, Baltimore.

2) ALDERMAN, D.J. 1988. Fisheries chemotherapy: A review. In: *Recent Advances in Aquaculture* 3 (eds. J.F. Muir & R.J. Roberts), pp. 161. Croom Helm, London.

3). AOKI, T., KITAO, T., LEMURA, N., MITOMA, Y. & NOMURA, T. 1983. The susceptibility of *Aeromonas salmonicida* strains isolated in cultured and wild salmonids to various chemotherapeutics. *Bull. Jap. Soc. Sci. Fish.* **49**, 17-22.

4). BAST, L., DALY, J.G., DEGRANDIS, S.A. & STEVENSON, R.M.W. 1988. Evaluation of profiles of *Aeromonas salmonicida* as epidemiological markers of furunculosis infections in fish. *J. Fish Dis.*, **11**,133-145.

5). BAUER, A.U., KIRKBY, W.M.M., SHERRIS, J.C. & TURCH, M. 1966. Antibiotic susceptibility testing by a standardised single disk method. *Amer. J. Clin. Path.*, **45**, 493-496.

6). DAFS 1989. *Marine Laboratory Aberdeen Annual Review 1987-88.* Her Majesty's Stationary Office.

7). INGLIS, V., FRERICHS, G.N., MILLAR, S. & RICHARDS, R.H. 1991. Antibiotic resistance of *Aeromonas salmonicida* isolated from Atlantic salmon, *Salmo salar* L. in Scotland. *J. Fish Diseases* (in press).

8). MURRAY, P.R., ZEITINGER, J.R. & KROGSTAD, D.J. (1982). Reliability of disc diffusion susceptibility testing. *Infection Control and Hospital Epidemiology,* **3**, 230-237.

9). VON GRAEVENTIZ, A. 1986. Use of Antimicrobial Agents as Tools in Epidemiology. In: *Antibiotics in Laboratory Medicine, 2nd ed.* (ed.by V. Lorian), pp. 810-814. Williams & Wilkins, Baltimore

Discussion

WITHNELL Dr Richards, can we draw any hope from the figures for oxytetracycline showing a reduction in resistant strains.

RICHARDS I think we can get some hope from the fact that I think that the figures don't seem to have changed at all. It is more or less a 50% number of strains that are resistant at present and it is still a very widely used and useful drug. I think it is very fortunate with its very extensive use that we haven't seen a major increase in resistance, so, hopefully we can keep on using it. Our biggest problem at the moment is the small numbers of strains that are developing that have multiple resistance which, of course, without effective treatment or without eradication of those stocks, will eventually lead to spread of those strains to affect a larger part of the industry. That is our biggest concern I think.

LEWIN Could I make a point on that. Based on your methods I don't think you can actually say that, because you are using disk sensitivities, so that your sensitive strains could actually be gradually creeping up in resistance and then you get a breakthrough of high level resistance. Quinolone resistance and other resistance is often a two step thing, of low level and then acquisition of high level, so disk sensitivity would not actually tell you that information, it will just tell you that your strains are sensitive, it won't tell you the level.

RICHARDS I think that that is fair comment, but we have been looking at a large number of strains with quite a degree of treatment going on for a long period of time. It is interesting that as far as the quinolones are concerned we saw a very rapid development of resistance to quinolones, during treatment very often. We suddenly found that on isolating strains during a treatment period, resistance developed during that period, to a high degree. I think it is encouraging that despite this tetracycline treatment having been carried out for some time, we have not seen that rapid an increase in resistance. These are seen to be trends, very clearly and I was trying to show you figures based on the use of these methods in a diagnostic service which still I think gives useful information.

AOKI How did you cross reference drug resistance on the sensitive strains, particularly nitrofurantoin and furazolidone, because are you sure of the number that are different.

RICHARDS With those two agents we were using disks that were readily available at the start of our screening period. In most cases, resistance seems to be present to both, if it was present at all, but there were some cases certainly of resistance to one or the other.

AOKI There is usually cross resistance you know.

RICHARDS Yes.

SCHLOTFELDT If you carry out during an outbreak several antibiogrammes, how long does it take to come to decision as to which drug to use?

RICHARDS In the clinical situation, if we have an outbreak of furunculosis, treatment must begin that very same day, before we have had the antibiogramme from isolates from those currently infected fish. So we monitor the situation on an on-going basis to know the best first choice of drug. The results we get from the next form of isolation will take two to three days. We hope that we will have made the right choice on that basis, we then if necessary would change that treatment. You have to be careful of course, if you're sampling fish during a outbreak. By treatment, we are really trying to stop the healthy fish becoming infected. So if you continually take samples from sick fish, you will not see the influence of that antibiotic treatment on them as they will not have taken the treatment concerned. So what we are trying to do is respond quickly before appetite goes, hope to get the right treatment and follow that up with a second treatment if necessary. In 95% of cases I would say, the treatment that we choose before we have the full results from that outbreak is the right one.

THOMEN I would like to make a small point to something that you said. You are the first person so far to mention something about finance. I think that if the fish farming community to a greater extent supported the companies which have licensed products they would stimulate an interest in developing more and better products for fish farming.

RICHARDS As far as the UK is concerned, we are only allowed to use licensed compounds, there is strict control. I should also point out that the Scottish Salmon Growers Association, based on a levy of production is in fact financing a search for new compounds with these big pharmaceutical companies, both in terms of new antibiotics and new lice treatments.

GRIVIERS How could you explain the dosage of the antibiotic disk, because for oxolinic acid you use 2 mg per disk, in France I think it is 10 mg. how do you explain this difference?

RICHARDS The dosage is based on MIC work showing that the MIC's used reflect the amount in that disk. It maybe depends on the dose rate used in France for treatment. I don't know what dose rate is used in France for treatment.

GRIVIERS The dosage is 12 mg/kg.

RICHARDS We use 10mg/kg, and it certainly seems to be parallel to the findings we get.

INGLIS The zone sizes simply indicate the kind of positive results we get in treatment.

CHAPMAN You commented on palatabilty problems with surface coating with potentiated sulphonamides. Have you much experience of in pellet incorporation of potentiated sulphonamides and, if so, what problems if any were you getting?

RICHARDS I think that the problem you should be aware of perhaps, is that the majority of feeds used for salmon are expanded feeds, using a lot of heat and pressure and the manufacturers do not in fact incorporate those antibiotics within the pellets, but surface coat them with the antibiotic. So whether the antibiotics prescribed for mixing at the mill or mixing at the fish farm, in most cases these are surface coated antibiotic treatments. Perhaps Mr Withnell would like to comment.

WITHNELL That is true, you get a fair degree of loss on expanded products. The other factor is that on expanded products you have very large runs and it is quite difficult to clean down in between runs and therefore the change over from medicated to non medicated could be quite expensive.

LAFONT What is the proportion of strains of *Aeromonas* that are resistant to both oxytetracycline and oxolinic acid.

RICHARDS It is shown in the table

INGLIS As a guess I think 30%.

RICHARDS A number of strains are certainly resistant to both.

WITHNELL Dr Richards, what are your feelings on combined drug usage which has not been applied in aquaculture very frequently, but is applied in other animal treatment.

RICHARDS Well I think simply, I would be very concerned about it. We have little enough to use at the moment, without using them all at once.

RODGER A question for Dr Richards or Dr Munro if he is here. With regard to dichlorvos. Is there any evidence for resistance with sea lice in Scotland at present?

MUNRO Yes.

RODGER Is there actual experimental evidence, is anybody actually working on it at present?

MUNRO Yes, Dr Wootton and Dr Sommerville at Stirling have a paper in draft, I'm not clear whether it is published yet, which indicates resistance as one would predict after repeated use of a pesticide.

RICHARDS I think, if you look at organophosphate use in general terms, you do expect resistance to build up. It took a long time in the salmon farming industry in Scotland and we assumed this was because we were continually getting reinfection from the wild state. In other words that we weren't getting multiplication of resistant organisms within a farm and I think that was probably the case. We have seen resistance building up especially now in areas which have a high number of sites and a high density of fish within them, suggesting that there is repopulation from that very fish farm source.

MECHANISMS OF RESISTANCE DEVELOPMENT IN AQUATIC MICROORGANISMS

C.S. Lewin

Department of Medical Microbiology, University of Edinburgh, Medical School, Teviot Place, Edinburgh, EH8 9AG

Summary: *The efficacy of antimicrobial chemotherapy, commonly used to control bacterial disease in aquaculture, is increasingly being compromised by the development of resistance in fish pathogens due either to plasmid carriage or chromosomal mutation. Plasmid-mediated resistance to chloramphenicol, trimethoprim, sulphonamides and tetracyclines have been identified in fish pathogens involving either drug modification, active efflux of the drug or modification of the target enzyme so that it is resistant to the antimicrobial. Although plasmid-mediated resistance to the 4-quinolones has not yet been identified bacterial resistance occurs due to chromosomal mutation. In* A. salmonicida *low level quinolone resistance appears to be due to alterations in outer membrane porins while high level resistance is probably due to alterations in DNA gyrase.*

***Résumé*: Mécanismes de développement des résistances dans les microorganismes aquatiques.**

Les conditions de l'aquacuture intensive, associant de fortes densités de population à des stress qui affaiblissent les poissons, sont propices au développement d'infections microbiennes contre lesquelles les antibiotiques constituent une arme efficace. Malheureusement leur emploi immodéré a favorisé l'émergence de nombreuses résistances chez les agents les plus dangereux. On connaît 4 sortes de mécanismes aboutissant à la résistance, caractérisés par la modification de l'antibiotique lui-même, la modification de sa cible moléculaire, la production accrue d'une enzyme cible ou d'un analogue structural, ou l'altération de la pénétration et de l'accumulation de l'antibiotique dans la cellule bactérienne. Du point de vue génétique l'apparition d'une résistance peut être due à une mutation chromosomique ou à l'acquisition d'un plasmide capable de circuler entre les différentes espèces de bactéries. Dans le dernier cas la propagation est rapide, et d'autant plus inquiétante que ces résistances transmissibles, souvent multiples, ne dépendent pas forcément de l'emploi des antibiotiques correspondants.

Les plasmides constituent un réservoir de résistances contre la plupart des antibiotiques utilisés ou non en aquaculture (Tableau I). Trois des types de mécanismes cités ont été observés. Il existe par exemple plusieurs classes de chloramphenicol-acetyl-transférases (CAT), dont 3 ont été identifiées chez des bactéries de poissons. La résistance aux tétracyclines est commandée par *plusieurs types de déterminants (*tet A-G, tet F, tet L-P*) agissant par inactivation de la molécule, modification de la perméabilité membranaire, ou production accrue d'une enzyme cible. Seul le premier est attesté chez les poissons. Par contre, la production d'enzymes cibles résistant aux actions des sulfamides et du triméthoprime, bien que classique, n'a pas été étudiée chez les animaux aquatiques.*

Des résistances de nature chromosomique sont récemment apparues contre les 4-quinolones, très populaires en aquaculture. Les deux types de mutations connus portent sur la modification de l'enzyme cible, l'ADN-gyrase, ou des protéines membranaires (porines) qui contrôlent la pénétration du produit dans la cellule. C'est le dernier mécanisme qui serait responsable des résistances d'Aeromonas salmonicida, *non seulement aux quinolones, mais aussi à l'amoxicilline, au chloramphenicol et aux tétracyclines, de sorte que l'emploi conjoint de l'acide oxolinique et de l'oxytétracycline accroît les risques de sélectionner de telles résistances. L'altération de l'ADN-gyrase ne peut cependant être exclue dans le cas d'autres bactéries pathogènes de poissons.*

On voit qu'il existe donc de grandes similitudes entre les mécanismes de résistance décrits chez les poissons et chez les autres espèces. Des travaux complémentaires seront cependant indispensables pour établir des stratégies de traitement et sélectionner de nouvelles molécules. Ces recherches spécifiques peuvent d'ailleurs conduire à l'identification de déterminants insoupçonnés, comme l'a montré la description de CAT IV et tet G *chez* Vibrio anguillarum.

Introduction

Aquaculture by its nature involves the intensive rearing of fish under conditions that will maximize the spread of bacterial disease through the population. Furthermore, the stress that is inherent in intensive rearing leads to a increased susceptibility to many diseases including those caused by bacteria. The economic consequences of uncontrolled outbreaks of bacterial disease would be disasterous and therefore various stratagems are employed in an attempt to control and limit the damage caused by such outbreaks. One of the most common methods used in attempts to control bacterial infections in fish is treatment with antibacterial agents. The efficacy of this stratagem depends on the infecting organisms being susceptible to the chemotherapeutic agents that are currently available for treatment. However, the wide use, perhaps even abuse, of chemotherapy for the control of bacterial infections in fish farms has led to the development of resistance to the antibacterials currently available for use in aquaculture in most fish pathogens. Cultured fish have been found to be infected by strains of *Aeromonas hydrophila, Aeromonas salmonicida, Edwardsiella tarda, Pasteurella piscicida, Vibrio anguillarum* and *Yersinia ruckeri* and Streptococci resistant to various antibacterials including nalidixic acid, oxolinic acid, tetracycline, trimethoprim, sulphonamides and chloramphenicol which are currently available in various parts of the world for the treatment of bacterial diseases in fish (2, 3, 4, 15). Bacterial drug resistance is therefore a major cause of concern in almost all types of aquaculture due to the economic consequences of an inability to control disease.

In general, there are four mechanisms which enable bacteria to resist antimicrobial agents:

1. Modification of the antimicrobial so that it is no longer active,
2. Modification of the drug's target so that it is no longer susceptible to the antimicrobial's action,
3. Increased production of the target enzyme so as to reduce the drug's effect,
4. A reduction in bacterial accumulation of the drug.

These resistance mechanisms can arise in bacteria by one of two means; chromosomal mutation or acquisition of plasmids. Chromosomal mutations are alterations in the bacterial chromosome which cause bacterial resistance to antimicrobials. This type of resistance can be transferred to daughter cells on cell division but cannot be passed on to other bacteria (Fig. 1).

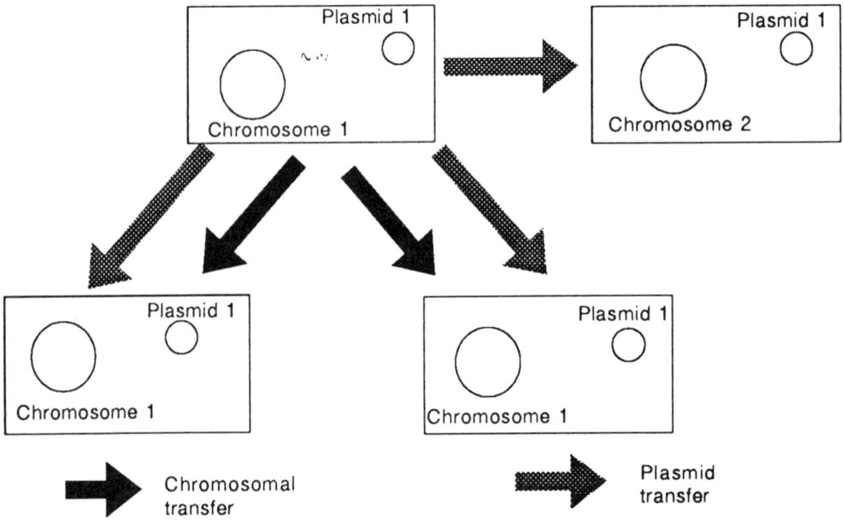

Figure 1 Transfer of genetic information coding for antibiotic resistance

Plasmids, on the other hand, are molecules of extrachromosomal DNA which can replicate independently of the chromosome and may be freely transferred between bacteria including those of different species. Plasmid-mediated drug resistance, unlike chromosomal resistance, can therefore spread resistance horizontally (to members of the same species and other species) as well as vertically (to daughter cells) (Fig. 1). Usually plasmids carry a number of genes conferring resistance to more than one antibiotic. Furthermore, persistence of plasmid carriage by bacteria is not always related to the use of antibacterials and therefore once introduced into the population the plasmid tends to remain. Indeed, this has been shown to be the case in plasmid carriage by *A. salmonicida*, the causative agent of furunculosis (13). As a result of these characteristics plasmids are usually the main carriers of resistance in many clinical bacteria and are a major cause of concern as they can lead to the rapid world-wide spread of resistance to chemotherapeutic agents.

1. Plasmid-mediated resistance in aquatic mircoorganisms

In aquaculture plasmids appear to be a significant reservoir for resistance to antimicrobial agents as plasmid-mediated resistance to the sulphonamides, trimethoprim, tetracycline and chloramphenicol has been identified in most common fish pathogens (Table I). Transferable resistance to macrolides has also recently been identified in non-haemolytic streptoccocus isolated from cultured yellowtail (9). It is also interesting to note that, despite aminoglycosides not being used in aquaculture, plasmid-mediated resistance to these antibacterials has been reported in various fish pathogens. Transferable resistance to streptomycin or kanamycin has been reported in *A. hydrophila*, *P. piscicida*, *E. tarda* and *A. salmonicida* isolates from Japan, Ireland and the U.S.A. (8, 13, 32). There does not appear to be a clear explanation for this phenomena although it has been suggested that the aminoglycoside resistance genes may be linked with the other drug-resistant genes (7). The proposal of Aoki *et al* is supported by the observation that in *A. salmonicida* the trimethoprim, spectinomycin and streptomycin determinants on an IncU plasmid transposed together (13). This suggests that plasmids carried by fish pathogens may potentially serve as a reservoir for resistance to antimicrobials that are not currently used in aquaculture. A worrying scenario, therefore, emerges if the introduction of an antimicrobial agent previously not available for aquaculture were to be compromised by the carriage of plasmid-mediated resistance in fish pathogens prior to its introduction.

Table I: Transferable resistance in fish pathogens

Strain	CM	TC	SA	TP	Reference
V.anguillarum	+	+	+	+	3,7
P.piscicida	+	+	+	-	4
E.tarda	+	+	+	-	8,7
Streptococci	+	+	-	-	9
A.salmonicida	+	+	+	+	13
A.hydrophila	+	+	+	-	2,8,32

CM = chloramphenicol, TC = tetracycline, SA = sulphonamide, TP = trimethoprim

2. Drug resistance determinants of R-plasmids

Three potential mechanisms of antibacterial resistance have been identified in plasmids:

(1) Inactivation or destruction of the drug,
(2) Production of an additional target resistant to the antimicrobial,
(3) A reduction in drug accumulation often due to active efflux of the drug from the cell.

All three of these mechanisms are associated with the antibacterials used in aquaculture.

. Chloramphenicol resistance:

Chloramphenicol is a bacteriostatic inhibitor of protein synthesis that acts by binding to the 50S subunit of the bacterial ribosome. Plasmid-mediated chloramphenicol resistance is due to the inactivation of the drug by an enzyme chloramphenicol acetyl transferase (CAT) which catalyzes the acetylation of chloramphenicol in the presence of acetyl-Co-A. The O-acetoxy derivatives of chloramphenicol are no longer able to bind to the bacterial ribosomes and therefore fail to inhibit bacterial protein synthesis (36). Two distinct groups of CAT variants compromising seven classes have been described: the constitutive group found in Gram negative bacteria and the inducible group associated with Gram positive bacteria (30, 36). CATS coded for by R-plasmids from Gram negative bacteria can be divided into three enzyme types CAT-I, CAT-II and CAT-III. Each enzyme compromises four identical subunits which catalyze the o-acetylation of chloramphenicol but can be differentiated by biochemical characteristics (18, 29, 36) and genetic techniques (29).

In fish pathogens plasmid-mediated chloramphenicol resistance has been shown by DNA-DNA hybridization to be due to a CAT-I enzyme in *P. piscicida* and a CAT-II type enzyme in *E. tarda, A. salmonicida* and *V. anguillarum* (2). A new class of CAT has been identified in *V. anguillarum.* Aoki (2) found that plasmid-mediated chloramphenicol resistance in *V. anguillarum* detected in Japan before 1977 was not classified into type I, II or III by DNA hybridization and therefore proposed that this chloramphenicol resistance was mediated by a new type of CAT, CAT-IV.

. Tetracycline resistance:

Tetracyclines are bacteriostatic antibiotics which inhibit protein synthesis by preventing aminoacyl tRNA from binding to the A site on the ribosome during elongation. In contrast to chloramphenicol where resistance is due to a single mechanism, three different mechanisms of plasmid-mediated tetracycline resistance have been identified. The first type, which is exclusive to Gram negative bacteria, consists of seven genetically distinct determinants termed *tetA-G* distinguishable by DNA-DNA hybridization (20, 21). Six of the determinants (*tetA-F*) have been found to reduce tetracycline accumulation by active efflux of the antibiotic at the inner membrane. The *tetF* determinant found in anaerobic bacteria also appears to involve a second mechanism, drug destruction, in addition to active efflux (20, 23, 26, 27). The second class of tetracycline resistant determinants, *tetL-P,* are found mainly in Gram positive bacteria but also in Gram negative bacteria (20, 21) and can be distinguished from each other and the *tetA-F* determinants by DNA-DNA hybridization. Unlike the *tetA-F* determinants, the *tetL-P* determinants do not mediate resistance by decreasing tetracycline accumulation or inactivating the drug but rather by coding for a protein which has a partial sequence similarity to the ribosome elongation factor. This protein protects ribosomes from inhibition by tetracycline (14, 20, 21).

The tetracycline resistant determinants of R-plasmids that have been identified in fish pathogens vary. In *A. hydrophila, E. tarda* and *P. piscicida* tetracycline resistance has been shown to be due to a class D determinant (8). A *tetC* determinant was identified in a single non-transferable plasmid of *A. salmonicida* (8). In Japanese strains of *V. anguillarum* plasmid-mediated resistance due to the *tetB* determinant was prevalent between 1973 and 1977 (7). However, since 1981 tetracycline resistance in this species has been primarily due to a tetracycline resistant determinant that could not be classified as *tetA-E* (2, 7) and it was suggested that resistance was coded for by a new determinant *tetG* (2, 21). The molecular mechanism(s) involved in the *tetG* determinant have yet to be determined (21) although it is presumed that they are similar to those found in other Gram negative bacteria. The mechanisms of transferable tetracycline resistance in the non-haemolytic streptococci isolated from yellowtail (9) have not yet been elucidated. It is assumed that the resistance determinants are similar to those found in other streptococci and involve protein synthesis resistant to tetracycline rather than active efflux of the drug.

. Inhibitors of folic acid synthesis:

The sulphonamides and trimethoprim are compounds which interfere with the synthesis of folic acid and therefore inhibit nucleotide biosynthesis as interruption of the supply of tetrahydrofolate soon brings nucleotide and nucleic acid biosynthesis to a halt. The sulphonamides inhibit the enzyme dihydropteroate synthetase blocking the formation of dihydropteroic acid from para-aminobenzoate and dihydropteroate. Trimethoprim and ormethoprim inhibit dihydrofolate reductase activity blocking the conversion of dihydrofolate to tetrahydrofolate. The two classes of drugs are often used together as the combination is synergistic. Both drugs block folic acid biosynthesis but at different points and the double blockage therefore seems to be especially effective in cutting off the bacterial supply of tetrahydrofolate. The two drugs in combination are known as a potentiated sulphonamide.

The biochemical mechanism of plasmid-mediated resistance to both the sulphonamides and dihydrofolate reductase inhibitors involves target enzymes resistant to the action of the antimicrobial. In the case of the sulphononamides, this is a drug resistant dihydropteroate synthetase while for trimethoprim and ormethoprim it is a drug-resistant dihydrofolate reductase (1, 31, 34). There are two known genes coding for plasmid-borne sulphonamide resistant dihydropteroate synthetases and these have been termed *sul-I* and *sul-II* (28). There are over ten different plasmid-borne trimethoprim-resistant dihydrofolate reductases that can be differentiated by DNA-DNA hybridization and biochemical characteristics (1).

There appears to have been very little work on the biochemical mechanisms by which plasmids mediate resistance to trimethoprim or sulphonamides in fish pathogens. Due to this lack of information it must be assumed therefore that the mechanisms of plasmid-mediated resistance to the inhibitors of folic acid synthesis are similar to those found in other species and due to the plasmid coding for a drug resistant dihydrofolate reductase or dihydropteroate synthase. Investigations of the biochemical mechanisms of resistance to this class of drugs would appear worthwhile as a potentiated sulphonamide, Romet (a combination of ormethoprim and sulphadimethoxine), is presently used in aquaculture (6, 24).

3. Chromosomally mediated antibacterial resistance

Chromosomally-mediated antimicrobial resistance is particularly significant as a resistant determinant in cases where plasmid-mediated resistance does not exist. This is the case for the 4-quinolones, as plasmid-mediated resistance to this synthetic class of antimicrobial agents has yet to be identified (16). The only pathway by which bacteria are therefore able to develop resistance to these drugs is via chromosomal mutation. The use of the 4-quinolones in aquaculture has been promoted due to the absence of plasmid-mediated resistance. However, the increasing use of these drugs in aquaculture has led to the development of chromosomal resistance which is becoming an increasing cause for concern. Resistance to the 4-quinolones has been identified in *A. salmonicida* (11), *V. anguillarum* (3), *P.piscicida*, *E. tarda* (25) and Y. ruckeri (33) isolated from diseased fish.

The increasing medicinal use of the 4-quinolones, particularly fluoroquinolones, has led to an improved understanding of the mechanisms by which bacteria are able to develop resistance to these drugs (22). Two types of chromosomal mutations are associated with resistance to these drugs: alterations in the target enzyme DNA gyrase and mutations that reduce bacterial accumulation of the 4-quinolones (22).

DNA gyrase (E.C. number 5.99.1.3) is the only bacterial enzyme capable of negatively supercoiling DNA. It is a tetramer consisting of two A and two B subunits. The A subunit cuts both strands of DNA simultaneously at intervals four base pairs apart and holds the strands apart but covalently bound to the enzyme. The B subunit, using ATP for energy, then introduces a negative supercoil into the DNA and the A strand then reseals the two strands. The gyrase genes that have been sequenced reveal close homology, suggesting that this enzyme is highly conserved between different species. High level resistance to the 4-quinolones in both Gram negative and Gram positive bacteria appears to be due to alterations in the A subunit of DNA gyrase coded for by the *gyrA* gene (22). Nucleotide sequence analysis of the resistant *gyrA* genes has found that quinolone resistance appears to be conferred by point mutations within a small region between amino acids 67 and 106 near the N. terminus of the A subunit situated close to the tyrosine at amino acid 122 that is covalently bound to DNA when the enzyme breaks the phosphodiester bonds of DNA.

The 4-quinolones must permeate into bacteria to reach their target DNA gyrase and therefore a reduction in bacterial drug accumulation will lead to resistance. All 4-quinolones penetrate bacteria through the outer membrane porins of Gram negative bacteria. The more hydrophobic drugs, such as nalidixic acid and oxolinic acid, (but not the hydrophilic drugs) are also able to penetrate through the phospholipid bilayer. In Gram negative bacteria resistance to the 4-quinolones due to reduced drug accumulation is often due to alterations in outer membrane proteins (OMPs) and is often associated with cross-resistance to antibacterials that are unrelated to the 4-quinolones (22).

There appear to have been relatively few investigations in to the mechanisms of resistance to the 4-quinolones in fish pathogens. However, it is clear that the mechanisms of resistance involve chromosomal mutations rather than plasmids as, in common with other bacteria, 4-quinolone resistance in fish pathogens is non-transferable. *A. salmonicida* is the species most investigated; probably because furunculosis caused by oxolinic acid resistant *A. salmonicida* is a major problem in salmon farms, particularly in Scotland and Norway. Reduced drug accumulation has been identified as a potential mechanism of resistance to the 4-quinolones in *A. salmonicida. In vitro* experiments have found that low level resistance to oxolinic acid and other 4-quinolones, including fluoroquinolones, in *A. salmonicida* is associated with alterations in OMPs which probably function as porins (12, 17, 19, 35). Two changes in OMPs have been associated with resistance: increased expression of a 37kDa protein which can occur alone or be accompanied by decreased expression of a 43kDa protein. The increased expression of the 37kDa OMP caused a 3-4-fold increase in MIC to oxolinic acid while both OMP alterations caused a ten-fold increase in MIC to the 4-quinolone (Table II). These alterations in OM expression not only caused resistance to the 4-quinolones but also to ampicillin, amoxycillin, chloramphenicol and tetracycline (Table II), (12, 19).

Table II: MICs (mg/L) of *A. salmonicida*

Strain	Oxolinic acid	Oxytetracycline	Amoxycillin
MT744#	0.03	0.2	0.2
AB176	0.1	1.0	0.75
MT363	0.03	0.2	0.2
AB174	1.0	1.5	0.75

#MT744 is the parent strain of AB176, a mutant exhibiting increased expression of a 37kDa OMP. *MT363 is the parent strain of AB174 exhibiting increased expression of a 37kDa OMP and decreased expression of a 43kDa OMP. Both mutants were selected on Tryptone soy agar containing oxytetracycline 1 mg/l.

It can be seen from Table 2 that in *A. salmonicida* selecting for resistance to oxytetracycline can also simultaneously select for resistance to oxolinic acid. The observation that oxytetracycline can select for oxolinic acid resistance and *vice versa* is a cause for concern as oxolinic acid and oxytetracycline are the two most commonly used drugs for furunculosis therapy in Europe. Until recently this phenomenon had only been reported in resistant mutants of *A. salmonicida* created in the laboratory and it might therefore be argued that such mutations do not occur in the wild. We have recently, however, identified a wild-type oxolinic acid resistant strain which showed increased expression of a

37kDa OMP (C.S. Lewin, A.C. Barnes, unpublished results) suggesting that these OMP alterations can occur in wild-type resistant isolates of *A. salmonicida*. In salmon farms the use of oxytetracycline may therefore select for low-level 4-quinolone resistance. This observation is of particular concern at it has been suggested in other Gram negative bacteria that the development of high level resistance to the 4-quinolones is a two-step process with the second step occurring more easily than the first. Thus it can be envisaged that tetracycline used for the treatment of furunculosis may facilitate the development of oxolinic acid resistance.

As mentioned previously, alterations in the A subunit of DNA gyrase are the second mechanism usually associated with resistance to the 4-quinolones. Little information actually exists on the DNA gyrases of fish pathogens as none appear to have been purified. However, as the DNA gyrases that have been sequenced all display a high degree of sequence homology it seems reasonable to speculate that the DNA gyrase of fish pathogens will be similar to those of other species. The antimicrobial resistance profile of fish pathogens therefore can give clues as to which mechanism of quinolone resistance is involved. *A. salmonicida*, *V. anguillarum*, *E. tarda* and *P. piscicida* displaying resistance to other 4-quinolones but not to unrelated antibacterials have been identified (11, 25, 33) suggesting that the mechanism of resistance is an altered DNA gyrase because alterations in the A subunit of DNA gyrase are the only mutation known to cause high level resistance to all 4-quinolones but not to other unrelated antibacterials

Conclusion

Fish pathogens which are resistant, due to plasmid carriage or chromosomal mutation, to the antimicrobials currently available for use in aquaculture, are a major cause of concern. Resistance to trimethoprim, sulphonamides, chloramphenicol and tetracyclines is transferable and appears to be mainly plasmid mediated. In contrast, resistance to the 4-quinolones is non-transferable and therefore chromosomally mediated. The antibiotic resistance determinants that have been investigated in fish pathogens appear to be similar to those found in other bacteria. There is, however, considerable scope for further investigation into the precise mechanisms by which antimicrobial resistance is mediated in fish pathogens as relatively little work has been carried out: particularly with reference to the 4-quinolones, sulphonamides and trimethoprim. These investigations are necessary for several reasons. Firstly, such studies can provide information which may enable the efficacy of therapy with the drugs currently available for use in aquaculture to be improved. For example, strategies that may serve to limit the emergence of resistance might be devised. Secondly, such investigations can aid in assessing the potential of new chemotherapeutic agents for the control of bacterial diseases. For example, one of the factors that will determine the suitability of the fluoroquinolones, whose potential use in aquaculture is being investigated, is their activity against oxolinic acid-resistant isolates. Finally, further investigations may identify novel resistance determinants as previously unknown resistance determinants have been found in fish pathogens: two such examples being the CATIV and *tetG* resistance determinant identified in *V. anguillarum*.

References

1). AMYES S.G.B. & TOWNER K.J., 1990. Trimethoprim resistance; epidemiology and molecular aspects. *J. Med. Microbiol.*, **31**, 1-19.

2). AOKI T., 1988. Drug-resistant plasmids from fish pathogens. *Microbiol. Sci.*, **5**, 219-223.

3). AOKI T., EGUSA S. & ARAI T., 1974. Detection of R factors in naturally occurring *Vibrio anguillarum* strains. *Antimicrob. Agents Chemother.*, **6**, 534-538.

4). AOKI T. & KITAO T., 1985. Detection of transferable R plasmids in strains of the fish pathogenic bacterium, *Pasteurella piscicida. J. Fish Dis.*, **8**, 345-350.

5). AOKI T., KITAO T., IEMURA N., MITOMA & NOMURA T., 1983. The susceptibility of *Aeromonas salmonicida* strains isolated in cultured and wild salmonids to various chemotherapeutics. *Bull. Jap. Soc. Sci. Fis heries*, **49**, 17-22.

6). AOKI T., KITAO T. & FUKUDOME M., 1989. Chemotherapy against infection with multiple drug resistant strains of *Edwardsiella tarda* in cultured eels. *Fish. Pathol.*, **24**, 161-168.

7). AOKI T., SATOH T. & KITAO T., 1987. New tetracycline resistance determinant on R plasmids from *Vibrio anguillarum. Antimicrob. Agents Chemother.*, 31, 1446-1449.

8). AOKI T. & TAKAHASHI A., 1987. Class D tetracycline resistance determinants of R plasmids from the fish pathogens Aeromonas hydrophila, Edwardsiella tarda and *Pasteurella piscicida. Antimicrob. Agents Chemother.*, **31**, 1278-1280.

9). AOKI T., TAKAMI K. & KITAO T., 1990. Drug resistance in a non-haemolytic *Streptococcus* sp isolated from cultured yellowtail *Seriola quinqueradiata. Dis. Aquatic Orgs.*, **8**, 171-177.

10). BAST L., DALY J.G., DeGRANDIS S.A. & STEVENSON R.M.W., 1988. Evaluation of profiles of *Aeromonas salmonicida* as epidemiological markers of furunculosis infections in fish. *J. Fish Dis.*, **11**, 133-145.

11). BARNES A.C., LEWIN C.S., HASTINGS T.S. & AMYES S.G.B., 1990. *In vitro* activities of 4-quinolones against the fish pathogen *Aeromonas salmonicida. Antimicrob. Agents Chemother.*, 1819-1820.

12). BARNES A.C., LEWIN C.S., HASTINGS T.S. & AMYES S.G.B., 1990. Cross-resistance between oxytetracycline and oxolinic acid in *Aeromonas salmonicida* associated with alterations in outer membrane proteins. *FEMS Microb. Letts.*, **72**, 337-340.

13). BRAZIL G., CURLEY D., GANNON F. & SMITH P., 1986. Persistence and acquisition of antibiotic resistance plasmids in *Aeromonas salmonicida*. In. *Banbury report 24*. Antibiotic resistance genes: ecology, transfer and expression. Cold Spring Harbor Laboratory, 107-114.

14). BURDETT, V., 1986. Streptococcal resistance mediated at the level of protein synthesis. *J. Bacteriol.*, **165**, 564-569.

15). CESCHIA G., GIORGETTI G., BERTOLDINI G. & FONTEBASSO S.,1987. The *in vitro* sensitivity of *Yersinia ruckeri* to specific antibiotics. *J. Fish Dis.*, **10**, 65-67.

16). COURVALIN P., 1990. Plasmid-mediated 4-quinolone resistance: a real of apparent absence? *Antimicrob. Agents Chemother.*, **34**, 681-684.

17). DARVEAU R.P., MacINTYRE S., BUCKLEY J.T. & HANCOCK R.E.W., 1983. Purification and reconstitution in lipid bilayer of an outer membrane, pore forming protein of *Aeromonas salmonicida. J. Bacteriol.*, **156**, 1006-1011.

18). GAFFNEY D.F., FOSTER T.J. & SHAW W.V., 1978. Chloramphenicol acetyltransferases determined by R plasmids from Gram negative bacteria. *J. Gen. Microbiol.*, **109**, 351-358.

19). GRIFFITHS S.G. & LYNCH W.H., 1989. Characterization of *Aeromonas salmonicida* mutants with low level resistance to multiple antibiotics. *Antimicrob. Agents Chemother.*, **33**, 19-26.

20). LEVY S.B., 1988. Tetracycline resistance determinants are widespread. *ASM News*, **54**, 418-421.

21). LEVY S.B., 198). Evolution and spread of tetracycline resistant determinants. *J. Antimicrob. Chemother.*, 24, 1-3.

22). LEWIN C.S., ALLEN R.A. & AMYES S.G.B., 1990. Potential mechanisms of resistance to the modern fluorinated 4-quinolones. *J. Med. Microbiol.*, **31**, 153-161.

23). McMURRY L., PETRUCCI R.E. & LEVY S.B., 1980. Active efflux of tetracycline encoded by four genetically different tetracycline resistance determinants in *Escherichia coli. Proc. Natl. Acad. Sci. USA*, **77**, 3974-3977.

24). MEYER F.P., 1989. Solutions to the shortage of approved fish therapeutants. *J. Aquatic An. Health* , **1**, 78-80.

25). NAKANO S., AOKI T., KITAO T. 1989. *In vitro* antimicrobial activity of pyridonecarboxylic acids against fish pathogens. *J. Aquatic An. Health*, 1, 43-50.

26). PARK B.H., HENDRICKS M., MALAMY M.H., TALLY F.P. & LEVY S.B., 1987. Cryptic tetracycline resistance determinant (class F) from *Bacteroides fragilis* mediates resistance in *Escherichia coli* by actively reducing tetracycline accumulation. *Antimicrob. Agents Chemother.*, **31**, 1739-1743.

27). PARK B.H. & LEVY S.B., 1988. The cryptic tetracycline resistance determinant on Tn4400 mediates tetracycline degradation as well as tetracycline efflux. *Antimicrob. Agents Chemother.* **32**, 1797-1800.

28). RADSTROM P. & SWEDBERG G., 1988. RSF1010 and a conjugative plasmid contain *sulII*, one of two known genes for plasmid-borne sulfonamide resistance dihydropteroate synthase. *Antimicrob. Agent Chemother.*, **32**, 1684-1692.

29). ROBERTS M.C., ACTIS L.A. & CROSA J.H., 1985. Molecular characterization of chloramphenicol-resistant *Haemophilus parainfluenzae* and *Haemophilus ducreyi*. *Antimicrob. Agents Chemother.*, **28**, 176-180.

30). ROBERTS M.C., CORNEY A. & SHAW. W.V., 1982. Molecular characterization of three chloramphenicol acetyltransferase isolated from *Haemophilus influenzae*. *J. Bacteriol.*, **151**, 737-741.

31. SKOLD P., 1976. R-factor-mediated resistance to sulfonamides by a plasmid borne, drug resistant, dihydropteroate synthase. *Antimicrob. Agents Chemother.*, **9**, 49-54.

32). TORANZO A.E., BARJA J.L., COLWELL R.R. & HETRICK F.M., 1983. Characterization of plasmids in bacterial fish pathogens. *Infect. Immun.*, **39**, 184-192.

33). TSOUMAS A., ALDERMAN D.J., RODGERS C.J., 1989. *Aeromonas salmonicida* development of resistance to 4-quinolone antimicrobials. *J. Fish Dis.*, 12, 493-507.

34). WISE, E.M. & ABOU-DONIA M.M., 1975. Sulfonamide resistance mechanism in *Escherichia coli* R plasmids can determine sulfonamide-resistant dihydropteroate synthases. *Proc. Natl. Acad. Sci. USA.*, **72**, 2621-2625.

35). WOOD S.C., McCASHION R.N. & LYNCH W.H., 1986. Multiple low level antibiotic resistance in *Aeromonas salmonicida*. *Antimicrobial. Agents Chemother.*, **29**, 992-996.

36). ZAIDENZAG Y., FITTON J.E. & PACKMAN L.C., 1979. Characterization and comparison of chloramphenicol acetyltransferase variants. *Eur. J. Biochem.*, **100**, 609-618.

PUTATIVE PUBLIC HEALTH RISKS OF ANTIBIOTIC RESISTANCE DEVELOPMENT IN AQUATIC BACTERIA

T. Midtvedt [*] & E. Lingaas [**]

[*] Department of Medical Microbial Ecology, Karolinska Institute, Box 60 400, S-10401 Stockholm, Sweden.

[**] Kaptein W. Wilhelmsen og Frues Bacteriological Institut, National Hospital, N-0027, Oslo, Norway.

Summary: *The results of modern epidemiological studies indicate that emergence and persistence of resistance to antimicrobial agents in all groups of ecosystems studied so far, from human and veterinary medicine, in agri- and aquaculture, follows a one-compartment model. The consequence of this is as follows: any use of any drug in any "sub-compartment" may create problems of resistance in other "sub-compartments". Therefore, a prudent use of antimicrobial agents in fish farming has to follow the same strategic and tactical considerations as established in human medicine. The veterinarians responsible for the present misuse of antibiotics in fish farming should calculate upon increasing "resistance" from the medical profession. A straight tip is that, from a medical point of view, tetracyclines and quinolones, e.g. the two groups of drugs most commonly used, should be banned from use in fish farming.*

Résumé : **Risques potentiels pour la santé publique du développement des microorganismes antibiorésistants en milieu aquatique.**

Depuis 4 ou 5 décennies l'usage croissant des antibiotiques en médecine humaine et en élevage, puis en agronomie et en aquaculture, a rompu l'équilibre des populations microbiennes naturellement résistantes. L'épidémiologie descriptive et analytique permet d'étudier l'apparition, le développement et la stabilité des antibiorésistances et d'apprécier leurs conséquences pour la santé humaine et animale. Cette démarche peut s'appliquer aux problèmes posés par l'emploi des antimicrobiens en aquaculture.

Les résistances sont très redoutées en milieu hospitalier, où elles peuvent s'établir durablement et évoluer, après apparition spontanée ou après introduction par les patients et par le personnel. La propagation en milieu médical de souches résistantes trouvant leur origine dans le traitement à doses subthérapeutiques d'animaux domestiques a été démontrée à plusieurs reprises, notamment pour les salmonelles, dont les résistances participeraient davantage de sources animales que d'une sélection par les traitements humains.

Les antibiotiques sont employés en aquaculture depuis plusieurs années. On admet qu'une cinquantaine de formules ont été préconisées, les plus répandues étant actuellement les quinolones et les tétracyclines. De nombreuses résistances sont apparues corrélativement. Des facteurs R, isolés de souches d'Aeromonas hydrophila, *ont pu être expérimentalement transmis à des entérobactéries. En Norvège, l'apparition de la vibriose d'eau froide, causée par* V. salmonicida *a contribué accroître l'emploi de l'oxytétracycline, qui de 3 000 kgs en 1981 est passé à 27 000 en 1987. Dès 1986 l'impact des résistances a entraîné l'adoption de l'acide oxolinique et d'autres quinolones, dont la consommation en 1990 aurait dépassé celle de la médecine humaine.*

Or, le milieu aquatique est très propice à la propagation des résistances. Il serait temps d'imposer des règles qui ne devraient admettre que l'emploi de composés rapidement dégradables, peu susceptibles de favoriser les résistances plasmidiques ou les résistances croisées, et dépourvus d'intérêt médical pour l'homme. L'oxytétracycline n'obéit à aucune de ces conditions. Les quinolones, n'induisent pas de résistance transmissible, mais des mécanismes de résistance croisée avec les tétracyclines ont été reconnus, leur persistance dans l'environnement est particulièrement longue, et nul ne peut nier leur utilité en pratique médicale. Ces considérations sont d'autant plus préoccupantes que les bactéries pathogènes de poissons ne sont pas seules concernées et qu'il faudrait tenir compte des nombreux agents potentiellement pathogènes pour l'homme qui colonisent les eaux ou sont propagés par les oiseaux aquatiques.

"I knew six honest service men
(They taught me all I know)
Their names are What and Why and When
and How and Where and Who"

These lyric interrogatives by Rudyard Kipling are still the basis of all epidemiology; even when evaluating putative public health risks of antibiotic resistance development in an aquatic environment.

It goes without saying that antimicrobial agents are some of Mother Nature's own weapons for establishment and maintenance of all microbial ecological systems. Consequently, antimicrobial resistance is a natural part of the regulatory factors in any ecosystem and has existed as long as there have been such ecosystems. During the last 4 or 5 decades however, a steadily increasing usage of antimicrobial agents, in human and veterinary medicine as well as in agri- and aqua-culture, has created opportunities for influencing Mother Nature's ecological balance.

The key questions are, whether, and to what extent, this increased and extended use of antimicrobial drugs has created any risk for the treatment of human diseases of microbial origin. In order to answer these questions, results from two basic types of epidemiological studies, *e.g.* descriptive and analytical, have to be utilized. Descriptive epidemiology defines the time, place, and microbial strains associated with a certain pattern of resistance. Analytical epidemiology is more focused on the fęatures of resistance, as what determines the persistance of the resistant microbial strains in a particular ecosystem, what are the risk factors that determine why some exposed individuals, fish as well as man, became ill while others do not, and so on.

Using terminology from pharmacology, the key questions can be slightly rephrased as follows: can the development, spread and maintenance of antimicrobial resistance in the microbial world be studied, utilizing a one-compartment model or do we have to utilize a multi-compartment model? If the development and maintenance of resistance is a multi-compartment story, antimicrobial drugs can be widely used in one compartment with little influence upon resistance patterns in the other compartments. However, if the one-compartment model is the most correct one, any usage of microbial drugs in one compartment will have influence upon resistance patterns in the other compartments.

In order to evaluate the putative public health risks following use of antimicrobials in aquaculture, results of a few specific investigations in some selected ecosystems from human and veterinary medicine will be commented on. In these investigations, descriptive and analytic epidemiology is combined in order to define the origin and persistance of resistant microbes in some inter-connecting ecosystems.

In human medicine, a hospital represents a relatively convenient ecosystem when examining the epidemiology of antimicrobial resistance. Essentially, there are three major sources for antibiotic-resistant organisms in a hospital. The organisms may emerge within the hospital itself, or may be introduced from the community, or from other hospitals.

The results from an American study may be convenient to elucidate some inter-hospital epidemiological questions. In 1981, methicillin-resistant *S. aureus* strains were isolated from a Seattle hospital over a 9-month period (10). The epidemic strain had been introduced to the hospital burns-unit from a patient who was transferred from a Houston hospital. It had a characteristic gentamicin resistance plasmid. In the burns-unit, chloramphenicol was used in an attempt to eradicate the resistant *S. aureus* strain. However, after a while, the organism acquired a chloramphenicol resistance plasmid. The organism also spread to the surgical intensive care unit with different plasmid to the one in the burns unit. During this time, there was an ongoing study in the intensive care unit in which chloramphenicol was used in patients with abdominal abscesses. The strain continued to cause infections in the surgical intensive care unit for several months. One patient was thought to be an important reservoir. He was colonized at multiple sites, and although he was isolated, the organism spread from him to a number of patients within the unit. When he was transferred to a rehabilitation unit, the organism spread to other patients within that unit also. The organism disappeared from the burns unit for 8 months, and, when it reappeared, it had a slightly different plasmid profile than at its first occurrence. Further epidemiological studies identified a nurse who had an external otitis that had persisted since the first outbreak. The epidemic strain with its characteristic plasmid profile was isolated from her lesion.

As summarized by Cohen (4), this outbreak demonstrates the introduction of a highly resistant organism from another hospital as well as the evolution of resistance by acquisition of additional antibiotic resistance determinants following use of antibiotics within the hospital. The likelihood and frequency of genetic exchange in the development of such resistance are demonstrated by the acquisition of different plasmids in different parts of the hospital. The persistence of the strain was attributable to patient and health-care worker reservoirs. The existence of such reservoirs may explain why certain strains became epidemic.

During the last few years, numerous reports have appeared demonstrating transmission of drug-resistant organisms from farms into the community and into the hospitals.

An ampicillin- and tetracycline-resistant strain of *Salmonella newport*, with an unique plasmid profile, could be traced from a herd of beef cattle, which had been given sub-therapeutic doses of antimicrobial drugs, through meat processing to supermarkets (8). The strain was identified from sick people both on the farm and in several states in the US. It was also spread into a hospital. A member of the farm family was hospitalized and underwent colonoscopy. The organism was then transmitted to another patient, also undergoing the same procedure. It was found to be likely that the selective pressures of the sub therapeutic antibiotics were responsible for the persistence of the organism in the herd. Antibiotics used for other infections in humans were found to be a risk factor for disease in infected persons.

A similar outbreak of a multi-resistant *S. newport* occurred in California in 1987 (14). The epidemic strain had a distinctive plasmid profile. The strain could be traced from humans back to a farm and was isolated at every step in the path of transmission. It was found that antibiotics had been used illegally in dairy cows.

The results of these two, as well as of several other combined epidemiological and laboratory studies, utilizing new methods in molecular technology, clearly demonstrate the path of transmission, from the sick persons back to the primary source-food of animal origin. It seems justifiable to assume that the antimicrobial drugs to which food animals are exposed provide selective pressure that can lead to the appearance and persistence of resistant strains - which subsequently may cause disease in humans. In fact, data from the United States as well as other countries indicate that antimicrobial use by humans does not play a major role in the emergence or persistance of resistant *Salmonella* (9).

The use of antimicrobial drugs in fisheries started decades ago (6). As early as in 1946, Gutsell recognized the potential of sulphonamides for combatting furunculosis. In the years to follow, the range of antimicrobial compounds rapidly expanded to encompass chloramphenicol, tetracyclines, kanamycin and others. Austin listed close to 50 antimicrobial compounds which have been used in fisheries for shorter or longer periods of time (2). At present, tetracyclines and various quinolones are the two groups of drugs most commonly used, at least in the European countries.

In parallel with the increase use of antimicrobials in fish farming, there has been an increasing number of reports dealing with the development of antibiotic resistance in fish pathogens. Below will be focused upon two pathogens, *Aeromonas hydrophila*, causing diseases in several fish species all over the world, and *Vibrio salmonicida* which is the causative agent in cold-water vibriosis.

The existence of resistance plasmids in *Aeromonas hydrophila* has been extensively studied by several scientific groups (1, 7). Some of the plasmid groups have been found to be transferable to enterobacteriacea as *Edwardsiella tarda* and *E. oli* (7). The high incidence of resistance factors in *A. hydrophila* was "assumed to be due to the selective pressure exerted by chemotherapeutics used in fish culturing" (1).

Cold water vibriosis was first described in Norway in 1977 (5). For the next decade, there were outbreaks of the disease in hundreds of Norwegian fish farms. Oxytetracycline was the most commonly used antibiotic, and the consumption rose sharply, from about 3.000 kg in 1981 to over 27.000 kg in 1987 (11), and, as could be predicted and expected strains of tetracycline-resistant *V. salmonicida* started to appear. The first strain was isolated in the spring 1985, and from the autumn 1986, such resistant strains became a serious problem in a wave of disease outbreaks that swept all over the Norwegian coast (15). The fading effect of oxytetracycline led to the use of oxolinic acid and other quinolones in Norwegian fisheries. In the years to follow, the consumption of quinolones increased dramatically, reaching a temporary top in 1990. During the three months July - September, approximately 19.500 kg of quinolones were used in fish farming in Norway. It is estimated that the total amount of kilogrammes quinolones used in fish farming in Norway in 1990 will be higher than the total amount of all antibiotics used in Norwegian human medicine that year (13). In parallel with increased use, resistant strains have started to occur, although no major report has been published.

Although type and amount of drug used over the years may vary, similar data are reported from most other countries having fish farms. Tetracyclines and quinolones are the two groups of drugs most commonly in use, and development of resistance does occur - if ever investigated upon.

Accepting the mere fact that the aquatic environment represents an unique milieu for spreading of resistance to other ecosystems, time has come, to list a set of standards for drugs to be utilized in fish farming. Austin has listed 12 commitments in order to ensure their safety and efficacy (2). Concentrating upon putative ecological and public health risks, the list can be highlighted as follows:

1. The compound must be rapidly broken down to non-toxic components.
2. It should not give rise to a plasmid-mediated resistance.
3. It should not give rise to any cross-resistance to other groups of antimicrobial drugs.
4. It should not be medically important.

The tetracyclines do not fulfil any of these points. They are not easily broken down and the microbes as well as the workers at and around a fish farm may be exposed to the drugs for a considerable period of time. Further, they are heavily involved in plasmid-mediated resistance and if so, the plasmid often demonstrates resistance to other groups of antimicrobial drugs. Last, but not least, they still are of considerable medical importance.

The quinolones fulfil one of these four points, e. g. they are - so far at least - not giving rise to plasmid-mediated resistance. On the other points, however, they are far from being ideal. They are not easily broken down and can be demonstrated in the environment months after they have been used at a fish farm. In fact, no microbial "quinolonidase" is so far isolated, and the environmental fate of these drugs are completely unknown - years after we started to pour out tons of these drugs into the sea (12). There is an increasing number of reports showing cross-resistance to other groups of drugs (via porin alterations). The results of some recent studies indicate that quinolones may be eliminated from the microbes by an efflux mechanism, similar to that found for tetracyclines. From an ecological point of view, efflux mechanisms represent a clear disadvantage as they increase the number of organisms being exposed to the drugs. Lastly, the quinolones are of considerable - and increasing - medical importance.

The question of possible public health hazards in the development of resistance is of course not restricted to the fish pathogens *per se*, but to all microbial taxa being exposed to the drugs used. Austin has listed approximately 20 groups of microbes which have been recovered from fish farms and are known to be potentially pathogenic for humans. In addition, Chang and Pien (3) listed several more species isolated from human infections associated with marine activities. The species commented on in these two reports include staphylococci, streptococci, enterobacteria and pseudomonas. In addition, microbial species harboured and spread by seabirds have to be taken into consideration. *Campylobacter jejuni*, already reported to have developed resistance to quinolones, is just one of several species with an established medical importance.

References

1). AOKI T., EGUSA R., OGATA Y. & WATANABE T., 1971. Detection of resistance factors in the fish pathogen *Aeromonas liquefaciens*. *J. Gen. Microbiol.*, **65**, 343-349.

2.) AUSTIN B. & AUSTIN D.A., 1987. *Bacterial fish pathogens. Disease in farmed and wild fish*. Ellis Horwood Ltd., Chichester, England, 364 pp.

3). CHANG J.W., & PIEN F.D.,1986. Marine-acquired infections Hazards of the ocean environment. *Marine Infect.*, **80**, pp.30, 32, 37, 41.

4). COHEN M.L., 1990. Epidemiologic studies of the emergence and transmission of drug-resistant bacteria. In, *Application of molecular biology in diagnosis of Infectious diseases*. Olsvik OE. & Bukholm G. eds, Norwegian College of Veterinary Medicine, Oslo, Norway, pp. 33-38.

5). EGIDIUS E., ANDERSEN K., CLAUSEN E. & RAA J., 1981. Coldwater vibriosis or "Hitra disease" in Norwegian salmonid farming. *J. Fish Dis.*, **4**, 353-354.

6). GUTSELL J., 1946. Sulfa drugs and the treatment of furuncuulosis in trout. *Science*, **104**, 85-86.

7). HEDGES R.W., SMITH P. & BRAZIL G., 1985. Resistance plasmids of aeromonads. *J. Gen. Microbiol.*, **131**, 2091-2095.

8). HOLMBERG S.D., OSTERHOLM M.T., SENGER R.A., COHEN M.L., 1984. Drug-resistant *Salmonella* from animals fed antimicrobials. *N. Engl. J. Med.*, **311**, 617-622.

9). HOLMBERG S.D., WELLS J.G. & COHEN M.L., 1984. Animal-to-Man transmission of antimicrobial-resistant *Salmonella*: Investigations of U.S. outbreaks, 1971-1983. *Science*, **225**, 833-835.

10). LOCKSLEY R.M., COHEN M.L., QUINN T.C., TOMPKINS L.S., COYLE M.B., KIRIHARA, J.M. & COUNTS G.W., 1982. Multiply antibiotic-resistant I: introduction,transmission, and evolution of nosocomial infection. *Ann. Intern. Med.*, **97**, 317-324.

11). MIDTVEDT T., 1988. Antibiotikabruk og fiskeoppdrett. Use of antibiotics and fish farming. *J. Med. Norwegian Ass.*, **108**, 1795-1796.

12). MIDTVEDT T., 1989. Quinolones and the environment. *Lancet*, October 28, 1040.

13). MIDTVEDT T., 1991. Quinolones - ecological aspects. *J. Med. Norwegian Ass.*, in press.

14). SPIKA J.S., WATERMAN M.H., SOO HOO G.W., ST. LOUIS M.E., PACER R.E., JAMES S.M., BISSETT M.L., MAYER L.W., CHIU J.Y. HALL, R., GREENE R., POTTER M.E., COHEN M.L. & BLAKE P.A., 1987. Chloramphenicol-resistant *Salmonella newport* traced through hamburger to dairy farms. *N. Engl. J. Med.*, **316**, 565-570.

15). SØRUM H., ROBERTS M.C. & CROSA J.H., 1990. Isolation and characterization of a tetracycline resistance gene from the fish pathogen *Vibrio salmonicida*. In, *Applicatin of molecular biology in diagnosis of infectious diseases*. Olsvik OE. & Bukholm G. eds, Norwegian College of Veterinary Medicine, Oslo, Norway, pp. 33-38.

Discussion

HASTEIN I think it is always interesting to discuss these things with Professor Midtvedt. We quite often discuss the use of drugs in Norwegian aquaculture. Today you say that we should ban the use of oxytetracycline and that we should ban the use of oxolinic acid. What do you suggest as compounds that we could use in the future, so that the audience, which includes registrators and pharmaceutical companies, could take your proposal home and start producing a drug that fulfilled the requirements that you have suggested in your paper?

MIDTVEDT You saw the criteria I listed earlier and you can start from them. I think time has come for the Norwegians to start to discuss very seriously which compounds they will use in the future. I think you have to accept usage for a year or two for the compounds you already have on the market and maybe they will still be effective. But for the future, you cannot continue and if you then start to go through the lists of antimicrobials (and to do that fully we would take the rest of the evening !) you do have groups of compounds which have not been used very much.

From an <u>ecological</u> point of view, talking just about ecology, some of the newer monobactams may be expected to be effective against the gram negatives, which is relevant to your problems in Norway. From a medical point of view, you may have some objections of using monobactams, but from an ecological point of view, they are safe to use. They are very rapidly broken down in the environment. You may have other groups, and, I may tell you my friend the Norwegian, that I have tried to talk to some of your companies about drugs to be used. It was of no interest to them. Thankyou very much, please go back to Sweden, was the message they told me.

LEWIN Professor Midtvedt, I think that you have raised some interesting questions, but I would like to point out something to you. For example, you said that with all of the fluoroquinolones, we don't know what happens to them. You are correct in that there is no drug destruction enzyme system, but most of the fluoroquinolones are actually fairly photosensitive and if you leave them out in light they just degrade. Secondly, you mentioned that you weren't happy about the use of fluoroquinolones in aquaculture. I do share some of your concerns, because I think you have raised a very valid point, but is it not better perhaps to use a fluoroquinolone which has a much lower mutation frequency i.e. less of a chance of resistance developing, than the current compounds that are being used, such oxolinic acid and nalidixic acid, which we know have very high frequency of mutation from both in vitro and in clinical studies? Would it not be wiser to perhaps switch into a newer fluoroquinolone being aware that there is a disadvantage to medicinal use, but weigh up the cost benefit of the equation? I think it is a very black picture that you paint.

MIDTVEDT Trying to answer your two main questions. The first one was about the breakdown of the quinolones out in nature. Of course most of them are sensitive to light, but underneath a fish farm and in the sediment there isn't much light. We saw that the half life of the oxolinic acid underneath a Norwegian fish farm was a year or so.

LEWIN But oxolinic is particularly stable, I would point out.

MIDTVEDT Yes, that is correct, however the other ones are also rather stable, but they have not been that much investigated, I admit. It is still rather dark underneath a fish farm. There is very little light coming down there to do the job. Then about the question about resistance and development. Certainly you have a quantitative aspect in the resistance, but nevertheless, you are just working at two or three logs and the drugs will still be there for a considerable period of time and will meet new bacteria. As long as you have quinolones in the environment, you can expect them to initiate a development of microbial resistance, whichever quinolone you are using. But you do have a quantitative aspect in that, I agree.

TREVES BROWN The speaker talked about four compartments and pointed out the correlation of resistance between the medical compartment and the veterinary compartment. The important question is, which of those compartments is the donor and which is the recipient? Of the four compartments he mentioned, those two must be the two closest together. Tetracyclines have been used in the veterinary compartment for what 35 years and they remain much more effective in the veterinary compartment than in the medical. I suggest the real problem is medical misuse and not veterinary misuse and I suggest that the aquaculture compartment is a long way away from the medical one.

MIDTVEDT I agree with you, but my statement is that, wherever the use or misuse, has taken place or is going to take place, we do have to talk about one compartment model for the development, spread and maintenance of resistance. Certainly the medical profession have a lot of problems ahead in their usage of quinolone. I can forsee one of them, the use of quinolones to prevent and treat travel diarrhoea. There are about 100 million people going north-south every year and these people are going on vacation to a very concentrated area in the Mediterranean or in subtropical climates. So far the quinolones are wonderful drugs for the prevention and treatment of diarrhoea If 100 million people are going to a very few places, you are going to create problems of resistance and alterations in the flora there. The medical profession is by no means better than any other field. We have just one compartment we have to talk about in the future.

SHEPHERD You said that there were various reports correlating increased drug resistance in microbial strains involved in aquaculture with increased disease in humans. Can you give us one example please?

MIDTVEDT I can give you one example. It is difficult to trace a strain from aquaculture to a human being, but returning to Norway, I will quote one story and that is a story with *Campylobacter jejuni* in the north. This bacterium was believed to be put into the water by seagulls and was resistant to various antibiotics. Certainly, you have no infections with *Campylobacter jejuni* in any fish farms in Norway, but you do have a resistance pattern rather similar to the patterns found in the humans.

LEWIN I feel obliged to comment on the last two points that you have made. Firstly, in regard to the *Campylobacter*, quinolones are also used in chickens, so it would not be surprising if you got resistance due to the use of quinolones in chicken farming or poultry farming rather than in aquaculture. I think that your premise is somewhat far fetched. Secondly, your first comment was on fluoroquinolones for traveller's diarrhoea. I agree with you that there are some problems with that, and, in fact, we have just published a paper on the subject which shows that there seems to be an association with heavy use of nalidixic acid that creates resistance problems. In the third world enteric pathogens tend to be resistant to almost any antibiotic you use due to abuse. Therefore I don't actually think the fluoroquinolones are the problem, it is the unregulated use of these drugs in those countries that is the cause.

MIDTVEDT Certainly, all the things you mentioned are part of the problem, however, taking the quinolones in Scandinavia, we are not using any quinolones to treat poultry in Scandinavia, but we are using quinolones in aquaculture. In Norway, the quinolones (except nalidixic acid) were not on the market for medical use until three or four months ago. They are quite new for human use, it is quite another story. Going back to the broader use of antimicrobials all around, I agree completely with you, but we do have to look upon the question as a one compartment model. That is my key message

ANON I would still like to comment on the one compartment model. If you look at the bacteria themselves and consider them just the bacteria, they just travel from one compartment to the other. I think that you have to take into consideration the hosts of these bacteria and I disagree that the fish pathogens use human beings or other animals as a normal host and can survive there, unless the human being or the other animal is treated with a similar drug. Then it can be selected out and then can cause problems, otherwise it will just die off in the other compartment.

MIDTVEDT This partly will depend also about the fate of the drugs and, if you return to the story of the quinolones, the quinolones will stay in the environment for a considerable period of time and will reach not only the fish bacteria, but it will also reach the bacteria in the birds, in the environment and in the fish farmers themselves and so on. Therefore you do have the possibility to have alteration in various hosts.

ANON I agree with you in that form, because I consider that as pollution and we have to get rid of pollution. From that point of view I totally agree with you.

MIDTVEDT Thankyou.

ANON 2 In the face of such heavy concerns, how is Swedish aquaculture performing? Do they not use these antibiotics, or what?

MIDTVEDT Aquaculture in Sweden is not as important as in Norway. You have a different situation there. Fish farm production in Sweden is next to nothing and the use of antibiotics is also very slight. That is not due to Sweden's antipollution controls, but they have neither the production nor the problems.

LAFONT You have advocated the use of monobactams as the only group of antibiotics which could be used in aquaculture. However I am afraid medical people would disagree and object, because they could be afraid of the risk of selecting for new, extended spectrum betalactamases, which cause concern in hospitals with the use of such drugs. So I am afraid even this group could not be allowed.

MIDTVEDT I said from a strict ecological point of view, monobactams could be suitable for use, but I also said that, from a medical point of view, you will have to talk to the medical profession about the usage. Going back to the Scottish story I just want to comment upon their usage of amoxycillin. I think you are all aware that amoxycillin is a betalactam. It is in use all around Europe for medical purposes and is inactivated by betalactamases. By using it in aquaculture in Scotland you will certainly create a problem with betalactamases. A question was asked earlier about drug hypersensitivity; so far no hypersensitivity to amoxycillin has been reported. However, it may be to early to expect to see that. If a fish producer is using a betalactam, it will not be during use that he will be hypersensitive, he won't have any reaction then, but he may be on later exposure to the drug. Going back to the monobacatam, I want to state that from a strict ecological point of view, they might be suitable, but from a medical point of view you do have to discuss the problem and I think I said that.

LEWIN I just can't understand how you can justify the use of a monobactam or even a penicillin, because these drugs have plasmid mediated resistance. I think my French colleagues will agree that you will tend to select multiple antibiotic resistance so that all you are going to do is create a very big problem for yourself. The use of amoxycillin may be efficient for a short time, but the experience that we have had in the medical field, as I am certain you well know, is that you get a huge amount of resistance. So why should we go around creating resistance problems for ourselves for the sake of ecology? I think you have to balance it out, its not a rational way. We need to devise a rational drug use policy rather than saying that "we are not doing this for ecological reasons", or "we are going to use this", which will create huge problems for industry and for human medicine. I am certain my French colleagues would agree with me on that.

MIDTVEDT And I also agree on that. Let me just make myself clear, that from a strict ecological point of view, it can be used, from a medical point of view you will create problems, if you use it. You have to discuss pro and contra, in order to solve the problems of the fish industry.

MICHEL A good debate and discussion, One of the purposes of this meeting was to confront and perhaps combine the ecological, medical, practical and commercial point of view and we have the promise of an active discussion for the last day.

EVALUATION OF THE SENSITIVITY OF BACTERIAL FISH PATHOGENS TO DIFFERENT ANTIMICROBIAL COMPOUNDS

A.E. Toranzo, J.L. Romalde, I. Bandín, Y. Santos & J.L. Barja

Departamento de Microbiología y Parasitología, Facultad de Biología, Universidad de Santiago, Santiago de Compostela 15706, Spain

Summary: *In this work the susceptibility of several Gram-negative (motile* Aeromonas, Aeromonas salmonicida, Vibrio *and* Yersinia *species) and Gram-positive (*Renibacterium salmoninarum*) bacterial fish pathogens to different antimicrobial compounds was studied in order to obtain useful information for therapeutic purpose. Sixteen antibacterial compounds including tetracyclines, macrolides, ß-lactam, aminoglycoside and polypeptide antibiotics, sulphonamides, quinolones and nitrofurans were evaluated by the agar plate dilution technique. Only tetracycline showed strong inhibitory effect against all the strains, with MIC values ranging from < 0.19 to 3.12 µg/ml. For the other drugs great differences were observed between Gram-negative and Gram-positive pathogens. So, whilst quinolones proved to be quite useful against* Vibrio, Yersinia *and* Aeromonas *species (MIC between < 0.019 and 0.15 µg/ml), erythromycin was the most effective against* R. salmoninarum. *Cefoxitine (a drug not commonly employed in fisheries) proved to be quite active against most of the Gram-negative isolates tested excepting* Y. ruckeri *strains.*

Résumé: **Evaluation de la sensibilité de souches bactriennes de poissons à différents composés antimicrobiens.**

Les infections bactériennes sont responsables de pertes importantes chez les poissons et nécessitent un fréquent recours à l'antibiothérapie malgré l'existence de vaccins contre la vibriose et la yersiniose. Depuis les premières tentatives avec les sulfamides de nombreuses formules ont été employées, mais peu d'études comparatives ont été menées jusqu'à présent. On a comparé les performancess de 16 agents antimicrobiens contre 71 souches bactériennes d'espèces variées. Les concentrations minimales inhibitrices (CMI) étaient établies par la technique de dilution en gélose.

La tétracycline présente le spectre d'activité le plus large, à la fois contre les souches Gram + et Gram -. Les Aeromonas *mobiles sont très sensibles aux quinolones et au chloramphenicol, bien que des résistances à la tétracycline et au chloramphenicol soient souvent exprimées par* A.sobria. A.salmonicida *est parfois résistant à l'acide oxolinique et Y.ruckeri au chloramphenicol, alors que les* Vibrio *spp. sont sensibles aux mêmes composés et au Romet (association sulfamide-orméthoprime). En revanche,* Renibacterium salmoninarum *ne répond qu'à la tétracycline, et surtout à l'érythromycine. Exception faite des quinolones il existe souvent des différences marquées d'efficacité à l'intérieur d'une même famille d'antibiotiques (supériorité de la tétracycline sur l'oxytétracycline, de la furazolidone sur le nifurpirinol, du Romet sur le Tribrissen). La céfoxitine présente également une activité intéressante à l'égard des Gram -,* Y.ruckeri *excepté. D'une manière générale la tétracycline, compte-tenu de son large spectre, apparaît comme l'antibiotique le plus intéressant.*

Introduction

Bacterial diseases are responsible for big losses in firshfarming production. Although vaccination procedures are currently employed in the prevention of vibriosis and enteric redmouth diseases, at present there are not suitable vaccines to prevent furunculosis, bacterial kidney disease (BKD) or septicaemia by motile *Aeromonas* and, therefore, antibacterial therapy is still essential in the control of clinical cases.

Following the introduction of sulphonamides, the first antimicrobial agent used in fisheries (4), a wide range of antibacterial compounds has been employed in aquaculture with different success (2,6,10). However, comparative studies of these chemo-therapeutants are practically absent at present and, consequently, it is difficult to assess the value of one drug over another. Previous report of our laboratory showed that the quinolones were the most effective compounds to control important bacterial diseases in aquaculture (8). In the present work, we have extended this study using both drugs commonly used in fisheries and other antimicrobial compounds which until now their efficacy against bacteria causing diseases in fresh and sea water fish had not been evaluated.

Material and methods

A total of 71 strains of Gram-negative and Gram-positive bacterial fish pathogens, obtained from private collections or isolated in our laboratory during epizootic outbreaks, were used in the drug susceptibility test. The strains belonged to the following species: *Yersinia ruckeri* (19 isolates), *Aeromonas hydrophila* (12), *A. sobria* (5), *A. salmonicida* (7), *Vibrio anguillarum* (12), *V. splendidus/ V. pelagius* group (3), *V. ordalii* (1), *V. tubiashii* (1) and *Renibacterium salmoninarum* (11). Stock cultures were stored frozen at -80°C in Tryptic Soy Broth (TSB, Difco) or Mueller-Hinton Broth (MHB, Difco) with 15% (v/v) of glycerol. Working cultures were grown in Mueller-Hinton Agar

(MHA) (Difco). For the *Vibrio* and *Renibacterium* strains all the media were supplemented with 1% NaCl and 0.1% L-cysteine hydrochloride (Sigma), respectively.

The chemotherapeutic agents used were: tetracycline (Te), oxytetracycline (OTe), erythromycin (E), cefoxitine (Cef), kanamycin (K), streptomycin (S), chloramphenicol (Cl), vancomycin (V), polymixin B (PB), sulphadiazine (Su), Romet (Ro), Tribrissen (Tri), nifurpirinol (Nf), furazolidone (F), oxolinic acid (OA) and Imequil (I). Soluble compounds were dissolved in distilled water or phosphate buffer, and 0.1 M NaOH or ethanol were added when it was necessary (1). Insoluble compounds were suspended in sterile distilled water.

The Minimum Inhibitory Concentrations (MICs) were determined according to the agar plate dilution technique (12). Brain Heart Infusion Broth (BHI, Difco) with 1% NaCl added was used for preparing inocula of the Gram-negative isolates and the cultures were incubated at 22°C for 24 h. *Renibacterium* isolates were grown in MHA with 0.1% L-cysteine hydrochloride (MHA-C) for 8 days at 15°C before experimental use. Inocula were adjusted with PBS to contain about 1 $x10^8$ colony-forming units (CFU)/ml. The diluted cultures were inoculated using an automatic multipoint inoculator (Denley) on plates of MHA (Difco) with 1% of Na Cl, containing the different concentrations of the drugs. The cultures of *R. salmoninarum* were spreaded onto MHA-C. All media were stored at 4°C and used within 48 h. The lowest concentration of antimicrobics (µg/ml) resulting in complete inhibition of visible growth after 48 h for the Gram-negative isolates and 15 days for *Renibacterium* was considered the MIC.

Results and Discussion

In order to obtain useful information for controlling epizootics we have evaluated a total of 16 antimicrobial compounds including drugs commonly used in fisheries and other chemotherapeutants which efficacy against bacterial fish pathogens had not been determined until present. The MIC values obtained indicate that only the tetracycline was highly effective against most of the strains tested. For the other drugs great differences were observed between Gram-negative and Gram-positive pathogens. In addition, within the Gram-negative isolates some variation was detected according to the bacterial species studied. These findings are in agreement with those of Ledo *et al* (8).

As shown in Table I, the motile *Aeromonas* strains were very sensitive to oxolinic acid and Imequil (MIC < 0.019 µg/ml). Whereas most of the A. hydrophila strains showed a high susceptibility to chloramphenicol (MIC < 0.19 µg/ml) and tetracycline (MIC ranging from 0.19 to 0.39 µg/ml), these compounds were poorly effective against *A. sobria* isolates. Although some

Table I. MIC values (µg/ml) of the 16 antibacterial compounds tested against motile *Aeromonas* strains

Chemotherapeutants

Bacterial strains	Cef	Cl	Te	OTe	OA	I	Su	Ro	Tri	V	PB	Nf	F	E	K	S
A.hydrophila																
P-192 (O1)	<1.88	1.56	1.56	25	<0.019	<0.019	37.8	0.2	9.39	250	<0.97	<1.17	1.17	1.56	10	31.2
81-83 (O2)	<1.88	<0.19	0.19	1.56	<0.019	<0.019	37.8	0.2	9.39	250	3.9	<1.17	<1.17	6.25	10	7.8
80-A2 (O2)	3.75	3.12	0.19	3.12	<0.019	<0.019	1200	9.39	9.39	500	500	4.69	<1.17	25	10	15.6
B-51 (O2)	1.88	<0.19	0.19	3.12	<0.019	<0.019	600	0.2	9.39	500	31.2	<1.17	<1.17	6.25	10	7.8
B-32 (O3)	1.88	<0.19	0.19	3.12	<0.019	<0.019	150	0.2	9.39	500	1.95	4.69	<1.17	1.56	10	15.6
B-35 (O3)	3.75	0.19	0.39	3.12	<0.019	<0.019	1200	2.35	9.39	500	3.9	9.39	1.17	25	10	7.8
B-36 (O3)	3.75	<0.19	0.19	3.12	<0.019	<0.019	150	0.2	9.39	250	1.95	9.39	<1.17	25	10	7.8
P-226 (O6)	1.88	0.19	0.19	3.12	<0.019	<0.019	1200	37.8	37.8	500	250	9.39	<1.17	25	10	31.2
T-80 (O6)	3.75	<0.19	0.39	3.12	<0.019	<0.019	1200	37.8	9.39	500	1.95	9.39	1.17	25	10	7.8
T-12 (O6)	3.75	0.19	0.39	3.12	<0.019	<0.019	1200	37.8	37.8	500	250	4.69	<1.17	25	10	31.2
Y.62 (O19)	<1.88	<0.19	0.19	3.12	<0.019	<0.019	600	0.2	9.39	250	15.6	2.34	<1.17	1.56	10	15.6
80-A1 (O24)	3.75	<0.19	0.39	3.12	<0.019	<0.019	1200	9.39	9.39	500	0.97	18.8	1.17	12.5	10	15.6
A. sobria																
ATCC9071 (O2)	30	1.56	0.78	25	0.15	1.25	1200	37.8	9.39	500	1.95	150	2.34	25	10	7.8
P-261 (O25)	<1.88	6.25	1.56	25	<0.019	<0.019	150	9.39	9.39	250	1.95	18.8	1.17	1.56	10	31.2
P-355 (O11)	<1.88	6.25	1.56	25	<0.019	<0.019	37.8	<0.2	9.39	125	1.95	2.34	<1.17	0.78	10	15.6
A7 (024)	<1.88	3.12	1.56	25	<0.019	<0.019	150	9.39	9.39	125	<0.97	2.34	<1.17	1.56	10	7.8
P-252 (019)	<1.88	<0.19	1.56	25	<0.019	<0.019	37.8	<0.2	9.39	250	1.95	1.17	<1.17	3.12	10	31.2

Key : See p.322

Table II.- MIC values (μg/ml) of the antibacterial compounds tested against *A. salmonicida*

	Chemotherapeutant															
Bacterial strains *A. salmonicida*	**Cef**	**Cl**	**Te**	**OTe**	**OA**	**I**	**Su**	**Ro**	**Tri**	**V**	**PB**	**Nf**	**F**	**E**	**K**	**S**
V5/80	<1.88	<0.19	<0.19	1.56	0.019	<0.019	600	4.69	9.39	500	0.97	37.50	4.69	3.12	2.5	15.6
416	<1.88	<0.19	6.25	25.00	0.310	0.150	1200	9.39	37.80	500	<0.97	18.80	2.34	3.12	2.5	500
PP-72	<1.88	<0.19	<0.19	1.56	1.250	0.150	9.39	0.20	4.69	500	0.97	2.34	<1.17	1.56	2.5	15.6
H22/77	<1.88	0.19	0.19	3.12	<0.019	<0.019	1200	2.35	9.39	500	0.97	2.34	<1.17	3.12	2.5	15.6
NF14.2	<1.88	0.19	0.19	3.12	<0.019	<0.019	1200	4.69	9.39	500	0.97	37.50	4.69	3.12	2.5	15.6
PP-43	<1.88	<0.19	0.19	3.12	<0.019	<0.019	600	4.69	9.39	500	0.97	4.69	<1.17	3.12	5	15.6
NF 16.1.	<1.88	0.19	0.19	3.12	0.019	0.037	600	4.69	9.39	500	0.97	18.80	4.69	1.56	5	7.8

Table III.- MIC values (µg/ml) of the chemotherapeutic agents evaluated against different *Vibrio* species

	Chemotherapeutant															
Bacterial strains (serotype)	**Cef**	**Cl**	**Te**	**OTe**	**OA**	**I**	**Su**	**Ro**	**Tri**	**V**	**PB**	**Nf**	**F**	**E**	**K**	**S**
V. anguillarum																
TM-14 (O1)	3.75	<0.19	0.19	3.12	0.019	0.019	600	<0.2	9.39	500	15.6	2.34	<1.17	6.25	10	31.2
RT-32 (O1)	3.75	<0.19	0.19	3.12	0.019	0.019	600	<0.2	9.39	500	125	4.69	1.17	6.25	10	15.6
96F (O1)	15	<0.19	0.19	3.12	0.019	0.019	600	<0.2	4.69	500	250	<1.17	<1.17	6.25	10	15.6
775 (O1)	3.75	<0.19	0.19	3.12	<0.019	<0.019	150	<0.2	4.69	125	125	2.34	<1.17	<0.39	10	3.9
R82 (O1)	15	0.19	0.39	3.12	0.019	0.019	150	<0.2	4.69	250	500	<1.17	<1.17	6.25	10	15.6
R83c (O1)	30	0.39	0.19	3.12	0.019	0.019	150	<0.2	4.69	125	500	<1.17	<1.17	12.5	10	31.2
RG111 (O2)	15	<0.19	0.39	3.12	<0.019	<0.019	150	<0.2	4.69	125	1.95	<1.17	<1.17	6.25	5	15.6
RV22 (O2)	60	<0.19	0.19	1.56	<0.019	<0.019	9.39	<0.2	4.69	125	1.95	<1.17	<1.17	6.25	5	7.8
19264 (O2)	3.75	<0.19	0.19	3.12	<0.019	<0.019	37.8	<0.2	9.39	250	1.95	1.17	<1.17	6.25	10	15.6
43F (O2)	60	<0.19	0.39	3.12	<0.019	<0.019	150	<0.2	4.69	250	0.97	<1.17	<1.17	6.25	5	15.6
14181 (O2)	120	<0.19	0.19	1.56	<0.019	<0.019	9.39	<0.2	4.69	125	0.97	<1.17	<1.17	3.12	2.5	3.9
RG111(O2)	15	<0.19	0.19	1.56	<0.019	<0.019	37.8	<0.2	4.69	250	0.97	<1.17	<1.17	6.25	5	15.6
V. splendidus / V. pelagius																
ATCC 25914	3.75	0.39	0.19	3.12	0.037	0.037	37.8	<0.2	4.69	31.2	15.6	37.5	<1.17	6.25	2.5	3.9
RG-91	<1.88	0.19	<0.19	3.12	<0.019	<0.019	150	<0.2	2.35	250	<0.97	2.34	1.17	1.56	10	15.6
RM 51	<1.88	<0.19	0.39	3.12	0.037	0.075	2.35	<0.2	4.69	15.6	<0.97	2.34	<1.17	1.56	2.5	1.95
V. ordalii																
NCMB2167	<1.88	0.19	0.19	3.12	0.075	0.075	37.8	<0.2	4.69	15.6	0.97	2.34	1.17	1.56	10	31.2
V. tubiashii																
ATCC 19109	1.88	0.19	0.39	3.12	0.075	0.075	37.8	<0.2	4.69	15.6	0.97	2.34	1.17	1.56	10	31.2

Table IV. MIC values (µg/ml) of the antibacterial compounds assayed against the *Y. ruckeri* strains.

	Chemotherapeutant															
Bacterial strains (serotype)	**Cef**	**Cl**	**Te**	**OTe**	**OA**	**I**	**Su**	**Ro**	**Tri**	**V**	**PB**	**Nf**	**F**	**E**	**K**	**S**
Y. ruckeri																
PP31 (O1)	30	1.56	0.78	12.5	<0.019	0.019	>1200	0.2	9.39	500	<0.97	150	4.69	50	5	7.8
RS 96(O1)	30	1.56	0.78	12.5	<0.019	0.019	>1200	0.2	9.39	500	500	150	4.69	50	2.5	7.8
11.4 (O1)	30	0.78	0.78	12.5	<0.019	0.019	>1200	0.2	9.39	500	125	75	4.69	50	2.5	7.8
87109 (O1)	30	1.56	0.78	12.5	0.037	0.075	>1200	0.2	9.39	500	500	150	4.69	50	10	15.6
FP13 (O1)	30	1.56	0.78	25	<0.019	0.019	>1200	0.2	9.39	500	<0.97	150	2.34	25	2.5	7.8
AG71 (O1)	30	3.12	3.12	25	<0.075	0.15	>1200	0.2	9.39	500	500	150	4.69	50	10	500
YR64 (O1)	30	1.56	0.78	12.5	<0.019	0.019	>1200	0.2	9.39	500	<0.97	150	4.69	50	5	7.8
1638 (O2)	30	0.78	0.78	12.5	<0.019	0.019	>1200	0.2	9.39	500	500	150	4.69	50	5	7.8
RS2 (O2)	30	1.56	0.78	12.5	0.075	0.15	>1200	4.69	9.39	500	125	37.5	4.69	50	10	31.2
RS6 (O2)	30	0.39	0.39	6.25	<0.019	0.019	>1200	0.2	9.39	500	500	75	2.34	50	5	7.8
11.31 (O2)	30	1.56	0.78	12.5	<0.019	0.019	>1200	0.2	9.39	500	500	75	2.34	50	5	7.8
RS54 (O3)	30	0.78	0.39	6.25	<0.019	0.019	>1200	0.2	9.39	500	31.2	75	9.39	50	2.5	7.8
1533(O3)	30	1.56	0.78	12.5	0.037	0.075	>1200	0.2	9.39	500	500	150	4.69	50	10	31.2
RS79 (O3)	30	0.78	0.39	6.25	<0.019	0.019	>1200	0.2	9.39	500	500	150	4.69	25	5	7.8
RS24 (O5)	30	1.56	0.78	12.5	0.037	0.037	>1200	2.35	9.39	500	125	150	4.69	50	5	7.8
RS25 (O5)	30	1.56	0.78	12.5	<0.019	0.019	>1200	0.2	9.39	500	500	75	2.34	50	5	7.8
11.47 (O5)	30	0.78	0.39	6.25	<0.019	0.019	>1200	0.2	9.39	500	31.2	37.5	4.69	50	5	15.6
11.73(O6)	30	1.56	0.78	12.5	0.019	0.037	>1200	0.2	9.39	500	31.2	150	4.69	50	10	15.6
RS80 (O6)	30	1.56	0.78	12.5	<0.019	0.019	>1200	0.2	9.39	500	500	75	2.34	50	5	7.8

Table V.- MIC values (µg/ml) of the antibacterial compounds tested against *R. salmoninarum*

	Chemotherapeutant															
Bacterial strains *R. salmoninarum*	**Cef**	**Cl**	**Te**	**OTe**	**OA**	**I**	**Su**	**Ro**	**Tri**	**V**	**PB**	**Nf**	**F**	**E**	**K**	**S**
ATCC 33209	30	0.78	0.78	6.25	1.25	10	150	600	>1200	500	>500	>150	37.5	0.39	>10	31.7
ATCC 33739	30	12.5	0.78	6.25	1.25	10	150	600	>1200	500	>500	75	37.5	0.39	>10	31.7
K46	>240	25	25	25	10	>10	600	600	600	125	250	>150	>150	0.39	>10	>500
K76	>240	25	25	25	10	>10	600	600	600	500	250	>150	>150	0.39	>10	>500
417	60	250	3.12	3.12	10	2.5	150	600	>1200	7.8	>500	>150	>150	0.39	>10	250
419	60	250	1.56	3.12	10	>10	150	600	600	7.8	>500	>150	>150	0.39	>10	500
420.	30	250	1.56	3.12	10	>10	150	600	600	7.8	>500	>150	>150	0.39	>10	>500
421	60	250	1.56	3.12	10	>10	150	600	600	125	>500	>150	37.5	0.39	>10	>500
422	120	>250	1.56	3.12	>10	>10	150	600	600	125	>500	37.5	>150	0.39	>10	>500
R1	>240	0.78	12.5	25	10	>10	600	600	>1200	500	>500	>150	18.8	0.39	>10	·>500
466	30	6.25	0.78	3.12	1.25	1.25	600	600	600	125	>500	75	37.5	0.39	>10	250

Key to tables I to V

Cl	Chloramphenicol	I	Imequil	OTe	Oxytetracycline	Su	Sulphadiazine
Cef	Cefoxitine	K	Kanamycin	PB	Polymyxin B	Te	Tetracycline
E	Erythromycin	Nf	Nifurpirinol	Ro	Romet	Tri	Tribrissen
F	Furazolidone	OA	Oxolinic acid	S	Streptomycin	V	Vancomycin

variability was observed among the strains in their susceptibility to some compounds such as cefoxitine, Romet, polymyxin B, and nifurpirinol, these differences were not related to the serotype.

The drug susceptibility pattern of *A. salmonicida* (Table II) was very similar to that of motile *Aeromonas*. However, only in this species we have detected some strains resistant to quinolones (i.e., strain PP-72 showed a MIC for oxolinic acid about 60-fold higher than the other strains). This finding is comparable with that reported by other authors in different countries (5,9,11), where oxolinic acid resistant strains were described after extensive use of the drug for the treatment of furunculosis.

Similarly to the *Aeromonas* sp, the quinolones were the most active substance against *Vibrio* strains (Table III). The MIC values ranged from < 0.019 µg/ml for *V. anguillarum* to < 0.019-0.075 µg/ml for the other *Vibrio* species. Most of the *Vibrio* isolates were also sensitive to Romet (MIC < 0.2 µg/ml), chloramphenicol and tetracycline (MIC between < 0.19 and 0.39 µg/ml).

Quinolones were also the most effective compounds agains *Y. ruckeri* strains (Table IV) with MIC values ranging from <0.019 to 0.15 µg/ml, efficacy reported by other authors (2). The *Y. ruckeri* isolates were also very sensitive to tetracycline (MICs from 0.39 to 0.78 µg/ml) and the two potentiated sulphonamides. In general, the 19 strains of *Y. ruckeri* displayed a very similar susceptibility pattern to chemotherapeutants and no significative differences were detected among serotypes.

R. salmoninarum isolates displayed, as shown in Table V, a drug susceptibility pattern different to that observed in Gram-negative fish pathogens. In general, natural antibiotics were much more effective than the synthetic antibacterial compounds assayed. In fact, neither quinolones nor nitrofurancs or sulphonamides showed inhibitory effect on the growth of *Renibacterium*, and tetracycline and erythromycin, were the most effective compounds against all the isolates.

It is noteworthy that in all the bacterial groups differences in the inhibitory spectrum were found within chemotherapeutants chemically related, being tetracycline clearly more effective than oxytetracycline, furazolidone than nifurpirinol, and Romet than Tribrissen. Only both quinolones tested showed quite similar high effectiveness against Gram-negative fish pathogens.

In regard to compounds not commonly employed in fisheries, cefoxitine (a cephalosporin) was found to be quite active against all the Gram-negative isolates tested, excepting *Y. ruckeri* strains. In addition, although the use of chloramphenicol is forbidden in many Countries, *Vibrio* and motile *Aeromonas* strains (excepting *A. sobria*) were highly susceptible to this drug (MIC <0.19 μg/ml). By contrast vancomycin and sulphadiazine were ineffective against most of the isolates.

According to these results tetracycline seems to be the antibiotic of choice for combating the main infections in aquaculture. However caution must be taken in the indiscriminate use of tetracycline because a high persistance of this antibacterial compound has been demonstrated in sediments from fishfarms (7), which can constitute a future risk of increasing the resistance to this drug.

Acknowledgements: This study was supported by grants PB-87-1027 from the Dirección General de Investigación Científica y Técnica (DGICYT), Spain. J. L. Romalde and I. Bandín acknowledge the Ministerio de Educación y Ciencia (Spain) for a Research fellowship

References

1). ANHALT, J.P. & WASHINGTON, J.A., 1980. Preparation and storage of antimicrobic solutions. In, *Manual of clinical microbiology*, 3th ed., H. Lennette, A. Balows, W. J. Hauser and J. P. Truant eds, American Society for Microbiology, Washington, D.C., pp.453-457.

2). AUSTIN, B.,1984. The control of bacterial fish diseases by antimicrobial compounds. In, *Antimicrobial and aquaculture, benefits and malefits*, M. Woodbine ed., Butterworths, London, pp.255-286.

3). CESCHIA, G., GIORGETTI, G., BERTOLDINI, G and FONTEBASSO, S., 1987. The *in vitro* sensitivity of *Yersinia ruckeri* to specific antibiotics. *J. Fish Dis.*, **10**, 65-67.

4). GUTSELL, J., 1946. Sulfa drugs and the treatment of furunculosis in trout. *Science*, **104**, 85-86.

5). HASTINGS, T. S., & McKAY, A., 1987. Resistance of *Aeromonas salmonicida* to oxolinic acid. *Aquaculture*, **61**, 165-171.

6). HERWIG, N., 1979. *Handbook of drugs and chemicals used in the treatment of fish diseases*. Thomas Books, Springfield, Illinois.

7). JACOBSEN, P. & BERGLIND, L., 1988. Persistence of oxytetracycline in sediments from fish farms. *Aquaculture*, **70**, 365-379

8). LEDO A., DOPAZO, C. P., ROMALDE, J. L, TORANZO, A. E and BARJA, J. L., 1987. Effectiveness of different chemotherapeutic agents for controlling bacterial fish diseases. *Bull. Eur. Ass. Fish Pathol.*, **7**, 20-22.

9). O'GRADY, P., PALMER, R., HICKEY, C., & SMITH, P. R., 1986. Antibiotic therapy of furunculosis in freshwater and seawater. *J. Fish Dis.*, **4**, 397-410.

10). SNIEZSKO, S. F., 1978. Control of Fish diseases. *Mar. Fish. Rev.*, **40**, 65-68.

11). TSOUMAS, A.; ALDERMAN, D. J., & RODGERS, C. J., 1989. *Aeromonas salmonicida*: development of resistance to 4-quinolone antimicrobials. *J. Fish Dis.*, 12, 493-507.

12). WASHINGTON, J. A., & V. L. SUTTER., 1980. Dilution susceptibility test: agar and macro-broth dilution procedures. In, *Manual of clinical microbiology*, 3th ed., E. H. Lennette, A. Balows, W. J. Hauser and J. P. Truant eds, American Society for Microbiology, Washington, D.C., pp.453-457.

BACTERICIDAL ACTIVITY OF 4-QUINOLONES, INCLUDING FLUMEQUINE, AGAINST *AEROMONAS SALMONICIDA*

A.C. Barnes, C.S. Lewin, T.S. Hastings, & S.G.B. Amyes

Department of Medical Microbiology, University of Edinburgh Medical School, Teviot Place, Edinburgh EH8 9AG.

Summary: *The activity against* Aeromonas salmonicida *of flumequine and five newer fluoroquinolones was investigated in terms of MIC and bactericidal activity and compared to oxolinic acid. In terms of MIC, enrofloxacin, sarafloxacin, PD117,596 and PD127,391 were more active than oxolinic acid against both oxolinic acid susceptible and resistant isolates of* A. salmonicida. *Flumequine was slightly more active in terms of MIC than oxolinic acid against oxolinic acid resistant isolates whereas CI934 was less active than oxolinic acid against all of the isolates tested. In addition to improved MICs, enrofloxacin, sarafloxacin, PD117,596 and PD127,391 were also rapidly bactericidal, with less than 1% of the bacteria surviving after 1 h exposure. In contrast, oxolinic acid, CI934 and flumequine were not bactericidal after 1 h, indeed, oxolinic acid was merely bacteriostatic even after 6 exposure. However, flumequine displayed bactericidal activity after 3 h, killing 90% of the bacteria. This bactericidal activity increased with duration of exposure with less than 1% of the bacteria surviving after 6 h exposure.*

Résumé: **Activité bactéricide de la fluméquine et d'autres quinolones contre *Aeromonas salmonicida*.**

La furonculose des saumons est souvent traitée à l'aide des 4-quinolones, qu'il s'agisse de l'acide oxolinique en Ecosse ou de la fluméquine en Norvège. Cependant les résistances tendent à se multiplier à l'encontre du premier. Aussi une analyse d'efficacité, comparant les concentrations minimales inhibitrices (CMI) et les effets bactéricides, a-t'elle été menée sur 5 fluoroquinolones: sarafloxacine, enrofloxacine, PD 127, PD 117,596 et CI 934. En termes de CMI, l'enrofloxacine, la sarafloxacine, le PD 127,391 et le PD 117 apparaissent nettement plus actifs que l'acide oxolinique (Tableau I), la fluméquine ne présentant de supériorité que contre les souches résistantes à l'acide oxolinique, et le CI 934 s'avérant constamment inférieur. Alors que l'acide oxolinique n'est pas bactéricide, les 4 premiers composés inactivent les bactéries à 91 % en 1 heure, la fluméquine réquérant 3 heures pour un résultat identique. Le CI 934 excepté, l'emploi des fluoroquinolones pourrait donc permettre d'éviter les rechutes, d'espacer les traitements,et peut-être de blanchir les poissons porteurs de germes. Bien que ses effets soient plus lents (mais des comparaisons pharmacocinétiques sont évidemment requises pour en juger définitivement) la fluméquine constitue également une alternative intéressante à l'usage de l'acide oxolinique et justifierait une demande d'autorisation pour l'Ecosse.

Introduction

Currently the control of furunculosis depends upon the use of antibiotics. The 4-quinolone oxolinic acid is one of the antibacterials most commonly used to treat infections caused by *Aeromonas salmonicida* (1,3). However, the effectiveness of oxolinic acid treatment has been hampered by the development of resistance in *A. salmonicida* which seems to be occurring with increasing frequency. Flumequine is another 4-quinolone which is currently used in the aquaculture industry in Norway, but not in Scotland.

In human medicine, the use of these quinolones has been superseded by newer compounds, which are more active against a wide range of pathogenic bacteria (7). We have therefore examined the activity of five new fluoroquinolones (sarafloxacin, enrofloxacin, PD127,391 and PD117,596, and CI934) and flumequine against 72 oxolinic acid susceptible and resistant Scottish isolates of *A.salmonicida* in comparison to oxolinic acid. This was done by determination of minimum inhibitory concentration (MIC) by agar dilution.

Previous studies have shown that there may be qualitative differences in lethality between 4-quinolones against *A.salmonicida* as Lewin and Hastings (5) demonstrated that oxolinic acid was unable to kill *A.salmonicida* after 3 h exposure. In contrast, two fluoroquinolones in medicinal use, ciprofloxacin and norfloxacin, were bactericidal (5). As MICs only provide information on the inhibition of bacterial multiplication, the bactericidal activity of the fluoroquinolones against selected isolates of *A.salmonicida* was also investigated.

Materials and Methods

1. Antibacterial preparation

Oxolinic acid, sarafloxacin, CI934 and flumequine were dissolved in 0.5M NaOH (0.2 ml/mg) prior to being dilute to the required concentration in sterile distilled water. Enrofloxacin, PD127,391 and PD117,596 were dissolved directly in sterile distilled water. Drug solutions were prepared fresh for all experiments.

2. Bacterial strains

The MICs of 72 oxolinic acid susceptible and resistant isolates of *A.salmonicida*, predominantly from Scottish salmon farms, were determined in this study. Two oxolinic acid resistant (MIC > 1.0 mg/l) and two oxolinic acid susceptible (MIC < 1.0 mg/l) strains of *A.salmonicida* subsp. *salmonicida* were used in the investigation into bactericidal activity. Stock cultures were stored in tryptic soy broth (TSB) + 10% glycerol at -70°C. Working cultures were maintained on drug free tryptone soy agar (TSA) plates (Oxoid, UK) at 4°C. Overnight cultures used in the bactericidal activity experiments were prepared by inoculating colonies into TSB and incubating at 22°C.

3. Determination of minimum inhibitory concentrations

MICs were determined on TSA plates as described previously (2).

4. Estimation of bactericidal activity

Table I: *In vitro* activities of 4-quinolones against 83 *A. salmonicida* strains

		[a]MIC at 22°C (µg/ml)		
Strain Type (no of isolates	Test agent	50%	90%	Range
Oxolinic acid resistant (38)	Oxolinic acid	3.00	7.50	1.00-15.00
	Sarafloxacin	1.50	4.00	0.20-4.00
	Enrofloxacin	0.50	1.00	0.05-5.00
	CI934	3.00	7.50	0.30-10.00
	PD127391	0.10	0.15	0.015-0.30
	PD117596	0.15	0.50	0.075-0.75
	Flumequine	2.00	4.00	0.10-20.00
Oxolinic acid susceptible (45)	Oxolinic acid	0.03	0.40	0.01-0.75
	Sarafloxacin	0.05	0.20	0.01-0.75
	Enrofloxacin	0.02	0.15	0.004-0.75
	CI934	0.40	1.00	0.04-3.00
	PD12739	0.0075	0.03	0.002-0.10
	PD117596	0.015	0.05	0.003-0.50
	Flumequine	0.075	0.10	0.075-0.50

[a] 50% and 90%, MIC for 50 and 90% isolates tested, respectively.

Killing efficacy was investigated essentially as described by Lewin and Hastings (1990) using two oxolinic acid resistant strains, MT364 and MT472, and two oxolinic acid susceptible strains, MT363 and MT736. A 1 in 50 dilution of a TSB culture containing approximately 10^8 cfu was made into fresh TSB containing various antibacterial concentrations. The antibacterial concentrations followed the arithmetic progression 1.5, 3, 5, 9. After 1 hr incubation at 22°C, samples were taken and serially diluted in TSB. Percent survival was then estimated by viable counting on TSA.

Results and Conclusions

In terms of MIC, enrofloxacin, sarafloxacin, PD127,391 and PD117,596 were significantly more active than oxolinic acid, with PD117,596 and PD127,391 being up to 100 times more active than oxolinic acid against resistant isolates (Table 1). Flumequine was also slightly more active than oxolinic acid against oxolinic acid resistant isolates, but it was less active against oxolinic acid susceptible isolates (Table 1). In the case of CI934, higher MICs were recorded against both oxolinic acid susceptible and resistant isolates of *A. salmonicida*.

In agreement with the previously reported results, oxolinic acid was not bactericidal against either the sensitive or the resistant isolates of *A. salmonicida* at concentrations up to 100 times the MIC after 3 h (Fig.1). Furthermore, even when the exposure time was increased to 6 h, no bactericidal activity was detected for oxolinic acid against *A.salmonicida* (Fig.1). Flumequine and CI934 were also not bactericidal after 1 h exposure. However, sarafloxacin, enrofloxacin, PD127,391 and PD117,596 were rapidly bactericidal, killing 99% of the bacteria after 1 h exposure (Fig.2). After 3 h exposure, flumequine displayed bactericidal activity, killing approximately 90% of the bacteria (Fig.3). The bactericidal activity of flumequine was observed to increase with duration of exposure as greater than 99% of the bacteria were killed after 6 h exposure (Fig.3). The lethality occurred at concentrations which should be achievable clinically for flumequine in salmon. However, with oxolinic acid resistant isolates high levels of flumequine would probably have to be maintained for 6 h or more for significant bactericidal activity to occur.

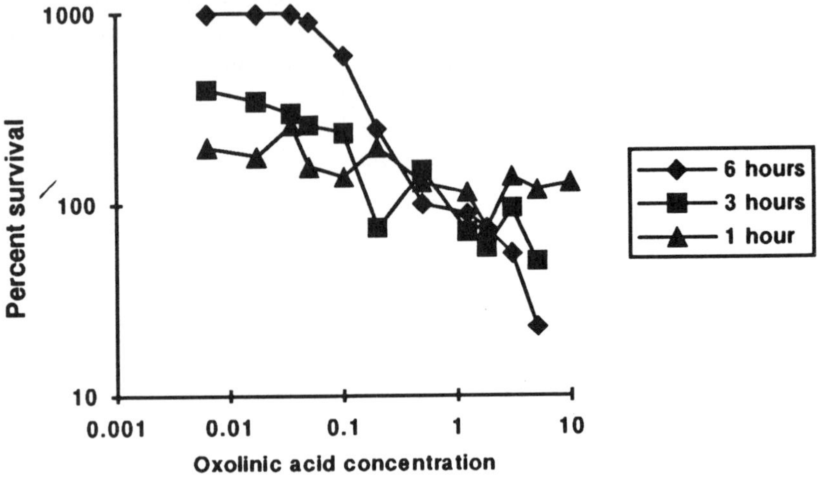

Fig 1 Bactericidal activity of oxolinic acid against *Aeromonas salmonicida* MI363

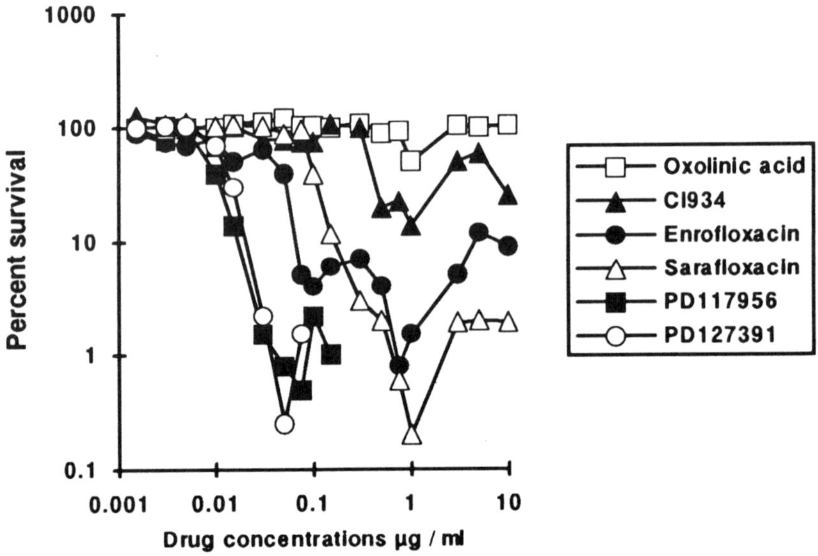

Fig 2 Bactericidal activity of 4-quinolones against *Aeromonas salmonicida* MI363 after 1h exposure

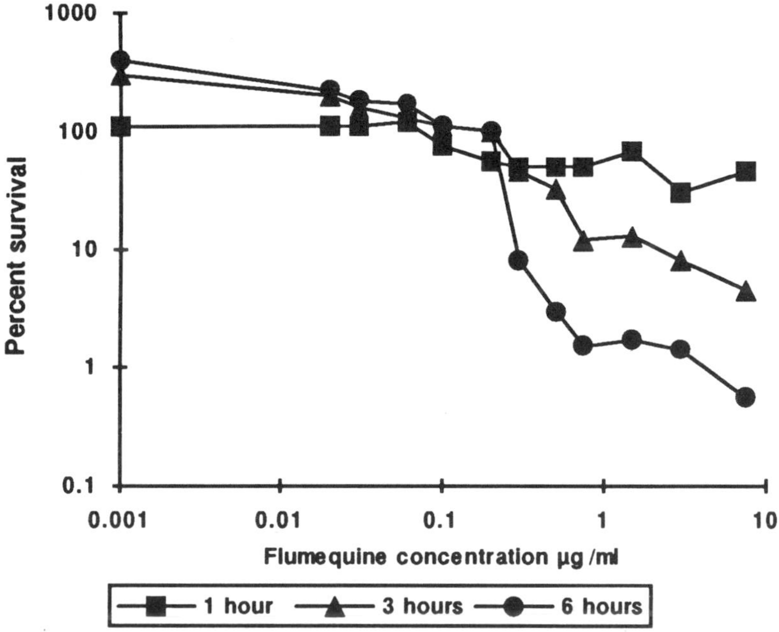

Fig 3 Bactericidal activity of flumequine against *A. salmonicida* MI363

In summary, with the exception of CI934, the fluoroquinolones were bactericidal against both oxolinic acid susceptible and resistant isolates of *A.salmonicida,* with bactericidal activity detected at concentrations slightly above the MIC. This bactericidal activity represents a qualitative difference between these fluoroquinolones and oxolinic acid as, in contrast, the latter compound was not bactericidal. This investigation found that 100% of the bacteria survived 6 h exposure to oxolinic acid at concentrations up to and including 100 times the MIC suggesting that this drug will not display significant bactericidal activity *in vivo.* These findings may explain why furunculosis frequently recurs after treatment with oxolinic acid, with relatively short periods of respite between outbreaks. The bactericidal activity of the fluoroquinolones may lead to longer periods between treatments which would have economic and environmental benefits for the aquaculture industry. Furthermore, the fluoroquinolones ability to kill bacteria may also be of use in eliminating bacteria in the asymptomatic carrier state of furunculosis. In this context, it is interesting to note that Scallan and Smith (1984) obtained promising results in reducing asymptomatic carriers of furunculosis in trout and salmon with flumequine.

Although the bactericidal activity of flumequine was not as rapid as that of the other fluoroquinolones, the *in vivo* significance of this qualitative difference should be investigated further. Factors such as pharmacokinetics, bioavailability and half life will also be involved in determining clinical efficacy and this qualitative difference may not be significant. It would appear that, with the exception of CI934, these fluoroquinolones merit further investigation as they may be more effective than oxolinic acid in the control of furunculosis. Based on these *in vitro* studies, CI934 would have to exhibit extremely good pharmacokinetics and have greatly increased bioavailability to be of use in furunculosis therapy. Flumequine, in addition to being more active than oxolinic acid in terms of both MIC and bactericidal activity, also exhibits good pharmacokinetics and bioavailability (O. Peterson, Apothokernes Laboratorium, A.S. Oslo, Norway, personal communication). Furthermore, flumequine is currently licensed for use in the aquaculture industry in Norway and must, therefore, be worthy of further study as a potential new drug for the treatment of furunculosis in Scotland.

Acknowledgements: We thank the Science and Engineering Research Council for the CASE Award to Andrew C. Barnes and we are grateful to the Scottish Salmon Growers Association for their financial support of this project.

References

1). AUSTIN B., RAYMENT J. & ALDERMAN D.J., 1983. Control of furunculosis by oxolinic acid. *Aquaculture,* 31, 101-108.

2). BARNES A.C., LEWIN C.S., HASTINGS T.S. & AMYES S.G.B., 1990. *In vitro* activities of 4-quinolones against the fish pathogen *Aeromonas salmonicida. Antimicrob. Agents Chemother.* 34, 1819-1820.

3). ENDO T., OGISHIMA K., HAYASAKI H., KANEKO S. & OSHIMA S., 1973 Application of oxolinic acid as a chemotherapeutic agent against infectious diseases in fish. *Bull. Jpn. Soc. Sci. Fish,* 39, 165-171.

4). HASTINGS T.S. & McKAY A., 1987. Resistance of *Aeromonas salmonicida* to oxolinic acid. *Aquaculture,* 31, 165-171.

5). LEWIN S.C. & HASTINGS T.S., 1990. *In vitro* activities of oxolinic acid, ciprofloxacin and norfloxacin against *Aeromonas salmonicida. J. Fish Dis.,* 13, 377-384.

6). SCALLAN A. & SMITH P.R., 1984. Control of asymptomatic carriage of *Aeromonas salmonicida* in Atlantic salmon smolts with flumequine. In, *Fish and Shellfish Pathology,* Academic Press, London.

7). SMITH J.T. & LEWIN C.S., 1988 Chemistry and mechanisms of action of the quinolone antibacterials. In, *The Quinolones,* V.T.Andriole ed., Academic Press, London, pp. 23-82.

IN VITRO ACTIVITY AND RESISTANCE DEVELOPMENT FOR SARAFLOXACIN (A-56620): AN AQUACULTURE ANTIBACTERIAL

J. Stamm

Abbott Laboratories, North Chicago, IL, 60064, USA

***Summary** : Sarafloxacin (A-56620) is a new aryl-fluoroquinolone antibacterial which has been shown to be extremely potent in vitro and in vivo against a broad spectrum of bacteria including the most common aquaculture pathogens. Sarafloxacin is being developed for therapeutic use in the worldwide aquaculture industry.*

Sarafloxacin has been compared with oxolinic acid, oxytetracycline, and ormetoprim-sulfadimethoxine (RometR) in in vitro studies of activity and resistance development using fish pathogens. Sarafloxacin was the most potent of the drugs with an MIC90 of < 1 mcg/ml. Bactericidal activity of sarafloxacin was rapid and extensive. The frequency of spontaneous resistance was low with sarafloxacin indicating poor ability to select resistant mutants. Organisms were also serially cultured in increasing concentrations of each drug. Susceptibility to each drug decreased gradually in a step-wise fashion. After 15 transfers, MICs of all drugs but sarafloxacin were above achievable in vivo concentra- tions for some organisms.

The above results plus the known absence of plasmid mediated transfer of quinolone resistance indicate that resistance to sarafloxacin will not develop easily or spread among aquaculture pathogens.

Résumé : **Etude in vitro de l'activité de la sarafloxacine et des risques de résitances bactériennes liés à son emploi en pisciculture.**

La sarafloxacine est un nouvel antibiotique à large spectre, très actif in vitro *et* in vivo *contre la plupart des bactéries rencontrés en pisciculture, dont on peut prévoir un emploi rapidement généralisé dans les élevages du monde entier. Des études* in vitro *sur des agents pathogènes de poissons ont permis de comparer l'activité antimicrobienne et la tendance à susciter l'apparition de résistances entre la sarafloxacine et d'autres molécules, incluant l'acide oxolinique, l'oxytétracycline et le Romet (association ormethoprime-sulfadiméthoxine autorisée aux Etats-Unis). La sarafloxacine s'est constamment montrée plus active, avec des* CMI_{90} *inférieures ou égales à 1 µg/ml. Cette activité, rapide et générale, est associée à une fréquence de résistances bien moindre que pour les autres molécules. La subculture de souches bactériennes en présence de concentrations croissantes d'antibiotiques révèle en effet qu'après 15 passages les CMI sont notablement augmentées pour tous les produits, excepté la sarafloxacine. En pratique le risque d'induire des mutants résistants est donc limité. Ces résultats, joints à la réputation des quinolones de ne pas susciter de résistances à support plasmidique, apportent une certaine garantie pour l'usage de la sarafloxacine en pisciculture.*

Introduction

Antibacterials are commonly used in the commercial production of fish to treat bacterial diseases. Oxytetracycline and ormetoprim-sulfadimethoxine (RometR) are currently approved by the U.S. Food and Drug Administration for use in aquaculture. Oxolinic acid, a quinolone antibacterial, is approved in Europe to combat furunculosis.

Sarafloxacin (A-56620) is a new aryl-fluoroquinolone antibacterial which has been identified as a potentially useful therapeutic agent against bac- terial diseases of fish. The compound has broad spectrum, potent antibacterial activity *in vitro*(5) and *in vivo* (1); it is bactericidal and pathogens have a low propensity for developing resistance to it (2,6). The mode of action is interference with the essential bacterial enzyme DNA-gyrase.

Abbott Laboratories is developing sarafloxacin for use in the aquaculture industry. Numerous in vitro and in vivo studies have been completed to support this development.

In Vitro Antibacterial Activity

1. Susceptibility to Sarafloxacin of *Aeromonas salmonicida* from Scotland and Ireland

Forty-seven *Aeromonas salmonicida* isolated from Atlantic Salmon during the last two years were tested for susceptibility to sarafloxacin. The strains were obtained from the Department of Fisheries, Dublin, Department of Fisheries and Agriculture for Scotland, and the Institute of Aquaculture, The University of Stirling.

A multipoint inoculator was used to place approximately 5 x 104 organisms on the surface of Isosensitest agar. Plates were incubated at 22°C for 48 hours.

The minimum inhibitory concentration (MIC) was defined as the lowest concentration which completely inhibited growth.

Results:-

Sarafloxacin MICs for these *A. salmonicida* strains ranged from 0.031 to 2.0 mcg/ml with MIC50 and MIC90 of 0.146 and 0.885 mcg/ml, respectively. All these strains were considered susceptible to sarafloxacin.

2. Comparative in vitro activity of sarafloxacin against fish pathogens

Studies have been performed to compare the in vitro activity of sarafloxacin, oxolinic acid, oxytetra- cycline, and the combination of sulfadimethoxine- ormetoprim (Romet 30R) against the significant aquaculture pathogens. The strains tested have been recently isolated from various locations, e.g. United States, Ireland, Canada, and Norway. MICs were determined using a standard agar dilution procedure including multipoint inoculation, Mueller Hinton agar, and 105 colony forming units inoculum.

Results:-

Aeromonas salmonicida (87 strains)

Compound	MIC Range	MIC 50	MIC 90
Sarafloxacin	<0.008-2	0.06	1
Oxolinic Acid	<0.008-8	1	2
Oxytetracycline	0.25->128	2	128
Romet 30	<0.25->128	4	>128

Based on mean MICs in this study, sarafloxacin is substantially more potent than oxolinic acid for these strains of A. salmonicida. If resistance to sarafloxacin is defined as MICs of 2 mcg/ml and above and to oxolinic acid as MICs of 1 mcg/ml and above (7), then 53% of this group of *A. salmonicida* are resistant to oxolinic acid compared with 15% to sarafloxacin. A large percentage of A. salmonicida were resistant to both oxytetracycline and a potentiated sulfonamide (Romet 30R).

Vibrio anguillarum (14 strains)

Compound	MIC Range	MIC 50	MIC 90
Sarafloxacin	0.015-0.5	0.12	0.25
Oxolinic Acid	0.015-2.0	0.06	0.25
Oxytetracycline	<0.25-1.0	0.25	0.5
Romet 30	0.06-1.0	0.25	1.0

All four antibacterials showed good activity against these organisms.

Yersinia ruckeri (17 strains)

Compound	MIC Range	MIC 50	MIC 90
Sarafloxacin	<0.008-0.5	0.015	0.06
Oxolinic Acid	0.015-8.0	0.03	0.06
Oxytetracycline	0.25-8.0	0.5	2.0
Romet 30	0.25->128	0.5	2.0

Sarafloxacin and oxolinic acid demonstrated excellent activity against these organisms. A small percentage (<10%) of strains were resistant to oxolinic acid, oxytetracycline, and/or Romet 30.

Resistance Development In Vitro

Sarafloxacin, oxolinic acid, oxytetracycline, and ormetoprim-sulfadimethoxine were compared to determine (1) the frequencies of in vitro resistance at eight-times the MIC, (2) the rates and extents of decrease in susceptibility when organisms were serially transferred to increasing concentrationsof drug, (3) the stability of the decreased susceptibility, and (4) cross-resistance to other antibacterials by organisms with developed resistance. Organisms tested included *Aeromonas salmonicida, A. hydrophila, Yersinia ruckeri, Vibrio ordalii,* and *V. anguillarum*.

1. Resistance Frequency

Resistance frequency experiments were performed to determine the number of spontaneous, single-step resistant mutants that were present in bacterial populations and that were selected by the antibacterial.

Overnight broth cultures were concentrated by centrifugation, resuspended in a small volume of broth, and plated on agar containing eight times the antibacterial MIC. Control cell counts were performed on drug-free agar. Plates were incubated at 25oC for 48 hours and resulting colonies were counted.

Results:-

Resistance frequencies were low for both sarafloxacin and oxolinic acid, i.e. <6.9 x 10-8, for isolates of *A. hydrophila, A. salmonicida, Yersinia ruckeri,*

Vibrio ordalii, and *V. anguillarum.* Similar frequencies were seen with oxytetracycline; those for ormetoprim- sulfadimethoxine were higher, i.e. 10-7-10-8.Mutants selected with the quinolones generally showed a low level of resistance to quinolones, i.e. they were 8 to 32-fold less susceptible than the corresponding parent. These mutants were considered still susceptible to quinolones.

2. Resistance Development with Serial Transfer

Serial transfer experiments were performed to determine the ability of organisms to develop resistance during contact with increasing levels of an antibacterial over time. The rapidity and extent of resistance development indicate whether or not resistance to a particular antibacterial is likely to develop during therapeutic use.

A series of up to 15 MICs was performed; organisms that showed growth at the highest concentration (one-half MIC) in one determination were used as the inoculum for the next determination.

Results:- During the serial transfers, the activities of the antibacterials against the test organisms typically decreased slowly in a stepwise fashion and many MICs were substantially higher than at the beginning. Final MICs of sarafloxacin for all organisms were at or below achievable in vivo concentrations; this was not true for oxolinic acid, ormetoprim-sulfadimethoxine and, especially, oxytetracycline.

3. Resistance Stability

Organisms from the final serial transfer were subcultured 10 times in drug-free broth to determine the stability of the observed resistance.The MICs of all antibacterials were similar for resistant mutants before and after 10 transfers in drug-free broth.

4. Cross-resistance to Other Antibacterials

Cross-resistance of developed mutants was tested with gentamicin, cefoxitin, and amoxicillin.Cross-resistance was present among the quinolones but not among the other antibacterials. Mutants selected with oxytetracycline and ormetoprim-sulfadimethoxine showed no significant change in susceptibility to any other antibacterial.

Discussion

The resistance frequency decreases with increased antibacterial concentration, and development of resistance depends on growth in the presence of subinhibitory concentrations of antibacterial. Therefore, the attainment and maintenance of adequate antibacterial levels in vivo during therapy, at least four times the MIC, is very important in minimizing the selection and survival of resistant organisms. Achievable in vivo concentrations of sarafloxacin in fish are not yet definitively known. However, studies to date indicate that levels should be well above MICs for all mutants selected in the resistance frequency experiments. Therefore, if resistant strains appeared during therapy, they would be susceptible and would not survive.Although several mutants developed in the transfer experiments showed greater resistance to quinolones, attainment of adequate in vivo quinolone levels during therapy would eliminate these mutants before several sequential mutations could occur.

It is important to note that many of the resistant mutants were slow growing, survived poorly in storage, or both. These mutants probably would be less able to compete in the natural environment and less virulent than the parent organism (2,4). To date, mutations that confer quinolone resistance have been found only on the bacterial chromosome, not on plasmids. Because quinolones appear to effectively inhibit the process of conjugative plasmid transfer (3), the spread of quinolone resistance by gene transfer is unlikely.

Resistance to sarafloxacin is not expected to develop easily during treatment of common bacterial diseases of fish, especially if in vivo levels of the compound are maintained at adequate levels. This advantage, plus excellent efficacy against fish pathogens, makes this compound a promising candidate for development as an aquaculture therapeutic.

References

1). FERNANDES P., CHU D., BOWER R., JARVIS K., RAMER N., & SHIPKOWITZ N. (1986). In vivo evaluation of A-56619 (difloxacin) and A-56620: new aryl-fluoro- quinolones. *Antimicrob. Agents Chemother.*, 29, 201-208.

2). FERNANDES P., HANSON D., STAMM J., VOJTKO C., SHIPKOWITZ N., & ST. MARTIN E. (1987). The frequency of in-vitro resistance development to fluoroquinolones and the use of a murine pyelonephritis model to demonstrate selection of resistance in vivo. *J. Antimicrob. Chemother.*, 9, 449-465.

3). NAKAMURA S., INOUE S., SHIMIZU M., IYOBE S., & MITSUHASHI S. (1976). Inhibition of conjugal transfer of R plasmids by pipemidic acid and related compounds. *Antimicrob. Agents Chemother.*, 10, 779-785.

4). SMITH J. (1984). Mutational resistance to 4-quinolone antibacterial agents. *Eur. J. Clin. Microbiol.*, 3, 347-350.

5). STAMM J., HANSON C., CHU D., BAILER R., VOJTKO C., & FERNANDES P. (1986). In vitro evaluation of A-56619 (difloxacin) and A-56620:new aryl-fluoroquinolones. *Antimicrob. Agents Chemother.*, 29, 193-200.

6). STAMM J. (1989). In vitro resistance by fish pathogens to aquaculture antibacterials, including the quinolones difloxacin (A-56619) and sarafloxacin (A-56620). *J. Aquatic Animal Health*, 1, 135-141.

7). TSOUMAS A., ALDERMAN D.J.., & RODGERS C.J. (1989). *Aeromonas salmonicida:* development of resistance to 4-quinolone antimicrobials. *J. Fish Diseases*, 12, 493-507.

SESSION 4

Residues

REVIEW OF THE METHODS USED IN THE DETERMINATION OF ANTIBIOTICS AND THEIR METABOLITES IN FARMED FISH

A. S. McGill & R. Hardy

Ministry of Agriculture; Fisheries & Food, Torry Research Station, PO Box 31, 135 Abbey Road, Aberdeen, AB9 8DG,

Summary: *The methods used for the determination of antibiotics and their metabolites in farmed fish are reviewed and the data presented in tabular form. They can be divided into two general types namely those that use microbiological assessment and those that use physiochemical techniques.*

Microbiological methods have certain advantages in that they are relatively inexpensive, simple and can provide an assessment of total antimicrobial activity. Conversely, they lack specificity, are relatively insensitive, do not differentiate between natural antimicrobial substances and synthetic antibiotics that are present, are time consuming using present methods, often lack reproducibility and are only semiquantitative.

Physiochemical methods using chromatographic techniques coupled with various detectors do not suffer from these disadvantages but unfortunately the equipment is complex and expensive, requires skilled operators and, except where it is known that only a limited range of antibiotics have been used, do not provide an assessment of total antimicrobial activity.

***Résumé*: Aperçu des méthodes utilisées pour la détection des antibiotiques et de leurs métabolites chez les poissons d'élevage.**

L'augmentation récente des productions aquacoles a principalement concerné les élevages traditionnels en eaux continentales. Cependant, dans certains pays, l'adoption de méthodes intensives s'est accompagnée d'une telle consommation de médicaments antimicrobiens qu'il a fallu en contrôler l'emploi, et surtout viller à ce que la qualité des produits livrés à la consommation ne s'en trouve pas compromise. L'analyste s'est vu confronté à des problèmes spécifiques (faiblesse des concentrations, métabolisme mal compris des molécules et de leurs métabolites, techniques d'extraction particu-lières) pour lesquels n'existaient que peu de références. De plus, la notion d'absence de résidus détectables est ambigüe et a conduit à s'interroger sur la définition de seuils acceptables et sur la sensibilité des techniques. La C.E.E. se penche actuellement sur ces questions.

Le dosage microbiologique est classique et encore très employé (Tableau I). Quelques qu'en soient les variantes, la technique est simple, peu coûteuse, et adaptée au traitement direct des tissus. Par contre elle est peu sensible (0.1 µg/g), peu précise, et ne permet pas de distinguer l'action d'espèces moléculaires différentes, ou d'inhibiteurs physiologiques présents dans les tissus. Elle requiert également beaucoup de temps. Des techniques rapides ont été proposées, mais n'ont pas encore été appliquées aux antibiotiques.

Les méthodes physicochimiques se sont récemment multipliées (Tableaux II à VII). La grande variété des protocoles décrits ne facilite cependant pas les comparaisons. L'extraction est souvent assurée par des solvants polaires qui ne rendent pas compte de l'existence de résidus liés. Les techniques de concentration et de mesure sont extrêmement diversifiées. Face à la fluorescence et à la chromatographie en phase gazeuse (peu employée) la chromatographie liquide à haute performance (HPLC) s'est imposée de plus en plus nettement, bien qu'aucune standardisation n'existe à ce jour. Précise, sensible, facile à automatiser, elle réclame des équipements coûteux. Il convient enfin de remarquer qu'on ne s'est guère intéressé jusqu'a présent à la recherche des métabolites (Tableaux VIII et IX).

On dispose donc pour la recherche des résidus d'un arsenal technique considérable, dont la sensibilité apparaît satisfaisante ou améliorable. Mais des protocoles standards sont encore à définir, et beaucoup de problèmes analytiques ne sont pas résolus. Surtout, il serait important de savoir différencier la mesure des résidus totaux et celle des éventuels résidus métaboliques.

Introduction

In the last few decades there has been a considerable increase in the global production of fish in inland waters; for instance the production in 1958 and 1988 was 4.8 x 10^6 tonnes and 13.4 x 10^6 tonnes respectively (23). Fish farming, particularly low technology farming, is responsible for most of this production and much of the fish produced is exposed only to natural antibiotic substances present in the water and feed and rarely, if ever, to anthropogenic antibiotics. However, in an increasing number of countries, large scale intensive fish farming is being practised and there is a growing use of chemicals, many of them antibiotics, to prevent or curb disease problems inherent in high intensity fish husbandry, to increase efficiency of feed utilisation and to promote growth. In view of the biological activity of these compounds there is concern about their use both from environmental and food viewpoints.

As a consequence of this concern most countries control the use of these compounds but even so there is a need to monitor their presence so as to ensure that the wholesomeness of the food is not compromised, micro-organisms harmful to man are not developed, the acceptable daily intake is not exceeded and the regulations are being complied with. Such monitoring poses a number of problems for the analyst because, the expected concentrations of the antibiotics are low, the types present are often unknown and biologically active metabolites may be present. Furthermore, the residues are usually present within the cell and sometimes combined with the cell components so that simple extraction procedures may be inefficient. This latter problem causes particular difficulties which are not resolved by simple recovery experiments using added antibiotics.

In addition to these difficulties there is a major problem when it is specified that no residues are acceptable. It is by no means certain that complete depuration from a chemical view-point is obtained and thus if an antibiotic has been used it is likely that it will be detected using sensitive equipment. The questions then arise concerning what concentrations pose an acceptable risk and the lowest limit of detection required of the analyst. These latter questions have been addressed to some extent by the Veterinary Medicinal Products Committee of the EC (personal communication, E. Good, VMD, UK).

Methodology

In essence the methods used fall into two categories, microbiological and physicochemical.

Microbiological assays

A summary of the papers in which reports on the use of microbiological assays have been described are presented in Table I. The assays rely to a large extent on the use of agar plates containing an appropriate growth medium and microorganism. The most commonly used microorganism is *B.cereus* although not surprisingly some microorganisms are more sensitive to specific antibiotics, eg *B.subtilis* appeared to be most suitable for the determination of sulphonamide drugs (45,77).

The tissue sample or extract is applied either to wells cut or formed in the agar or is used to impregnate a small disk of filter paper which is then placed in the agar surface. After incubation of the agar plates for the appropriate time and temperature, usually 24 hrs at 30 C, the diameter of the zone in which no growth occurs is measured. The amount of antibiotic present is then estimated from appropriate calibration curves.

The technique is relatively simple, it can integrate total antimicrobial activity, use the whole tissue sample and thus overcome problems associated with sample extraction and is inexpensive to perform, but it does suffer from a number of disadvantages. First of all, many of the methods used lack sensitivity. The detection limits quoted are often >0.1 $\mu g\ g^{-1}$ which may be greater than the maximum residue limit (MRL). Related problems are, the poor differentiation between different antibiotics and also between these and naturally occurring inhibitors of microorganisms, such as the lysozymes, and the accuracy and precision of the methods.

The method is also time consuming in that the total analysis time is in excess of 24 hours.

Many of these drawbacks could be overcome by the use of modern rapid methods now developed for the estimation of bacterial numbers and growth and indeed have been used to demonstrate the presence of bacterial inhibitors in milk (17). Such methods however do not appear to be used for general assays of antibiotic residues.

Physicochemical analysis

A precis of the recent publications on physicochemical assays is presented in Tables II to VII. It is immediately apparent that a variety of extraction, concentration and analytical conditions have been used even for the same antibiotic and consequently the reported recoveries have varied quite markedly. A number of authors have reported recoveries for different concentrations of an analyte so in Tables II to VII the lowest concentration used by the author has been selected for brevity and to represent the most demanding conditions. Furthermore, in many of the reports recoveries for several species or tissue samples have been given and some of this detail is also presented in the tables along with the statistical data.

In nearly all instances the extractions of the residues have been carried out on the homogenised sample using polar solvents. In so doing most workers have chosen to ignore the possibility of bound residues being present whose extraction cannot be mimicked by simply extracting added antibiotics to the tissue sample.

After extraction the samples are usually passed through a concentration and clean up procedure and for this purpose a number of different proprietary solid phase cartridges have been used. As can be seen in Table VIII the choice of cartridge is important because even apparently similar types have markedly different retention characteristics, particularly when fish extracts are used.

A variety of chromatographic methods have been used in the final analysis of the extracts most commonly in association with UV or fluorescence detectors. Gas chromatographic analyses are not common in the determination of fish antibiotics even though various forms of this technique have been used in the analysis of sulphonamides (33,82,84) and chloramphenicol (12,26,33) in meat.

There are a number of reports on the use of thin layer chromatography (TLC), see Table VII, but notwithstanding the simplicity of the technique it does not seem to be a favoured method even though very low detection levels have been quoted.

The preferred method of chromatographic analysis is HPLC and the methods used are presented in Tables II to VI. The HPLC conditions vary from one group of workers to the next and the quoted accuracy and sensitivities differ accordingly. In contrast to the microbiological methods, the physicochemical methods are relatively complex and tend to require several analyses if complex mixtures are suspected as being present. HPLC methods also require expensive equipment and high grade technical support. The advantages of these methods are that they can be automated and they have the potential to identify in a specific manner the antibiotics present with considerable sensitivity and accuracy.

Metabolites

The determination of antibiotic metabolites in fish have been largely ignored and only a few reports have been published (Table IX) despite the fact that some such as sulphonamide metabolites have antimicrobial activity (63).

Conclusions

A wide array of methods are used in the analysis of antibiotic residues in fish. Most of the techniques are, or in the case of the microbiological ones could be made, sufficiently sensitive and accurate for safety and legislative purposes. There is a need to define standard methods and for this purpose more inter-laboratory calibration exercises are required to compare methods and select the best. A number of analytical problems remain. Perhaps the most important is the assessment of total residues and metabolites. This can be done in an experimental situation by feeding radioactive drugs to the fish. The assumption is then made that the findings from such experiments hold for all situations in which the drug may be used and thus the measurement of any component of the drug such as the unbound residue will provide measures of total residues and metabolites. Such an assumption is not satisfactory and improved procedures are required so as to ensure that the analyses represent accurately the total amount of antibiotics and biologically active metabolites present.

List of abbreviations used for antibiotics

AMP: Ampicillin
CAP: Chloramphenicol
CTC: Chlortetracycline
DC: Doxycycline
DFZ: Difurazone
EF: Enrofloxacine
EM: Erythromycin
F: Flumequine
FM: Framycetin
FMZ: Furamizole
FP: Furpyrinol
FZ: Furazolidone
NA: Nalidixic acid
NF: Nitrofurazone
NaNFS: Sodium nifurstyrate
OA: Oxolinic acid
OMP: Ormetoprim
OTC: Oxytetracycline
PA: Piromidic acid
SDZ: Sulphadiazine
SDM: Sulphadimidine
SDMX: Sulphadimethoxine
SMMX: Sulphamonomethoxine
SMR: Sulphamerazine
SIZ: Sulphisozole
SQX: Sulphaquinoxaline
TC: Tetracycline
TMP: Trimethoprim
NFP: Nifurpirinol

Table I

Microbiological assays

Fish	Drug	Tissue	Test organism	Method		Ref
Salmon, coho, sockeye	CAP	K,L,M	*A. liquifaciens*	Turbidity	NS	**20**
Rainbow trout	TC SMR	M B, K, L	*Sarcina lutea* (SI) *Staph. aureus*	Disc plate Tissue dir.ect	NS	**83**
Channel catfish	OTC	L, M, S	*B. cereus* (ATCC 9634)	Cylinder plate	L =0.1 - 0.15 M =0.15-0.25 S =0.09	**24, 25**
Trout - rainbow, brook, brown	OTC	K, L, M, P	*B. cereus* (Pfizer)	Cylinder plate	0.1-0.5	**32**
Haddock, cod, flounder	CTC	M	*B. cereus* (ATCC 1778)	Paper chromatog.raphy	0.04	**9**
Carp	OA	K, M, S	*E. coli* (`11346)	Cup plate	K, M = 1.0 S= 0.2	**18**
Rainbow trout	CTC	B, K, L, M	*B. subtilis* Bremen	Cyl plate Electrophoresis	NS	**27**
Rainbow trout	TMP	B, K, M	*B. pumilis* (CN 60 Wellcome)	Plate	NS	**54**
Rainbow trout	CAP FM OTC TMP	M	*Sarcinia lutea* (ATCC 3491) *Staph. epidermis* (ATCC 12228) *B. cereus* (ATCC 11778) *B. pumilis* (CN 607 Wellcome)	Cyl inder plate	NS	**55**
Goldfish, eels	PA	B, K, L, M, Skin	*E. coli* kp	Cup plate	L=0.6,K=0.72 M, B=0.25,	**50**
Rainbow trout	TMP	M, L, P	*B. pumilis* (Wellcome)	Plate	NS	**6**
Carp	AMP NF OTC	S	*Staph. aureus*(Naval biosci ences lab) *E.coli* (ATCC266-76N) *Staph. aureus* (Naval biosciences lab)	Cylinder plate	NS	**10**
Yellowtail	EM	B, K, L, M & spleen	*M. luteus* (ATCC9341)	Cup plate	0.03-0.08	**51**
Channel catfish	CAP EM,FP OTC	Total visceral fluids	*B. subtilis* (ATCC6633)	Cylinder plate	NS	**64**

Rainbow trout	F	L, M	*B. subtilis* (CIP 5262)	Tiss.ue plate. direct	2.5	**15**
Rainbow trout	OTC	M	*B. cereus* (ATCC11778)	Cylinder plate	0.92	**76**
Rainbow trout	OTC SDZ TMP	M	*B.cereus* (ATCC 11778) *B. subti;lis* (ATCC 6633) *B. pumilis* (CN607)	Cylinder plate	0.04 0.04 0.0002	**77**
Rainbow trout, ayu yellowtail, , carp	SMMX SDMX SIZ	M	*B .subtilis*	Disc plate	NS	**45**
Rainbow trout	OTC	M	*B .cereus* (ATCC11778) *B .subtilis*	Cylinder plate	0.1 0.4	**4**
Rainbow trout	F	M	*Proteus morgani*	Disc plate	0.25	**89**
Chinook salmon	EM	Eggs, fry	*M . luteus*	Disc plate	NS	**14**
Salmon	F,OA OTC	S	*Vibrio anguillarum*	Nunc plate	0.8	**65**
Carp	OTC	L,M,P skin & bone	*B .subtilis* BGA *B. cereus* (ATCC 6941)	Plate	NS	**29, 30**
Brown trout, Salmon	F	S	*Vibrio anguillarum* (775)	Cup plate	0.8	**66**
Rainbow trout	CTC OTC	M	*B. cereus* (ATCC 1178) *B. subtilis*(ATCC6633) *M. luteus* (ATCC 9341)	Disc plate	0.01 0.05 <0.05	**44**
Rainbow trout African catfish	OTC	P	*B. cereus* (ATCC 6941) *B .subtilis* BGA	Plate	NS	**31**
Salmon	OTC	L, S	*B. cereus* (NCIB8849)	Nunc pl	0.3, 0.5	**13**
Salmon, trout	OTC NF SDZ OA F, EF NF	L, M	*B. cereus* (ATCC 1178) *B. cereus* (ATCC 1178) *M. luteus* *E. coli* *E. coli* *E.coli*	Tissue direct plate	0.02 0.04 0.08 0.04 0.02	**70**

Table II

HPLC Analysis of Tetracylines

Sample, Extraction and isolation	HPLC Conditions	Detector	% Recovery, (CV or SD quoted) Amount added (AA) and limit of detection (LD) (μg^{-1} or cm^{-3})	Reference
Yellowtail, 20g, 1M HCl, centrifuged, filtered (2x), XAD 2 Column	25 cm x 4,mm i.d. a) dimethyl silica 10µm or b) Polystyrene gel 10µm a) 0.05 M phosphate buffer pH 2.5, acetonitrile (90:10) or b) 0.04M potassium phosphate - 0.08M EDTA - methanol (10:10:80)	UV 370nm	OTC = : 79 (5.4) TC= : 84 (6.4) AA= : 1.0 CTC = : 70 AA = : 3.0 LD= : 1.0 OTC and TC : 3.0 CTC	70
Yellowtail, rainbow trout, red sea bream, eel, 5g muscle 0.5% metaphosphoric acid - methanol (8:2) Filtered, concentrated Sep-Pak C_{18} cartridge	25 cm x 4.6 mm i.d. Nucleosil $5C_{18}$ Acetonitrile - 0.5M monobasic sodium phosphate (35 : 65) pH 2.5	UV 265nm	OTC = 75-82 (±3.1-5.6), not eel CTC= 73-78 (±1-3.2) AA= 1.0 LD= 0.05 OTC, 0.1 CTC	35
Yellowtail, eel, 5g sample 0.1M Na EDTA -McIlvaine buffer (pH4.0) Centrifuged, filtered, (2x), Baker $10C_{18}$ cartridge	25cm x 4 mm i.d. Lichrosorb RP.8 10µm Aqueous 0.01M oxalic acid pH2.0 - methanol -acetonitrile (2.5 : 1 : 1.5)	UV 350nm	OTC = 91(3.5), 94(1.1) TC = 88(2.5), 89(0.8) CTC = 93(2.9), 88(3.1) DC = 83(0.9), 85(1.5) AA= 1.0 LD= 0.05 OTC, TC, 0.1 CTC, DC	69
Rainbow trout, 10g muscle, 2g liver 1M HCl - 50% Trichloracetic acid (15:1) Centrifuged, filtered Sep-Pak C_{18} cartridge	25cm x 4.6 mm i.d. ODS Hypersil 5µm Diammonium hydrophosphate (5g), diethanolamine (5ml), water (850ml) Acetonitrile (190ml) Dimethylformamide (60ml), pH 2.5)	UV 365nm	Muscle / Liver OTC= 53(3.8) / 69(3.8) AA = 0.05 / 0.25 LD = 0.005 / 0.01	61

Rainbow trout, 25g muscle 0.5M HCl/0.37M $HClO_4$ Filtered (2x), Amberlite XAD 2 resin	15cm x 4.6 i.d. Hypersil sas 5µm , 0.1M citric acid (340ml), 0.1M trisodium citrate (5ml), 0.1MNa$_2$ EDTA (5ml), to 500ml with acetonitrile	UV 370nm	OTC = 72 (±7.9) TC = 76 (±5.5) CTC= 57 (±8.3) AA= 0.1 LD= 0.005	57
Rainbow trout, 5g muscle, 2g liver, 500ml serum 0.1M citric acid, 0.2M disodium hydrogen phosphate (62:38) pH4.0 : Sonicated, centrifuged, filtered, Bond-Elut C_{18} cartridge	10cm x 4.4 ,, i.d. ODS 5µm Acetonitrile - N,N-dimethyl formamide - 0.01M aqueous oxalic acid pH2.1	UV 355nm	Muscle Serum Liver OTC= 88(3.0) 91(4.1) 95(2.3) TC = 78(3.8) 80(4.1) 77(4.3) CTC= 78(4.3) 94(3.3) 83(3.4) DC = 87(3.2) 97(2.5) 85(2.7) AA = 1.0 LD= 0.05 0.05 0.1	7
Rainbow trout 100ml plasma + trifluoracetic acid (10:1) Centrifuged	10cm x 4.6 mm i.d. Cyanosphei 5 cartridge HPLC 5µm 0.02M oxalic acid (pH2.)) - methanol - N,N-dimethylformamide (950:50:50)	UV 350 nm	OTC= 98-99 (4.2-5.2) AA = 0.4, 0.7 and 1.2 LD = 0.04	42
Rainbow trout as Onji et al (70)	but C_{18} column Acetonitrile (195ml) -dimethylformamide (65ml) + 0.05M ammonium sulphate to 1 litre	UV 370 nm	Muscle Blood Skin Gutted Whole OTC = 70-84 98-107 70-84 70-84 AA= 0.5 1.0 0.5 0.5 LD = 0.05 0.05 0.1 0.05	43
Rainbow trout 10g muscle, 4ml serum, 3g liver & kidney 5g skin & bone 5% trichloroacetic acid - 0.5% Na$_2$ EDTA Centrifuged, rotary evaporated Sep-Pak C_{18} cartridge	25 cm x 4 mm i.d. M&S C_{18} Methanol - acetonitrile - 0.2M oxalic acid (1:1:3.5) pH 2.0	UV 360 nm	Muscle Serum Liver OTC= 74 (5.7) 82 (4.8) 75 (5.1) Kidney Skin OTC = 72 (7.3) 65 (4.3) AA= 2.0 LD= 0.05, muscle & serum 0.1 others	94

Salmon 5g muscle McIlvaine buffer Filtered Chelating sepharose column, centrifuge (3x), filtered Bond Elut C_{18} cartridge	25 cm x 4 mm i.d. Lichcrocart RP 18 7µm Aqueous 0.01M oxalic acid acetonitrile - methanol (75:17:10)	UV 335 nm	OTC = 82 (±5) TC = 66 (±7) CTC = 0 AA= 0.2 LD= OTC 0.08, CTC 0.09	73
Rainbow trout 10 g muscle succinate buffer, Filtered Chelating Spharose column XAD2 column	20 cm x 3 mm i.d. Lichrosorb RP6 0.01M oxalic acid - acetonitrile (1:1)	UV 350 nm	OTC= 94 (3) TC = 60 (4) CTC = 61 (14) AA = 0.01 LD = 0.01	21
Catfish 0.5 g muscle Bulk C18 silica, blended, hexane extracted, Acetonitrile - methanol elution	10cm x 1.4 i.d. ODS silica 10µm 0.02M aqueous oxalic acid - acetonitrile - methanol (70:27.5:2.5)	UV 365nm	OTC = 79 (±13.3) AA = 0.2 LD= <0.05	52
Rainbow trout 10g muscle 10% phosphoric acid, saturated Na_2 EDTA - 40% methanol (1:1:100) Rotary evaporated Bond Elut C_{18} cartridge	15 cm x 4.6 i.d. Rogel RP 5µm 0.025 KH2 PO4 (60ml) - 1M NaOH (16ml) -0.1M tetrabutylammonium hydrogen sulphate (10ml) - dieionised water (7.8ml) -acetonitrile (226ml)	UV 370 nm	OTC = 90 (4.5) AA= 0.2 LD = 0.005	Author's unpublished results

Table III

HPLC Analysis of quinolones

Sample, extraction and isolation	**HPLC conditions**	**Detector**	**% Recovery (CV or SD quoted), Amount added and limit of detection (LD) (ug g^{-1} or cm^{-3})**	**Reference**
Rainbow trout sweet smelt 10g muscle .2M phosphate buffer (pH6) Ethyl acetate extraction centrifuged rotary evaporated solvent partition	50cm x 2.6mm i.d. Zipax SAX 0.1 tetraborate (pH9.1) containing 0.006M Na_2SO_4	UV 258 nm	NA = 98 (0.9-2.0) AA = 0.1 LD = 0.05	**46**
Rainbow trout sweet smelt yellowtail amago salmon eel 20g muscle 0.2M phosphate buffer (pH6) Ethyl acetate extraction centrifuged rotary evaporated solvent partition	50cm x 2.6mm i.d. Zipax SAX 0.1 tetraborate (pH9.1) containing 0.006M Na_2SO_4	UV 335 nm	OA = 88-89 (0.8-2.2) NA = 85, 85 (0.8, 2.0) PA = 75, 76 (0.9, 1.8) AA = 0.1 LD 0.3 OA, 0.04 NA, 0.1 PA	**47**
Eel $1cm^3$ plasma 7g tissue McIlvaine buffer Several chloroform/buffer partitions Centrifugation steps Evaporation to dryness	30cm x 4mm i.d. Bondapak C_{18}	UV 280 nm	PA and metabolites - not reported LD 0.2	**67**

Fish, not specified 20g muscle 0.2M Na_2PO_4 buffer (pH6.0) Ethyl acetate shaken and centrifuged Evaporated to dryness Acetonitrile hexane NaCl solution Chloroform partition steps	25cm x 4mm i.d. Lichrosorb RP-18 10um Phosphate buffer (pH6.4) - methanol (14:11) Containing 0.05% TMCA-Br	UV 330 nm	OA not reported PA metabolites not reported	**48**
Rainbow trout yellowtail sweet fish carp eel 5g muscle 0.1% metaphosphoric acid - methanol (6:4)	25 cm x 4.6 mm i.d. Kaseisorb LC ODS 300-5 5mM sodium dihydrogen phosphate - acetonitrile (6:4)	Fluorescence Emission 325nm Exitation 365nm UV 280nm	OA = 84-89 (±2.1-4.6) NA = 82-85 (± 2.0-2.7) PA = 81-85 (±1.5-3.3) AA = 0.2 OA 0.4 NA, PA LD = 0.01	**36**
Yellowtail 0.5cm^3 0.5g muscle , liver, kidney 0.2M sodium acetate - HCl buffer (pH2) Ethyl acetate extraction and partition into 0.1N NaOH and finally chloroform Centrifuged between stages	25 cm x 4.6mm i.d. TSK gel DEAE-2 SW 0.015M phosphate buffer (pH7) - acetonitrile (65:35)	Fluorescence Exitation 337nm Emission 365nm	OA = muscle 76 (3.8) serum 77 (3.4) kidney 61 (7.2) liver 61 (5.6) AA = 1.0 LD = 0.1-0.3 Internal STD Naladixic acid	**19**
Rainbow trout yellowtail eel 10g muscle Acetone extraction filtered (2x) + propanol, rotary evaporated Hexane extraction and partition into chloroform Alumina column	25 cm x 4.6 mm i.d. Nucleosil C_{18} THF - acetonitrile - phosphoric acid - water (29+1.0+0.6+69.94)	UV 260nm	OA = 77-91 (± 3.4-4) NA = 85-90 (±2.0-3) PA = 75-82 (±1.3-3.3) AA = 0.5 LD = 0.02 OA, 0.04 NA, 0.08 NA	**62**

Eel 5g tissue Acetonitrile, centrifuged (2x) Rotary evaporated to dryness and taken up in 1% TCA. Bond-Elut C_{18} Cartridge	25 cm x 4.6 mm i.d. Nucleosil C_{18} 5 um 0.02M citric acid(0.01M tetrabutyl ammonium bromide) - acetonitrile (6:4)	UV 254 nm 280nm	OA = 82 (4.8) NA = 83 (2.2) PA = 78 (2.1) AA = 0.2 LD = 7	**37**
Yellowtail 10g muscle & liver 10cm^3 blood 1-4g kidney As Kasuga et al 1981 (46)	25 cm x 4mm i.d. Tayo gel DEAE-2SW 0.015M phosphate buffer (pH7.0) - acetonitrile (65:35)	UV 335nm	Muscle Liver Kidney Blood OA = 86 78 84 77 AA = 2.0 LD = 0.02 0.05- 1.0 0.02	**92**
Rainbow trout amago salmon 10g muscle 1.5g muscle 0.5g kidney 2cm^3 blood As Kasuga et al 1981 (46)	As Ueno et al 1988 (96)	UV 335nm	Muscle Liver Kidney Blood OA = 77.79 82.79 90.80 89.83 AA = 2.0 LD = 0.02	**93**
Rainbow trout 100mm^3 serum Chloroform 5g muscle/skin NHCl - ethyl acetate Filtered and evaporated to dryness, further partition centrifugation and evaporation steps	15cm x 3.9mm i.d. Waters Novapak C_{18} 5um Methanol (420cm^3) - water 580cm^3) - Na_2HPO_4 (1.88g) - N acetyl, NNN trimethylammonium bromide (0.75g)	UV 265nm	Muscle/skin Serum OA = 80 92 AA = 0.01 LD = 0.01	**3**
Rainbow trout sweet fish red sea bream yellowtail eel 5g n hexane ethyl acetate (1:3) Centrifuged (2x) Baker 10 amino Cartridge	7.5cm x 4.6mm i.d. Nucleosil C_{18} 3um Acetonitrile - methanol - 0.01M aqueous oxalic acid (3:1:6)	UV 295 nm	OA = 76-92 (2.2-9.8) NA = 87-95 (1.0-6.8) PA = 74-91 (1.5-8.3) AA = 0.1 LD = 0.05	**38**

Rainbow trout	As Kasuga et al 1982 (47), but 100cm column		Muscle Blood Skin Gutted Kidney OA = 92-105 82-85 91-105 80-87 92-105 AA = 1.0 LD = 0.05 0.1 0.05 0.05 0.05	
Salmon $250mm^3$ plasma + $250mm^3$ internal standard Bond Elut C2 cartridge or $20mm^3$ plasma Precolumn Polystyrene divinylbenzene 36um	15cm x 4.6 i.d. Polystyrene divinylbenzene 5um Acetonitrile- tetrahydrofuran - 0.02M orthophosphoric acid (20:15:65) OA (22.5:17.5:60)	Fluorescence Exitation 262nm Exitation 389nm	Direct column switching OA = 104 (4.2) 98 (2.7) F = 103 (2.8) 95 (5.3)* AA = 0.5 0.05, 0.1* LD = 0.005, 0.01*	**72**
Salmon rainbow trout 5g muscle 5g liver NaOH - acetone centrifuged (3x) Liquid/liquid extraction	15cm x 4.6mm i.d.PLRP-S Polymer absorbent 5um 0.002M phosphoric acid acetonitrile - tetrahydrofuran (60:20:15)	Fluorescence Exitation 260nm Emission 380nm	Muscle Liver OA = 97 (7.2) 98 (7.2) F* = 89 (3.9) 88(4.0) AA = 0.01 0.1* 0.01 0.05*	**74**
Salmon rainbow trout 5g muscle Homogenised as above Chloroform extract, dried Hexane & NaOH extract C_2 Bondwell column	15cm x 4.6mm i.d.PLRP-S Polymer absorbent 5um 0.002M phosphoric acid acetonitrile - tetrahydrofuran (60:20:15)	Fluorescence Exitation 260nm Emission 380nm	Muscle OA = 78 (2.2) F* = 72 (3.8) AA = 0.1 0.5* LD =0.0005 OA 0.002F*	**74**

Salmon 2g muscle McIlvaine buffer (pH3.6) - methanol (55:45) Centrifuged vacuum reduced DCM extraction washed to dryness with N_2 + buffer, centrifuged	10cm x 5mm i.d. ODS Hypersil 3um 0.1 citric acid (pH3.2) - methanol - acetonitrile - tetrahydrofuran (60:30:5.5)	Fluorescence Exitation 374nm Emission 363nm	OA = 90-93 (1.3) F = 84-87 (2.0) AA = 0.1 OA 0.05 F LD = 0.005	**78**
Salmon 100mm^3 serum Centrifuged 5g muscle 2g liver 0.1M $K_2H_2PO_4$, pH7.0, 4.5 Sonicated centrifuged filtered (3x) Bond Elut C_{18} cartridge	15cm x 4.6mm i.d. Pinkerton GFF 5um ISRF 0.1M aqueous $K_2H_2PO_4$ (pH2.0) - acetonitrile (90:10)	UV 254nm	Serum / Muscle / Liver OA = 100 (3.9) / 88 (2.4) / 84 (4.6) NA = / 88 (3.0) / 87 (3.5) AA = 1.0 LD = 0.01	8
Salmon 100mm^3 plasma Zinc sulphate - acetonitrile Shaken centrifugation	15cm x 4.6mm i.d. MOS Hypersil (C_8) 3um 0.025 oxalic acid (pH3.2) - acetonitrile - methanol - tetrahydrofuran (80:2.5:15:2.5) or (50:20:25:5)	Fluorescence Exitation 325nm Emission 360nm	OA = 90 (3.5) F = 94 (5.6) AA= 0.05 LD = 0.003 OA 0.005 F	**80**
Salmon 5g muscle 0.05M phosphate buffer (pH9) - hexane. Homogenised soicated centrifuged Aqueous phase (1.5cm^3) Dialysis precolumn Precolumn only	As Rasmussen et al 1988 (72), but mobile phase (20:14:75)	Fluorescence Exitation 325nm Emission 365nm	Precolumn HPLC / Dialysis Precolumn HPLC OA = 96 (±2.2) / 104 (±3.5) F = 105 (±5.7) / 101 (±1.9) AA = 0.5 / 0.05 LD = 0.002 OA / 0.003 F	

Salmon 2g liver 0.4M phosphate buffer (pH10) - hexane Homogenised sonicated centrifuged Dialysis precolumn	As Rasmussen et al 1988 (72), but mobile phase (20:14:66)	Fluorescence Exitation 325nm Emission 365nm	OA = 96 (+/-1.5) F = 107 (+/-1.9) AA = 0.05 LD = 0.004 OA, 0.007 F	2
Salmon $100mm^3$ blood, plasma Dialysis precolumn	As Rasmussen et al 1988 (72), but mobile phase (20:14:66) Internal standard naladixic acid	UV 325nm	Plasma Blood OA = 104 (3.1) 108 (2) F = 102 (5.3) 106 (3.7) AA = 0.5 LD = 0.05	1
Salmon 2g muscle 0.5-1g kidney Na_2SO_4 ethylacetate (2x) Solvent partition hexane : 0.1M aqueous oxalic acid Centrifuged	Partisil ODS - 3um Acetonitrile - methanol - 0.01M oxalic acid (3:1:6)	Fluorescence Exitation 327nm Emission 369nm	OA = 71-83 AA = 0.01-0.1 LD = 0.01	88
Rainbow trout salmon as OTC table	7cm x 4.6mm i.d. Nucleosil 5um 0.025M $K_2H_2PO_4$ ($450cm^3$) - acetonitrile ($50cm^3$) - nonylamine ($0.4cm^3$) - orthophosphoric acid ($0.4cm^3$)	Fluorescence Excitation 355nm Emission 380nm	OA = 75 (2.3) AA = 0.2 LD = 0.002	Authors unpublished results

Table IV

HPLC analysis of nitrofurans

Sample, extraction and isolation	HPLC conditions	Detector	% Recovery (CV or SD quoted), Amount added and limit of detection (LD) (ug g^{-1} or cm^{-3})	Reference
Rainbow trout carp seet smelt yellowtail red sea bream tilapia eel 10g Methanol - NN dimethyl formamide (49:1) Filtered (2x) Rotary evaporated Methanol Sep-Pak C_{18} (not cartridge)	25cm x 4mm i.d. PC8-10 Acetonitrile 0.05M NaH_2PO_4 (3:7) 1% TEA pH6 with N_3PO_4	UV 380nm	FZ = 93-101 NaNFS = 75-98 NFP = 81-100 DFZ = 54-72 AA = 0.4 LD = 0.05 FS, NaFS, NFP, 0.1 DFZ	58
Rainbow trout sweet smelt eel 10g muscle DCM - Na_2O_4 Filtered (2x) Rotary evaporated Chloroform Florisil column	15cm x 4mm i.d. Lichrosorb RP18 5um Acetonitrile - water (45:55) - acetic acid 2.5 x 10^{-5}%	UV 390nm	NFP = 82-85 (3.0-3.7) AA = 0.1 LD = <0.001 Internal standard 2-nitro-*p*-cresol	71
Red sea bream yellowtail sweet smelt eel 10g muscle Methanol - dimethylformamide (49:1) Centrified filtered (2x) Solvent extraction Sep-Pak C_{18} (not cartridge)	25cm x 2.6mm i.d. TSK gel ODS-120T Water - methanol (60:40) Containing 0.01M tetrabutyl ammonium bromide + 0.04M NaH_2O_4 (pH5.2)	UV 260nm	FZ = 82-88 FMZ = 86 DFZ = 64-72 AA = 1.0 LD = 0.05 FZ, 0.1 FMZ, DFZ	56

Rainbow trout carp eel 10g Acetone filtered + n propanol and concentrated Partition with acetone hexane 3% NaCl solution finally into chloroform and concentrated Alumina column	25cm x 4mm i.d. Nucleosil 5 C_{18} Acetonitrile - acetic acid - water (70:0.1:30)	UV 380nm	FZ = 98-100 (±5.2-7.9) DFZ = 84-89 (±.9-2.0) AA = 0.1 FZ 0.2 DFS LD = 0.3	34
Rainbow trout carp 10g muscle Homogenised with buffer (pH5.4) then ethyl acetate centrifuged. Ethyl acetate (2x) evaporated to dryness, dissolved in DCM silica gel column	25cm x 4mm i.d. Lichrosorb RP18 5um 0.06M sodium acetate - acetonitrile (7:3)	Electrochemical - 550mv 50na	FZ not reported LD = 0.0005	81
Rainbow trout yellowtail red sea bream eel 5g muscle 0.5% metaphsphoric acid - methanol (8:2) Filtered reduced under vacuum Sep-Pak C_{18} cartridge.	25cm x 4.6mm i.d. Nucleosil 5c Acetonitrile - 0.05M NaH_2O_4 (35:65) pH 2.5	UV 265nm	FZ = 69-81 (±1.5-6.6) DFZ = 6.3-9.1 (±0.6-4.8) AA = 1.0 LD = 0.05 Fz 1.0 DFZ	35

Rainbow trout yellowtail eel 10g muscle Acetone extraction, filtered (2x) + propanol, rotary evaporated Hexane extraction and partition into chloroform Alumina column (neutral)	25cm x 4.6mm i.d. Nucleosil C_{18} THF - acetonitrile - phosphoric acid - water (29:1:0.6:69.95)	UV 260 nm	FZ = 76-83 (±2.7-3.5) NFP = 74-81 (±1.7-8.0) DFZ = 71-78 (±1.9-3.8) FMZ = 83-86 (±0.4-2.9) AA = 0.9 LD = 0.04 FZ, 0.06 DFZ, 0.08 FMZ, NFP	**62**
Salmon 2g liver muscle +NaCl McIlvaine buffer (pH3.6) - Methanol (70:30), centrifuged (2x) Evaporated -15cm^3, partitioned into dichloromethane and evaporated to 2cm^3 + hexane Bond Elut NH_2 column	10cm x 5mm i.d. ODS Hypersil 3um Acetonitrile - water (16:84) containing 0.001M Na_2EDTA and 0.1M KNO_3 (pH3.2)	UV 400nm	Muscle Liver FZ = 95 (±3.3) 94 (±4.4) AA = 0.025 LD = 0.005	**79**
Trout 10g muscle 1g liver Ethanol, centrifuged (2x) Evaporated, solvent partition steps Extralut 3 column	25cm x 4.6mm i.d. Cs Multisorb Si 60 5um Chloroform - acetonitrile - ethyl acetate - glacial acetic acid (80+30+10+1)	UV 265nm 365nm	Muscle Liver FZ = 88 (±15) 78 (±13) LD = 0.02 0.05 Metabolite = 75 (±2) 80 (±) LD = 0.01 0.03	

Table V

HPLC analysis ofsulphonamides and synergists

Sample, extraction and isolation	HPLC conditions	Detector	% Recovery (CV or SD quoted), Amount added and limit of detection (LD) (ug g^{-1} or cm^{-3})	Reference
Rainbow trout 250mm^3 blood Water + acetonitrile Mix and centrifuge	Redpac A C_{18} Cartridge Acetonitrile - distilled water (24:76) + 0.4% triethylamine pH3.8 with H_3PO_4	UV 254nm	SDM not specified	11
As nitrofurans			Muscle SDMX = 90-96 SMMX = 90-96 SQX = 90-96 AA = 1.0 LD = 0.05	56
As nitrofurans but mobile phase (10:0.05:70)		UV 272nm	Tissue SMMX= 97-101 (±1.0-1.2) SDMX = 95-99 (±1.2-1.7) AA = 0.2 LD = 0.03	34
As tetracyclines			Muscle SMMX = 83-92 (±1.8-6.6) SDMX = 73-80 (±1.5-5.6) AA = 1.0 LD = 0.05	35

As nitrofurans	25cm x 4mm i.d. Lichrosorb RP18 5um 0.034M ammonium sulphate + acetic acid methanol (7:3)	Electro chemical +1100nV 50nA	Muscle DM Not specified LD = 0.002	**28**
Carp 200mm^3 blood Heperinised, centrifuged	15cm x 4.6mm i.d. Lichrosorb RP8 5um Methanol - acetonitrile - distilled water - 0.2M sodium acetate - 0.2M acetic acid (15:4:27.4:50:3.6) pH4.8	UV 272nm	SDM and acetyl metabolite Hydroxy metabolite also LD + 0.01-0.04	**28**
Rainbow trout yellowtail eel 10g muscle Acetone extraction, filtered (2x) + propanol , rotary evaporated Hexane extraction and partition into chloroform Alumina column	25 cm x 4.6mm i.d. Nucleosil C_{18} THF - acetonitrile - phosphoric acid - water (29+1+0.06+69.94)	UV 260nm	SMR = 73-79 (±2.1-4.1) SMMX = 71-78 (±.6-4.8) SIZ = 71-81 (±2.2-3.0) SDMX = 88-92 (±2.8-3.4) Fort = 0.05 LD = 0.04 SMR 0.06 others	**62**
Catfish 2.5g muscle, liver, kidney 0.5M TBAH - 0.05M buffer pH10 - N NaOH (0.2+1+0.5) DCM mixed and centrifuged DCM layered and filtered	30cm x 3.9mm i.d. u porasil column Chloroform - methanol - water - concentrated ammonium hydroxide (1000:28:2:0.6)	UV 288nm	Muscle Liver Kidney SDMX = 101 (±6.5) 118 (±15) 114 (±26) OMP = 108 (±12) 128 (±8) 115 (±13) Fort = 0.05 LD = 0.05	**98**

Rainbow trout As Weidenfield & Macasieb (97)	30cm x 3.9mm i.d. u porasil silica gel 10um Chloroform - (methanol - water - ammonium hydroxide (150:92:1) 500:25	UV 280nm	Muscle Liver Kidney TMP = 95-100 88-97 96-98 AA = 1.0 0.5 1.0 1.0 LD = 0.01	**43**
Chinook salmon 5g muscle Acetonitrile - trichloracetic acid Centrifuged evaporated/N_2 + water membrane filtered Sep-Pak C_{18} cartridge	25cm x 4.6mm i.d. Ultrasphere ion pair 5um Acetonitrile - methanol - 0.1M phosphate buffer (pH4.0) (17:10:73)	UV 280nm	SDMX = 54 OMP = 57 AA = 0.5 LD = 0.2	**96**
Rainbow trout 0.5g muscle Bulk C_{18} silica blended hexane extracted DCM elution	30cm x 4mm i.d. ODS silica 10um 0.017M aqueous H_3PO_4 - acetonitrile (65:35)	UV 270nm	SDMX = 106 (±22.6) AA = 0.05 LD = 0.005	**53**

Table VI

HPLC analysis of chloramphenicol and ampicillin

Sample, Extraction & Isolation	HPLC conditions	Detector	% Recovery (CV or SD quoted) Amount added (AA) and limit of detection (LD) ($\mu g\ g^{-1}$ or cm^{-3})	Reference
As tetracyclines	As tetracyclines	As tetracylcines	CAP=75-88 (±4.3-9.2) AA= 1.0 LD= 0.05	35
As tetracylcines	As tetracyclines	Electrochemical 780 mV, 50nA	CAP= not reported LD= 0.0008	81
10 g Acetate buffer -$Na_2\ So_4$ Ethyl acetate, filtered (2x) Washed, evaporated Solvent partitiion Centrifugation, filtration	10 cm x 4mm i.d. Hypersil H5 ODS Water - acetonitrile - acetate buffer (80:20:1)	UV 276 nm	CAP = 62 (+/-12) AA= 0.05 LD= 0.01	5
Yellowtail 10g muscle Methanol, centrifuged, filtered (x2) n-propyl alcohol, Evaporated Florisil cartridge Several other steps	15 cm x 4.6 mm i.d. Nucleosil C18 Methanol - buffer (pH 6.0) (15:85)	UV 222 nm	AMP= 62 (9) AA= 0.1 LD= 0.03	59

Table VII
Thin layer chromatography

Agent	Fish, Extraction, isolation	Recovery , % (CV or SD quoted) Amount added (AA) and limit of detection (LD)(μg g^{-1} or cm^{-3})	Reference
OTC, CTC, TC	Haddock muscle HCl Amberlite resin Reverse phase plate	20 AA = 1.0 LD = 0.025 Semi quantative above 0.5	**75**
OTC, CTC, TC	Rainbow trout, carp, eel, amago 20g muscle 0.1M Na_2 EDTA - McIlvaine buffer (pH4.0) Seppak C_{18} cartridge, Silica gel HPTLC plate - UV densitometry	60-79 (±1.1-4.6) AA = 1.0 LD = 0.1	**68**
CAP	Rainbow trout, salmon 2g muscle Phosphate buffer (pH6.0), Chloroform-ethyl acetate (2:1),Direct or silica gel TLC Colour reaction reduction/diazotisation	LD = 0.5	**85**
FZ	Carp, eel 20g muscle, Elaborate work up Silica gel plate - fluorescence	77 (8.4, 14.4) AA = 0.001, LD = 0.0002	**87**
F	Rainbow trout 10g muscle Chloroform HPTLC 60F UV 254 nm	LD = 30.0	**90**
11 sulphas including SDZ, SDM, SDMX, & SMR	Rainbow trout 2.5g muscle Ethyl acetate, HPTC Kieselghur 60 UV 254 nm without indicator	40-50% AA = 0.04 LD = 0.005	**22**

Table VIII

Retention characteristics of different batches of cartridges * (methanol elution)

Type	**Recovery %**	
	Addition of an aqueous solution containing 2µg OTC	**Addition of fish extract containing 2 µg OTC**
Bond Elut C_{18}	95	85
	40	0
	95	90
	71	0
	86	42
Baker 10 C_{18}	95	81
	0	0
Sep-Pak C_{18}	90	50
	72	60

* Pretreated with aqueous Na_2 EDTA solution

Table IX

Summary of metabolite and binding studies

Antibiotic	Fish	Metabolite	Technique	Reference
SDM	Rainbow trout Blood	N_4 acetyl	HPLC (Table V)	11
SDM	Carp Blood	N4 acetyl 5-OH and 6-OH detected	HPLC (Table V)	28
SDMX	Channel catfish : plasma	Labelled ^{14}C binding expt.	TLC, HPLC	86
PA	Eel tissues	2-OH and 3-OH	HPLC(Table III)	67
PA	Fish not specified	2-OH and 3-OH	HPLC(Table III)	48.
PA, NA, OA	Rainbow trout Bile, urine	2-OH and 3-OH conjugated and unconjugated	TLC Qualitative	39
OA	Yellow tail : All tissues	Glucuronide in all tiisues	HPLC (Table III)	92
OA	Rainbow trout, tilapia, red sea bream,mackerel, eel yellowtail, flounder, Bile	6-OH and 7-OH. No free 6-OH or 7-OH detected	TLC Qualitative	40
OA	Rainbow trout	As above	TLC Qualitative	41
CAP	Rainbow trout	^{3}H study : Glucuronide Base and alcohol Evidence of binding	HPLC	16
FZ	Eel	3-(4-cyano-2-oxobutylidene-amino)-2-oxazolidone	TLC Qualitative	60
FZ	Trout, : Muscle, liver	As above	HPLC (Table IV)	99

References

1). AGASØSTER, T. & RASMUSSEN, K.E., 1991. Automated analysis of oxolinic acid and flumequine in salmon whole blood and plasma using dialysis combined with trace enrichment (ASTED) as on line sample preparation for high performance liquid chromatography. *J. Chromatogr.*, **564** (1), 171-179.

2). ANDRESEN, A.T. & RASMUSSEN, K.E., 1991. Automated on-line dialysis and column-switching HPLC determination of flumequine and oxolinic acid in fish liver. *J. Liquid Chromatogr.*, in press.

3). ARCHIMBAULT, P., AMBROGGI, G. & NICOLAS, S., 1988. Oxolinic acid in the trout: bioavailability and tissue residues. *Ann Rech Vet.*, **19**, 39-43.

4). ASPLUND, K. & HIRN, J., 1985. Detection of oxytetracycline from farmed fish. *Suomen Eläinlääkarilehti*, **91** (3), 115-120.

5). BECHEIRAZ, M., HALDEMANN, A. & ETTER, R. (1983). Bestimmung von Chloramphenicol in tierischen Lebensmitteln. *Mitt. Gebiete Lebensm. Hyg.*, **74**, 147-155.

6). BERGSJØ, T. & SØGNEN, E., 1980. Plasma and tissue levels of trimethoprim in the rainbow trout, *Salmo gairdneri*, after absorption from fresh and salt water. *Acta Vet. Scand.*, **21**, 18-25.

7). BJÖRKLUND, H., 1988. Determination of oxytetracycline in fish by high performance liquid chromatography. *J. Chromatogr.*, **432**, 381-387.

8). BJÖRKLUND, H., 1990. Analysis of oxolinic acid in fish by high-performance liquid chromatography. *J. Chromatogr.*, **530**, 75-82.

9). BLAKELY, J., KRAMER, J. & SELZER, G.B., 1969. Sensitive identity test for chlortetracycline in fish. *J. Ass. Off. Anal. Chem.*, **52** (5), 935-938.

10). BLASIOLA, G.C. Jr, DEMPSTER, R.P & MORALES, P., 1980. Uptake studies of ampicillin, nitrofurazone and oxytetracycline in carp, *Cyprinus carpio*, with specialreference to use of antibacterials in aquaria and notes on heparin as an anticoagulant. *J. Aquaricult.*, **1** (1), 1-5.

11). BORGAN, A., ØDEGAARD, S. & BERGSJØ, T., 1981. Temperature related absorption and excretion of sulphadimidine in rainbow trout, *Salmo gairdneri. Acta Vet. Scand.*, **22**, 211-217.

12). BORIES, G.S.F., PELERAN, J.C. & WAL, J.M., 1983. Liquid chromatographic determination and mass spectrometric confirmation of chloramphenicol residues in animal tissues. *J. Assoc. Off. Chem.*, **66** (6), 1521-1525.

13). BRUNO, D.W., 1989. An investigation into oxytetracycline residues in Atlantic salmon, *Salmo* salar L. *J. Fish Dis.*, **12**, 77-86.

14). BULLOCK, G.L. & LEEK, S.L., 1986. Use of erythromycin in reducing vertical transmission of bacterial kidney disease. *Vet. Human Toxicol.*, **28**, 18-20.

15). CHEVALIER, R., GERARD, J.P. & MICHEL, C., 1981. Distribution et cinétique tissulaire de la fluméquine chez la truite arc-en-ciel (*Salmo gairdneri*, Richardson). *Revue Med. Vet.*, **132** (12), 831-837.

16). CRAVEDI, J.P., HEUILLET, G., PELERAN, J.C. & WAL, J.M., 1985. Disposition and metabolism of chloramphenicol in trout. *Xenobiotica*, **15** (2), 115-121.

17). EASTER, M.C. & GIBSON, D.M., 1989. Determination of microorganisms by electrical measurements. In *Rapid methods in food microbiology. Progress in Industrial Microbiology*, Vol **26**, M.R. Adams & C.F.A. Hope ed., Elsevier, pp 57-100.

18). ENDO, T., OGISHIMA, K., HAYASADA, H., KANEDO, S. & HOSHIMA, S., 1973. Application of oxolinic acid as chemotherapeutic agent against infectious diseases in fish (1) Antibacterial activity, chemotherapeutic effects and pharmacokinetics of oxolinic acid in fishes. *Bull. Jap. Soc. Sci. Fish.*, **39** (2), 165-171.

19). ENDO, T., ONOZAWA, M., HAMAGUCHI, M. & KUSUDA, R., 1987. Enhanced bioavailability of oxolinic acid by ultra-fine size reduction in yellowtail. *Nippon Suisan Gakkaishi*, **53** (10), 1711-1716.

20). EVELYN, T.P.T., 1968. Tissue levels of Chloramphenicol attained in sockeye (*Oncorhynchus nerka*) and Coho (*O. kisutch*) salmon by feeding. *Bull. Off. int. Epiz.*, **69** (9-10), 1453-1463.

21). FARRINGTON, W.H., TARBIN, J., BYGRAVE, J. & SHEARER, G., 1991. Analysis of trace residues of tetracyclines in animal tissues and fluids using metal chelate affinity chromatography/HPLC. *Food Addit. Contam.*, **8** (1), 55-64.

22). FARRINGTON, W.H. Unpublished results. MAFF, Food Science Laboratory, Norwich.

23). FISHERIES STATISTICS, CATCHES AND LANDINGS (1958-88), FAO, Rome.

24). FRIBOROUGH, J.H., ROBINSON, J.A. & MEYER, F.P., 1969. Oxytetracycline residues in tissues of blue and channel catfishes. US Fish and Wildlife Service, Tech. Paper 38, 1-7.

25). FRIBOROUGH, J.H., ROBINSON, J.A. & MEYER, F.P., 1969. Oxytetracycline levels produced in catfish serum by three methods of treatment. US Fish and Wildlife Service, Tech. Paper 39, 1-6.

26). GINKEL, L.A. van, ROSSUM, H.J. van, ZOONTJES, P.W., BLITTERSWIJK, H. van, ELLEN, G., HEEFT, E. van der, DE JONG, M., & ZOMER, G., 1990. Development and validation of a gas chromatographic - mass spectrometric procedure for the identification and quantification of residues of chloramphenicol. *Analytica Chimica Acta*, **237**, 61-69.

27). GLENDE, W., WENZEL, S. & ROTH, L., 1974. Untersuchungen uber die Antibiotikaverweildauer und den Antibiotikagehalt in Regenbogenforellen (*Salmo gairdneri*) nach peroraler Verabreichung verschiedener Chlortetracyclindosen. *Arch. Lebensmittelhygiene*, **6**, 124-129.

28). GRONDEL, J.L., NOUWS, J.F.M. & HAENEN, O.L.M. 1986. Fish and antibiotics: pharmacokinetics of sulphadimidine in carp (*Cyprinus carpio*). *Vet. Immunol. Immunopathol.*, **12**, 281-286.

29). GRONDEL, J.L., NOUWS, J.F.M. & VAN MUISWINKEL, W.B., 1987. The influence of antibiotics on the immune system: immuno-pharmokinetic investigations on the primary anti-SRBC response in carp, *Cyprinus carpio* L., after oxytetracycline injection. *J. Fish Dis.*, **6**, 35-43.

30). GRONDEL, J.L., NOUWS, J.F.M., DE JONG, M., SCHUTTE, A.R. & DRIESSENS, F., 1987. Pharmacokinetics and tissue distribution of oxytetracycline in carp, *Cyprinus carpio* L., following different routes of administration. *J. Fish Dis.*, **10**, 153-163.

31). GRONDEL, J.L., NOUWS, J.F.M., SCHUTTE, A.R. & DRIESSENS, F., 1989. Comparative pharmacokinetics of oxytetracycline in rainbow trout (*Salmo gairdneri*) and African catfish (*Clarias gariepinus*). *J. Vet. Phar. Ther.*, **12**, 157-162.

32). HERMAN, R.L., COLLIS, D. & BULLOCK, G.L., 1969. Oxytetracycline residues in different tissues in trout. US Fish and Wildlife Service. Tech. Paper 37, 1-6.

33). HOLTMANNSPÖTTER, H. & THIER, H.P., 1982. Analysenmethode für Rückstände von sechs Sulfonamiden und Chloramphenicol durch Gaschromatographie an Glaskapillaren. *Dtsch. Lebensmittel-Rundschau*, **78** (10), 347-350.

34). HORI, Y., 1984. Determination of synthetic antibacterials in cultured fish by high performance liquid chromatography. *J. Food Hyg. Soc. Jap.*, **25** (2), 158-162.

35). HORIE, M., HOSHINO, M., NOSE, N., IWASAKI, H. & NAKAZAWA, H., 1985. Simultaneous determination of antibiotics and synthetic antibacterials in fish by high performance liquid chromatography. *Eisei Kagaku*, **31** (6), 371-376.

36). HORIE, M., SAITO, K., HOSHINO, Y., NOSE, N., MOCHIZUKI, E. & NAKAZAWA, H., 1987. Simultaneous determinations of nalidixic acid, oxolinic acid and piromidic acid in fish by high performance liquid chromatography with fluorescence and UV detection. *J. Chromatogr.*, **402**, 301-308.

37). HORII, S., MONMA, C. & MATSUMOTO, M., 1987. High performance liquid chromatographic method for the simultaneous determination of oxolinic, nalidixic and piromidic acids in eel. *Ann. Rep. Tokyo Metr. Res. Lab. P.H.*, **38**, 182-184.

38). IKAI, Y., OKA, H., KAWAMURA, N., YAMADA, M. & NAKAYAMA, H., 1989. Improvement of chemical analysis of antibiotics. XVI. Simple and rapid determination of residual pyridone-carboxylic acid antibacterials in fish using a prepacked amino cartridge. *J. Chromatogr.*, **477**, 397-406.

39). ISHIDA, N. & TAKESHI, H., 1985. TLC analysis of nalidixic acid type synthetic antibacterial agents and their metabolites in salmonids. *Bull. Natl. Res. Inst. Aquaculture*, **7**, 97-104.

40). ISHIDA, N., 1990. Comparison of biliary metabolites of oxolinic acid in seven species of teleost. *Nippon Suisan Gakkaishi*, **56** (1), 55-59.

41). ISHIDA, N., 1990. Metabolites of oxolinic acid in rainbow trout (*Salmo gairdneri*). *Nippon Suisan Gakkaishi*, **56** (1), 159.

42). IVERSEN, B., AANESRUD, A., KOLSTAD, A.K. & RASMUSSEN, K.E., 1989. Determination of oxytetracycline in plasma from rainbow trout using high-performance liquid chromatgraphy with ultraviolet detection. *J. Chromatogr.*, **493**, 217-221.

43). JACOBSEN, M.D., 1989. Withdrawal times of freshwater rainbow trout, *Salmo gairdneri* Richardson after treatment with oxolinic acid, oxytetracycline and trimethoprim. *J. Fish Dis.*, **12**, 29-36.

44). JINBO, K., MOMOSE, R., MARUYAMA, T. & MATSUMOTO, M., 1988. An improvement on the detection method for residual tetracyclines in cultured fish. *Ann. Rep. Tokyo Metr. Res. Lab. P.H.*, **39**, 108-111.

45). KANZAKI, M., TAKEBA, K., MURAKANI, F., MARUYAMA, T. & MATSUMOTO, M., 1984. An improvement of simplified detection method for residual sulphonamides in meat and fish. *Ann. Rep. Tokyo Metr. Res. Lab. P.H.*, **35**, 202-206.

46). KASUGA, Y., OTSUKA, K., SUGITANI, A. & YANADA, F., 1981. Determination of nalidixic acid in fishes by high performance liquid chromatography. *J. Food Hyg. Soc. Jap.*, **22** (6), 479-483.

47). KASUGA, Y., SUGITANI, A. & YAMADA, F., 1982. Simultaneous determination of oxolinic, nalidixic and piromidic acids in fishes by high performance liquid chromatography. *J. Food Hyg. Soc. Jap.*, **23** (4), 344-347.

48). KASUGA, Y., SUGITANI, A. & YAMADA, F., 1983. Identification of oxolinic acid and two major metabolites of piromidic acid in fishes. *J. Food Hyg. Soc. Jap.*, **24** (5), 484-487.

49). KASUGA, Y., SUGITANI, A., YAMADA, F., ARAI, M. & MORIKAWA, S., 1984. Oxolinic acid residues in tissues of cultured rainbow trout and ayu fish. *J. Food Hyg. Soc. Jap.*, **25** (6), 512-516.

50). KATAE, H., KOUNO, K., TAKASE, Y., MIYAZAKI, H., HASHIMOTO, M. & SHIMIZU, M., 1979. The evaluation of piromidic acid as an antibotic in fish: an *in vitro* and *in vivo* study. *J. Fish Dis.*, **2**, 321-335.

51). KATAE, H., KOUNO, K., SHIMIZU, M., KUSUDA, R., TANIGUCHI, M., SHIOMITSU, K. & HASEGAWA, H., 1980. Studies on chemotherapy of fish disease with erythromycin-I, absorption, distribution, excretion and residue of erythromycin in cultured yellowtails. *Fish Pathol.*, **15** (1), 7-16.

52). LONG, A.R., HSIEH, L.C., MALBROUGH, M.S., SHORT, C.R. & BARKER, S.A. 1990. Matrix solid phase dispersion, isolation and liquid chromatographic determination of oxytetracycline in catfish (*Ictalurus punctatus*) muscle tissue. *J. Assoc. Off. Anal. Chem.*, **73**, 864-867.

53). LONG, A.R., HSIEH, L.C., MALBROUGH, M.S., SHORT, C.R. & BARKER, S.A, 1990. Matrix solid phase dispersion, isolation and liquid chromatographic determination of sulfadimethoxine in catfish (*Ictalurus punctatus*) muscle tissue. *J. Assoc. Off. Anal. Chem.*, **73**, 868-871.

54). McCARTHY, D.H., STEVENSEN, J.P. & SALSBURY, A.W, 1974. A comparative pharmaco-kinetic study of seven sulphonamides and a sulphonamide potentiator, trimethoprim in rainbow trout. *Salmo gairdneri*, Richardson). *Aquaculture*, **4**, 299-303.

55). McCRACKEN, A., FIDGEON, S., O'BRIEN, J.J. & ANDERSON, D., 1976. An investigation of antibiotic and drug residues in fish. *J. Appl. Bact.*, **40**, 61-66.

56). MOCHIIKE, A., 1983. Simultaneous determination of sulfanilamides and nitrofurans in fish meat by high performance liquid chromatography. Report of the Hiroshima Prefectural Institute of Public Health, No 30, 1-5.

57). MURRAY, J., McGILL, A.S. & HARDY, R., 1988. Development of a method for the determination of oxytetracycline in trout. *Food Addit. Contam.*, 5 (1), 77-83.

58). NAGATA, T., MIYAMOTO, F. & SAEKI, M., 1982. Determination of nitrofurans in cultured fishes by high performance liquid chromatography. *J. Food Hyg. Soc. Jap.*, **23** (3), 278-282.

59). NAGATA,T. & SAEKI, M., 1986. Determination of ampicillin residues in fish tissues by liquid chromatography. *J. Assoc. Off. Anal. Chem.*, **69** (3), 448-450.

60). NAKABEPPU, H. & TATSUMI, K.,1984. Metabolism of furazolidone in eels. *Chem. Pharm. Bull.*, **32** (10), 4193-4196.

61). NORDLANDER, I., JOHNSSON, H. & ÖSTERDAHL, B., 1987. Oxytetracycline residues in rainbow trout analysed by a rapid HPLC method. *Food Addit. Contam.*, **4** (3), 291-296.

62). NOSE, N., HOSHINO, Y., KIKUCHI, Y., HORIE, M., SAITOH, K., KAWACHI, T. & NAKAZAWA, H., 1987. Simultaneous liquid chromatographic determination of residual synthetic antibacterials in cultured fish. *J. Assoc. Off. Anal. Chem.*, **70** (4), 714-717.

63). NOUWS, J.F.M., VREE, T.B. & HEKSTER, Y.A., 1985. *In vitro* antimicrobial activity of hydroxy and N_4-acetyl sulphonamide metabolites. *Vet. Q.*, **7**, 70-72.

64). NUSBAUM, K.E. & SHOTTS, E.B. Jr, 1981. Absorption of selected antimicrobial drugs from water by channel catfish, *Ictalurus punctatus*. *Can. J. Fish Aquat. Sci.*, **38**, 993-996.

65). O'GRADY, P., PALMER, R., HICKEY, C. & SMITH, P.R., 1986. Antibiotic therapy of furunculosis in freshwater and sea water. In *Pathology in marine aquaculture*, C.P. Vivarès, J-R Bonami & E. Jaspers eds., European Aquaculture Society. Special Publication No. 9, Bredene, Belgium.

66). O'GRADY, P., MOLONEY, M. & SMITH, P.R., 1988. Bath administration of the quinolone antibiotic flumequine to brown trout (*Salmo trutta*) and Atlantic salmon (*S. salar*). *Dis. Aquat. Org.*, 4, 27-33.

67). OIDA, T., KOUNO, K., KATAE, H., NAKAMURA, S., SEKINE, Y. & HASHIMOTO, M., 1982. Levels of piromidic acid and its metabolites in eel tissues. *Bull. Jap. Soc. Sci. Fish*, **48** (11), 1599-1608.

68). OKA, H., UNO, K., HARADA, K. & SUZUKI, M., 1983. Improvement of chemical analysis of antibiotics. II -Evaluation of prepared C_{18} cartridge for extraction of tetracyclines. *Yakugaku Zasshi*, **103** (5), 531-537.

69). OKA, H., MATSUMOTO, H., UNO, K., HARADA, K., KADOWAKI, S. & SUZUKI, M., 1985. Improvement of chemical analysis of antibiotics. VIII -Application of prepacked C_{18} cartridge for the analysis of tetracycline residues in animal liver. *J. Chromatogr.*, **325**, 265-274.

70). ONJI, Y., UNO, M., & TANIGAWA, K., 1984. Liquid chromatographic determination of tetracycline residues in meat and fish. *J. Assoc. Off. Anal. Chem.*, **67** (6), 1135-1137.

71). OTSUKA, K., HORIBE, K., SUGITANI, A. & YAMADA, F., 1982. High performance liquid chromatographic determination of nifurpirinol in fish. *J. Food Hyg. Soc. Jap.*, **23** (5), 373-376.

72). RASMUSSEN, K.E., TØNNESEN, F., HUUTHANH, H., ROGSTAD, A. & AANESRUD, A., 1989. Solid-phase extraction and high performance liquid chromatographic determination of flumequine and oxolinic acid in salmon plasma. *J. Chromatogr.*, **496**, 355-364.

73). REIMER, G.J. & YOUNG, L.M., 1990. Validation of a method for determination of tetracycline antibiotics in salmon muscle tissue. *J. Assoc. Off. Anal. Chem.*, **73** (5), 813-816.

74). ROGSTAD, A., HORMAZABAL, V. & YNDESTAD, M.,1989. Simultaneous extraction and determination of oxolinic acid and flumequine in fish tissues by high performance liquid chromatography. *J. Liquid Chromatogr.*, **12** (15), 3073-3086.

75). RYAN, J.J. & DUPONT, J.A., 1974. Chemical analysis of tetracycline residues in animal tissues. *J. Assoc. Off. Anal. Chem.*, **57**, 828-831.

76). SALTE, R.,1982. Oxytetracycline residues in rainbow trout (*Salmo gairdneri*) fed a commercial medicated feed. *Acta Vet. Scand.*, **23**, 150-152.

77). SALTE, R. & LIESTØ K., 1983. Drug withdrawal from farmed fish. Depletion of oxytetracycline, sulfadiazine, and trimethoprim from muscular tissue of rainbow trout (*Salmo gairdneri*). *Acta Vet. Scand.*, **24**, 418-430.

78). SAMUELSEN, O.B., 1989. Determination of flumequine in fish by high performance liquid chromatography and fluorescence detection. *J. Chromatogr.*, **497**, 355-359.

79). SAMUELSEN, O.B., 1990. High performance liquid chromatographic determination of furazolidone in Atlantic salmon (*Salmo salar*) tissue. *J. Chromatogr.*, **528**, 495-500.

80). SAMUELSEN, O.B., 1990. Simple and rapid method for the determination of flumequine and oxolinic acid in salmon (*Salmo salar*) plasma by high performance liquid chromatography and fluorescence detection. *J. Chromatogr.*, **530**, 452-457.

81). SCHMIDT, T. & BUNING-PFAUE, H., 1985. Rückstandsverhalten von Arzneistoffen in der Intensivhaltung von Nutzfischen. *Deutsche Lebensmittel Rundschau*, **81** (8), 239-243.

82). SIEGERT, K., 1985. Nachweis von Sulfonamiden und Antibiotika mit Methoden der gekoppelten Gaschromatographie-Massenspektrometrie. *Fleischwirtschaft*, 65, 1496-1497.

83). SILVEN, L., JOHANSSON, N. & LJUNGBERG, O., 1968. Terramycin and sulphamerazine residues in muscle tissue and interior organs of rainbow trout after oral drug treatment. *Bull. Off. Int. Epiz.*, **69** (9-10), 1465-1474.

84). SIMPSON, R.M., SUHRE, F.B. & SHAFER, J.W., 1985. Quantitative gas chromatographic - Mass spectrometric assay of five sulfonamide residues in animal tissue. *J. Assoc. Off. Anal. Chem.*, **68** (1), 23-26.

85). SKARE, J., HASTEIN, T. & FRASLIE, 1974. Investigation of residual concentrations after chloramphenicol treatment of rainbow trout (*Salmo gairdneri*) and Atlantic salmon (*Salmo salar*). *Nordisk Veterinaer Medicin*, **26**, 108-115.

86). SQUIBB, S.K., MICHEL, C.M.F., ZELIKOFF, J.T. & O'CONNOR, J.M., 1988. Sulfadimethoxine pharmacokinetics and metabolism in channel catfish (*Ictalurus punctatus*). *Vet. Hum. Toxicol.*, 30, 31-35.

87). SUGIMOTO, N., KASHIWAGI, S. & MATSUDA, T., 1979. Research on the change in concentration of furazolidone (N-(5 nitro-2 furyliden)-3 amino -2 oxazolidone) in fish muscle using a fluorescent spectrometric determination. *Bull. Japan. Soc. Sci. Fish.*, **45**, 353-362.

88). SVED, S., LAROCQUE, L., WENINGER, A., SCHNURR, M. & VAS, D., 1990. The assay of oxolinic acid in salmon tissues by HPLC with fluorescence detection. Proceedings of Conference on residues of veterinary drugs in foods. Noordwejkerhout, Netherlands. ISBN 90-61519-011-6, 356-360.

89). TAO, S.H., NICOLAS, M. & BONNOT, M., 1985. Methode de recherche de residus de la flumequine dans la chair des truites. *Revue Med. Vet.*, **136** (8-9), 623-625.

90). TAO, S.H. & POUMEYROL, M., 1989. Méthode rapide d'identification de la flumequine dans la chair des truites. *Revue Med. Vet.*, **140** (4), 321-323.

91). THANH, H.H., ANDRESEN, A.T., AGASØSTER, T. & RASMUSSEN, K.E., 1990. Automated column-switching high-performance liquid chromatographic determination of flumequine and oxolinic acid in extracts from fish. *J. Chromatogr.*, **532** (2), 363-373.

92). UENO, R., HORIGUCHI, Y. & KUBOTA, S., 1988. Levels of oxolinic acid in cultured yellowtail after oral administration. *Nippon Suisan Gakkaishi*, **54** (3), 479-484.

93). UENO, R., OKUMURA, M., HORIGUCHI, Y. & KUBOTA, S.S., 1988. Levels of oxolinic acid in cultured rainbow trout and amago salmon after oral administration. *Nippon Suisan Gakkaishi*. **54** (3), 485-489.

94). UENO, R., UNO, K., KUBOTA, S.S. & HORIGUCHI, Y., 1989. Determination of oxytetracycline in fish tissues by high performance liquid chromatography. *Nippon Suisan Gakkaishi*, **55** (7), 1273-1276.

95). VALSET, G., LOSNEGARD, N. & YNDESTAD, M., 1989. Routine microbiological method for the detection of antibiotics or chemotherapeutics in fish. Directorate of Fisheries, Norway.

96). WALISSER, J.A., BURT, H.M., VALAG, T.A., KITTS, D.D. & McERLANE, K.M., 1990. High performance liquid chromatographic analysis of Romet-30 in salmon following administration of medicated feed. *J. Chromatogr.*, **518**, 179-188.

97). WEINFELD, R.E. & MACASIEB, T.C., 1979. Determination of trimethoprim in biological fluids by high performance liquid chromatography. *J. Chromatogr.*, **164**, 73-84.

98). WEISS, G., DUKE, P.D. & GONZALES, L., 1987. HPLC method for the simultaneous analysis of sulfadimethoxine and ormetoprim in tissues and blood of cattle, chickens and catfish. *J. Agric. Food Chem.*, **35**, 905-909.

99). WÖLWER-RIECK, U. & BUNING-PFAUE, H., 1990. Das Rückstandsverhalten des Furazolidonmetaboliten 3-(4-cyano-2-oxo-butylideneamino)-2-oxazolidon in Forellen. *Z. Lebensm Unters Forsch.*, **191**, 319-321.

Discussion

ALDERMAN Dr McGill has pointed out many of the problems associated with the use and analysis of antimicrobials in fish. In particular his comments on the variations between different batches of solid phase extraction columns is a distinct warning to everyone, particularly when processing large numbers of samples for residue analysis which is going to happen as monitoring of fish products for drug and other residues increases.

WITHNELL What are the chances of false positives occurring in laboratories not accustomed to analysing for residues in fish flesh?

MCGILL With HPLC techniques there should be no false positives, but with the microbial assay which, as I pointed out is not as specific, there are materials other than antibiotics there which can cause inhibition. Microbial assays are very useful however for a quick screening, but most people, if they get an indication from the screening of the presence of possible antimicrobial residues, carry out a confirmatory re-analysis by other techniques, probably HPLC.

MUNRO Are any antibiotics likely to be bound more than any other?

MCGILL Yes, there is evidence for binding, Cravedi has done work in France on chloramphenicol. Furazolidone is bound. However it is extremely difficult to find any information on the binding of antibiotics to fish tissue or the way in which it occurs and a lot of new work is going to be needed. Indeed the same is true for other meat, there doesn't seem to be any available information as to the extent of binding.

DOSAGE DU CHLORAMPHENICOL DANS LA CHAIR DE POISSON PAR CHROMATOGRAPHIE LIQUIDE HAUTE PERFORMANCE. APPLICATIONS DANS DES ETUDES CINETIQUES

Brigitte Joseph-Enriquez

Service de Pharmacie et Toxicologie, Ecole Nationale Vétérinaire d'Alfort, 94704 Maisons-Alfort Cedex, France

Résumé: *Une méthode de dosage du chloramphénicol par chromatographie liquide haute performance a été mise au point pour déceler la présence de cet antibiotique dans la chair de poisson. Cette méthode peut être utilisée pour des études cinétiques, elle demande à être améliorée dans sa limite de détection pour être applicable à la détermination de niveaux résiduels de l'ordre de quelques dizaines de ppb.*

Summary : **Application of an HPLC method for detection of chloramphenicol residues in fish tissues to kinetic studies.**

A method of determining concentrations of chloramphenicol by liquid chromatography has been developed to enable the detection of this antibiotic in fish flesh. This method is simple and specific, but requires further development before its limit of detection is low enough to allow its use for pharmacokinetic studies. Currently limit of detection is in the order of 100 ppb with a recovery of 20% from spiked samples.

Introduction

Le chloramphénicol (C.A.P.)

Le chloramphénicol est un antibiotique produit naturellement par un microorganisme (*Streptomyces venezuelae*), et que l'on sait produire par synthèse depuis 1948.

C'est en 1950 que l'on découvrit ses effets délétères sur l'hématopoïèse en clinique humaine.

Aujourd'hui, en médecine humaine, le chloramphénicol n'est usité qu'en cas de nécessité absolue pendant une durée limitée et en général pour des traitements locaux : on dénombre 8 spécialités commerciales dont 7 à usage oculaire (collyres), 1 à usage général (comprimés). Dans ce dernier cas, le fabricant ne recommande l'emploi que si d'autres médicaments connus pour être moins dangereux se sont révélés inefficaces.

Le thiamphénicol, de structure proche (Fig. 1) mais non responsable d'aplasie médullaire, est utilisé notamment pour lutter contre la fièvre typhoïde ou d'autres salmonelloses. On lui reconnaît néanmoins des effets hématologiques réversibles à l'arrêt du traitement (déficit des lignées rouge, granuleuse et plaquettaire).

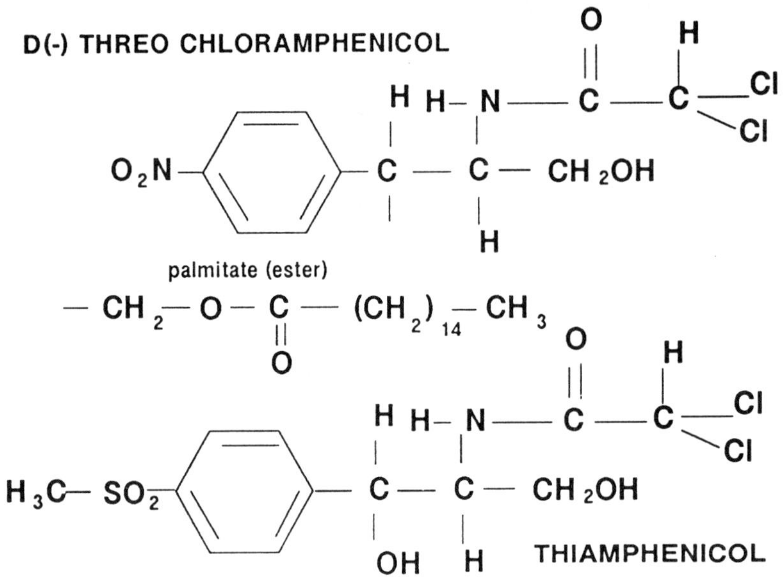

Fig 1 Structures moléculaires du chloramphénicol et du thiamphénicol

En médecine vétérinaire, le chloramphénicol connaît encore de nombreuses applications du fait de son large spectre, de son efficacité particulière sur les salmonelles, pasteurelles, coliformes, sur les genres *Aeromonas* et *Vibrio* (en aquaculture), enfin par son coût modéré.

Les nombreuses spécialités présentées dans le Dictionnaire des Médicaments Vétérinaires (11) en témoignent, bien qu'encore en sursis d'obtention d'AMM : plus de 70 médicaments dont environ 60 destinés à un effet général (comprimés, solutions buvables, poudres orales, prémélanges, solutés injectables) à base de chloramphénicol (Fig. 1) ou de son ester palmitique associé ou non à d'autres antiinfectieux ou des antiinflammatoires.

Le législateur français n'a en effet interdit d'emploi cet antibiotique chez les animaux destinés à la consommation humaine (également le législateur américain) que dans le cas de poules pondeuses. Un arrêté (A.M. du 9 Janvier 1987 paru au J.O. du 21 Janvier 1987) a interdit la commercialisation des aérosols contenant des préparations à usage topique afin de protéger l'utilisateur lors des pulvérisations. Or le risque pour la santé humaine est lié à la présence de résidus de chloramphénicol dans les denrées alimentaires de toute origine animale (poissons y compris), susceptibles d'engendrer le syndrome d'aplasie médullaire. L'aplasie peut en effet survenir après ingestion du chloramphénicol (et réduction en dérivé aminé dans le tube digestif), à une dose faible. Bien que tous les mécanismes d'apparition de ce syndrome ne soient pas élucidés, il apparaît que des taux résiduels dans les denrées alimentaires de l'ordre de 50 ppb ont peu de chances d'avoir un effet nocif à long terme pour le consommateur. Il était donc opportun de choisir une méthode de dosage qui ait une sensibilité de cet ordre pour la caractérisation des résidus de chloramphénicol chez le poisson, afin d'être en mesure de vérifier le bon respect du temps d'attente par l'utilisateur.

D'autre part, le dosage du chloramphénicol dans la chair du poisson permet aussi de vérifier la cinétique d'incorporation et de déplétion et donc de confirmer l'efficacité d'une présentation originale de chloramphénicol dans cette espèce. C'est ce point de vue que nous développerons ici.

Le choix de la chromatographie liquide haute performance

Les méthodes de dosage du chloramphénicol sont nombreuses. Elles ont suivi les progrès technologiques dans le domaine analytique.

Certaines sont à l'heure actuelle dépassées car souvent peu sensibles. Ainsi les dosages microbiologiques ont une sensibilité qui varie entre 1 et 10 ppm selon les modalités pratiques de réalisation des tests et la nature des prélèvements (sérum, urine, chair d'animaux traités).

Différentes techniques de colorimétrie ont été proposées de 1949 à 1980. Elles utilisent dans un premier temps la réduction du groupement nitré (par le chlorure de titane - méthode de GLAZKO (5) - par le zinc en présence de chlorure de calcium, à chaud en milieu éthanolique - méthode d'Auterhoff et Beinroth, (1)). Dans l'ensemble ces méthodes sont moins spécifiques mais plus sensibles que les méthodes microbiologiques (détection d'environ 0.1 ppm).

La fluorimétrie reprend la technique de réduction du groupement nitré (4) avec la poudre de zinc en milieu chlorhydrique. L'amine aromatique obtenue est couplée à la fluorescamine et la fluorescence est lue et comparée à celle obtenue à partir d'un échantillon non réduit.

La sensibilité de la méthode serait de 60 ppb pour le plasma, 100 ppb pour le muscle. Comme la méthode colorimétrique, elle dose le chloramphénicol et l'ensemble des métabolites possédant un groupement p.nitrophényle, y compris les dérivés glycuronoconjugués.

Proposée par Summa, (9) pour l'analyse du chloramphénicol dans les formulations, la polarographie a été développée par la suite pour doser les résidus de chloramphénicol dans le sérum. La sensibilité de cette méthode avoisine 20 ppb.

Les méthodes enzymatiques sont basées sur l'acétylation du chloramphénicol par l'enzyme chloramphénicol-acétyltransférase à l'aide d'acétyl-coenzyme A (marqué (8) ou non (2) au ^{14}C). Cette enzyme est isolée de souches *d'Escherichia coli* résistantes au chloramphénicol.

Dans la technique de Lietman et coll. (8) l'acétyl-coenzyme A est maqué au ^{14}C et le ^{14}C-acétoxy-chloramphénicol est séparé du précurseur marqué par extraction sélective dans le benzène et dosé ensuite par scintillation liquide.

Ce type de dosage ne peut donc être réalisé que dans des laboratoires spécialisés. De même, les méthodes immunoenzymatiques nécessitent du personnel spécialisé.

Ainsi, nous nous sommes orientés vers les méthodes spécifiques et suffisamment sensibles que sont les méthodes chromatographiques, et en particulier l'HPLC du fait de notre expérience précédente dans le cadre du dosage du chloramphénicol dans le lait (7).

Nous avons dans un premier temps pratiqué l'extraction du CAP par des mélanges de solvants, en particulier le méthanol, puis dosé spécifiquement le CAP libre.

Notre méthode permet de détecter pour un échantillon de 100 mg des concentrations de l'ordre de 500 ppb. Sa sensibilité devrait être prochainement améliorée.

Matériel et méthodes

Réactifs

Tous les solvants utilisés sont purs pour analyse (méthanol, acétonitrile, chloroforme, isooctane - Prolabo HPLC). Le chloramphénicol standard est le Chloramphénicol Sigma^{R}.

Extraction

50 à 100 mg de chair de poisson (muscle dorsal) sont mis en suspension puis broyés dans un milieu contenant de l'acétate d'éthyle et du chloroforme (1/2). Après centrifugation à 4000 tours/mn pendant 10 mn, le surnageant est récupéré puis mis à évaporer à seccité à température ambiante pendant 24 heures.

Le résidu est repris par un mélange de 0,75 ml de chloroforme, 0,5 ml d'isooctane, 0,5 ml du solvant d'élution (méthanol/eau 65%/35%). Après centrifugation à 4000 tours/mn pendant 5 mn, le surnageant (0,5 ml de méthanol/eau) sert directement à l'analyse.

Conditions chromatographiques

Cent µl de l'extrait sont analysés dans les conditions suivantes :

- Le chromatographe utilisé (HPLC Perkin-Elmer) est équipé d'une colonne Merck-Hiber (25 cm x 4 mm ID) 250-4 lichrosorb RP18. Les particules de remplissage ont un diamètre de 7 µm. Il s'agit donc d'une chromatographie en phase inversée.

- La colonne est précédée d'une précolonne de 5 cm de longueur, remplie de particules identiques à celles composant la colonne mais de granulométrie supérieure (25 à 40 µm).

L'éluant est composé à 65% de méthanol et à 35% d'eau. Il est utilisé après dégazage d'une quinzaine de minutes par barbotage d'azote. Le débit est fixé à 10 ml/mn avec une pression en tête de colonne d'environ 3000 PSI.

L'injection de l'échantillon est effectuée au moyen d'une boucle d'injection d'une capacité de 50 µl.

Après séparation, la détection est assurée par spectrophotométrie (Perkin-Elmer LC85) à 280 nm. Les pics composant le chromatogramme sont identifiables par leur temps de rétention. Un intégrateur (Hewlett-Packard 3390 A) réalise le calcul des aires. Les concentrations en CAP des échantillons sont obtenues par comparaison avec une solution de référence standard (1 ppm) dans du méthanol-eau (65/35).

Pour chaque extrait, deux injections ont été réalisées; la moyenne des deux aires est retenue.

Résultats et discussion

Le chromatogramme obtenu à partir de la solution de référence du CAP révèle un temps de rétention est de 3,70 mn dans les conditions chromatographiques précitées.

La linéarité de la réponse a été vérifiée pour des concentrations de 0,5 à 10 µg/ml (soit 25 à 500 ng).

La reproductibilité de l'analyse chromatographique est estimée par 6 injections successives de 100 µl/ml (50 ng). Le coefficient de variation mesuré est de 4,4%. Pour une solution de référence à 0,1 µg/ml, ce coefficient atteint 8,4%.

Les chromatogrammes obtenus avec des extraits, d'une part de muscle de poisson tout venant (témoin), d'autre part du même échantillon surchargé à 2 mg/kg, montrent que le temps de rétention est inchangé par rapport à celui du standard. De plus, dans la chair de poisson n'existe pas de composé qui interfèrerait dans la détection avec le CAP.

Le rendement moyen d'extraction obtenu à partir d'échantillon de chairs de poisson supplémentées à 1, 2, 5 et 10 mg/kg atteint 30%.

Des études sont en cours pour déterminer avec exactitude la limite de détection de la méthode.

Une expérimentation menée dans notre service sur de jeunes poissons (3) montre que l'on détecte facilement des pics correspondant à des concentrations de quelques centaines de ppb, ce qui est apparemment suffisant pour des études cinétiques menées sur le muscle. En effet, d'après Lasserre (6), des concentrations de l'ordre de 30 ppm de chloramphénicol se maintiennent dans le muscle de truites arc-en-ciel de 6 à 18 heures après l'ingestion d'une dose de chloramphénicol base de 100 mg/kg. Par contre, nous travaillerons à l'abaissement de la limite de détection pour être à même de contrôler les niveaux résiduels jusqu'au seuil de 50 ppb.

Conclusion

La méthode de dosage choisie a montré sa spécificité et sa simplicité de mise en oeuvre. Il est intéressant de pouvoir travailler sans interférence entre le chloramphénicol et les constituants normaux de la chair de poisson également entraînés lors de l'extraction. Des études restent à mettre en oeuvre pour améliorer la méthode et notamment son seuil de détection.

Références

1). AUTERHOFF H. & BEINROTH A.S., 1980. Nachweiss aromatischer Nitrogruppen über die Hydroxylaminstufe. *Arch. Pharm.*, **313**, 807-809.

2). BOECKX R.L. & BRETT E.M., 1981. A bioluminescence micromethod for measuring chloramphenicol in serum. *Clin. Chem.*, **27**, 819-822.

3). CHEREL P.A. & NIN F., 1989. *Contribution à l'étude des capacités de bioencapsulation de principes actifs médicamenteux par Artemia. Intérêt en Ichtyopathologie.* Thèse de Doctorat Vétérinaire, Ecole Nationale Vétérinaire d'Alfort, n°100.

4). CLARENBURG R. & RAO V.R., 1977. A fluorometric method to assay chloramphenicol. *Drug Metab. Dispos.*, **5**, 246-252.

5). GLAZKO A.J., WOLF L.M. & DILL W.A., 1949. Biochemical studies on chloramphenicol (chloromycetin). I. Colorimetric methods for the determination of chloramphenicol on related nitrocompounds. *Arch. Biochem.*, **23**, 411-418.

6). LASSERRE J., 1972. *Recherche du chloramphénicol utilisé en pisciculture dans différents tissus de la truite arc-en-ciel Salmo gairdneri Richardson, conséquences thérapeutiques et sanitaires.* Thèse de Doctorat Vétérinaire, Ecole Nationale Vétérinaire d'Alfort, n° 79.

7). LIBEAU Véronique, 1985. Les résidus de chloramphénicol ; leur dosage dans le lait par HPLC. DEA de Toxicologie fondamentale et appliquée à l'environnement et aux industries pharmaceutiques et agroalimentaires, Université Paris VII.

8) ETMAN P.S., WHITE T.J. & SHAW W.V., 1976. Chloramphenicol : an enzymological microassay. *Antimicrob. Ag. Chemother.*, **10**, 347-353.

9). SUMMA A.F., 1965. Polarographic determination of chloramphenicol preparations. *J. Pharm. Sci.*, **54**, 442-444.

10). VAN DER LEE J.J., NOUWS J.F.R. & BLOEMENDAL F.W.R., 1982. Physicochemical methods for pharmacokinetics and residue analysis of chloramphenicol and degradation products in dairy cows. *J. Vet. Pharmacol. Therap.*, **5**, 161-175.

11). MEISSONNIER, E., DEVISME P., JOIN-LAMBERT, P., 1991. *Dictionnaire des Médicaments Vétérinaires.* 6e Ed., Editions du Point Vétérinaire, Paris, 1050 pp.

ANTIBIOTHERAPY USING BIOCARRIERS (*Artemia salina*) IN HATCHERIES

P. Chérel & F. Nin

Service de Pharmacie et Toxicologie, Ecole Nationale Vétérinaire d'Alfort, 94704 Maisons-Alfort Cedex, France

Summary : *A method for the bioencapsulation of therapeutic products for those aquaculture species which will only accept live foods is described. Nauplius larvae of* Artemia salina *were exposed to a 3g/l microglobular preparation of chloramphenicol for 48h before feeding them to young chiclids (*Chiclasoma nigrofascitum*). The resulting tissue levels of chloramphenicol achieved in the chichlids were determined by HPLC after a fixed period. The chloramphenicol concentrations attained inthe muscled were greater than those attained by normal methods and indicate that the approach has practical possibilities.*

Résumé : **Bioencapsulation des produits thérapeutiques pour les espèces aquicoles n'acceptant que des proies vivantes.**

Une méthode de bioencapsulation des produits thérapeutiques destinés aux espèces aquicoles n'acceptant que des proies vivantes a été expérimentée. Des larves d'Artemia salina *au stade nauplius, exposées 48 heures à une concentration de 0,3 g/l de chloramphenicol sous forme microglobulaire, ont été distribuées à de jeunes cichlidés (*Cichlasoma nigrofascitum*). Les concentrations tissulaires étaient mesurées à temps fixes par HPLC. Les résultats indiquent que les concentrations musculaires atteintes sont supérieures aux CMI usuelles et que le procédé peut présenter un intérêt pratique.*

Here we report the results of veterinarian thesis work carried out to test the feasibility of antibiotic treatments of fish fry in hatcheries using the oral route.

The hatcheries breeding water is highly contaminated due to the intensive confining of the young animals. Thus disease outbreaks are very frequent encouraging the use and overuse of antibiotics. Usually and originally the antibiotic (often as active ingredient) is poured directly into the breeding medium which is very detrimental to the environment in which the waste water from the hatcheries is discharged. For example, chloramphenicol resistant bacterial strains have been found on shrimp spawners caught in the wild off the coast of Ecuador (Cahu, personal communication) where many hatcheries are set-up.

The possible use of the oral route for chiclid culture which appears to be a much less polluting therapy is hindered by the strictly carnivorous behavior of the larvae, eating only living and moving preys. The nauplii of *Artemia salina,* a small branchiopod crustacean, are intensively used to feed the larvae. The nauplii hatch from dormant eggs (cysts) which can be kept in cans for years.

It is possible to modify the nutritional value of the nauplii by dipping them in media containing particles in suspension which they filter and concentrate.

This enrichment method was been originally developed to enhance the unsaturated fatty acid content of the larvae feed.

Using lipidic microglobules of 2 µm diameter containing 10% of chloramphenicol palmitate, we obtained chloramphenicol enriched nauplii or to be more 'fashionable' "bioencapsulated chloramphenicol".

The bioencapsulation protocols employed are illustrated in (Figs.1 and 2). After cyst hatching, free surviving nauplii were separated from the cyst shells and then transferred to the enrichment bath containing 0.3 g/l of microglobules. Eighteen hours later a further 0.3 g/l was added to the bath. After 24 hours nauplii were rinsed with fresh water and fed to the fish.

To evaluate the efficacy of this original galenic form we chose cichlids fingerlings, *Cichlasoma nigrofasciatum,* as model. We indirectly evaluated the nauplii chloramphenicol content by measuring the chloramphenicol kinetic in the fish muscle using the HPLC procedure previously described by Joseph-Enriquez.

Enriched nauplii were fed *ad libitum* to 48 hours fasted fingerlings. After digestion of the nauplii and hydrolysis of the chloramphenicol palmitate, 5 fishes were sampled at regular intervals frozen until analysis (Fig.3)

After feeeding with the enriched nauplii, the muscle chloramphenicol content (Fig.4) of the chichlid larvae was above 2 ppm for 7 hours (a much higher concentration in liver can be ignored). These concentrations are in the range of the MIC (minimal inhibitory concentration) of many fish pathogens.

By increasing the antibiotic content of the enrichment bath and/or the number of medicated feeds, it seems possible to attain higher and longer drug concentrations in the fish.

The evident advantages of this therapeutic route are the savings in active ingredient (allowing the use of more expensive drugs, if necessary) and the reduction in the pollution (only a small volume of water, the nauplii enrichment bath is contaminated).

Moreover the same enrichment technique may be used with different biocarriers such as *Brachionus plicatilis*, filtering rotifer, also used for larval feeding.

	EXPERIENCE 1	EXPERIENCE 2
NAUPLIES		
DENSITE	35,000/l	20,000/l
VOLUME D' ENRICHISSEMENT	10 l	50 l
MILEU	EAU DE MER RECONSTITUEE A 27·C, OXYGENEE	
ALEVINS		
AGE	30 j	45 j
POIDS MOYEN	220 mg	290 mg
MILEU	EAU DOUCE A 25C, OXYGENEE	
PRELEVEMENTS	T A T + 18 h	T A T + 8 h

Fig.1 Conditions of production and maintenance of experimental animals

Fig.2 Treatment of *Artemia* nauplii with chloramphenicol

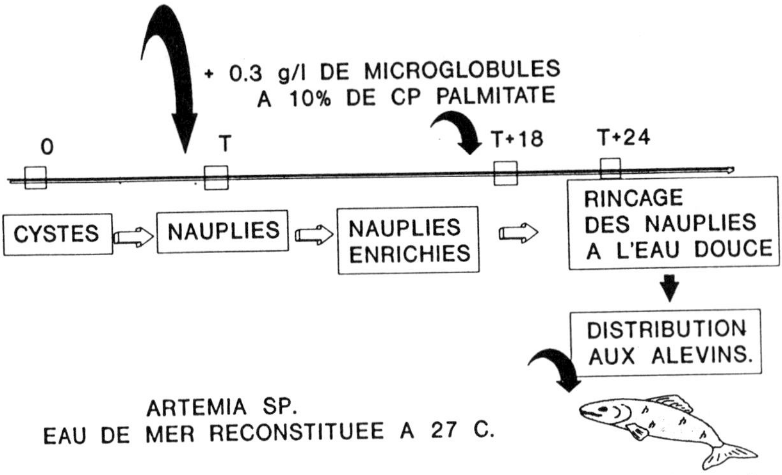

Fig.3 Experimental schedule for fish treatment and control

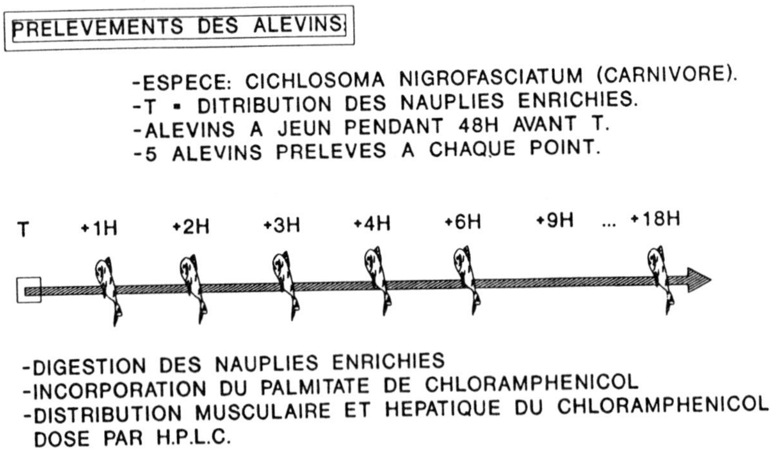

Fig.4 Kinetics of chloramphenicol in fish muscle after oral ingestion of medicated *Artemia* larvae

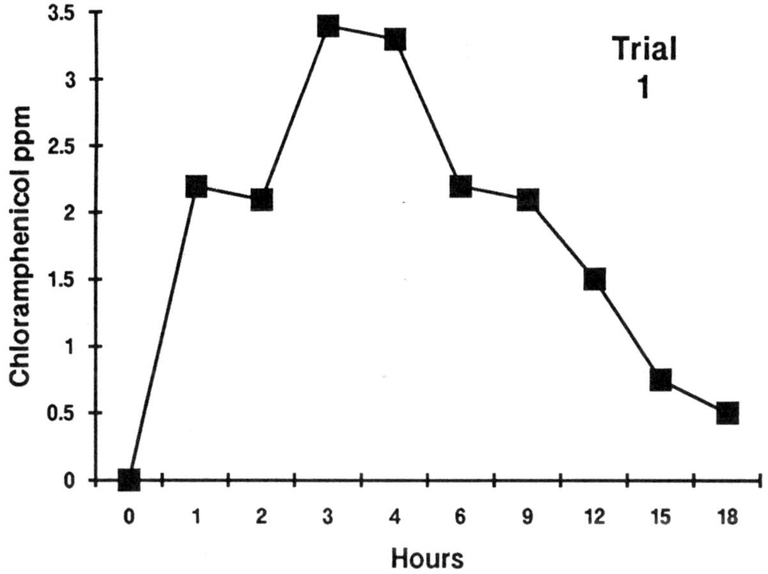

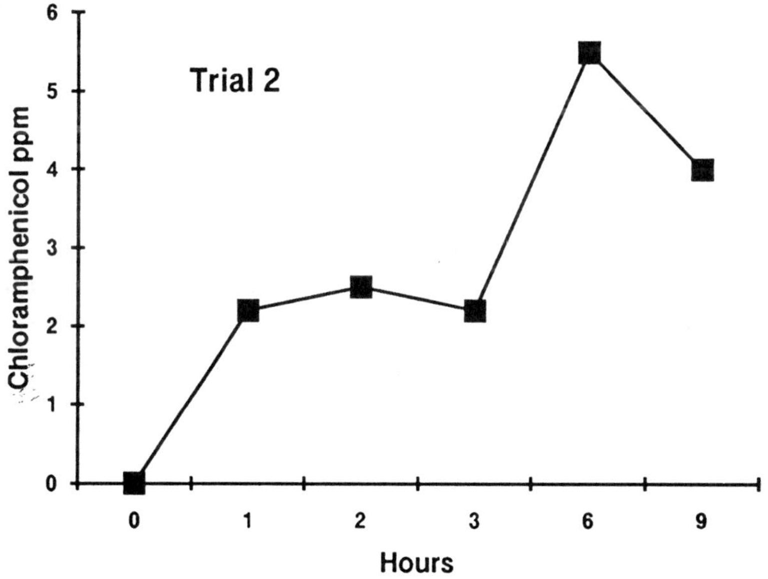

ANTIBIOTIC RESIDUES IN FARMED SHRIMP - A DEVELOPING PROBLEM?

J.H.Brown & I. Higuera - Ciapara

Institute of Aquaculture, University of Stirling,
Stirling, Scotland, FK9 4LA, U.K.

Summary : *The dramatic growth in world production of farmed shrimp has led to increasing pressure in some areas to further intensify farming methods. Problems of disease often accompany these developments as environmental conditions deteriorate. Pressure to ensure production can lead to reliance on antibiotics, both under controlled veterinary supervision and "prophylactically". The question as to whether antibiotic residues are likely to be a problem in shrimp can be approached in two ways. First by looking at what drugs are administered and, by calculating residence times, determine whether residues are likely to be present. However, information on which to base this for shrimp is scanty. The other approach is by direct measurement of residues in marketed shrimp and a preliminary survey, reported here, resulted in four positive samples of tetracyclines out of 136 samples examined.*

***Résumé*: Les résidus d'antibiotiques dans les crevettes d'élevage: un nouveau problème?**

L'élevage des crevettes connaît une grande faveur en zone tropicale depuis quelques années, puisque la production mondiale est passée de 90 000 t en 1980 à 565 000 en 1989. Les méthodes, intensifiées de manière parfois incontrôlée,ont favorisé l'impact croissant des maladies, comme l'illustre le quasi-anéantissement de la production taïwanaise en 1988. Des vibrioses à agents opportunistes en particulier sont favorisées par les stress, la surdensité, et les maladresses de gestion. Redoutées dans toutes les phases d'élevage, elles sont prétexte à des suradministrations d'antibiotiques, parfois préventives, et à des abus auxquels contribuent certainement des facteurs psychologiques. Ces pratiques ont entraîné la multiplication des résistances, notamment à l'égard de l'oxytétracycline. Elles font surtout surgir de nouveaux problèmes liés à l'apparition de résidus, et à leurs effets dans des écosystèmes et des contextes d'exploitation aux caractéristiques encore peu connues.

On ne dispose que d'une étude expérimentale sur le devenir et la persistance de l'oxytétracycline chez les crevettes. Selon les doses cette persistance varie entre 3 et 15 jours, mais on n'est pas sûr d'atteindre des concentrations tissulaires suffisantes pour affecter Vibrio *spp. si les traitements ne sont pas maintenus au moins 3 semaines. Il en est de même pour la sulfaméthazine. En théorie, les délais d'attente imposés aux poisson sont applicables, bien que la physiologie des crustacés rende toute extrapolation hasardeuse. En pratique, aucune législation sur les délais d'attente n'existe encore dans les pays intéressés.*

Une recherche de résidus, sur des crevettes surgelées ou précuisinées de provenances diverses, a été entreprise à l'Institut d'Aquaculture de Stirling. La technique de Charm, qui mesure par dosage microbiologique la compétition entre les résidus recherchés et un antibiotique radiomarqué, a été employée. Quatre lots congelés se sont avérés contenir des traces d'oxytétracycline, la recherche de chloramphenicol et les mesures effectuées sur plats conditionnés demeurantnégatives. Il ne s'agit cependant que d'un travail préliminaire mené à faible échelle (136 échantillons), et ne permettant pas de conclusions définitives.

La présence possible de résidus dans les crevettes commence à inquiéter sérieusement les responsables sanitaires de nombreux pays. Le Japon a décidé de contrôler systématiquement les lots de crevettes importées de Taïwan, sensibilisant ainsi les autorités chinoises. La tendance est donc à l'amélioration. Il n'en faut pas pour autant perdre de vue que c'est souvent la pertinence des traitements qui devrait être remise en question.

Introduction

Shrimp farming has been a major success in tropical aquaculture in recent years. The growth in world aquaculture production of shrimp has been remarkable, from 90,000 t in 1980 to about 565,000 t in 1989. This success has increased the pressure to intensify farming practice to maximize profits. Lack of control over farm developments has led to huge areas becoming unsuitable for shrimp production (e.g. the Samut Sakorn area southwest of Bangkok, Thailand). This intensification of farms has also led to increasing disease problems, as exemplified by the dramatic loss of production seen in Taiwan since 1988 (3). One of the major causes of mortality in ponds appears to be due to infection with *Vibrio* spp. Vibriosis is a disease particularly related to stress. *Vibrio* bacteria are ubiquitous and are not primary pathogens in shrimp (8) but there are many factors that come into play in the expression of disease. There is considerable evidence that disease is closely related to management practices and poor environmental conditions(6,10). It is now considered that avoidance of disease by using stocking densities of not more than 30 post larvae m^2; providing good water management (including maintenance of stable algal blooms) and adequate nutritional inputs is a more reliable control method than treatment with antibiotics (11). Antibiotics can be effective if treatment is given early enough in an outbreak of vibriosis and continued over the requisite time period for treatment (10).

Concern about the prophylactic use of antibiotics has already been expressed in relation to shrimp hatchery practice (1) but there is increasing evidence for concern over this practice in shrimp pond growout. Reliable sources indicate that there is pressure on many small farmers to try antibiotics, or, what in some cases merely purport to be antibiotics for "treating" their ponds, particularly for the "one - month syndrome" when heavy mortalities can occur for a number of not fully explicable reasons within six weeks of stocking at densities above 25 m^2. There is also evidence of some placebo effect (as far as local farmers are concerned) from the use of any substance when this condition arises (Dan Fegan pers. comm. 1990). This raises doubts as to the value of using antibiotics in ponds. A particular example of misuse of antibiotics is given by one farmer in Indonesia who uses antibiotic-treated feed once a week "prophylactically" right up to harvest time and claims that this improves his yields. (In comparison with neighbouring farmers his results are indeed better but this could be for a number of reasons). Another example from Indonesia is of a farm where it is considered good if oxytetracycline medicated food is given over only 10% of the 130 day growing period. Antibiotics are usually presented for a greater percentage of this period. Antibiotic treatment can come to be considered an essential part of the husbandry and this overdependance may lead to the development of antibiotic resistant strains of bacteria. Indiscriminate use of antibiotics in Central Thailand has resulted in very low sensitivity of *Vibrio* spp. to oxytetracycline (11).

Antibiotic treatments undertaken in a controlled manner are not going to be of such concern for the purpose of this paper, provided there is adequate withdrawal time before marketing but the use of antibiotics is not a procedure to be undertaken lightly. One reason is that the expense of medicated feed adds significantly to production costs where feed already constitutes at least 50-60% of the operational costs on an intensive shrimp farm. Use of antibiotics raises a very important question on the safety of their use in environmental and public health terms. The association of *Vibrio* spp. with shellfish and the implications of antibiotic resistant bacteria becoming associated with shrimp are dealt with in (1). Also, although the role of bacteria within pond systems is poorly understood they could be contributing directly to prawn productivity (9). There is no information available on how the prolonged use of antibiotic will affect the pond ecosystem.

Thus there are practices in shrimp farming that could lead to the presence of antibiotic residues in shrimp. The question to be addressed in this paper is to what extent these residues are actually present? There are two different methods to answer the question . The first is to calculate the amounts of antibiotics administered, the amount that will be taken up into the shrimp and then, with reference to residence times, determine how likely it is that residues will be a problem. The second method is by direct measurement of residues in marketable shrimp.

a) Uptake and residence time studies of antibiotics in shrimp

For the first method there is a dearth of information on both uptake and residence times in shrimp. Corliss (4) performed trials measuring the uptake of oxytetracycline in *Penaeus setiferus* at three different dose rates (1000 mg kg^{-1}, 5000 mg kg^{-1}, and 10,000 mg kg^{-1}) monitoring the resulting levels in shrimp, reared at 25 ±2°C. She also measured the retention times of antibiotic after feeding had ceased. Residues were undetectable within three days for the 1000 mg kg^{-1} group and within two weeks for groups fed at 5000 mg kg^{-1} and 10,000 mg kg^{-1}. In a separate study (14) the minimum inhibitory concentration (MIC) for the *Vibrio* spp. causing vibriosis in Kuruma shrimp (*Penaeus japonicus*) ranged from 200 to 400 mg kg^{-1} in 75.6% of the 49 strains tested. Tissue levels obtained from the lowest dosage level in the Corliss study (4) did not consistently come within this effective range (14). Using a much lower dosing level, [oxytetracycline (250 mg kg^{-1} and sulphamethazine (250 mg kg^{-1}) the resulting tissue levels in shrimp infected with *Vibrio* spp. were found to be within this range at just 200 mg kg^{1} oxytetracycline (14) only after three weeks (7). These two studies raise important questions as to what actually happens in ponds and particularly highlights the need for more work in this area. The study (4) is the only one that has looked at residence times in shrimp. On the basis of this information the withdrawal times for oxytetracycline supplied under license for treatment of fish in the UK (400 degree days) would be adequate if applied to shrimp but there could be other factors at work in the case of shrimp. For example, does the moult cycle affect the ability of the shrimp to rid itself of residues? Preliminary studies currently underway (January 1991) at the Shrimp Culture Research Centre in Thailand have shown that residence times for shrimp differ markedly between laboratory and pond trials; no detectable residues were found after a week in laboratory trials following therapeutic treatment but were still detectable up to thirty days in shrimp treated in ponds (G. Nash, pers.comm. 1991). It must be remembered that most shrimp growing areas do not have strict drug licensing so the finer points of degree day withdrawal times will not be clearly stated or enforced.

b) Measurements of antibiotic residues in farmed shrimp

At the Institute of Aquaculture in Stirling we have screened for residues in shrimp from India, Ecuador, Thailand, China and Pakistan obtained from several commercial outlets in the U.K.. These assays were carried out in three series. For the first and third series the sources of the shrimp were not known and were not necessarily farmed shrimp. In the second series the sources of the shrimp were from one or more farms in either Thailand or Ecuador.

Material and methods.

The screenings were carried out using the Charm receptor assay. This is a biological assay using microbial cells containing binding sites for specific antibiotic families on or within the cells (2). Any antibiotic in the tissue being screened will take up sites in the special microbial cells and prevent the full uptake of radioactively-labelled antibiotic. Subsequent counting of the sample using a scintillation counter results in a lowered count if antibiotic was present in the sample. Samples are counted together with both negative and positive controls.

The first series of screenings were carried out on headless frozen shrimp from China, India, Thailand and on samples of ready cooked headless shrimp from Ecuador/Thailand.

Forty grams of peeled shrimp was homogenized with 40 ml of extraction buffer and incubated for 30 min at 60°C. After centrifuging for 10 min at 1600 g the clear supernatant was collected for testing. The Charm 7000 test for beta-lactams, sulphonamides, tetracyclines and chloramphenicol was carried out in the first series (Penicillin Assays Inc). In all screenings negative kidney extract (NKE) controls and positive controls supplied by Penicillin Assays Inc. were carried out at the same time. In the second series of tests, 14 batches of shrimp from different, but known, locations in Thailand or Ecuador were obtained and of these, nine were also supplied in cooked form. Samples from these were tested for chloramphenicol and tetracyclines. In the third series samples from Pakistan and Thailand were tested for the presence of chloramphenicol. A total of 12 tests were carried out.

Results

The results for the first series are given in Table I. below. Of the 44 samples two were shown to be positive for tetracyclines in the batch from Ecuador/Thailand.

In the second series (Table II), two of the samples from one of the 14 different batches were positive for tetracyclines and were judged to contain just under 500 mg kg^{-1} by comparison with the standard control. Further testing on other samples from this particular batch did not contain tetracyclines and nor did the cooked samples from this batch. Information about this batch suggests it could be the produce from more than one farm. In all 3 series of tests, samples were negative for chloramphenicol and the tests for the other antibiotics in the first series were also negative.

Table I Results from the first series of sceenings

	ß-lactams	Sulphonamides	Tetracyclines	Chloramphenicol
CHINA				
range cpm	1589-1652	4220-4550	1121-1362	1670-1724
no. of samples	3	2	3	4
positives	0	0	0	0
INDIA				
range cpm	1523-1666	5865-6854	1719-1870	1495-1668
no. of sample	3	2	3	3
positives	0	0	0	0
THAILAND				
range cpm	1420-1569	3706-4622	1165-2023	1216-1748
no. of samples	3	2	3	4
positives	0	0	0	0
ECUADOR/ THAILAND				
range cpm	1210-1560	6122-7473	775-987*	1988
no. of samples	3	2	3	1
positives	0	0	2	0
NKE	1518-1544	2018-2182	1232-1268	1216-1519
Standard		735-815 ($40 \mu g\ kg^{-1}$)	859-915 ($500 \mu g\ kg^{-1}$)	

* Counts for two further samples of same extract confirmed positive by comparison with 500 $\mu g\ kg^{-1}$ standard.

Discussion

If chloramphenicol were present in shrimp it would be cause for great concern due to the known adverse affects of this drug in humans, particularly its association with the development of aplastic anaemia from very small doses (15), so it is reassuring that were no positive results for this drug. Finding the four positive samples of tetracyclines is not surprising in view of what is known of antibiotic use in many farms and the fact that oxytetracycline is probably the most commonly used of these drugs. Nothing can be concluded from the fact that cooked samples were negative since there were also other samples from that particular batch that were clear. There is evidence that cooking has minimal effect on the presence of antibiotics (12). This survey is however a preliminary study only and was carried out in response to the question posed in the title of this paper. The sample is very small and even the significance of the positive results is unclear when many countries are still discussing tolerance levels. The positive levels are however above the 0.1 mg kg^{-1} level set for salmonids, catfish and lobsters by the U.S. Food and Drug Administration.

Table II Results from the second series of screenings

	Tetracyclines	Chloramphenicol
Cooked samples		
THAILAND		
Range cpm	1264-1356	1551-1883
no of samples	3	3
positives	0	0
ECUADOR		
range cpm	968-1336	1514-1894
no. of samples	8	6
positives	0	0
NKE	1219-1399	1598-1642
standard	374-440 (500μg kg^{-1})	731-924 (10μg kg^{-1})
Uncooked samples		
THAILAND		
range cpm	592-1450	1422-2114
no. of samples	27	3
positives	2	0
ECUADOR		
range cpm	758-1339	1680-2142
no. of samples	19	11
positives	0	0
NKE	1636	-
standard	586 (500μg kg^{-1})	-

There is now increasing concern over the presence of residues all over the shrimp producing world. Japanese health authorities were reported to have found residues of oxytetracycline and oxolinic acid in shipments of shrimp from Thailand (5). From the time of that discovery the Japanese authorities announced that imports of farmed shrimp from Thailand would no longer be just randomly checked but that every shipment would be screened for residues. The immediate result of this announcement was a 27% drop in price for Thai shrimp. This has resulted in Government initiatives within Thailand to control the use of antibiotics and other chemicals and to monitor residues in shrimp and sediments. Shrimp farms will be required to register with the Thai Department of Fisheries which will do the monitoring (G. Nash pers. comm. 1991). This will be extremely beneficial in the long term and highlights the importance of Governments taking responsibility. Whilst large producers can set up their own links with overseas importers it is difficult to avoid a bad reputation affecting the market for all the produce from a particular country, although possibly only deserved through a small proportion of producers. There are many ways in which the increasing production of shrimp world-wide will engender competition and the presence of unnecessary chemicals is surely one negative factor that all producers should endeavour to avoid.

The levels of residues that are permitted or acceptable have not been considered in this paper. For many countries no guidelines have been set or they are still under discussion. Evidence presented above serves to show that there could be a potential problem but not an insuperable one in that antibiotics are by no means essential for shrimp production; indeed they may be harmful for long term production. This paper also serves to emphasise that too little is known about the behaviour of antibiotics within shrimp, the environment and their effect on the consumer.

References

1). BROWN.H. (1989) - Antibiotics: their use and abuse in aquaculture. *World Aquaculture*, 20(2), 34-35,38-39,42-43.

2). CHARM, S.E. & CHI, R. (1988). - Microbial receptor assay for rapid detection and identification of seven families of antimicrobial drugs in milk: a collaborative study. *J. Assoc. Off. Anal. Chem.*, 7(2), 304-316

3). CHEN S.N. (1989) - Investigation and resolution of mass mortality in cultured giant tiger prawn *Penaeus monodon*, in Taiwan. *Report to USDA* 26pp.

4). CORLISS J.P. (1979) - Accumulation and depletion of oxytetracycline in juvenile white shrimp (*Penaeus setiferus*). *Aquaculture*, 16, 1-6.

5). DHANASETTAKORN A. 1990. - "Find sparks full check on Thai prawns." *The Nation, Bangkok,* Nov 27 1990.

6). FEGAN D.F., FLEGEL T.W., Sriurairatana S. & Waiyakruttha M. (1991) - The occurrence, development and histopathology of monodon baculovirus in *Penaeus monodon* in Southern Thailand. Aquaculture (In Press).

7). HIGUERA - CIAPARA I., BROWN J.H. & JAUNCEY K. (1991) - Effect of oxytetracycline and sulphamethazine on weight gain and survival of *Penaeus monodon* under stress. Poster paper given at O.I.E. Conference "Problems of chemotherapy in Aquaculture: from theory to reality". Paris, March 12-15, 1991 (This publication)

8). LIGHTNER D.V. (1988) - Vibrio disease of penaeid shrimp. In *"Disease Diagnosis and Control in North American Marine Aquaculture."* p 44. Eds. C.J.Sindermann and D.V.Lightner. Elsevier. 431pp

9). MORIARTY D.J.W. (1986) - Bacterial productivity in ponds used for culture of penaeid prawns. *Microb Ecol.* 12, 259-269.

10). NASH G.L. (1990) - *Penaeus monodon* growout diseases. *Proc Aquatech '90. Putra World Trade Centre, Kuala Lumpur, 11-14 June, 1990.*

11). NASH G.L., NITHIMATHACHOKE C., TUNGMANDI C., ARKAJAMORA A., PRATHANPIPAT P., & RUAMTHAVEESUB P. (1991) - Vibriosis in pond reared *Penaeus monodon* in Thailand. *Proc. of Symposium on diseases in Asian aquaculture, 26-29 November 1990, Bali, Indonesia.* (In Press).

12). O'BRIEN J.J., CAMPBELL N. & CONAGHAN T. (1981) - Effect of cooking and cold storage on biologically active antibiotic residues in meat. *J.Hyg., Camb.* (87) 511-523

13). PENICILLIN ASSAYS INC. (1990) - The Charm System for Monitoring Antibiotics in Tissue. Operators's manual. Malden, Ma. 12pp.

14). TAKAHASHI Y., ITAMI T., NAKAGAWA A., NISHIMURA H., & ABE T. (1985) - Therapeutic effects of oxytetracycline trial tablets against vibriosis in cultured Kuruma prawns *Penaeus japonicus* Bate. *Bull. Japan. Soc. Sci. Fish.*, 51 (10), 1639-1643.

15). WORLD HEALTH ORGANISATION. (1988). - Joint FAO/WHO Expert Committee on Food Additives. Toxicological evaluation of certain veterinary drug residues in food. (WHO food additive series: 23)

Discussion

ALDERMAN There have been suspicions that farmed shrimp may be a problem in regard to residues and as we have seen, confirmation is beginning to appear. Are there questions for Dr Brown?

LE GOUVELLO Does anybody know how to carry out pharmacokinetic studies on shrimp?

JANET BROWN We don't even know what the residence times for shrimp are.

LE GOUVELLO Has nobody worked on the haemolymph?

JANET BROWN As far as I know there has been no work done on the behaviour of antibiotics in shrimp. Nobody has the basic information on residence times, but I don't see any reason why it shouldn't be done

LE GOUVELLO So nobody has tried to work out a protocol?

JANET BROWN As far as I know, no.

MIDTVEDT Will your Charm method find metabolites as well as original compound?

JANET BROWN The method distinguishes between antibiotic families and I don't really know whether it will pick up metabolites.

THOMEN Do you know anything about false negative levels?

JANET BROWN The method is best used as a rapid screening method, it is quite sensitive although sensitivity differs for different antibiotic families. For example the standard for chloramphenicol is 10ppb whilst for tetracycline it is around 500ppb. Thus false negatives might occur for the tetracycline family, but would be very unlikely for choramphenicol. Also not all antibiotics can be screened for by this method, for example there is no method yet available for oxolinic acid although one is under development.

PHARMACOKINETICS OF CHEMOTHERAPEUTANTS IN FISH AND SHELLFISH

T. Endo

Biological Research Laboratory, Tanabe Seiyaku Co., Ltd., 2-2-50, Kawagishi, Toda, Saitama 335, Japan

Summary: *Many antibiotics and antimicrobials are used in the treatment of diseased cultured fish and shellfish. The pharmacokinetics of these drugs are important in evaluation of their effects and side effects. Drug concentrations in the animal body fluids are determined by methods such as HPLC with high sensitivity, accuracy and specificity. Recently, the comparative pharmacokinetics of various chemotherapeutic agents in several fish have been compared, and the characteristics of pharmacokinetics in fish and shellfish were clarified using compartment models. Now, pharmacokinetic studies in fish and shellfish are necessary to determine withdrawal periods and suitable uses of chemotherapeutic agents.*

Résumé **: Pharmacocinétique des agents thérapeutiques chez les poissons et les crustacés.**

Chez les poissons comme chez les mammifères, l'étude pharmacocinétique des médicaments repose sur des dosages biologiques effectués dans les liquides corporels. On peut ainsi évaluer leur efficacité, leurs effets secondaires et leur persistance en tant que résidus. Ces propriétés sont conditionnées par l'absorbtion, la distribution, le métabolisme et l'excrétion du produit, et exprimées par la demi-vie, la clairance et le volume de distribution. Au niveau tissulaire, on applique les méthodes immunologiques et la chromatographie HPLC, qui présentent l'avantage d'une spécificité et d'une sensibilité plus grandes. Dans le cas présent on s'est intéressé à l'oxytétracycline, à l'acide diamino-pyrimidique et aux 4-quinolones, dont l'administration et l'absorbtion sont le plus souvent orale et balnéatoire.

Métabolisme et élimination: les molécules et leurs métabolites sont généralement éliminés dans l'urine et la bile, parfois par les branchies et le mucus. Peu d'études existent encore à ce sujet, mais des métabolites secondaires de l'acide piromidique et des sulfamides ont été identifiés.

Liaison aux protéines plasmatiques: cette propriété, qui augmente la clairance, semble nettement moins marquée chez les poissons que chez les mammifères.

Pharmacocinétique: l'emploi des modèles de compartiments a permis d'estimer des délais d'attente pour plusieurs antibiotiques. De nombreux facteurs influencent la cinétique. La température agit selon la loi du Q 10 (le

métabolisme est doublé pour un accroissement de température de 10°C). Bien que les délais d'attente préconisés varient selon les auteurs, on compte 60-90 j pour les tétracyclines, 40-60 j pour le triméthoprime, 20 j pour l'acide oxolinique. Les doses, les durées d'administration, et d'autres facteurs peuvent moduler ces données. Par voie orale la forme micronisée de l'acide oxolinique est absorbée plus efficacement. La biodisponibilité varie selon la nature des sels employés (cas des sulfamides) et parfois selon la fréquence de distribution. L'espèce surtout semble importante, les particularités métaboliques et physiologiques propres à chacune pouvant induire de grandes différences dans la demi-vie, les concentrations tissulaires et le temps de disparition des molécules. Le poids joue également, l'élimination étant plus lente chez les jeunes animaux dans le cas de la truite. La voie d'administration est bien sûr un facteur déterminant. En bain, le pH ambiant, la concentration saline de l'eau, le pKa et le degré d'ionisation de la molécule peuvent conditionner des différences d'absorbtion. Par exemple les sulfamides sont mieux absorbés en eau de mer qu'en eau douce, l'acide oxolinique moins bien. L'absorbtion de la fluméquine est gênée par un pH élévé, une température basse et une forte dureté.

Crustacés: des études menées avec l'oxytétracycline et la sulfadiméthoxine montrent que cette dernière s'accumule fortement dans la carapace.

Le modèle de compartiments a donc apporté un précieux concours dans la conception et la stratégie des traitements chez les animaux aquatiques. La diversité des facteurs pouvant moduler la pharmacocinétique des produits laisse pourtant encore beaucoup de travail en perspective.

Introduction

Because concentrations of antibiotics and antimicrobials in the tissue fluids are easily determined using bioassay, the pharmacokinetics of these drugs have been studied. Direct actions of antibiotics and antimicrobials such as antibacterial and bactericidal activity as well as the antibacterial spectrum, are very important in chemotherapeutic agents. However, pharmacokinetics including absorption, distribution, metabolism and excretion are also necessary for evaluation of the effects of these drugs. For example, pharmacokinetic data are necessary to decide when to withdraw chemotherapeutics used against infectious diseases of domesticated animals, fowls and cultured fish. Recently, pharmacokinetics in the serum have been expressed in terms of several parameters such as half-life, clearance and volume of distribution. The use of these parameters enables prediction of the plasma concentration irrespective of the administration routes at any dose. Once the concentrations of chemotherapeutics in the blood and tissues are determined, whether or not the tissue levels were beyond the minimum inhibitory concentration (MIC) against bacteria and how long the tissue levels remained beyond the MIC can be ascertained. Furthermore, the pharmacokinetic parameters from the serum concentrations show whether or not the concentration of drugs were beyond the standard values provided for settlement of the residue analysis.

Drug concentrations in the tissues are usually determined by high performance liquid chromatography (HPLC), immunological methods and bioassay. HPLC has higher sensitivity, accuracy and specificity in comparison with bioassay. Because of the development of many detectors, HPLC is the best method to determine drug concentrations in tissue fluids. These developments in the method for determination have improved pharmacokinetic investigations of antibiotics and antimicrobials in animal tissue fluids.

This paper reviews the pharmacokinetics of chemotherapeutic agents in fish and shellfish. Although many drugs have been used in these fields, oxytetracycline, sulfonamides, diaminopyrimidine derivatives and 4-quinolones are focused because of their wide use.

Administration route and absorption

Chemotherapeutics used in fish farming are usually administered orally or by bathing. Intramuscular and intravenous administration, which are commonly used in human and large domesticated animals, are rarely used in cultured fish regimens. Oral administration through diet is generally convenient in fish cultured in groups. When chemotherapeutants are administered to juveniles or to fish with poor appetites, bath administration provides a more uniform drug uptake in comparison to oral administration.

Elimination and metabolism

An administered drug is largely excreted in the urine and/or bile as the parent compound and/or as metabolites. Some of the administered drug is occasionally excreted through the gills, and also from mucus of the body surface. However, since no excretion rate of chemotherapeutants used in fish and shellfish cultures has been reported, little is known of the rate of excretion by each elimination route. The elimination rate of xenobiotics used in fish farming was commonly estimated by comparing the drug levels between the kidney and liver or bile. For example, Ueno *et al.* reported that because of the high ratio of the drug concentrations in both liver and kidney to those in the serum, oxolinic acid in yellowtail might be eliminated from the liver to the bile and/or from the kidney to the urine (30). A few reports concerning metabolites of chemotherapeutics used in fish cultures have been presented. For example, two metabolites of piromidic acid (PA), 2-OHPA and 3-OHPA, were detected in eels (25). As for sulfonamides, a N_4-acetyl metabolite of sulfadimethoxine was determined in rainbow trouts and channel catfish (21, 29).

Protein binding

Plasma protein binding of drugs affects their pharmacokinetics and probably influences their activity. Since protein bound drug can not penetrate the blood vessel wall, it is suggested that only free antibiotics have antibacterial activity in the infected lesions. The extent of antibiotic protein binding varies with animal species. Recently, in a report on the plasma protein binding of chemotherapeutics in fish and shellfish, the relationships between protein binding and pharmacokinetics are discussed. The plasma protein binding of ciprofloxacin in carp, African catfish and trout were 21.6, 20.2 and 23.1%, respectively (23). The percentage of plasma bound oxytetracycline in carp was 25.1% (14). The protein binding rate of sulfadimethoxine in rainbow trout was 17%, which might cause rapid elimination from the body fluids and a large distribution to the tissues in these fish (21). The binding rate of sulfadimethoxine to catfish plasma proteins was 18.4%. This is far lower than that of mammals (60-88%), causing rapid blood clearance of sulfadimethoxine in catfish. The authors suggested that general differences in blood parameters such as pH, fatty acid composition and/or quantity or quality of plasma proteins, might be responsible for the different clearance rates observed between fishes and mammals (29). The binding rate of ormethoprim to rainbow trout plasma protein was about 30%, and the *in vitro* plasma protein binding of sulfadimethoxine was not altered by an increase in ormethoprim concentration and *vice versa* (5, 6).

Analysis of pharmacokinetics

Pharmacokinetics is the study of the fate of drugs in the body. The pharmacokinetics of drugs is described by compartment models using differential equations. Several parameters obtained from a compartment model are enough to estimate the pharmacokinetics in the body. The pharmacokinetics of intravenous, intramuscular and oral sulfadimethoxine, ormethoprim, oxytetracycline and ciprofloxacin in fish have been investigated using compartment models (5, 6, 15, 16, 21, 23, 29).

The withdrawal time for chemotherapeutics used in fish farming has been statistically calculated from drug concentrations in edible tissues sampled serially. The withdrawal period is defined as the time intercept between the detection limit of the applied method of analysis and the 90% upper prediction limit of linear regression on the logarithm of the tissue drug concentration (26, 27).

Factors influencing pharmacokinetics in fish

Because fish are poikilothermic and aquatic, the pharmacokinetics of mammals can not be applied to fish. Although chemotherapeutics have been used successfully in the treatment of the diseases of cultured fishes over a long period, only a few pharmacokinetic studies have been performed in fish. However, these studies so far have addressed several factors supposedly affecting the pharmacokinetics of chemotherapeutics after oral, intravenous, intramuscular and bathing administration in fish.

1. Oral, intravenous and intramuscular administration

- Temperature

The metabolic rate of chemotherapeutics in fish doubles approximately with each 10°C rise in temperature showing that pharmacokinetics depend on water temperature (7).

The absorption and excretion of oxytetracycline, trimethoprim and oxolinic acid were temperature-dependent, and the recommended withdrawal times for fresh water rainbow trout treated with these drugs were 90 days below 6°C, 70 days between 6 and 12°C and 60 days above 12°C for oxytetracycline, 60 days below 8°C and 40 days above 8°C for trimethoprim, and 20 days at approximately 18°C for oxolinic acid (18). The withdrawal times of these chemotherapeutics used in aquaculture were obviously extended when the water temperature is lowered.

The serum and tissue concentrations of oxolinic acid in rainbow trout reached a maximum concentration within 2 days at 15°C and within 3 days at 10°C. They had disappeared within 12 days at 15°C and within 18 days at 10°C after administration with a dose of 40 mg/kg (31). After oral administration with a dose of 20 mg/kg for 5 days, persistence of oxolinic acid in the muscle and liver of rainbow trout was a little shorter at temperatures between 8.5 and 11.5°C than between 17.0 and 19.6°C (19).

Sohlberg reported that flumequine given 5 mg/kg orally and intra-arterially to rainbow trout was eliminated in river water when the temperature declined during the experimental period. A plateau was observed in the elimination curve of the intra-arterially administration group during the steepest fall in temperature (28).

Salte *et al.* also recognized that water temperature was an important determinant of pharmacokinetics in fish, and supposed that the withdrawal times were 60 days at temperatures above 10°C and 100 days between 7 and 10°C for fish treated with up to 75 mg/kg of oxytetracycline orally for up to ten consecutive days. They recommended that sulfadiazine-trimethoprim should be applied to slaughter fish during summer only, because these drugs persisted at a lower temperature (27).

Moreover, the absorption and elimination of oxytetracycline in the serum, muscle and liver of rainbow trouts was investigated at 5, 10 and 16°C after a single oral administration of 75 mg/kg. Oxytetracycline in the serum reached a maximum concentration within 1 h at 16°C, 12 h at 10°C and 24 h at 5°C. The half-life of oxytetracycline was 4.8 days at 16°C, 6.1 days at 10°C and 8.9 days at 5°C. Therefore drug elimination was markedly temperature-dependent. The authors reported that withdrawal times varied from 27 days to 135 days depending on the temperature and tissue studied. The pharmacokinetics of a drug in all organisms is dependent on several factors such as the total intake of medicated food (drug dose), gastric emptying time, absorption, biotransformation and excretion rate. In poikilothermic species all these parameters are temperature-dependent (3).

Pharmacokinetics in poikilothermic animals including fish is affected by water temperature, and the elimination rates of drugs in fish are in general faster at high temperature than at low temperature.

- Species

The pharmacokinetic differences of 4-quinolones among various fishes have been reported. The plasma disposition of ciprofloxacin has been investigated in carp, African catfish and trout after intravenous and intramuscular administration with a dose of 15 mg/kg, and pharmacokinetic parameters were calculated. This study shows striking differences in pharmacokinetic behaviour, especially in the distribution volume (Vd_{area}) and distribution rates (K_{12} and K_{21}) of ciprofloxacin among the three fish species involved. Pharmacokinetic differences in relation to physiological diversity among these species were suggested (23).

The pharmacokinetics of oxolinic acid as well as ciprofloxacin vary with fish species. Tissue levels of oxolinic acid in salmonids to which rainbow trout and amago salmon belong, have been determined by HPLC at 15°C after oral administration with a dose of 40 mg/kg. In rainbow trout, oxolinic acid was detected 9 days after administration, and in amago salmon 27 days after treatment. Therefore, the times required for disappearance of oxolinic acid from the tissues were different between two fish species. The absorption and elimination of oxolinic acid in rainbow trout and amago salmon after oral administration was slower than those in yellowtail. The authors supposed that this might be explained by species differences, water temperature, their movement, and other factors (30, 31).

Oxolinic acid residues in the tissues were compared between rainbow trout and ayu after administration of 20 mg/kg for 5 days at temperatures between 17.1 and 19.6°C. In 240 g rainbow trout, oxolinic acid was not detected in the muscle and liver at 21 days, whereas in 40 g ayu, the drug had disappeared from both tissues at 14 day, thus the persistence of oxolinic acid in these tissues was longer in rainbow trout than in ayu (19).

The pharmacokinetics of piromidic acid in ayu, amago salmon and eel were examined. At temperatures between 19.6 and 22.4°C, the concentration (13.35 µg/ml) in the blood of ayu reached a maximum 3 h after administration, whereas at temperatures between 13.0 and 13.7°C the maximum concentration (7.73 µg/ml) was reached at 24 h in blood of amago salmon (20).

A comparative pharmacokinetic study was conducted in rainbow trout and African catfish following intravenous and intramuscular administration of oxytetracycline of 60 mg/kg. A two (in case of catfish) or three compartment model was the best fit for the plasma concentration-time curves. The extrapolated zero time concentration obtained in trout was 8-fold higher than that of African catfish. A significant difference with respect to the relatively large apparent distribution volume (Vd_{area}) after intravenous administration was observed. This difference may have resulted from the differences in accumulation and/or tissue compositions (16).

Sulfisozole/trimethoprim residues were examined in goldfish, yellowtail and eel following oral administration with 50 mg/kg for 7 days (10 days for goldfish). Though the trimethoprim concentration in the kidney of goldfish had fallen to 0.05 µg/ml 5 days after administration at 25°C it was detected at 40 days in yellowtails at temperatures between 25 and 28°C and at 30 days for eels at 25°C (12).

Biliary metabolites of oxolinic acid were compared in seven fish species after oral administration. Only oxolinic acid was detected in the unconjugated form in all species, whereas there were some differences in the conjugated fractions among the freshwater fish (rainbow trout, Japanese eel, tilapia) and the seawater fish (red sea bream, Japanese mackerel, yellowtail, Japanese flounder) (17). It was suggested that the efficacy and the toxicity of chemotherapeutics used for cultured fishes were influenced by metabolic differences in fish species.

Fish can live in a relatively wide temperature range. The pharmacokinetics of most of fish species may be mainly temperature-dependent. Droy *et al.* and Grondel *et al.* suggested that the differences in pharmacokinetic behaviour among fish species may be due to differences in tissue composition, the type of muscle and variability in interspecies physiology and metabolism (6, 13, 15).

- <u>Food</u>

Marked differences were observed with respect to the bioavailability of oxytetracycline in carp following intramuscular and oral administration with a dose of 60 mg/kg. The bioavailability following intramuscular and oral administration were approximately 80% and 0.6%, respectively. The slow and incomplete absorption following oral administrations may be due to a tendency of tetracyclines to bind and form complexes with some substances in the alimentary tract. It may be useful to avoid food interaction by a change in the food composition or by increments of oxytetracycline in the diet (15).

- <u>Other factors</u>

Pharmacokinetic variations depending upon body weight were studied in rainbow trout. The decrease in the concentration of oxolinic acid in the liver and muscle was slower in rainbow trout weighing 10 g than in those of 160 g. Oxolinic acid disappeared within 13 days from the liver of rainbow trout weighing 160 g, whereas it was 19 days in 10 g fish. Thus the withdrawal period was longer in small than in large fish (19).

The bioavailability of a drug is influenced by its own characteristics such as solubility, diameter of crystals and chemical form. The bioavailability of oxolinic acid in red sea bream and yellowtail is enhanced by smaller particles. The pharmacokinetic parameters in yellowtail were compared between ultra-fine (1.0 µm in diameter) and ordinary (6.4 µm in diameter) oxolinic acid after a single oral administration of 30 mg/kg. These two preparations required almost the same time to reach the peak serum concentration, but the peak concentration and the area under serum concentration-time curve (AUC) of the ultra-fine type was 1.7 and 1.3 times higher than those of the ordinary type, respectively. The bioavailability of oxolinic acid in red sea bream was enhanced by micronization as observed with yellowtails. The enhanced bioavailability of oxolinic acid by micronization is probably due to the increase in absorption (8, 9).

The bioavailability of sulfadimethoxine varied with the chemical forms. A higher bioavailability at a dose of 20 mg/kg was demonstrated in the rainbow trout fed with the sodium salt (63%) rather than with the free base (34%). The bioavailability of the sodium salt decreased with an increase in the dose (21). On the other hand, both forms of sulfadimethoxine administered to channel catfishes had almost equal bioavailabilities (34.1% vs 31.4%) (29).

Pharmacokinetic parameters derived from intravenous elimination curves in rainbow trout differed between single and multiple administrations of sulfadimethoxine. The half-lives of the terminal elimination phase for the single and multiple administrations were 16.9 and 36.4 h, respectively. An alteration in the metabolic profile or elimination route by the multiple dose administration had probably occurred (21). On the other hand, no differences in the metabolism between single and multiple-dose administrations were noted in catfish (29).

2. Bath administration

The effect of pH on the absorption of sulfonamides in water was examined in rainbow trout. The absorption from fresh water of sulfanilamide (pKa 10.4) and sulfadimidine (pKa 7.4) at pH 4, 6, 8 and 10 over a period of 96 h was studied. The uptake of sulfonamide was significantly higher at pH 6 than at pH 10, and the increase in the absorption due to the alteration of pH was moderate (2).

Lo *et al.*, using several pharmacokinetic models to predict the accumulation of sulfonamide by fish, found that the extent of accumulation of sulfonamides (sulfapyridine, sulfisomidine and sulfadimethoxine) had two pH-independent regions, with the low pH region having the greater accumulation (22).

The pH effects of bathing water on the uptake of oxolinic acid by goldfish were examined. When the concentration of oxolinic acid in bathing water was 10 μg/ml, the equilibrium concentrations calculated using a compartment model in goldfish were 74.7, 23.4 and 11.5 μg/g at pH 6.0, 6.9 and 7.7, respectively. Thus, the accumulation of oxolinic acid was greater at pH below the pKa(6.9) of this compound. The absorption of these chemotherapeutics from fresh water depend on their pKa values and the pH of the bathing water (10). Thus, the uptake of xenobiotics such as weak electrolytes is affected by the degree of ionization.

The uptake of xenobiotics by fish depends on the salt concentrations of bathing water. Bergsjo *et al.* investigated the salt effects by exposing rainbow trouts to 650 mg/L sulfadimidine or 400 mg/L sulfanilamide in either fresh water or salt water. For both sulfonamides, the uptake was higher in salt water than in fresh water. The authors suggested that the salt effect could be partly explained by supplementary absorption from the gut (2). On the other hand, the uptake of oxolinic acid from bathing water by ayu was not affected by the presence of 0.85% NaCl, but was affected by 12.5% and 25% artificial seawater containing about 0.4% and 0.8% NaCl, respectively. The uptake of oxolinic acid from bathing water by ayu might be inhibited by calcium and/or magnesium ions in artificial sea water (11).

The temperature, pH and the hardness of water of the bath influenced the uptake of flumequine by brown trout. A progressive decline in the uptake occurred with an increase in the pH, *i.e.* the uptake decreased from 35 μg/ml at pH 7.0 to 5 μg/ml at pH 8 in the 100 ppm bath test. The relationship between serum levels of flumequine and temperature was approximately linear between 3 and 15°C. A decrease in flumequine absorption with increasing levels of calcium was pronounced at calcium levels to 75 mg/l $CaCO_3$ (24).

Since most of chemotherapeutics used for fish cultures are either weakly acidic or weakly basic, it is likely that the pH of water used for aquacultures influences their elimination. In fact, the uptake of chemotherapeutics from bathing water is influenced by the pH of the water. The excretion of chemotherapeutics from the gills has been little investigated despite its importance.

Pharmacokinetics of chemotherapeutants in shellfish

Corliss investigated the accumulation and depletion of oxytetracycline given in a formulated feed to juvenile white shrimp at concentrations ranging from 1,000 to 10,000 mg/kg of feed for 3 weeks. Oxytetracycline was detected at 24 h after the administration of 5,000 and 10,000 mg/kg doses, and at 48 h after 10,000 mg/kg. He recommended that the dose to protect against vibriosis was 5,000 mg/kg for 14 days and that the withdrawal period required 10 days (4).

The pharmacokinetics of intrapericardially administered sulfadimethoxine in American lobster has been analyzed in detail using a two-compartment model. The sulfadimethoxine disposition in the haemolymph was best described by a two compartment model, and neither the administered dose nor the sex of the lobster had any effects on the pharmacokinetic parameters. Total body clearance (Clb), apparent steady-state volume of distribution (Vd_{ss}) and half-life were 13.8 ml/h/kg, 1.369 ml/kg and 76.7 h, respectively, and the binding of sulfadimethoxine to haemolymph was 53.5%. Lobsters exhibited somewhat smaller Clb, larger Vd_{ss} and longer biological half-life compared with mammalian species. The shell was an important site of distribution of sulfadimethoxine in lobsters (1). The shell is very important to pharmacokinetics in crustaceans, since the ecdysis of the shell may be closely related to the elimination of xenobiotics.

Conclusion

Fish and shellfish cultures can not proceed without the use of chemotherapeutics. Recently, the pharmacokinetics in fish and shellfish have been analyzed using compartment models, and their characteristics have been clarified by comparing the pharmacokinetic parameters between these species and mammals. In addition to oral and bathing administration, intravenous, intraperitoneous and intramuscular administrations are also used in comparative pharmacokinetic investigations on fish and shellfish. The bioavailability is calculated by comparing the pharmacokinetics between intravenous and the other administrations. The best administration method for a regimen and the proper withdrawal period for chemotherapeutics can be determined by examining the pharmacokinetics, which are thereafter applied for planning preparation of chemotherapeutics used for fish cultures.

Although bacterial infections in domesticated animals and fowls is successfully reduced by the use of chemotherapeutics, antiseptics and vaccines, control of infection in cultured fish and shellfish is far from perfect. The pharmacokinetics in fish and shellfish are affected by several factors such as water temperature and animal species. In addition, it is altered by administration techniques. These factors influence the efficacy, safety and residue of chemotherapeutics. Investigations with respect to pharmacokinetics of chemotherapeutics and the continuing application of their results will be necessary for the safe and effective use of chemotherapeutics in fish and shellfish farming.

References

1). BARRON M.G., GETUDIS C. & JAMES M.O., 1988. Pharmacokinetics of sulphadimethoxine in the lobster, *Homarus americanus*, following intrapericardial administration. *Xenobiotica*, **18**, 269-276.

2). BERGJSØ T. & BERGJSØ H.T., 1978. Absorption from water as an alternative method for the administration of sulphonamides to rainbow trout, *Salmo gairdneri*. *Act. Vet. Scand.*, **19**. 102-109.

3). BJÖRKLUND H. & BYLUND G.,1990. Temperature-related absorption and excretion of oxytetracycline in rainbow trout (*Salmo gairdneri* R.). *Aquaculture*, **84**, 363- 372.

4). CORLISS J. P., 1979. Accumulation and depletion of oxytetracycline in juvenile white shrimp (*Penaeus setiferus*). *Aquaculture*, **16**, 1-6.

5). DROY B F , TATE T , LECH J J & KLEINOW KM, 1989. Influence of ormetoprim on the bioavailability, distribution, and pharmacokinetics of sulfadimethoxine in rainbow trout (*Oncorhynchus mykiss*). *Comp. Biochem. Physiol.*, **94**, 303-307.

6). DROY B F, GOODRICH M S, LECH J J & KLEINOW K M, 1990. Bioavailability, disposition and pharmacokinetics of ^{14}C-ormetoprim in rainbow trout (*Salmo gairdneri*). *Xenobiotica* **20**, 147-157.

7). ELLIS A.E., ROBERTS R.J. & TYTLER P., 1978. The anatomy and physiology of teleosts. in Roberts R.J. ed., *Fish Pathology*, Baillière Tindall, London, pp. 13-54.

8). ENDO T., ONOZAWA M., HAMAGUCHI M. & KUSUDA R.,1987. Bioavailability of ultra-fine preparation of oxolinic acid in red sea breams. *Nippon Suisan Gakkaishi*, **53**, 1493.

9). ENDO T., ONOZAWA M., HAMAGUCHI M. & KUSUDA R., 1987. Enhanced bioavailability of oxolinic acid by ultra-fine size reduction in yellowtail. *Nippon Suisan Gakkaishi*, **53**, 1711-1716.

10). ENDO T. & ONOZAWA M., 1987. Effects of pH and temperature on the uptake of oxolinic acid by goldfish. *Nippon suisan Gakkaishi*, **53**, 551-555.

11). ENDO T. & ONOZAWA M., 1987. Effects of bath salinity and number of fish on the uptake of oxolinic acid by ayu. *Nippon Suisan Gakkaishi*, **53**, 557-562.

12). FUJIHARA Y., KANO T. & FUKUI H., 1984. Sulfisozole-trimethoprim as a chemotherapeutic agent for bacterial infections in yellowtail and eel. *Gyobyo Kenkyu*, **19**, 35-44.

13). GRONDEL J.L., NOUWS J.F.M. & HAENEN O.L.M., 1986. Fish and antibiotics; Pharmacokinetics of sulphadimidine in carp (*Cyprinus carpio*). *Vet. Immunol. Immunopathol.*, **12**, 281-286.

14). GRONDEL J.L., NOUWS J.F.M. & VAN MUISWINKEL W.B., 1987. The influence of antibiotics on the immune system; immuno-pharmacokinetic investigations on the primary anti-SRBC response in carp, *Cyprinus carpio* L.; after oxytetracycline injection. *J. Fish Dis.*, **10**, 35-43.

15). GRONDEL J.L., NOUWS J.F.M., DE JONG M., SCHUTTE A.R. & DRIESSENS F., 1987. Pharmacokinetics and tissue distribution of oxytetracycline in carp, *Cyprinus carpio* L., following different routes of administration. *J. Fish Dis.*, **10**, 153-163.

16). GRONDEL J.L., NOUWS J.F.M., DE JONG M., SCHUTTE A.R. & DRIESSENS F., 1989. Comparative pharmacokinetics of oxytetracycline in rainbow trout (*Salmo gairdneri*) and African catfish (*Clarias gariepinus*). *J. Vet. Pharmacol Therap.*, **12**, 157-162.

17). ISHIDA N., 1990. Comparison of biliary metabolites of oxolinic acid in seven species of teleost. *Nippon Suisan Gakkaishi*, **56**, 55-59.

18). JACOBSEN M.D., 1989. Withdrawal times of freshwater rainbow trout, *Salmo gairdneri* Richardson, after treatment with oxolinic acid, oxytetracycline and trimetoprim. *J. Fish Dis.*, **12**, 29-36.

19). KASUGA Y., SUGITANI A., YAMADA F., ARAI M. & MORIKAWA S., 1984. Oxolinic acid residues in tissues of cultured rainbow trout and ayu fish. *Shokuhin Eiseigaku Zasshi*, **25**, 512-516.

20). KATAE H., KOUNO K., MORIKAWA S., TASHIRO F., TAKAHASHI S., MATUMARU Y., USHIYAMA M., MIYAZAKI T. & KUBOTA S.S., 1979. Studies on chemotherapy of fish diseases with piromidic acid - I. Absorption, distribution, excretion and toxicity of piromidic acid in some species of fishes. *Gyobyo Kenkyu*, **14**, 79-91.

21). KLEINOW K.M. & LECH J.J., 1988. A review of the pharmacokinetics and metabolism of sulfadimethoxine in the rainbow trout (*Salmo gairdneri*). *Vet. Hum. Toxicol.*, **30** (Supplement 1), 26-30.

22). LO I-H & HAYTON W.L, 1981. Effects of pH on the accumulation of sulfonamides by fish. *J. Pharmacokinet. Biopharm.*, **9**, 443-459.

23). NOUWS J.F.M., GRONDEL J.L., SCHUTTE A.R. & LAURENSEN J., 1988. Pharmacokinetics of ciprofloxacin in carp, African catfish and rainbow trout. *Vet. Quart.*, **10**, 211-216.

24). O'GRADY P., MOLONEY M. & SMITH P.R., 1988. Bath administration of the quinoline antibiotic flumequine to brown trout *Salmo trutta* and Atlantic salmon *S. salar*. *Dis. Aquat. Org.*, **4**, 27-33.

25). OIDA T., KOUNO K., KATAE H., NAKAMURA S., SEKINE Y. & HASHIMOTO M., 1982. Levels of piromidic acid and its metabolites in eel tissues. *Bull. Jpn. Soc. Socie. Fish.*, **48**, 1599-1608.

26). SALTE R., 1982. Oxytetracycline residues in rainbow trout (*Salmo gairdneri*) fed a commercial medicated feed. *Acta Vet. Scand.*, **23**, 150-152.

27). SALTE R. & LIESTØL K., 1983. Drug withdrawal from farmed fish: depletion of oxytetracycline, sulfadiazine, and trimethoprim from muscular tissue of rainbow trout (*Salmo gairdneri*). *Acta Vet. Scand.*, **24**, 418-430.

28). SOHLBERG S., CZERWINSKA K., RASMUSSEN K. & SOLI N.E., 1990. Plasma concentrations of flumequine after intra-arterial and oral administration to rainbow trout (*Salmo gairdneri*) exposed to low water temperatures. *Aquaculture*, **84**, 355-361.

29). SQUIBB K.S., MICHEL M.F., ZELIKOFF J.T. & O'CONNOR J., 1988. Sulfadimethoxine pharmacokinetics and metabolism in the channel catfish (*Ictalurus punctatus*). *Vet. Hum. Toxicol.*, **30** (Supplement 1), 31-35.

30). UENO R., HORIGUCHI Y. & KUBOTA S.S., 1988. Levels of oxolinic acid in cultured yellowtail after oral administration. *Nippon Suisan Gakkaishi*, **54**, 479-484.

31). UENO R., OKAMURA M., HORIGUCHI Y. & KUBOTA S.S., 1988. Levels of oxolinic acid in cultured rainbow trout and amago salmon after oral administration. *Nippon Suisan Gakkaishi*, **54**, 485-489.

Discussion

ALDERMAN I would like to thank Dr Endo for a thoroughly competent review of a difficult subject.

TREVES BROWN The normal procedure for administering antibacterial drugs to sick fish is to dose every day for several days, a procedure adopted from that traditionally used with warm blooded animals. Perhaps it might be just as correct to give larger doses say twice a week?

ENDO I have no data on such multiple applications.....the pharmacokinetics of drugs in fish is still in its early stages and the pharmacokinetics of multiple doses is difficult to analyse.

ALDERMAN You and other speakers have mentioned the problems of uptake of drugs such as oxytetracycline as a result of possible complexing with for example magnesium and other salts in the diet. Do you have any thoughts on a possible route by which the diet might be modified to mitigate that effect?

ENDO I am afraid I have no good ideas there.

THOMEN You referred to a study by Jacobsen, a comparative study. That study used two different qualities of oxolinic acid. If you go into that study you might be able to interpret differences in excretion time as a function of the quality of the product. Are you aware of any similar studies?

ENDO I am not aware of any evidence in that area.

FIJAN Mr Chairman, may I ask your opinion as a microbiologist whether we should consider also the influence of temperature on the replication rate of bacteria which are being treated as well as on the metabolism of the fish receiving the drug?

ALDERMAN Temperature plays a major role in everything, drug uptake, excretion and the metabolism of the fish and obviously also with the pathogen and the development of the disease. So yes, but you are then adding yet another complication; let us first try and understand how drugs behave in fish, then look at MICs, serum levels etc.

ENDO Fish temperature is also important in the area of metabolites.

EVALUATION DES RESIDUS DE MEDICAMENTS CHEZ LES POISSONS: LE ROLE DES ETUDES METABOLIQUES

J.P. Cravedi

INRA, Laboratoire des Xénobiotiques,
BP 3, 31931 Toulouse Cédex, France

Résumé : *Comme pour les autres espèces d'élevage, l'utilisation de médicaments en pisciculture peut entraîner la présence de résidus dans la chair des poissons. Avant 1975, l'analyse de ces résidus concernait essentiellement la substance initiale puisque les produits de biotransformation étaient prétendus sans risque toxicologique. Au cours des 15 dernières années cette affirmation s'est révélée fausse pour plusieurs produits chimiques. L'évaluation des résidus de médicaments est donc à présent basée non seulement sur le médicament de départ, mais également sur tous les composés qui en sont issus (métabolites libres ou conjugués, et résidus liés de façon covalente aux macromolécules.*

Les aspects multiformes des résidus sont décrits, explicités et illustrés par des exemples. Ceux-ci montrent en quoi les données sur le métabolisme jouent un rôle déterminant dans l'évaluation toxicologique des résidus chez le poisson et soulignent comment ces études permettent la mise en évidence de biomarqueurs, de tissus-cibles, ou de résidus-liés, pour aboutir finalement à une meilleure détermination du temps de retrait d'un médicament.

Summary: **Assessment of drug residues in fish: the role of metabolic studies**

As in other food-producing animals, the use of chemotherapeutants in fish farming has the potential to leave residues in edible tissues. Before 1975, the analysis of these residues were primarily concerned with the parent drug since it was thought that biotransformation products were without toxicological concern. Over the last 15 years, this assumption was proven to be false for many chemicals. Consequently, the assessment of drug residues in animals are now not only based on residues of the parent drug but also on all compounds derived from it (free metabolites, conjugated metabolites and residues covalently bound to macromolecules).

The multiform aspects of residues are described and subsequently clarified and illustrated by examples. These examples demonstrate the relevant points for which metabolism data play a key role in developing information on the safety assessment of drug residues in fish. This includes the helpfulness of a metabolic study in the selection of marker residue or a target tissue, as well as to discover bound residues and finally to set a safe withdrawal period.

Introduction

La production piscicole mondiale est actuellement proche de 10 millions de tonnes par an et l'on estime qu'elle pourrait atteindre celle de la pêche en l'an 2000, c'est-à-dire environ 75 millions de tonnes. Parallèlement au développement de la pisciculture intensive, l'apparition et la propagation très rapide des maladies des poissons ont rendu nécessaire l'utilisation d'antibiotiques et de fongicides dans les élevages.

Comme pour toute espèce animale d'élevage, l'administration au poisson de ces substances actives peut engendrer la présence de résidus dans les tissus. Il apparait donc indispensable de s'assurer de l'absence de risque pour l'homme à consommer les animaux traités.

La prise de conscience de ce problème de sécurité alimentaire lié à l'utilisation en élevage d'additifs et de médicaments, date des années 1960-1970, et a eu pour conséquence la mise en place d'organismes réglementaires et de structures de contrôle destinés à prévenir les risques et à protéger la santé publique. Durant cette période, le concept de résidu était limité à la substance originellement administrée présente dans les produits animaux. L'animal était considéré comme un filtre biologique à l'encontre des composés chimiques et il était classiquement admis que les produits de métabolisation de la substance active offraient un bien moindre risque toxicologique. Il a été montré depuis que si la plupart des métabolites étaient sans aucun doute moins toxiques que le produit de départ, il existait de nombreuses exceptions à la règle (4). Aujourd'hui on entend par résidu non seulement la substance chimique originale, mais également tous les composés qui en proviennent : métabolites libres, conjugués, ou encore liés de façon covalente aux macromolécules. L'élargissement de ce concept à des formes différentes du médicament initial n'a été possible qu'avec l'aide des traceurs radioactifs (^{3}H ou ^{14}C). Cet élément méthodologique ainsi que les progrès considérables accomplis en chimie analytique et notamment en chromatographie et en spectrométrie de masse ont été déterminants pour la recherche, l'isolement puis l'identification des résidus souvent présents à l'état de traces dans les tissus. La notion de résidu est par conséquent intimement liée aux limites des techniques analytiques mises en oeuvre pour le rechercher.

Cette présentation a pour objectif d'appréhender les différents aspects de la génèse des résidus chez le poisson en démontrant qu'une bonne connaissance du métabolisme de la substance étudiée est essentielle dans toute tentative d'analyse ou d'évaluation toxicologique. Pour illustrer nos propos et en raison du faible nombre d'études de métabolisme concernant le médicament chez le poisson, nous avons été amenés à choisir quelques exemples parmi des xénobiotiques autres que ceux utilisés en thérapeutique. Conceptuellement en effet, la notion de résidu ne diffère pas lorsque l'on passe d'un médicament à des contaminants de l'environnement tels que les pesticides, les hydrocarbures ou les PCB. Dans les deux cas, il s'agit d'entités chimiques capables de pénétrer dans l'animal et d'y persister plus ou moins longtemps en fonction de ses capacités métaboliques.

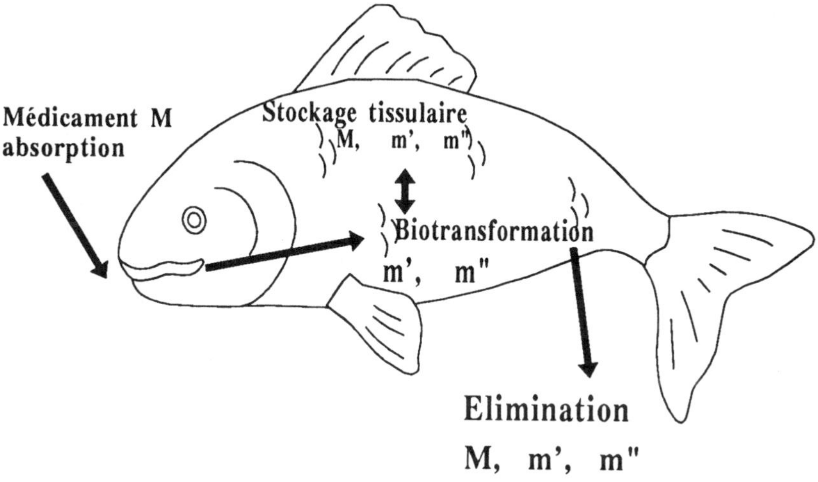

Figure 1: Destinée d'un médicament chez le poisson. m', m" sont des métabolites du médicament M.

Le devenir des médicaments chez le poisson

Bien que les premiers travaux faisant état des capacités de biotransformation des poissons datent de 1958-1960, il n'était pas rare, jusqu'en 1974, de lire dans certains articles de synthèse que cette espèce animale était démunie d'enzymes du métabolisme des xénobiotiques, l'élimination des agents exogènes se faisant par simple diffusion au travers des branchies (7). Depuis, il a été bien établi que les poissons effectuent de nombreuses biotransformations par les systèmes liés au cytochrome P450 ou par les enzymes de conjugaison.

La figure 1 schématise la destinée d'un médicament chez le poisson. Celui-ci est absorbé au travers du tractus digestif lorsque l'administration se fait par la voie alimentaire. Il pénètre par les branchies dans le cas de traitements balnéatoires. Après avoir pénétré dans la circulation sanguine par l'une de ces deux voies, ou encore par injection directe, le médicament atteint les différents tissus de l'organisme qui pourront constituer des lieux de stockage ou de biotransformation. Certains produits chimiques lipophiles tels les PCB sont stockés sans élimination apparente pendant plus de 100 jours chez la truite (13). Dans la plupart des cas néanmoins, le médicament subit au niveau du foie notamment, un certain nombre de biotransformations. Les métabolites ainsi qu'une partie de la substance originellement administrée sont excrétés hors de l'animal par les voies urinaire, branchiale et fécale (cette dernière comprenant l'élimination biliaire).

Les réactions de biotransformation chez le poisson sont multiples et comparables à celles décrites chez le mammifère. Des oxydations (hydroxylations, déalkylations), des hydrolyses, des réactions de conjugaisons (acétylations, glucuronidations, sulfatations, liaisons aux acides aminés ou au glutathion) ont été mises en évidence *in vitro* et *in vivo* chez de nombreuses espèces (12). Ces étapes sont bien souvent nécessaires à l'élimination d'un médicament surtout si celui-ci est liposoluble et par conséquent peu apte a être excrété sans modification dans un milieu aqueux tel que la bile, l'urine ou l'eau environnante. L'on conçoit donc aisément que les capacités d'un animal à métaboliser une substance constituent, avec les caractéristiques physico-chimiques de cette dernière les principaux facteurs déterminant sa demi-vie et donc sa concentration dans les tissus. Pour illustrer cet aspect, Kapoor et ses collaborateurs (11) ont montré que chez *Gambusia affinis*, le facteur de bioconcentration atteignait 84500 pour le DDT [2-bis (p-chlorophenyl)-1, 1, 1-tri-chloroéthane] alors qu'il n'était que de 5,5 pour une molécule très proche et de liposolubilité équivalente, le 2-bis (p-méthylthiophényl) 1, 1, 1-trichloroéthane. Cette différence est due au fait qu'à l'inverse du DDT, le second composé est promptement métabolisé par *Gambusia affinis*.

Apport des études de métabolisme dans la stratégie de contrôle des résidus

L'évaluation toxicologique des résidus, telle qu'elle a été définie par la Food and Drug Administration aux USA nécessite en premier lieu l'étude du métabolisme du médicament chez l'espèce cible (18). Cette étude constitue un élément principal des dossiers de demande d'autorisation d'emploi de ces substances dans la plupart des pays. Ayant déterminé la nature et la teneur des métabolites présents dans les tissus, ainsi que leur vitesse d'élimination, il devient possible de définir les conditions de bonne utilisation d'un médicament en élevage. Il convient en particulier de tenir compte, dans l'évaluation du temps de retrait, de la présence de métabolites ayant une signification au plan toxicologique.

Bien que faisant partie des contaminants et non des médicaments, l'aflatoxine sera prise comme exemple car elle illustre parfaitement ces aspects du problème. Cette mycotoxine est un des plus puissants hépatocancérogènes connus et la truite y est particulièrement sensible : après 12 mois d'un régime contaminé par 4 ppb d'aflatoxine B_1, 50 % des truites sont atteintes d'un cancer du foie (8). L'aflatoxicol, le métabolite le plus abondant de l'aflatoxine B_1 chez la truite, possède une activité carcinogène proche de cette dernière (1) et une demi-vie hépatique plus longue (16). Ces données font ressortir la nécessité d'inclure l'aflatoxicol dans tout programme de surveillance et d'évaluation des résidus d'aflatoxine chez la truite. Il apparait au travers de cet exemple une première notion importante du concept de résidu : l'identification des biomarqueurs d'une chimiothérapie. Il est en effet irréaliste en matière de stratégie de contrôle de vouloir analyser tous les métabolites tissulaires. Le choix se porte donc sur un composé (le biomarqueur) dont la présence dans les tissus témoigne du contact de l'animal avec le médicament, l'additif ou le contaminant. Chez le carassin, Inoue *et al.* (10) ont montré que lors d'un traitement aux sulfamides, la forme libre était éliminée plus rapidement que le métabolite acétylé. Dans ce cas il pourrait être envisagé d'utiliser la forme acétylée comme biomarqueur de ce type de traitement à condition de vérifier que les espèces cibles c'est-à-dire d'élevage, se comportent comme le carassin.

La majorité des médicaments utilisés en aquaculture disparaissent rapidement de l'organisme après traitement et il n'existe pas véritablement de tissu cible, c'est-à-dire de tissu dans lequel le médicament ou ses métabolites s'accumulent préférentiellement. Néanmoins, certaines substances échappent à cette règle. Outre le cas bien connu des tétracyclines en général et de l'oxytétracycline en particulier, qui se fixent de façon permanente au tissu osseux ce qui permet d'utiliser ces molécules pour marquer certaines populations de poissons (17), il a été rapporté chez la truite quelques exemples intéressants.

Douze jours après administration orale de Bacitracine, la concentration résiduelle dans le rein postérieur était respectivement 7 et 56 fois supérieure à celle mesurée dans le foie et le muscle (9). Le trimethoprim ou l'ormetoprim persistent très longtemps dans la peau des truites traitées par ces antibactériens. Droy *et al.* (6) ont montré que la demi-vie de l'ormetoprim dans la peau était de 99 jours. Le trimethoprim, en plus de son affinité pour la peau, se concentre et persiste de façon importante dans le tractus uvéal (2). Notons que cette dernière observation aurait probablement échappé à toute investigation menée avec d'autres techniques que l'autoradiographie, soulignant une fois de plus l'étroite association entre l'analyse des résidus et les moyens méthodologiques mis en oeuvre pour y parvenir.

Le développement des techniques analytiques et des moyens d'investigation a pour effet de repousser sans cesse les limites maximales admissibles pour les résidus de médicaments. Il permet aussi d'ouvrir des perspectives nouvelles sur l'étude de la génèse des résidus. Un exemple de ce dernier aspect est apporté par l'analyse des résidus de chloramphénicol chez la truite. Nous avons analysé les résidus hépatiques de chloramphénicol 24 h après administration *per os* d'une dose unique de ^{3}H-chloramphénicol (50 mg/kg). Les méthodes d'extraction et d'analyse des métabolites ont été décrites précédemment, Cravedi et al. (5). La mesure des résidus totaux, basée sur la quantité de radioactivité présente dans le foie au temps 24 h indique une concentration de 32 ppm d'équivalent chloramphénicol (Cravedi, résultats non publiés). Une analyse plus fine révèle la nature et la proportion des résidus hépatiques : le chloramphénicol inchangé ne représente que 3 % de la radioactivité totale, 1 % est dû aux métabolites libres, 8% aux métabolites conjugués, et le reste soit 88 % est inextractible aux solvants organiques. Seul un traitement par HCl 6N à reflux permet de libérer la totalité des résidus. Leur analyse indique alors qu'il s'agit de chloramphénicol-base, c'est-à-dire d'un produit d'hydrolyse du chloramphénicol ou de ses métabolites. L'acide chlorhydrique concentré a donc démasqué ces résidus tissulaires mais n'a pas préservé leur structure chimique. L'identification de ces résidus, liés de façon covalente aux macromolécules endogènes présente pourtant un intérêt évident dans l'estimation du risque qu'ils représentent. Ils font actuellement l'objet d'un programme de recherche au laboratoire qui visera à déterminer leur structure et a trouver les moyens de les rendre accessibles à l'analyse.

Conclusion

Il apparait au travers de ces exemples que la notion de résidu est parfois difficile à appréhender et qu'elle a beaucoup évolué depuis quelques années. De ce fait, les études qui ont pu être menées chez le poisson, espèce d'élevage de faible poids économique, sont peu nombreuses et souvent incomplètes, surtout en ce qui concerne les molécules génériques. Il est aujourd'hui parfaitement établi que les espèces aquatiques possèdent un équipement enzymatique capable de biotransformer les médicaments en de multiples métabolites . Pour un médicament donné, le profil métabolique est généralement voisin de celui que l'on peut obtenir chez les mammifères. Cette similitude n'est cependant pas suffisante pour que l'on puisse extrapoler des données obtenues chez le rat, le porc ou les bovins, à des poissons. Plus que les autres ces derniers sont sensibles aux paramètres environnementaux susceptibles de modifier les voies et les taux de métabolisation. Il convient naturellement d'en tenir compte dans toute étude visant à définir la cinétique d'élimination des résidus. Le principal de ces facteurs reste la température d'élevage. Ainsi la demi-vie hépatique de l'oxytétracycline chez la truite passe de 4,2 j. pour une température de 10°C à 9,5 j. pour 5°C, Björklund & Bylund (3). Cet effet a conduit certains auteurs à déconseiller l'utilisation de médicaments tels que l'oxytétracycline, la sulfadiazine ou le trimethoprim à des températures inférieures à 10°C, les temps de retrait pouvant dans ces conditions dépasser 100 jours, Salte & Liestøl (15).

Il existe un obstacle supplémentaire à l'estimation des taux résiduels d'un médicament en élevage piscicole : la diversité des espèces concernées. Si les données dont nous disposons actuellement sont loin d'être complètes en ce qui concerne la truite, elles sont bien souvent inexistantes pour d'autres espèces d'élevage. Que connait-on du devenir des antibiotiques chez le brochet, l'esturgeon ou le bar ? Quelques études de métabolisme comparé ont montré qu'il n'était pas toujours possible d'extrapoler des résultats obtenus chez une espèce à d'autres espèces parfois voisines. Pour reprendre un exemple déjà utilisé ici, l'accumulation et la persistance d'ormetoprim dans la peau ont été mises en évidence chez la truite, mais n'ont pas été observées chez le poisson-chat, Plakas *et al.* (14). Il semble donc nécessaire, compte-tenu de la diversification progressive des productions piscicoles, d'élargir à d'autres poissons que le "modèle truite" l'étude de la génèse des résidus de médicaments. La prise en compte des différents éléments qui ont été développés devrait permettre d'affiner les modalités d'évaluation toxicologique des résidus médicamenteux chez le poisson et de conserver à celui-ci l'image d'"aliment-santé" dont il bénéficie auprès du consommateur.

Références

1). BAILEY G.S., HENDRICKS J.D., NIXON J.E. & PAWLOWSKI N.E., 1984. The sensitivity of rainbow trout and other fish to carcinogens. *Drug Met. Rev.*, **15**, 725-750.

2). BERGSJØ T., NAFSTAD I. & INGEBRIGTSEN K., 1979. The distribution of 35S-sulfadiazine and 14C-trimethoprim in rainbow trout, *Salmo gairdneri.Acta Vet.Scand.*, **20**, 25-37.

3). BJÖRKLUND H. & BYLUND G., 1990. Temperature-related absorption and excretion of oxytetracycline in rainbow trout (*Salmo gairdneri* R.). *Aquaculture*, **84**, 363-372.

4) BUHLER D.R. & WILLIAMS D.E., 1988. The role of biotransformation in the toxicity of chemicals. *Aquat. Toxicol.*, **11**, 19-28.

5). CRAVEDI J.P., HEUILLET G., PELERAN J.C. & WAL J.M., 1985. Disposition and metabolism of chloramphenicol in trout. *Xenobiotica*, **15**,115-121.

6). DROY B.F., GOODRICH M.S., LECH J.J. & KLEINOW K.M., 1990. Bioavailability, disposition and pharmacokinetics of ^{14}C-ormetoprim in rainbow trout (*Salmo gairdneri*). *Xenobiotica*, **20**, 147-157.

7). GOLDSTEIN A., ARONOW L. & KAPLAN S., 1974. *Principles of drug action, the basis of pharmacology*. Wiley, New-York, 283-284.

8). HENDRICKS J.D., 1982. Chemical carcinogenesis in fish. In *Aquatic Toxicology*, L.R. Weber, ed.. Raven Press, New York, pp.149-211.

9). HØY T., HORSBERG T.E., NAFSTAD I. & BERGE G.N., 1989. The absorption and distribution of ^{14}C-bacitracin in rainbow trout (*Salmo gairdneri*) after a single oral dose, as observed by whole body autoradiography. *Pharmacol. Toxicol.*, **64**, 262-265.

10). INOUE S., YOSHIDA B. & IMAI S., 1966. Studies on sulfonamides concentration in fishes. I - Absorption of sulfonamides by goldfish from solution and their toxicity to the fish. *Bull. Jpn Soc. Sci. Fish*, **32**, 573-578.

11). KAPOOR I.P., METCALF R.L., HIRWE A.S., COATS J.R. & KHALSA M.S., 1973. Structure activity correlations of biodegradability of DDT analogs. *J. Agric. Food Chem.*, **21**, 310-315.

12). LECH J.J. & VODICNIK M.J., 1985. Biotransformation. In *Fundamentals of aquatic toxicology*, G.M. Rand & S.R. Petrocelli, ed., Hemisphere Publishing Corporation, Washington, pp.526-557.

13). NIIMI A.J. & OLIVER B.G., 1983. Biological half-lives of polychlorinated biphenyl (PCB) congeners in whole fish and muscle of rainbow trout (*Salmo gairdneri*). *Can. J. Fish Aquat. Sci.*, **40**, 1388-1394.

14). PLAKAS S.M., DICKEY R.W., BARRON M.G. & GUARINO A.M., 1990. Tissue distribution and renal excretion of ormetoprim after intravascular and oral administration in the channel catfish (*Ictalurus punctatus*). *Can. J. Fish. Aquat. Sci.*, **47**, 766-771.

15). SALTE R. & LIESTØL K., 1983. Drug withdrawal from farmed fish - Depletion of oxytetracycline, sulfadiazine and trimethoprim from muscular tissue of rainbow trout (*Salmo gairdneri*). *Acta Vet. Scand.* **24**, 418-430.

16). SHELTON D.W., GOEGER D.E., HENDRICKS J.D. & BAILEY G.S., 1986. Mechanisms of anti-carcinogenesis : the distribution and metabolism of aflatoxin B1 in rainbow trout fed Arochlor 1254, *Carcinogenesis*, **7**, 1065-1071.

17). WEBER D. & RIDGWAY G.J., 1967. Marking Pacific salmon with tetracycline antibiotics. *J. Fish Res. Bd. Can.*, **24**, 849-865.

18). WEBER N.E., 1983. - Metabolism and kinetics in the regulation of animal drugs. *J. Anim. Sci.*, **56**, 244-250.

Discussion

DUNIER Avez-vous une idée du devenir de l'oxytétracycline? Avez-vous fait des analyses au niveau des biotransformations, connaissez-vous les produits de dégradation et les organes-cibles pour l'oxytétracycline?

CRAVEDI Pour les organes-cibles, un certain nombre de travaux ont montré que l'oxytétracycline a une bonne affinité pour le calcium et qu'on en retrouve des traces importantes au niveau des os, des cartilages, et des écailles des poissons. C'est d'ailleurs une technique utilisée pour marquer les populations de poissons. En ce qui concerne les biotransformations, il n'y a aucune donnée sur le poisson à ma connaissance, mais il faut savoir que l'oxytétracycline n'est pratiquement pas métabolisée chez les mammifères. Avec le risque qu'on peut prendre en extrapolant, on pourrait penser que chez les poissons également la métabolisation de ce genre de composés est inexistante. Mais on n'a pas d'information précise.

DUNIER Et vous n'avez pas fait d'analyse au niveau du rein?..D'analyse de résidus?

CRAVEDI: Pas du tout.

DUNIER: La raison de cette question, c'est que l'oxytétracycline a été montrée immunosuppressive. M. Boon, de Hollande, devait faire un exposé hier. Il était absent, mais ç'est démontré depuis pas mal d'années. Je voulais donc savoir s'il existe une possibilité d'avoir une cible préfèrentielle au niveau du rein ou des organes lymphoïdes.

CRAVEDI: Je n'ai pas d'information. Je n'ai moi-même jamais vraiment travaillé sur les tétracyclines en dehors des problèmes d'absorption digestive de l'oxytétracycline.

PHARMACOKINETICS IN FARMED ATLANTIC SALMON

S.O. Hustvedt & R.Salte

AKVAFORSK (Institute of Aquaculture Research) P.O.Box 10
N-1432 Ås-NLH; Norway

Summary: *Pharmacokinetics of oxolinic acid, i.e. absorption, distribution, and elimination in Atlantic salmon (*Salmo salar *L.) are presented. These data are compared with those scarce data that are available in the literature.*

A cost-effective experimental design is also presented which has proved its value in pharmacokinetic experiments in designing effective treatments, and in obtaining data which are required for drug licensing both in Norway and in the EEC.

Résumé **: Pharmacocinétique chez le saumon atlantique d'élevage.**

Les thérapeutiques les plus employées chez le saumon mettent en jeu des quinolones et des organophosphorés. Dans le cas des agents antimicrobiens on admet une relation entre les concentrations tissulaires recherchées et les concentrations sanguines. C'est à l'évolution de ces dernières qu'on se réfère pour guider les traitements et choisir les présentations, les doses, les fréquences et les voies d'administration. De même, la pharmacocinétique est la clé de toutes les études sur l'efficacité, les résidus, et les effets sur l'environnement.

Grâce à une technique de canulation de l'aorte dorsale on a suivi les effets de l'administration d'acide oxolinique par voies orale et intraveineuse, les concentrations tissulaires étant parallèlement contrôlées par autoradiographie. L'analyse statistique des surfaces sous courbes de concentration permettait d'extrapoler les résultats à l'infini. On a ainsi estimé le temps de persistance moyen de l'antibiotique, sa demi-vie, sa clairance totale, son volume de distribution, sa constante d'absorbtion, et sa biodisponibilité.

La constante d'absorbtion détermine les concentrations maximales et le temps requis pour les atteindre. Pour l'acide oxolinique, une concentration de 1 à 3 µg/ml est atteinte en 9-20 h (8-10 h avec une présentation enrobée de type Aqualets). La biodisponibilité en cas de prise unique est de 20-25 % (valeur faible, bien que supérieure à celle de l'oxytétracycline), mais peut être améliorée par micronisation. La liaison aux protéines plasmatiques est faible (39 %) et la distribution tissulaire rapide. Enfin la pharmacocinétique de l'acide oxolinique ne paraît pas dépendre de la dose, la demi-vie variant de 51 à 78 h en milieu marin. Chez la truite cette demi-vie est 2 fois plus courte et impose des administrations plus rapprochées. Ces résultats ont permis de proposer pour le traitement des saumons une distribution de 15 mg/kg d'acide oxolinique pendant 3 jours consécutifs, puis toutes les 48 heures.

Il existe un besoin croissant de thérapeutiques nouvelles en aquaculture, et le modèle d'analyse présenté peut s'appliquer aux nombreuses études nécessaires à l'obtention des autorisations. Jusqu'à présent l'emploi des quinolones s'est généralisé dans l'ignorance de leur pharmacocinétique à basse température et de leurs effets pour les espèces non visées et pour l'environnement. D'autres composés prometteurs, comme l'enrofloxacine, la difloxacine, la sarafloxacine et les bromopyrroles, n'échapperont pas à des contrôles plus rigoureux. Enfin, l'amélioration génétique des poissons, qui affecte à différents degrés certains aspects de leur métabolisme, imposera sans doute pour sa part une évaluation continue des systèmes de délivrance thérapeutique.

Introduction

The ultimate objective of a given treatment is to control the disease in question, and the most commonly utilized drugs in Atlantic salmon (*Samo salar* L.) are the quinolones and the organophosphates. For bacterial infections this objective can be achieved by maintaining an adequate concentration of the antimicrobial agent at the site of infection. Usually a correlation between blood concentration of the drug and the concentration at the site of infection is assumed, and thereby a correlation with effect. Thus, blood drug concentration-time relationships in fish may be used to design the dose, dosage form, dose frequencies, and route of administration of a given drug.

Studies of the pharmacokinetics, the efficacy, residues, and the environmental fate of drugs which are of potential value in aquaculture are also required for drug licensing. The need for studies of future compounds to be used in aquaculture is, however, not met. In this presentation we focus on how this need can be met, using oxolinic acid (OA) as a model substance.

Methods

Study design

Pharmacokinetic studies in fish can be performed as cross-over studies where repeated blood samples from individual resting fish are taken via a cannula in the dorsal aorta. The fish are anaesthetized in tricaine methane sulphonate (MS-222), whereafter a cannula made of polyethylene tubing PE-50 (Clay Adams Intramedic®, Parsippany, New Jersey) is inserted into the dorsal aorta. The cannulation procedure should be performed as described by Soivio *et al.* (15,16), with the exception that each fish should be given an oral anticoagulant, i.e. aspirin, into the stomach together with heparin through the cannula as described by Salte and Hustvedt (14), to ensure a long-term functional cannula. The method is a reliable means of obtaining the pharmacokinetic data which are required for drug approval and drug therapy. In the cross-over study each fish are given both an intravascular injection, and after a washout period, an oral dosage of the drug. Rate and degree of absorption, distribution as well as rate of elimination of the drug can then be calculated. In addition withdrawal periods can be estimated (13), for later verification through clinical trials.

Methods such as autoradiography may, when properly designed, be useful in investigating tissue-selective metabolic processes, but does not *per se* give information about the chemical identity of a labelled compound. Such a method may, however be a valuable supplementation to pharmacokinetic studies.

Pharmacokinetic analysis

The following parameters are discussed:

- the area under the serum concentration versus time curve (AUC), calculated by means of the trapezoidal rule from time zero to the time of the concentration at the last sampling time or C' ($AUC_{0-c'}$) and extrapolated to infinity ($AUC_{0-\infty}$) according to the formula :

$$AUC_{0-\infty} = AUC_{0-c'} + C'/\beta$$

where ß is the final slope of the straight line of elimination in a two-compartment model calculated by means of nonlinear regression.

- the area under the first moment of drug concentration curve from time zero to infinity (AUMC), extrapolated to infinity according to the formula:

$$AUMC_{0-\infty} = AUMC_{0-c'} + t'C'/\beta + C'/(\beta)^2$$

where t' is the time of the last point which is measured;

- the mean residence time (MRT) evaluated according to the formula:

$$MRT = AUMC_{0-\infty}/AUC_{0-\infty}$$

- the terminal serum elimination half-life ($t_{1/2}$):

$$t_{1/2} = 0.693/\beta$$

- the total systemic clearance (Cl_T):

$$Cl_T = dose_{i.v.}/AUC_{0-\infty}$$

- the volume of distribution at steady state (V_{dss}):

$$V_{dss} = dose_{i.v.} * MRT/AUC_{0-\infty}$$

- the absorption rate constant (k_a), the maximum serum concentration (C_{max}), and the time to reach C_{max} (T_{max}) that are all calculated by means of nonlinear regression.

- the bioavailability (F) of an extravascular (e.v.) dose, as determined after oral or intraperitoneal injection:

$$F = AUC_{e.v.} * dose_{i.v.}/AUC_{i.v.} * dose_{e.v.}$$

Discussion

Absorption

The absorption rate constant is a major determinant of maximum drug concentration as well as the time required to reach this concentration. An early occurring concentration maximum can reasonably be expected to cause a more rapid onset of action. Peak serum concentrations (C_{max}) of OA incorporated into medicated feed occur 9-20 h after oral administration (10,11). Corresponding time with OA-containing Aqualets[P] were found to be 8-10 h (18). An oral dose of 10-20 mg/kg body weight results in a C_{max} of 1-3 µg/ml (10). Feed delays the time to reach peak serum concentrations (T_{max}) without affecting the absolute value of C_{max} or the area under the curve (AUC) (10). The estimated bioavailability (F) of drugs in Atlantic salmon is generally low; after a single oral dose of OA F was found to be 20-25 % (10,11). When comparing F for various drugs the ranking is: flumequin > OA >> oxytetracycline (18) (further pharmacokinetics of flumequin and OA in Aqualets® (Apothekernes Laboratorium A.S.) are to be published in near future, A. Rogstad, personal communication). Both first-pass metabolism and mal-absorption may account for the poor bioavailability of OA in fish; when "micronized" particles were used, F was shown to be as high as 40% (Hustvedt, unpublished data), which is in correspondence with data from other fish species (4).

Distribution

OA seems to exhibit low protein binding of about 39 % to serum proteins (2). An extensive apparent volume of the central compartment indicates that OA is initially rapidly distributed to tissues outside of the blood (11). This result is similar to that seen for ^{14}C-ormetoprim in rainbow trout (3). Furthermore, OA becomes extensively distributed in body fluids and tissues both in Atlantic salmon and rainbow trout, as is evidenced by a volume of distribution (V_{dss}) which exceeds total body water (9,11). V_{dss} of OA were estimated to be lower in Atlantic salmon than in rainbow trout. Variations in this parameter have been reported by several workers (3,6) depending both on type of drug, and fish species. An extensive distribution is also seen in autoradiography studies after bath administration of diflubenzuron (8).

Elimination

OA seems to exhibit dose-independent linear pharmacokinetics (11). Terminal elimination half-life is sufficiently long to permit dosing every 48 h as $t_{1/2}$ is 51-78 h in Atlantic salmon (11) kept in seawater. Corresponding figures for rainbow trout was found to be 20-43 h (9). The same study showed that Cl_T was more than twice as high in rainbow trout kept in seawater than in Atlantic salmon. Consequently, OA has to be administered with shorter dose intervals to rainbow trout.

Results which are presented in a case report (12), clearly demonstrate that dosage regimens can be designed to maintain concentrations within the therapeutic window, provided the pharmacokinetic parameters of a given compound is known. In this study the therapeutic schedule was based on previous estimates of pharmacokinetic parameters of OA (see above). According to these estimates the terminal elimination half-life would permit dosing every 48 h, and an oral dose of 15 mg per kg body weight incorporated into medicated feed should be sufficient to achieve an average serum concentration of 3.4 µg OA per ml; results showed that OA serum levels did range from 2.5-5.8 µg per ml from day 2 until day 10 after initiation of treatment, thus supporting this notion. To achieve this serum concentration as quickly as possible, the medicated feed had initially to be given for three consecutive days (loading interval) followed by maintenance intervals of 48 h.

The efficacy of a given treatment should depend on the minimum inhibitory concentration (MIC) of the drug at the site of infection and not on the frequency of dosing. The drug level at the site of infection should exceed the MIC during the major part of the dosage intervals. The peripheral levels are dependent both upon the site and the extensiveness of the infection. What levels of average serum concentrations to be chosen are obviously open to discussion. With the same total dose, however, "pulse dosing" is assumed to give higher peripheral concentrations than those achieved by "single doses".

The above results demonstrate that both application of pharmacokinetic principles, and studies of effects are needed to achieve optimal antimicrobial therapy in fish.

Concluding remarks

From our point of view, there is a registration crisis in todays commercial aquaculture. Several of the most serious fish pathogens have developed resistance to most of the few therapeutic products that are available. Thus, there is a pressing need for data on the pharmacokinetics, withdrawal periods, metabolism, and environmental fate of new compounds of potential value in aquaculture. It is important to note that variation in both V_{dss} and Cl_T are to be expected depending both on type of drug, fish species, water temperature and salinity. Moreover, one should stress the need for obtaining such data before the compounds are included into the therapeutic spectra.

The experimental design which is described above has proved its value in pharmacokinetic studies and in designing treatments under fish farming conditions. The mathematical characterization of the pharmacokinetic processes permits quantitative prediction of amounts and concentrations of a drug in the fish as a function of time and dosing regimen. In addition, the experimental design is adequate in obtaining data which are required for drug licensing both in Norway and in the EEC.

As regards the quinolones (and any other drug which is presently in use in commercial aquaculture for that matter), it is important to notice that neither mutagenicity, nor the carcinogenicity and toxicity to non-target organisms have been elucidated. Some other fundamental questions pertaining to the quinolones also remain unanswered: how do they actually work?, how do bacteria develop resistance to quinolones?, how (if at all) are they broken down in the environment? In our opinion, information about how drugs are broken down or concentrated through the food chains should have been elucidated before they were permitted for use in aquaculture. It is also a fact that we do not have sufficient pharmacokinetic data of the quinolones at low sea-water temperatures, i.e. below 5°C. This notwithstanding the quinolones OA and flumequine have demonstrated the advantages of this class of compounds. Other quinolones which have been identified as potentially useful therapeutic agents against bacterial diseases of fish are enrofloxacine (7), difloxacin and sarafloxacin (17). Now, attention is also drawn to another class of chemicals, i.e. the bromopyrroles which show pronounced activity against a wide range of Gram-negative and Gram-positive bacterial fish pathogens (1).

Finally, genetic improvements through fish breeding programs will affect growth rates together with various aspects of metabolism (5). This calls for a continuous evaluation of drug delivery systems, feeding practice, dose levels and intervals.

References

1). AUSTIN B., 1989. Developments in chemotherapy. In: *Diseases of fish and shellfish.* IV EAFP international conference. Santiago, Spain: 7.

2). BENJAMINSEN, T. & HUSTVEDT S.O., 1986. *Oxolinsyre til laksefisk; analysemetode og farmakokinetikk.* M.sc.thesis, Institute of Pharmacy, University of Oslo, 87 pp.

3). DROY, B.F., GOODRICH M.S., LECH J.J. & KLEINOW K.M., 1990. Bioavailability, disposition and pharmacokinetics of ^{14}C-ormetoprim in rainbow trout (*Salmo gairdneri*). *Xenobiotica*, **20**, 147-157.

4). ENDO T., ONOZAWA M., HAMAGUCHI M. & KUSUDA R., 1987. Enhanced bioavailability of oxolinic acid by ultra-fine size reduction in Yellowtail. *Nippon Suisan Gakkashi*, **53**, 1711-1716.

5). GJEDREM T., 1983. Genetic variation in quantitative traits and selective breeding in fish and shellfish. *Aquaculture*, **33**, 51-72.

6). GRONDELl J.L., NOUWS J.F.M., DE JONG M., SCHUTT E R. & DREISSENS F., 1987. Pharmacokinetics and tissue distribution of oxytetracycline in carp, (*Cyprinus carpio* L.), following different routes of administration. *J. Fish Dis.*, **10**, 153-163.

7). HORSBERG T.E. & HØY T., 1989 Medikamenter til fisk. Farmakologi og farmakoterapi. *Norsk Vet. Tidsskr.*, **101**, 497-510.

8). HORSBERG T.E. & HØY T.T., 1989. Diflubenzuron orally against salmon lice - clinically effects and pharmacokinetic properties. *Diseases of fish and shellfish*, IV EAFP international conference, Santiago, Spain, 119.

9). HUSTVEDT S.O. & SALTE R., 1991. Distribution and elimination of oxolinic acid in rainbow trout (*Oncorhynchus mykiss* Walbaum) after a single rapid intravascular injection. *Aquaculture*, **92**, 297-303.

10). HUSTVEDT S.O., SALTE R, KVENDSTO. & VASSVIK V. 1991 Bioavailability of oxolinic acid in Atlantic salmon (*Salmo salar* L.) with medicated feed. *Aquaculture*, submitted.

11). HUSTVEDT S.O., SALTE R., & VASSVIK V., 1991. Absorption, distribution and elimination of oxolinic acid in Atlantic salmon (*Salmo salar* L.) after various routes of administration. *Aquaculture*, **95**, 193-199..

12). HUSTVEDT S.O., SALTE R. & VASSVIKV., 1991. Combatting Cold-water vibriosis in Atlantic salmon (*Salmo salar* L.) with oxolinic acid: a case report. *Aquaculture*, submitted.

13). SALTE R. & LEISTØLI K., 1983. Drug withdrawal from farmed fish. *Acta vet. scand.*, **24**, 418-430.

14). SALTE R. & HUSTVEDT S.O., 1991. Pretreatment with oral anticoagulant ensure a long-term functional dorsal aorta cannula. (as manuscript)

15). SOIVIO A., WESTMAN. & NYHOLM K., 1972. Improved method of dorsal aorta catheterization: haematological effects followed for three weeks in rainbow trout (*Salmo gairdneri*). *Finnish Fish Res.*, **1**, 11-21.

16). SOIVIO A., NYHOLM K. & WESTMAN K., 1975. A technique for repeated sampling of the blood of individual resting fish. *J. exp. Biol.*, **62**, 207-217.

17). STAMM J.M., 1989. *In vitro* resistance by fish pathogens to aquaculture antibacterials. Including the quinolones Difloxacin (A-56619) and Sarafloxacin (A-56620). *J. Aquatic Animal Health*, **1**, 135-141.

18). SYVERTSEN C., ROGSTAD A. & BERGE G., 1989. Aqualets, a drug delivery system for farmed fish. *Diseases of fish and shellfish*, IV EAFP international conference, Santiago, Spain, p201.

Discussion

ALDERMAN The data presented in this paper perhaps give us an indication of the way in which we may begin to think about using loading doses and non continuous treatments to reduce the amount of drug used to treat fish and thereby perhaps reduce the impact on the environment.

WITHNELL I was wondering what speed the fish was swimming at in the cannula experiment.

HUSTVEDT In the cannula experiment the fish were in receptacles with a very low swimming speed. The water flow was increased to get the fish moving, but a very low swimming speed

WITHNELL Do you think that this might have influenced the rate of excretion?

HUSTVEDT Yes, but the data from the cannulated experiment are found to be very similar to those found in case reports from disease cases.

THOMEN I would like again to refer to the Jacobsen study and ask if you have compared different qualities of oxolinic acid, especially different levels of impurities?

HUSTVEDT No, the only study of different forms is the one that I have shown you with the micronised oxolinic acid.

PHARMACOKINETICS OF ANTIMICROBIALS IN SOME FRESH WATER FISH SPECIES

J.F.M. Nouws*, J.L. Grondel, J.H. Boon**
& V.J.Th. van Ginneken**,*****

* Farma Research B.V., Snelliusstraat 6, 6533 NV Nijmegen, The Netherlands
** Agricultural University, Zodiac, Wageningen, The Netherlands
*** Gorlaeus Laboratories, State University of Leiden, The Netherlands

Summary: *The pharmacokinetics of antimicrobial agents in fresh water fish were studied following intravenous, intramuscular, oral and intraperitoneal administration. Large species differences and a marked effect of temperature were demonstrated with respect to elimination half-lives, drug distribution and metabolism of antimicrobials. Sulphatroxazole, sulphadiazine and sulphadimidine were slightly metabolized (hydroxylated and/or acetylated) in the rainbow trout and/or carp; trimethoprim was not metabolized in the carp. No glucuronide conjugates of sulphonamides or their metabolites, trimethoprim, chloramphenicol and flumequine could be detected in the plasma of fish investigated. The oral bioavailability of oxytetracycline and doxycycline is poor in carp and trout.*

***Résumé*: Pharmacocinétique des antibiotiques chez quelques espèces de poissons d'eau douce.**

L'étude pharmacocinétique des produits thérapeutiques employés chez les poissons devient de plus en plus importante compte-tenu des problèmes croissants rencontrés en pratique et de la crainte des résidus. De grandes différences existent selon les espèces, les caractéristiques chimiques des produits, les conditions d'environnement, et les particularités physiologiques des poissons. Plusieurs études, menées sur des poissons d'eau douce par voies intraveineuse (IV), intramusculaire (IP), intrapéritonéale (IP) ou orale (PO), ont permis d'établir le volume de distribution, la clairance totale, la demie-vie et la biodisponibilité de différents produits. Les tétracyclines étaient dosées par méthode microbiologique, les sulfamides, triméthoprime, chloramphenicol et fluméquine par HPLC.

Chez la truite et la carpe injectées par voie IV la température influence la distribution des tétracyclines et des sulfamides (Tableaux I à III, Fig.1 à 4). La demi-vie est toujours diminuée à température élevée, mais la clairance, diminuée chez la truite, augmente chez la carpe. La voie IM (Tableau IV et Fig.5) permet de retrouver des différences selon l'espèce et la température. La demi-vie apparaît particulièrement longue pour l'oxytétracycline et la fluméquine chez l'anguille. Chez la truite et la carpe respectivement étudiées à 10 et 20°C, on constate dans le 2ème cas une élimination plus rapide du triméthoprime et du chloramphenicol. Lea grande hétérogènéité des situations constatés après injection IP (Tableau VI) conduisent à déconseiller cette voie d'administration.

La voie orale (Tableau V, Fig.7 et 8) montre que l'oxytétracycline et la doxycycline sont assez mal absorbées, les meilleurs taux plasmatiques étant, à l'inverse des mammifères, obtenus avec l'oxytétracycline, tant chez la carpe que chez la truite. Bien que l'absorption apparaisse diphasique chez la carpe, les demi-vies ne diffèrent chez les deux espèces que pour la doxycycline. Le triméthoprime, le chloramphenicol et les sulfamides ont également une bio-disponibilité médiocre, une grande part des produits se trouvant éliminée dans les fécès, qui avec la bile constituent la voie d'excrétion principale en eau douce.

Des métabolites acétylés et hydroxylés sont libérés en quantités mesurables chez la carpe et la truite, tandis que le triméthoprime ne paraît pas métabolisé par la carpe. Enfin on n'a observé aucune évidence de glucurono-conjugaison des substances antibiotiques employées dans cette étude.

Introduction

In recent years due to growing concerns about therapeutic treatments and about the presence of residues, more attention has been given to the pharmacokinetic behaviour of antimicrobials in various fish species (6,14,15). Recent studies (3,5,6,7,8,11) have pointed out that large differences in the pharmacokinetics of drugs exist between fish species. It was found that, besides the physico-chemical characteristics of the drug (lipophilicity, pK_a-value), external factors like water temperature, oxygen concentration, salinity, and food composition, and intrinsical factors like metabolic and excretion capacity of the fish influence the kinetics (6,14). Therefore the above mentioned external and intrinsical factors affect the therapeutic efficacy and withdrawal times of drugs administered to fish (14,15).

For fresh water fish, several experiments were performed by our group in the last 6 years in which the intravenous, intramuscular, oral and intraperitoneal route of drug administration were studied. The IV route of drug administration is the most suitable one for studying the pure pharmacokinetic behaviour of the drug in the body. No drug absorption occurs and the availability of the drug is 100%. Several pharmacokinetic parameters can be obtained like distribution volume, total body clearance, and elimination half-life (1). With these parameters, species differences and effect of environmental factors can be demonstrated. Moreover the area under the curve (UC) can be used for the calculation of drug bioavailability following other routes of administration. This paper deals with selected pharmacokinetic data obtained after intravenous, intramuscular, oral and intraperitoneal drug administration to fresh water fish.

Materials and methods

Details about fish, housing system, food, experimental set-up, intravenous, intramuscular, and intraperitoneal drug administration and blood sampling technique have been published previously (3,5,6,7). For oral drug administration a watery mixture of the drug, food and toluidine blue (as control on the gastro-intestinal reflux) was made and administered orally with a syringe annex stomach tube. In this way comparative oral pharmacokinetic experiments with oxytetracycline (OTC) and doxycycline (DOX) were performed in carp and trout.

The following drugs were used: Sulphadimidine[R]-33% (containing 330 mg sulphadimidine/ml exc.; Aesculaap BV, NL); Diatrim[R] (containing 200 mg sulphadiazine and 40 mg trimethoprim/ml exc.; AUV, NL); Leotrox[R] (containing 200 mg sulphatroxazole and 40 mg trimethoprim/ml exc.; Leo Ltd., DK); Engemycine[R]-5% (containing 50 mg oxytetracycline/ml exc.; Mycofarm BV, NL); Gloveticol[R] (containing 250 mg chloramphenicol/ml exc.; Mycofarm BV, NL); Flumequine 3% (Dopharma BV, NL); oxytetracycline and doxycycline powder (Sigma, US).

Oxytetracycline and doxycycline concentrations were determined microbiologically using *Bacillus mycoides* as test strain (7). Sulphadimidine, sulphatroxazole, sulphadiazine, trimethoprim, chloramphenicol and flumequine were determined with HPLC as previously described (3,5,8,10,12,13). Pharmacokinetic and statistical analysis (two sample t-test) were performed using standard methods (1,9).

Results and discussion

Intravenous route

In Table 1 the mean pharmacokinetic parameters for oxytetracycline in the carp and rainbow trout are summarised (see also Figs 1 and 2). The effect of temperature on the pharmacokinetic parameters is clearly demonstrated. In the trout a shorter elimination half-life and a smaller distribution volume (V_{Darea}) were observed at 19°C compared with those at 10°C. The total body clearance was similar at these two temperatures. In the carp a larger distribution volume, a shorter elimination half-life and higher total body clearance were found at 20°C than 8°C. The tissue concentration ratios reflected the observed alteration in distribution volume at these temperatures (Table 1). Thus the change in pharmacokinetic parameters had an obvious physiological significance.

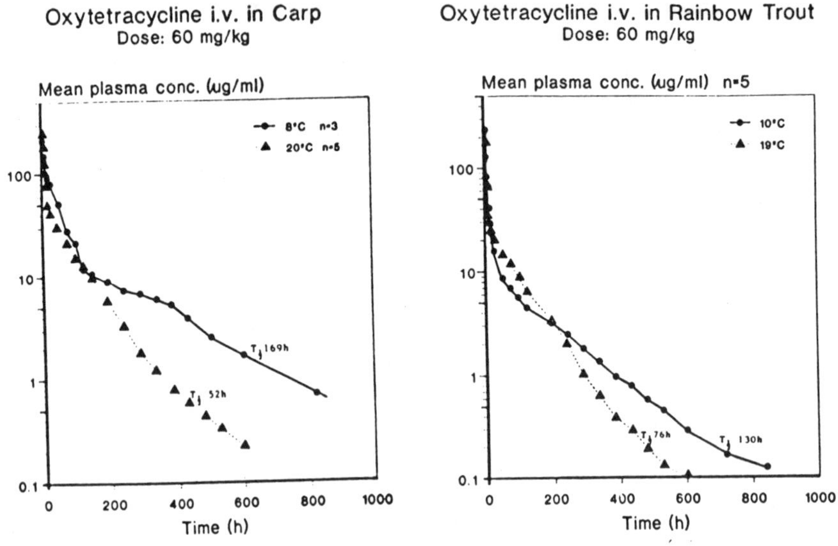

Figs. 1 & 2. Mean plasma concentration-time curves of oxytetracycline, at different temperature levels, in trout (Fig.1) and carp (Fig.2) injected intravenously. ($T_{1/2}$ elimination half-life).

Table 1. Selected mean pharmacokinetic parameters for oxytetracycline (Engemycine[R]-5%) in the carp and rainbow trout (Dose 60 mg/kg intravenously).

Species	Trout		Carp	
	10°C	19°C	8°C	20°C
Number	6	6	6	6
$V_{D area}$, litre/kg	4.0	2.66*	1.49	2.25*
V_1, litre/kg	0.29	0.36*	0.34	0.19*
Cl_B, 1/h/kg	0.0236	0.0245	0.0061	0.011*
B_0, µg/ml	8.5	17.8*	20.9	50*
$T_{1/2}$ β, h	130	76*	169	52*
Concentration ratios:				
Muscle/plasma	0.65	0.37*	0.15	0.81*
Fat/plasma	0.17	0.05*	0.36	x
Liver/plasma	3.1	2.4	1.33	1.92

* A temperature related significant difference ($P<0.05$)

x Not determined

Table 2. Selected mean pharmacokinetic parameters for sulfadimidine in the carp and rainbow trout (Dose 100 mg/kg intravenously).

Species	Trout		Carp	
Temperature	10°C	19°C	10°C	20°C
Number	6	6	6	6
V_{Darea}, litre/k	1.20	0.83*	1.15	0.90*
Cl_B, 1/h/kg	0.041	0.040	0.016	0.025*
B_0, µg/ml	107	168*	81	114*
$T_{1/2}\ \beta$, h	20.6	14.6*	50.3	25.6*
% CH_2OH	-	-	2.40	1.28*
% N_4-SDM	4.1	3.2	4.2	5.6*
C_{max}, µg/ml (N_4-SDM)	1.47	1.63	2.35	2.68
T_{max}, h (N_4-SDM)	24	14	40	24*

- Absent * A temperature related significant difference (P<0.05)

Table 2 shows the effect of temperature on the kinetics and metabolism of sulphadimidine (SDM) in trout and carp (see also ref. 5). The elimination half-life became shorter at the higher temperature in both species. In the trout no alteration in the total body clearance (Cl_B) occurred for SDM (the same feature as for OTC). For sulphadiazine (SDZ) and trimethoprim, an increase of CL_B and a shorter elimination half-life were noticed in the carp with the higher water temperature (Table 3; Figs. 3 and 4), which was also observed for OTC and SDM (Tables 1 and 2).

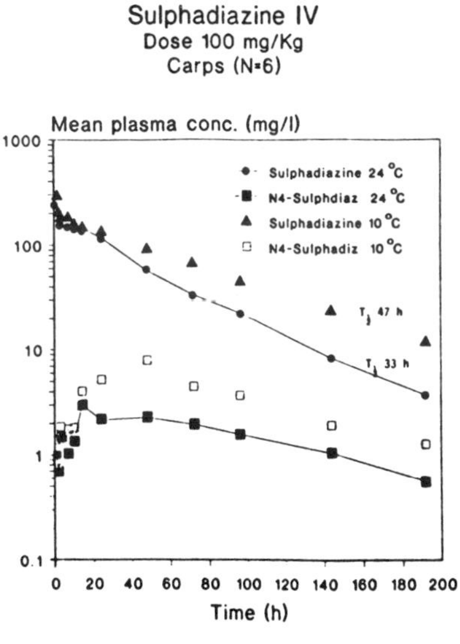

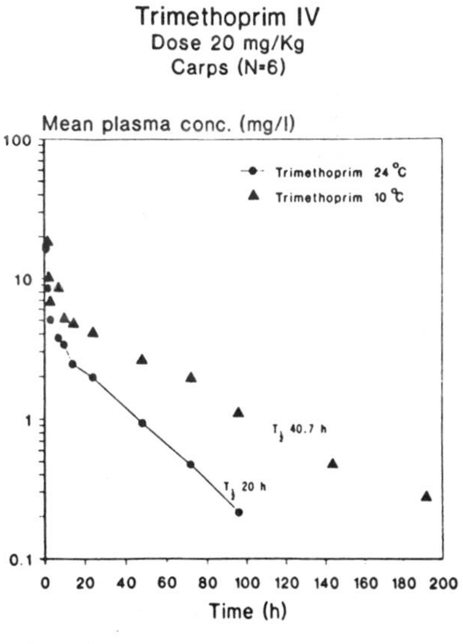

Figs 3 & 4. Mean plasma concentration-time curves of sulphadiazine and its N_4-acetyl metabolite (Fig.3) and of trimethoprim (Fig.4), at two different temperature levels, in carp injected intravenously ($T_{1/2}$ = elimination half-life)

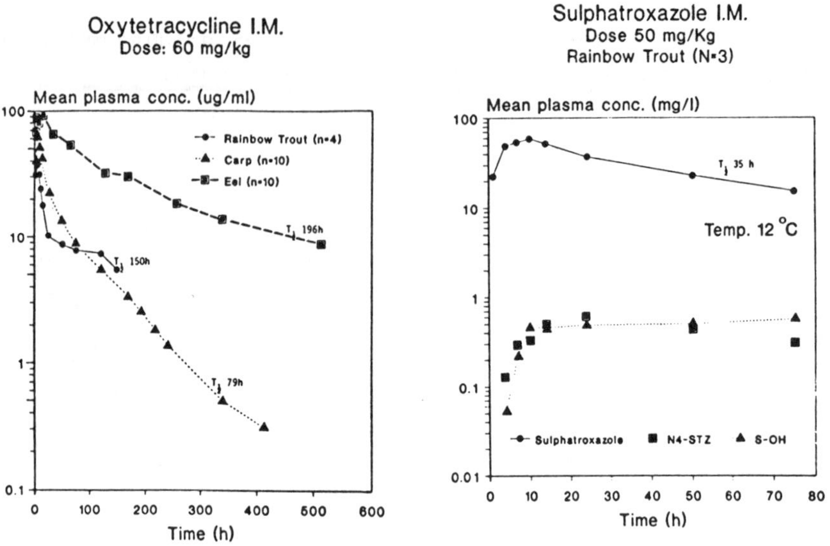

Fig. 5 & 6. Mean plasma concentration-time curves of oxytetracycline in trout, carp and eel, at 10, 20 and 22°C respectively (Fig.5), and of sulphatroxazole and its N_4-acetyl (N_4-STZ) and hydroxy (S-OH) metabolites in trout at 12°C (Fig.6), after single intramuscular injections ($T_{1/2}$= elimination half-life).

Table 3. Effect of temperature on selected mean pharmacokinetic parameters for sulfadiazine and trimethoprim (Diatrim[R]) in the carp.

Antibiotic	Sulphadiazine		Trimethoprim	
dose, mg/kg	100	100	20	20
Number	6	6	6	6
Temperature	10°C	24°C	10°C	24°C
V_{Darea}, litre/kg	0.53	0.60	3.1	4.0
Cl_B, 1/h/kg	0.0079	0.0122*	0.047	0.101*
B_0, µg/ml	190	173	6.5	4.5
$T_{1/2}\beta$ h	47.1	33.0*	40.7	20.0*
% N_4-SDZ(a)	5.1	3.8	x	x
% N_4-SDZ(b)	8.4	10.8	x	x
T_{max}, h (N_4-SDZ)	34	21	x	x
C_{max}, µg/ml (N_4-SDZ)	9.1	3.1*	x	x

x Not relevant

* A temperature related significant difference ($P<0.05$)

(a) Based on AUCs

(b) Based on actual plasma concentrations beyond 72 h p.i.

The percentage of N_4-acetyl metabolite in plasma increased gradually after SDM or SDZ administration. An equilibrium between the parent drug and its N_4-acetyl metabolite was not achieved in the 192 h sampling period (Fig. 3).

The calculated plasma percentage of N_4-acetyl metabolite vs the parent compound based on drug concentrations beyond 72 h p.i. did not alter significantly. However in the carp the plasma percentage of N_4-acetylsulphadimidine (N_4-SDM), based on AUC values, increased slightly at the higher temperature, while the percentage hydroxy metabolites (CH_2OH) decreased. This was not observed in the trout. In the carp no hydroxy metabolites of SDZ (10), no metabolites of TMP (13) and no glucuronides of these metabolites could be detected in plasma.

Intramuscular route

In Table 4 the pharmacokinetic parameters are summarized following IM administration of some drug formulations. Species differences (Fig. 5) and the influence of water temperature on pharmacokinetic parameters of OTC are presented. Most striking are the extremely long elimination half-lives found for OTC and flumequine in the eel. Taking the water temperature into account for TMP and chloramphenicol a shorter elimination half-life was found in the trout (10°C) than for the carp (20°C) (see also Table 3). The glucuronide conjugate of chloramphenicol was not detected in the plasma of the trout or the carp. Sulphatroxazole was acetylated and hydroxylated in the trout to the same extent (2.1 and 4.1 %, respectively; Fig. 6). It is noteworthy that following

Table 4. Selected mean pharmacokinetic parameters following intramuscular administration of antibiotics.

Antibiotic	Species	Number	Temp °C	Dose mg/kg	C_{max} µg/kg	T_{max} h	$T_{1/2}$ h
Oxytetracycline (Engemycin 5%)	Carp	6	8	60	48	8	157
Oxytetracycline (Engemycin 5%)	Carp	10	20	60	57	14	79
Oxytetracycline (Engemycin 5%)	Trout	4	10	60	35	10	150
Oxytetracycline (Engemycin 5%)	Eel	10	22	60	100	16	196
Trimethoprim (Leotrox^R)	Trout	4	12	10	0.7	4	8
Sulphatroxazole (Leotrox^R)	Trout	4	12	50	50	15	35
Chloramphenicol (Gloveticol^R)	Carp	4	21	60	22	4	10.5
Chloramphenicol (Gloveticol^R)	Trout	4	12	60	17	1	10.4
Flumequine (Flumequine-3%)	Eel	5	22	9	9	75	256

intraperitoneal sulphatroxazole administration in the carp (Table 6) the plasma percentage of the N4-acetylsulphatroxazole was 16.6%, while only traces of the hydroxy metabolite could be detected.

With respect to IM administration of drugs, one has to keep in mind that the composition of the formulation may have an extensive influence on drug pharmacokinetics. Large differences may exist between formulations, a feature well-known in mammalian animals.

Oral route

Comparative oral pharmacokinetic experiments with oxytetracycline (OTC) and doxycycline (DOX)) were performed in carp and trout. The following results were obtained (Table 5; Figs. 7 and 8):

a) In both species higher concentrations in plasma were achieved for OTC than DOX (0.4 vs 0.2 µg/ml, respectively), which is the opposite of what is found in mammalian species.

b) Both drugs were poorly absorbed as revealed in low drug concentrations in plasma; the bioavailability of orally administered OTC was low: for the carp 0.38% and for the trout 1.25%.

c) No difference in maximal OTC or DOX concentrations in plasma were observed between carp and trout (see Figs. 7 and 8).

d) The absorption phase of DOX was biphasic in the carp; the reason is unclear.

e) The elimination half-life of DOX differed between trout and carp; for OTC a similar elimination half-life was observed.

Table 5. Selected pharmacokinetic parameters for doxycycline and oxytetracycline after oral application (Dose 60 mg/kg).

Antibiotic	Doxycycline		Oxytetracycline	
Species	Carp	Trout	Carp	Trout
Temperature °C	20	10	20	10
Number of fish	6	6	6	6
$T_{1/2\,abs}$, h	2.0*	5.4	5.1	5.2
C_{max}, µg/ml	0.23	0.18	0.46	0.35
T_{max}, h	16	16	24	48
$T_{1/2}$ h	24.4	36.6	56	50
Bioavailability %	x	x	0.38	1.25

* Biphasic absorption pattern
x Not determined

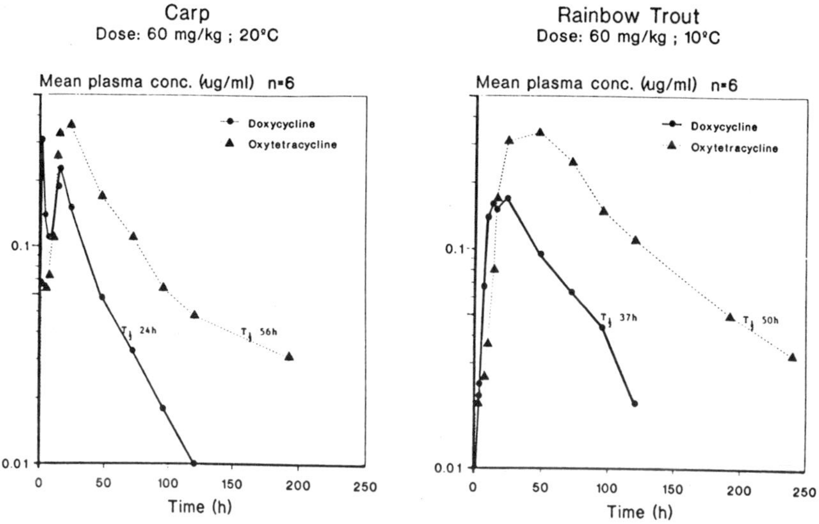

Fig. 7 & 8. Mean plasma concentration-time curves of oxytetracycline and doxycycline in carp (Fig.7) and trout (Fig.8)) after oral administration.

For chloramphenicol a low oral bioavailability (based on AUC values) of about 15% was observed in the carp. Also Cravedi *et al.* (4) demonstrated that only 16% of the orally administered labelled chloramphenicol was excreted by the kidney, and the rest was excreted with the faeces. Besides Bersjø *et al.* (2) demonstrated following oral administration a poor absorption of labelled sulphadiazine and trimethoprim in the trout. Drug excretion in carp and trout seems to occur mainly by the biliary and faecal route, while drug excretion routes by gills, skin and kidney were minor ones. Previously published results (2,4) support the present results indicating that the oral drug bioavailability in fresh water fish might be poor (certainly lower than in mammalian species). An explanation may be that the absorbing area in the gastro-intestinal tract is relatively small (e.g. the carp is stomachless). Besides the pH in the stomach and duodenum is almost neutral which would corroborate presumably the hindrance in the absorption of those drugs which have a low pK_a-value (like tetracyclines). Therefore it is urgent to elucidate the factors influencing oral drug absorption in fish and to initiate the development of those oral products characterised by excellent bioavailibility and palatability.

Intraperitoneal route

The problem with this route of administration is that large variations in plasma drug concentration-time profiles were observed (8). An explanation may be that a part of the dosage is deposited in the gastro-intestinal tract, which may result in poor drug absorption. This phenomenon is observed in particular with OTC formulations. With respect to trimethoprim, and the sulphonamides, the drug concentration-time curves in the carp were more uniform. Table 6 summarizes the experiments performed in the carp. It may be concluded that intraperitoneal route is not advisable.

Table 6. Selected pharmacokinetic parameters of some drugs after intraperitoneal administration to carp.

Drug	Dose (mg/kg)	C_{max} µg/ml	T_{max} (h)	$T_{1/2}$ (h)
Oxytetracycline	60	65	5	36
Trimethoprim	10	4	2	5
Sulphatroxazole	50	58	7	13
Sulphadimidine	60	41	5	13

Conclusions

More basic pharmacokinetic parameters obtained after IV and oral drug administration are needed for rational use of antimicrobials in aquaculture. Moreover improvement of oral drug formulations may be also a way to optimize oral drug therapy.

Acknowledgements: Thanks are due to Mrs. O. Haenen and Mrs. M. de Jong, for their assistance in some of these trials during their stagiare periods at Zodiac (Wageningen) and RVV-District 6 (Nijmegen).

References

1). BAGGOT J.D., 1977. *Principles of drug disposition in domestic animals.* Saunders Co, Philadelphia.

2). BERGSJØ T., NASFED I., INGEBRIGSTE K., 1979. The distribution of ^{35}S-sulfadiazine and ^{14}C-trimethoprim in rainbow trout, *Salmo gairdneri. Acta Vet. Scand.*, **20**, 25-37.

3). BOON J.H., NOUWS J.F.M., HEYDEN M.M.M.T. van, BOOMS G.H.R. and DEGEN M., 1991. Disposition of flumequine in plasma of European eel (*Anguilla anguilla*) after a single intramuscular injection. *J. Fish Diseases*, in press.

4. CRAVEDI J-P, HEUILLET G., PELERAN J-C, WAL J-M., 1985. Disposition and metabolism of chloramphenicol in trout. *Xenobiotica*, **15**, 115-121.

5. GINNEKEN V.J.Th., van, NOUWS J.F.M., GRONDEL JL, DRIESSENS F, DEGEN M., 1991. Pharmacokinetics of sulphadimidine in carp (*Cyprinus carpio* L) and rainbow trout (*Salmo gairdneri* Richardson) acclimated at two different temperature levels. *The Veterinary Quarterly*, in press.

6. GRONDEL J.L.,1986. The immune system as a target for antibiotics. Thesis, Agriculture University of Wageningen.

7. GRONDEL J.L., NOUWS J.F.M., SCHUTTE R., DRIESSENS F., 1989. Comparative pharmacokinetics of oxytetracycline in rainbow trout (*Salmo gairdneri*) and African catfish (*Clarias gariepinus*). *J. Vet. Pharmacol. Therap.*, 12, 157-162.

8. HAENEN O., GRONDEL J.L., NOUWS J.F.M., 1985. Pharmacokinetics of oxytetracycline, trimethoprim, sulphatroxazole and sulphadimidine in the carp. Abstracts 3rd EAVPT congress, Ghent, p.112.

9. MENDENHALL, 1971. *Introduction to probability and statistics*. Wadsworth Publ. Co., Belmont, USA.

10. NOUWS J.F.M., FIRTH E.C., VREE T.B., BRAAKMAN M., 1987. Pharmacokinetics and renal clearance of sulfamethazine, sulfamerazine, and sulfadiazine and their N4-acetyl and hydroxy metabolites in horses. *Am. J. Vet. Res.*, **48**; 392-402.

11. NOUWS J.F.M., GRONDEL J.L., SCHUTTE R., LAURENSEN J., 1988. Pharmacokinetics of ciprofloxacin in carp, African catfish and rainbow trout. *Vet. Quarterly*, **10**, 211-216.

12. NOUWS J.F.M., VREE T.B., MEVIUS D., DEGEN M.,1989. Pharmacokinetics, metabolism and renal clearance of sulphatroxazole in calves and cows. *J. Vet. Pharmacol.Therap.*, 12, 50-57.

13. NOUWS J.F.M., VREE T.B., DEGEN M., MEVIUS D., 1991. Pharmacokinetics of a sulphamethoxazole/trimethoprim formulation in pigs after intravenous administration. *Vet. Quarterly*, in press.

14. RASMUSSEN F., 1988. Therapeutics used in fish production. Pharmacokinetics, residues, withdrawal times. FAO-Report of the Working Party on Withdrawal Period of Drugs, European Inland Fisheries Advisory Commission, XV/88/inf.13.

15. SCHMID, 1980. Arzneimittelrückstände bei Fischen. *Tierärztl. Prax.*, **8**, 237-244.

STUDIES ON CHEMOTHERAPY FOR MAJOR CULTURE FISHES IN JAPAN

K. Inouye*, S. Morikawa & T. Hara*****

* Natl. Res. Inst. Aquaculture, Tamaki, Mie, 519-04, Japan
** Gifu Prefectural Office, Yabuta, Gifu, 500, Japan
*** Res.Dep., The Fisheries Agency, Kasumigaseki, Tokyo, 100, Japan

Summary: *Chemotherapy of fish diseases in Japan, including furunculosis, vibriosis, pseudotuberculosis and streptococcicosis was reviewed. When sulfa drugs were orally administered to rainbow trout, the absorption rate was low and the total absorbed amount was small compared to the case of mammals. Administration of sulfa drugs as sodium salt showed quicker absorption than their free form. When an antibiotic agent was administered orally to yellowtail, absorption was quick and the drug was retained in the tissue for a fairly long time. Sulfa drugs showed the highest effect on furunculosis and vibriosis, followed by antibiotics and synthetic antibacterial agents in the order, while ABPC and OA were effective on pseudotuberculosis, and macrolide antibiotics were effective on streptococcicosis. The therapeutic effect varied depending on the preparation of drug-containing feed. In salmonids, variation in the drug uptake among individuals was larger as the size of pellets became larger. When minced fish meat was used as drug-containing feed in marine fish culture, therapeutic effect was reduced by the dispersion of drugs into water.*

Résumé **: Recherches thérapeutiques chez les principaux poissons élevés au Japon.**

*De nombreuses espèces de poissons ont un intérêt économique au Japon, mais la plupart des études pharmacocinétiques ont été réalisées chez les salmonidés et la sériole (*Plecoglossus altivelis*). Les maladies les plus graves sont en effet, pour les salmonidés la furonculose et les vibrioses causées par des* Vibrio anguillarum *de type 2, pour la sériole la pseudotuberculose à* Pasteurella piscicida *et les streptococcies. Chez les salmonidés les sulfamides distribués par voie orale sont très lentement et relativement mal absorbés, mais l'emploi de sels sodiques en améliore les performances. Ce sont eux qui présentent les effets les plus marqués contre la furonculose et la vibriose. Chez la sériole par contre tous les antibiotiques sont rapidement distribués dans les tissus et y persistent longuement. L'ampicilline et l'acide oxolinique, surtout sous forme micronisée, sont très efficaces contre la pseudotuberculose, tandis que les streptococcies répondent bien aux macrolides.*

Les résultats thérapeutiques dépendent dans une large mesure de la préparation de l'aliment médicamenteux. Les variations individuelles dans la prise alimentaire des salmonidés sont plus marquées lorsqu'on utilise des granulés de grande taille, tandis que le recours à des viandes de poisson émincées favorise les pertes de produit actif par dissolution. Les effets toxiques des traitements restent assez mal connus si l'on excepte des observations consécutives à des surdosages de sulfamides.

Introduction

Major culture fishes in Japan include rainbow trout (*Oncorhynchus mykiss*), Yamame salmon (a landlocked salmon, *0. masou*), Amago salmon (a landlocked salmon, *O.rhodurus*), eel (*Anguilla japonica*), carp (*Cyprinus carpio*) and Ayu (*Plecoglossus altivelis*) as fresh water fishes, and yellowtail (*Seriola quinqueradiata*) and red sea bream (*Pagrus major*) as saltwater fishes.

With the advance of R&D studies on fish drugs, various chemotherapeutic studies such as drug sensitivity, safety, tissue level and clinical test have been performed on culture fishes, especially on salmonids which have a long history of culture. Among marine culture fishes, yellowtail have been studied most extensively because of their high productivity. In this communication, chemotherapy in Japan is reviewed for furunculosis caused by *Aeromonas salmonicida* and vibriosis by *Vibrio anguillarum* biotype 2 in salmonids and for pseudotuberculosis by *Pasteurella piscicida* and streptococcicosis by *Streptococcus* sp. in yellowtail.

1. *In vitro* antibacterial activities of drugs against fish disease bacteria

Aoki *et al.* (1) have measured MIC of drugs against 129 strains of *A. salmonicida* isolated from salmonids at culture farms in various parts of the country from 1979 to 1986 and reported that decreased sensitivity was found in various drugs. Also, they have measured MIC of drugs against 14 strains of *V. anguillarum* biotype 2 isolated at various parts of the country from 1976 to 1985 and reported that some of the strains showed decrease in sensitivity to drugs but the degree of the decrease was rather small compared to the case of *A. salmonicida*. Kitao *et al.* (14) and Kusuda *et al.* (19) have reported that MIC against *P. piscicida* varied depending on the kind of drugs. Takemaru and Kusuda (24) and Aoki *et al.* (1) have reported that *Streptococcus* sp. had high sensitivities to macrolide antibiotics.

During the period of from 1974 to 1986 the present authors have isolated 776 strains of *A. salmonicida* and 309 strains of *V. anguillarum* biotype 2 at various parts in Gifu Prefecture and examined their sensitivities to sulfamonometoxine (SMM), oxytetracycline hydrochloride (OTC), chloramphenicol (CP) and nalidixic acid (NA) by disk method. The sensitivities to these drugs decreased with years but were recovered after discontinuation of the application of these drugs. Some of the strains showed multiple drug resistances.

Takashima *et al.* (23) have examined drug sensitivities of 281 *P. piscicida* strains isolated at various districts during the period of from 1981 to 1983 and reported that 262 of them were R factor-related multiple drug resistant strains, most of which being triple drug resistant strains to CP, tetracycline hydrochloride (TC)and SMM. On the other hand, Kusuda *et al.* (18) examined drug sensitivities of 307 *P. piscicida* strains isolated from 1984 to 1985, and found that 49% of the strains isolated in 1984 and 34% of those isolated in 1985 were resistant against ampicillin (ABPC) and 13% of the 1984 isolates and 25% of the 1985 isolates were resistant against oxolinic acid (OA). According to another examination on 306 strains isolated during the period from 1986 to 1988 (19), 36% and 37% of them were resistant against ABPC and OA, respectively.

2. Tissue level of drugs

Hara and Inoue (4) have examined the tissue level of sulfa drugs in rainbow trout after the oral administration and found that the absorption rate of these drugs was low compared to the case of mammals and drug distribution in each tissue varied depending on the kind of sulfa drugs. They could not find type differences in sulfa groups in the action to rainbow trout such as long acting and short acting types generally observed in mammals. According to Hara *et al.* (5), administration of a sulfa drug as sodium salt showed quicker absorption than its free form.

Tanaka (1968) has examined the tissue level of SMM in yellowtail after oral administration of the drug and found that the blood level reached its maximum 24 hours after the administration and the results were almost the same as those in rainbow trout examined by Hara and Inoue (4). Hatai *et al.* (8) have reported that the tissue level of CP in yellowtail reached its maximum 48 hours after its oral administration and a tissue level of more than 10 times higher than the MIC value was maintained for over 24 hours when CP was administered at a dose of 50mg/kg. In the case of ABPC, its blood and muscular levels reached the maximum 1 hour after the administration and the maximum levels were 0.981 μg/g in the liver, 0.735 μg/g in kidney, 0.358 μg/g in blood and 0.197 μg/g in muscle, respectively (9). In the case of the oral administration of erythromycin (EM), its tissue and blood levels reached their maximum 1 to 3 hours after the administration and the levels in the liver, kidney and spleen were about four times higher than that in blood (12).

Regarding nalidixic acid-based synthetic antibacterial agents, Endo (2,3) orally administered a standard OA preparation and a finely powdered OA preparation to yellowtail and compared tissue levels of OA after the administration of these two preparations. He found that the period required to reach to the maximum level of OA and its biological half life in the serum and each organ tested were almost the same in both cases of the preparations. However, the maximum level in the fish administered finely powdered preparation was 1.3 to 2.0 times higher than that administered the standard one, thus showing the improvement of bioavailability of OA in powdered drug form, which has also been confirmed in red sea bream.

Hara (7) has reported that experimentally obtained effective muscular levels of SMM and sulfadimetoxine (SDM) for the therapy of furunculosis were 26 mg/kg and 29 mg/kg, respectively, and the serum level of SMM was 55 mg/l, which coincided with the generally considered effective blood level of sulfa drugs (50 mg/l) in mammals. In addition, Hara and Inouye (6) have reported that the effective therapeutic dose of OA calculated from the results of experiments using artificially infected furunculosis to rainbow trout was about 2.5 µg/g in the liver, 1.5 µg/g in the kidney and 0.5 µg/g in muscle respectively.

3. Therapeutic effects of drugs on fish diseases

Hara (7) examined therapeutic effects of various antibacterial agents by administering them to rainbow trout juveniles which had previously been inoculated with *A. salmonicida*. When the results were evaluated in terms of ED_{50} (mg/kg/day), synthetic antibacterial agents (NA and OA) showed the highest therapeutic effect followed by antibiotics and sulfa drugs in the order.

According to a study on the therapeutic effect of florfenicol (FF) on yellowtails experimentally infected with *P. piscicida*, the ED_{50} value was 0.82 g/kg/day when FF was administered by mixing with pellets but the therapeutic effect was lower when FF was mixed with minced fish meal due, probably, to dispersion of the drug into the water (25).

In the case of the natural onset of furunculosis in landlocked salmons (Yamame and Amago), the sulfa drug was effective when it was administered at a dose of 100 mg/kg/day for 5 to 7 days: the antibiotic agent was effective at a dose of 10-30 mg/kg/day for 10 days (Hara, 1978); a potentiated sulfa drug was effective at 30 mg/kg/day for 5 days; OA was effective at 10 mg/kg/day for 5 days; NA was effective at 25 mg/kg/day for 5 days (21).

In the case of the natural onset of pseudotuberculosis in yellowtails, ABPC was effective when it was administered at a dose of 12 mg/kg/day for 5 days

(17) and the therapeutic effect of a finely powdered OA preparation was higher than that of the standard OA preparation because of the quick absorption of the former preparation (22).

In the case of the natural onset of streptococcicosis in yellowtails, EM was effective when it was administered at a dose of 25-50 mg/kg/day for 5 days (13), and spiramycin (SP) was effective when administered at a dose of 25 mg/kg/day for 10 days (16).

4. Preparation of drug-containing feed

Because artificial formula feed has been used for salmonids, Hara (7) has prepared SMM-containing pellets of various sizes and measured the muscular level of SMM in rainbow trout after oral administration of each size of the pellets under the same feeding amount. He found that difference in the drug uptake among individuals became larger as the size of pellets was larger, though the maximum muscular level scarcely changed by the pellet size.

For the rearing of a yellowtail which has the largest aquaculture production in Japan, three types of feed preparations are currently used, that is, minced or properly cut fresh or frozen fish, the moist pellet prepared by mixing powdered feed and fish meat and the dry pellet. It is known that the feed made of fresh fish disperses into water at higher rate compared to moist pellets and other feed, thus arising a problem of decreased therapeutic effect of the drug contained in the fresh fish-based feed. Hatai *et al.* (10, 11) have examined the effect of feed types by measuring tissue levels of CP in yellowtail after administration of each CP-containing feed type and found that a feed made of fresh fish showed the lowest effect and the tissue level increased as the amount of the powdered feed added to the fresh fish increased. Therefore, it is recommended to use a binder when the fresh fish-based feed is used.

5. Side effects

Very little has been reported on the side effect of drug administration to culture fishes. Blackened body color and loss in appetite caused by overdose of sulfa drugs are known as examples of such side effects. A vertebral fracture was observed when sulfisomidin (SIM) was administered to yellowtail at a dose of 800 mg/kg (15) In the case of SMM, however, such a fracture did not occur even when it was administered at a dose of 1000 mg/kg, indicating that the fracture was not dose-dependent but related to the kind of sulfa drugs (20).

References

1). AOKI T., TAKESHITA S. & KITAO T.,1983. Antibacterial action of chemotherapeutic agents against non-hemolytic *Streptococcus* sp. isolated from cultured marine fish, yellowtail *Seriola quinqueradiata. Bull. Jap. Soc. Fish.*, **49**(1), 1673-1677.

2). ENDO T., ONOZAWA M., HAMAGUCHI M. & KUSUDA R., 1987a.
Bioavailability of ultra-fine preparation of oxolinic acid in red sea breams. *Nippon Suisan Gakkaishi*, **53**(8), 1493.

3). ENDO T., ONOZAWA M., HAMAGUCHI M. & KUSUDA R., 1987b. Enhanced bioavailability of oxolinic acid by ultra-fine size reduction in yellowtail. *Nippon Suisan Gakkaishi*, **53**(10), 1711-1716.

4). HARA T. & INOUE S., 1967. Concentration of various sulfonamides in the tissues of rainbow trout after oral dosage. *Bull. Jap. Soc. Sci. Fish.*, **33**(7), 618-523.

5). HARA T., INOUE S. & SAITO M., 1967. Concentration of sulfonamides in the tissues of rainbow trout after oral dosage of the sodium salts. *Bull. Jap. Soc. Sci. Fish.*, **33** (7), 624-627.

6). HARA T. & INOUYE K., 1974. Studies on therapy of fish disease with chemotherapeutic agents. Oxolinic acid. Summaries of the spring meeting of the Jap. Soc. Sci. Fish., p.85.

7). HARA T., 1980. Studies on control of furunculosis in salmonids. Ph. D dissertation of Tokyo Univ., Tokyo

8). HATAI K., HIRAKOA T., SAHASHI Y., MATSUSHIMA M., IWAHASI Y., SASAKI T., MIZUNO N., & EGUSA S., 1974. Absorption and excretion of chloramphenicol in fishes -I. Tissue levels of chloramphenicol in cultured yellowtail after oral administration. *Fish Pathol.*, **9**(1), 50-70.

9). HATAI K., IWAHASHI Y. & EGUSA S., 1978. Absorption and excretion of ampicillin in yellowtail. *Fish Pathol.*, **13** (2), 73-78.

10). HATAI K., YASUMOTO S. & YASUNAGA N., 1982. Absorption and excretion of drug in cultured young yellowtail, *Seriola quiqueradiata* -I Effects of dosing method on chloramphenicol levels in the tissues. *The Aquaculture*, **29** (4), 199-210.

11). HATAI K., YASUMOTO S. & YASUNAGA N., 1984. Effects of way of dosing on tissue levels of erythromycin in cultured young yellowtails, *Seriola quinqueradiata. Bull. Nagasaki Prefect. Inst. Fish.*, **10**, 79-83.

12). KATAE H., KOUNO K., SHIMIZU M., KUSUDA R., TANIGUCHI M., SHIOMITSU K. & HASEGAWA H., 1980. Studies on chemotherapy of fish disease with erythromycin -I. Absorption, distribution, excretion and residue of erythromycin in cultured yellowtails. *Fish Pathol.*, **15**(1), 7-16.

13). KATAE H., 1982. Erythromycin - The application to streptococcal infections in yellowtails. *Fish Pathol.*, **17**(1), 77-86.

14). KITAO T., NAKAUCHI R., SAITO R. & TANAKA I., 1989. *In vitro* studies of amoxicillin on antibacterial activity for some fish pathogens. *Fish Pathol.*, **24**(2), 83-88.

15). KUBOTA S., KOJIMA S. & ISHIDA A., 1970. A side-effect of sulfonamides in fish. *Fish Pathol.*, **4**(2), 98-102.

16). KUBOTA S. & MIYAZAKI T., 1980. Studies on toxicity of spiramycin -II. Therapy of streptococcicosis in yellowtail. Summaries of the spring meeting of the Jap. Soc. Sci. Fish., p.45.

17). KUSUDA R. & INOUE K., 1977. Studies on the application of ampicillin for pseudotuberculosis in cultured yellowtails -III. Therapeutic effect of ampicillin on yellowtails artificially and naturally infected with *Pasteurella piscicida. Fish Pathol.*, **12**(1), 7-10.

18). KUSUDA R., ITAOKA M. & KAWAI K., 1988. Drug sensitivity of *Pasteurella piscicida* strains isolated from cultured yellowtail from 1984 to 1985. *Nippon Suisan Gakkaishi*, **54**(9), 1521-1526.

19). KUSUDA R., SUGIURA H. & KAWAI K., 1990. Drug sensitivity of *Pasteurella piscicida* isolated from cultured yellowtail from 1986 to 1988. *Nippon Suisan Gakkaishi*, **56**(2), 239-242.

20). MATUSHIMA M., MATSUBARA S., AI T. & INOUE S., 1971. Toxicity of sulfonamides in yellowtail. *Fish Pathol.*, **5**(2), 107-115.

21). MORIKAWA S., 1980. Studies on the furunculosis in salmonid fishes -XIV. On the chemotherapy and toxicity of sulphadiazine, trimethoprim and A-TP. *Report Gifufx Prefec. Experim. Station.*, **27**, 49-56.

22). TAKAHASHI Y. & ENDO T., 1987. Evaluation of the efficacy of an ultrafine preparation of oxolinic acid in the treatment of pseudotuberculosis in yellowtail. *Nippon Suisan Gakkaishi*, **53**(12), 2175-2162.

23). TAKASIMA N., AOKI T. & KITAO T., 1985. Epidemiological surveillance of drug-resistant strains of *Pasteurella piscicida. Fish pathol.*, **20**(2/3), 209-217.

24). TAKEMARU I. & KUSUDA R., 1988. *In vitro* antibacterial activity of josamycin against *Streptococcus* sp. and various fish pathogenic bacteria. *Nippon Suisan Gakkaishi*, **54**(9), 1527-1531.

25). YASUNAGA N. & TSUKAHARA J., 1988. Dose titration study of florfenicol as a therapeutic agent in naturally occurring pseudotuberculosis. *Fish Pathol.*, **23**(1), 7-13

COMPARISON OF THE BIOAVAILABILITY OF MICRONIZED OXOLINIC ACID IN SEVERAL ANIMALS INCLUDING FISH

T. Endo

Biological Research Laboratory, Tanabe Seiyaku Co. Ltd.,
2-2-50, Kawagishi, Toda, Saitama 335, Japan

Summary: *The bioavailability of micronized oxolinic acid (OA) was evaluated in fish (yellowtail), chickens and dogs (beagles) by comparing pharmacokinetic parameters. In yellowtail after administration of 30 mg/kg, the maximum serum concentration (C_{max}) and area under concentration curve (AUC) of ultra-fine OA (1.0 µm in diameter) was 1.7 times and 1.4 times greater than that of ordinary OA, respectively. In chickens administered with 10 mg/kg, the C_{max} and AUC of ultra-fine OA was 2.0 and 2.9 times greater than that of ordinary OA. In beagles administered with 50 mg/animal, the AUC and the C_{max} of the micronized OA (1.3 µm in diameter) was 2.6 and 2.7 times higher than that of the mechanically pulverized OA (16 µm in diameter).*

Résumé: **Comparaison de la biodisponibilité de l'acide oxolinique sous forme micronisée chez le poisson et d'autres animaux.**

*L'administration orale de composés thérapeutiques chez les poissons est dépendante de nombreuses causes de variations, incluant la température, l'espèce traitée et la présentation du produit. La notion de biodisponibilité joue donc un rôle essentiel dans l'évaluation de l'efficacité des traitements. L'acide oxolinique, très employé en aquaculture, semble présenter des variations importantes du fait de sa faible solubilité dans l'eau. On a donc évalué sa biodisponibilité selon l'administration sous forme classique ou micronisée chez la sériole (*Plecoglossus altivelis*), le poulet, et des chiens beagles. Les paramètres cinétiques étaient évalués en HPLC après administration unique. Bien que des différences sensibles existent selon les espèces traitées, on peut conclure que la micronisation de l'acide oxolinique améliore sensiblement sa biodisponibilité.*

Introduction

Antibiotic therapy continues to play an important role in the control of bacterial diseases in cultured fish. Various factors affect the bioavailability of orally administered drugs in fishes. The concentration of oxolinic acid (OA) in fish body fluids was related to water temperature and fish species (5, 6, 8, 9). In rainbow trout, sulfadimethoxine in a sodium salt form resulted in higher tissue concentrations than sulfadimethoxine in a free form (7).

Recent developments in pharmacokinetics have led to the concept of bioavailability in the evaluation of drugs. For example, there are detailed guidelines for bioequivalency tests of medical drugs in Japan, and the evaluation of drug equivalency is based on the maximum serum concentration of a drug (C_{max}), time (T_{max}) for C_{max} and area under the serum concentration-time curve (AUC) obtained in crossover trials (1).

OA has high antibacterial activity against gram-negative bacteria and certain gram-positive bacteria. In Japan, this compound has been widely used as a chemotherapeutic agent for bacterial infections in domesticated animals, fowls and cultured fishes. However, OA has the disadvantage of use limitations because of sparing solubility in water. Accordingly, finely divided OA with a reduced particle diameter has been developed which assures improved bioavailability in beagles, red sea breams and yellowtail fish (2, 3, 4).This study compared the enhanced bioavailability of micronized OA among yellowtails, chickens and beagles.

Test antibacterials

A 5% suspension of ultra-fine OA with a particle diameter of 1.0 μm and, as a reference control, a commercial OA preparation containing 5% of OA crystals with a particle diameter of 6.4 μm (ordinary OA) were applied in the chicken and yellowtail trials (3). The particle diameters of OA used in the trials using beagles were 0.13, 1.3 and 16 μm (4).

Enhanced bioavailability of oxolinic acid by micronization

Yellowtail

Concentrations of OA in yellowtail serum were assayed by high performance liquid chromatography (HPLC) after a single oral administration of 30 mg/kg. Pharmacokinetic parameters obtained from the serum concentrations are shown in Table I.

Both drugs reached a maximum concentration at 6 h after administration, and the C_{max} of ultra-fine OA (1 μm in diameter) and ordinary OA (6.4 μm in diameter) were 2.38 and 1.39 μg/ml, respectively. Thus, the C_{max} of ultra-fine OA was higher than that of ordinary OA. Thereafter, the serum concentrations of both drugs gradually decreased. The AUC_{0-8} of ultra-fine OA was 11.8 μg.h/ml, which was greater than that (8.7 μg.h/ml) of ordinary OA. These results showed that an enhanced bioavailability was obtained by micronization.

Chicken

OA concentrations in chicken serum were assayed by HPLC after single oral administrations of ultra-fine OA (1.0 µm in diameter), ordinary OA (6.4 µm in diameter) and an aqueous solution of OA at a dose of 10 mg/kg, respectively, and pharmacokinetic parameters obtained from the serum concentrations are shown Table I.

Serum concentrations in this trial were always the highest with ultra-fine OA, followed by aqueous solution then ordinary OA. The C_{max} of ultra-fine and aqueous OA, which were 3.09 and 2.78 µg/ml respectively, were significantly higher ($p<0.001$) than that of ordinary OA, whereas there was no significant difference in the C_{max} between ultra-fine and aqueous OA. The AUC_{0-8} of ultra-fine and the aqueous solution of OA were significantly greater ($p<0.05$) than that of ordinary OA, and the AUC_{0-8} of ultra-fine was significantly greater ($p<0.05$) than that of aqueous solution of OA. Thus, the bioavailability of OA in chickens after a single oral administration depended on the particle diameter of the OA crystals and the liberation rate. Therefore, the bioavailability of OA in chicken was highest in ultra-fine OA, followed by aqueous then ordinary OA.

Table I: Maximum serum concentration (C_{max}), time (T_{max}) for C_{max} and area under the serum concentration-time curve (AUC) for various diameters of OA particles.

Animal	Diameter (µm)	T_{max}(h)	C_{max} (µg/ml)	AUC_{0-t}(1) (µg*h/ml)
Yellowtail	1.0	6.0	2.38	11.8
	6.4	6.0	1.39	8.70
Chicken	1.0	5.2	3.09	19.38
	6.4	6.4	1.56	6.73
	solution	5.2	2.78	15.56
Beagle	0.13	2.0	4.52	20.89
	1.3	1.1	5.67	23.00
	16.0	1.8	2.08	8.79
	solution	0.63	7.05	21.96

(1) AUC_{0-7}; beagle, AUC_{0-8}; chicken and yellowtail

Source: Endo *et al.* (1987) & Hirakawa *et al.* (1983).

Beagles

Beagles were administered with two kinds of ultra-fine OA (0.13 µm and 1.3 µm in diameter), mechanically pulverized OA crystals (16 µm in diameter) and an aqueous solution at a dose of 50 mg/animal. Hirakawa *et al.* have compared the absorbency among these preparations. T_{max}, C_{max} and AUC_{0-7} obtained from various particle sized OA are shown Table I. The absorbency of the two ultra-fine OAs was higher than that of mechanically pulverized OA. Although the AUC_{0-7} of the two ultra-fine OAs were almost equal, they were both greater than that of mechanically pulverized OA. Moreover, these two OAs and the aqueous OA were almost equal in AUC_{0-7}. Thus, the bioavailability of OA in beagles depended on the particle size of the preparations.

Comparison of enhanced bioavailability by micronization among animal species

It was concluded that the bioavailability of OA in yellowtails, chickens and beagles was enhanced by micronization of the crystals. But, there were differences in the increase in pharmacokinetic parameters due to micronization of the particles among animal species. The value of T_{max} in yellowtails did not depend on the diameter of OA crystals, whereas the T_{max} of ultra-fine OA (1.0 µm in diameter) in chickens was shorter than that of ordinary OA (6.4 µm in diameter). In beagles, the T_{max} of OA with a diameter of 0.13 µm was longer than that of micronized OA with a diameter of 1.3 µm. The C_{max} of ultra-fine size types in yellowtails and chickens were 1.7 and 2.0 times higher than those of ordinary OA, respectively. In beagles, the C_{max} of micronized types, with diameters of 0.13 and 1.3 µm, were 2.2 and 2.7 times higher than that of OA 16 µm in diameter. Moreover, marked differences were observed among three animal species with respect to the increment of AUC resulting from micronization. The AUC_{0-8} of ultra-fine in yellowtails was 1.4 times higher than that of ordinary OA, whereas the AUC_{0-8} of ultra-fine in chicken was 2.9 times higher than that of ordinary OA. The AUC_{0-7} of the two micronized OAs in beagles were 2.4 and 2.6 times higher than that of mechanically pulverized OA (16 µm in diameter). Therefore, the C_{max} increase due to micronization was highest in beagles, followed by chickens and yellowtails, and the AUC increment was lower in yellowtail than in chicken and beagle dog.

It is suggested that the differences in enhanced bioavailability of OA resulting from micronization is due to the degree of micronization, animal species and the properties of the substance administered.

References

1). EJIMA A., 1982. Seibutugakuteki Doutousei Shiken niokeru Mondaiten. *Rinsho Yakuri,* **13**, 199-203.

2). ENDO T., ONOZAWA M., HAMAGUCHI M. & KUSUDA R., 1987. Bioavailability of ultra-fine preparation of oxolinic acid in red sea breams. *Nippon Suisan Gakkaishi,* **53**, 1493.

3). ENDO T., ONOZAWA M., HAMAGUCHI M. & KUSUDA R., 1987. Enhanced bioavailability of oxolinic acid by ultra-fine size reduction in yellowtail. *Nippon Suisan Gakkaishi,* **53**, 1711-1716.

4). HIRAKAWA Y. & HARADA K., 1983. Studies on development of the ultra-fine size reduction method of slightly soluble medicinal crystals. IV. Enhanced bioavailability of oxolinic acid by ultra-fine size reduction. *Yakugaku Zasshi,* **103**, 1190-1194.

5). JACOBSEN M.D., 1989. Withdrawal times of fresh water rainbow trout, *Salmo gairdneri* Richardson, after treatment with oxolinic acid, oxytetracycline and trimethoprim. *J. Fish Dis*., **12**, 29-36.

6). KASUGA Y., SUGITANI A., YAMADA F., ARAI M. & MORIKAWA S., 1984). Oxolinic acid residues in tissues of cultured rainbow trout and ayu fish. *Shokuhin Eisei Zasshi,* **25**, 512-516.

7). KLEINOW K.M. & LECH J.J., 1988. A review of the pharmacokinetics and metabolism of sulfadimethoxine in the rainbow trout (*Salmo gairdneri*). *Vet Hum. Toxicol.*, **30** (Supplement 1), 26-30.

8). UENO R., HORIGUCHI Y. & KUBOTA S.S., 1988. Levels of oxolinic acid in cultured yellowtail after oral administration. *Nippon Suisan Gakkaishi,* **54**, 479-484.

9). UENO R., OKAMURA M., HORIGUCHI Y. & KUBOTA S.S., 1988. Levels of oxolinic acid in cultured rainbow trout and amago salmon after oral administration. *Nippon Suisan Gakkaishi,* **54**, 485-489.

THE DISPOSITION OF IVERMECTIN IN ATLANTIC SALMON (*SALMO SALAR*)

T. Høy, T. E. Horsberg & I. Nafstad

The Norwegian College of Veterinary Medicine, Department of Pharmacology and Toxicology, P.O.box 8146 Dep.,N-0033 Oslo 1, Norway

Summary: *The disposition of tritium labelled ivermectin (22,23 di-^{3}H-avermectin B1a) in Atlantic salmon was studied by whole-body autoradiography, liquid scintillation counting, and thin-layer chromatography. Ivermectin was slowly absorbed, the highest concentrations being found in lipid-containing organs. High concentrations were also found in the central nervous system. The total amount of radioactivity in blood, muscle, liver and kidney diminished by only 35 % from day 4 to day 28 after administration. Excretion was mainly by the biliary route. The drug was mainly excreted in the unchanged form.*

***Résumé*: Biodisponibilité de l'ivermectine chez le saumon atlantique.**

L'ivermectine est un dérivé synthétique de lactone monocyclique, très efficace contre certains agents parasitaires externes et internes. Elle bloque la transmission de l'influx nerveux en potentialisant le relargage de l'acide gamma-aminobutyrique (GABA). Cet effet est peu marqué chez les mammifères. Chez le saumon, on l'emploie pour lutter contre les copépodes parasites à la dose de 0,2 mg/kg PV, les posologies supérieures présentant des risques toxiques.

La distribution tissulaire de l'ivermectine a été étudiée en administrant le produit radiomarqué par l'intermédiaire de sondes gastriques, à des saumons élevés à 7°C. Des autoradiographies étaient pratiquées sur coupes après sacrifices à intervalles fixes. Parallèlement, la persistance du produit dans la bile était recherchée par chromatographie en couche mince. Deux heures après ingestion l'ivermectine est présente dans l'estomac et les coecca pyloriques, puis elle passe dans les différents organes. Les concentrations culminent à 4 jours, surtout dans les tissus riches en lipides (dont le système nerveux central). L'élimination est lente, et la persistance est supérieure à 28 jours (Tableau I). A ce moment l'intestin et la bile présentent encore une forte activité, et un métabolite majeur a pu y être identifié.

Le métabolisme de l'ivermectine favorise donc son accumulation dans les tissus gras, d'où elle s'élimine lentement. Chez le saumon, à l'inverse des mammifères, les concentrations sont plus élevées dans le tissu nerveux que dans le muscle. Celà pourrait expliquer les mortalités observée à des posologies seulement 2 fois supérieures à la dose thérapeutique usuelle. Bien qu'on n'ait pas observé de résidu non polaire ces propriétés rendent l'ivermectine peu recommandable pour le saumon, des exceptions pouvant être admises pour le traitement d'animaux de moins d'1 an.

Introduction

Ivermectin (22,23-dihydroavermectin B_1a) is a semi-synthetic derivative of a macrocyclic lactone produced by *Streptomyces avermitilis*. Ivermectin is highly efficient against a wide range of internal and external parasites in many host species. Its effect is due to increased release of the inhibitory neurotransmitter gamma-aminobutyric acid (GABA) from presynaptic nerve terminals, and enhanced GABA binding to postsynaptic receptors.

At therapeutic doses, ivermectin does not affect mammals significantly, as the GABA-system is restricted to the central nervous system and the drug does not readily pass the blood-brain barrier (1).

Palmer *et al.*(3) described preliminary trials on the use of ivermectin against salmon lice infestations. A single oral dose of 0.2 mg/kg b.wt., the recommended mammalian dose, produced significant reduction in the number of copepods on test fish, and no increased mortality. On increasing the dose to 0.4 mg/kg b.wt., a higher mortality rate was observed in the medicated group compared with the controls.

The aim of this investigation was, using whole-body autoradiography and scintillation counting, to examine the tissue distribution of 22,23 ^3H-avermectin B_1a (ivermectin) in Atlantic salmon after the administration of a single oral dose.

Materials and methods

Twenty healthy Atlantic salmon smolt, weighing 68 ± 11 g were used. They were kept in a 500 l fibreglass tank supplied with running salt water with a salinity of 28 - 34 $^o/_{oo}$. The temperature was kept at 7.1 ± 0.5 ° throughout the experiment.

Seventy four MBq of 22,23 di-^3H-avermectin B_1a (2220 MBq/mg) dissolved in methanol was kindly supplied by Merck Sharp & Dohme Research Laboratories, Rahway, New Jersey, USA. The solution was evaporated to a volume of 0.1 ml under a gentle stream of purified air. It was then mixed with 0.5 ml of ethanol, and distilled water was added to give a final volume of 5 ml. No precipitation of the drug occurred. Medication was carried out by inserting a stomach tube and administering 0.2 ml of the solution, corresponding to 2.96 MBq into the stomach of each fish, using a 1 ml syringe.

At fixed time intervals, 1-2 fish were killed by an overdose of CO_2, immediately frozen with liquid nitrogen (-196 °C) and transported to the laboratory. They were embedded in a 1 % solution of sodium carboxy methyl cellulose, frozen in a mixture of n-hexane and solid CO_2, and subjected to whole body autoradiography according to the method of Ullberg (4).

From the remaining material, samples of approx. 10 mg were taken from heart blood and bile, as well as kidney, liver, brain and muscle tissue. The samples were subjected to liquid scintillation counting according to standard procedures.

Calculations of the percentage of the administered radioactivity dose present in liver, kidney, blood and muscle at different sampling times were made using the total dose administered to each fish, the relative contribution of the organ to total bodyweight and the weight of the fish.

Bile samples from fish sacrificed after 1, 3 and 15 days were subjected to thin layer chromatography to obtain an approximation of the amount of unchanged drug excreted via the biliary route. The samples were diluted with water, mixed with equal parts of methanol to precipitate mucus and proteins, and centrifuged. The supernatant was collected and methanol was removed under a gentle stream of nitrogen. Aliquots were applied on pre-coated TLC plates (Sil G-25 UV_{254} Machery-Nagel, Düren, W. G.) together with standards of ivermectin. The TLC- plates were developed in a solvent system of chloroform:ethyl acetate:methanol:methylene chloride 9:9:1:2. Each zone was divided into 0.5 cm segments, the gel of each segment was scraped off with a scalpel into a separate liquid scintillation vial and scintillation cocktail was added to the vials. An approximation of the amount of drug excreted in unchanged form could be achieved by comparing the results of the different bile samples with the standards.

Results

Two hours after administration, the radioactive material was present in the stomach and pyloric caeca, but only negligible amounts were present in other organs. The activity then gradually increased in most organs.

Peak concentrations were found 4 days after administration. Radioactivity was present in all organs and tissues. The concentration of ^{3}H-ivermectin was particularly high in fat-containing tissues. High activity was found in the central nervous system (CNS), the highest levels being found in the cerebellum and cerebral cortex. The concentration of radioactivity in the brain was 1.7 times higher than the concentration in muscular tissue. It exceeded that of the muscular tissue from 2 days after administration and throughout the sampling period.

Later, the activity slowly decreased in all organs. At the final sampling time, 28 days after administration of the drug, the highest activity was detected in bile and the intestinal contents. Liver, kidney and brain also contained considerable levels of radioactivity at this time.

The results of the calculations of the total amount of radioactivity in blood, muscle, liver and kidney compared to the dose administered, are presented in Table I. Four days after administration, approx. 29 % of the dose was present in blood, muscle, liver and kidney. Twenty-eight days after administration, approx. 19 % of the administered dose was still present in these organs.

Table I: Percentage of administered dose of radioactivity present in blood, muscle, liver and kidney at different intervals after administration.

	Blood	Muscle	Liver	Kidney	SUM
2 h	1.1	2.1	1.2	0.5	4.9
12 h	1.4	5.0	1.1	0.6	8.1
1 d	1.6	10.7	1.2	0.5	14.0
2 d	2.21	4.2	1.6	0.6	18.6
4 d	1.5	24.7	2.5	0.6	29.3
7 d	2.1	23.2	1.2	0.8	27.3
15 d	0.9	8.9	0.8	0.3	10.9
21 d	1.1	16.0	1.3	0.6	19.0
28 d	1.5	15.9	1.3	0.4	19.1

Table II: Distribution of unchanged drug, the major metabolite, and the sum of other polar metabolites present in the bile at different sampling times.

time	unchanged drug	main metabolite	other polar metabolites
1 d	77*	12	11
3 d	44	25	31
15 d	42	26	32

*values = relative percentages in sample

One major metabolite, more polar than the unchanged drug, was detected. The amounts of unchanged drug, the major metabolite and the sum of other polar metabolites in the bile are presented in Table II.

Approx. 77 % of the radioactivity in the bile one day after administration of the drug had the same Rf-value as the unchanged drug, this proportion dropping to approx. 42 % after 15 days.

Discussion

The present data show that the drug is slowly absorbed from the intestine of salmons, reaching maximum concentrations in all organs approx. 4 days after administration. Elimination is very slow. The difference in total amount of radioactivity in blood, muscle, liver and kidney at day 4 after treatment compared with day 28 was only 35 %. The drug accumulates in fatty tissues, from which it is slowly released. Particular note should be made of the fact that the concentration of ivermectin in brain was higher than in muscular tissue at all sampling times except 12 and 24 hours after administration. This is in complete contrast to the results of Campbell *et al.* (1), who reported the concentration in the brain of male rats to be less than 1/10 of the concentration in muscular tissue. The present study shows that ivermectin easily crosses the blood-brain barrier in Atlantic salmon. In fact, the results indicate that this barrier is poorly developed in fish compared to mammals.

Palmer *et al.* (3) found that an oral dose of 0.4 mg/kg b.wt. resulted in increased mortality among treated salmon. This might be explained by passage of ivermectin across the blood-brain barrier and its effect on the central nervous system.

The results obtained from the thin-layer chromatographic study indicate that the major source of radioactivity in the bile is the unchanged drug. No metabolite less polar than the parent drug was detected in the bile. In steer and sheep fat, residues less polar than the parent compound have, however, been detected (2). This rather unusual finding was probably caused by fatty esters of a metabolite of ivermectin, and accounted for about 50 % of the residues in fat. Whether this is also the case in salmon has not been examined.

It can be concluded that ivermectin is not well suited for the oral treatment of salmon lice infestations or other parasitic diseases in Atlantic salmon. This is due to the high concentrations of the drug reached in the central nervous system and its very slow excretion. A certain use on salmon the first year in sea might, however, be considered.

References

1). CAMPBELL, W.C., FISHER, M.H. & STAPLEY, E.O 1983. Ivermectin: A potent new antiparasitic agent. *Science*, **221**, 823-828.

2). JACOB, T.A., BUHS, R.P.,. CARLIN, R.P., LEE-CHIU, S.H., MIWA, G.T., & ROSEGAY, A. 1983. The metabolism and tissue residue profile of ivermectin. In: *Recent developments in the control of animal parasites*. W.D.H. Leaning ed., Proc. MSD-AGVET Symp., Perth, Australia, pp 56-97.

3). PALMER, R., RODGER, H., DRINAN, H., DWYER, C. & SMITH, P.R., 1987. Preliminary trials on the efficacy of ivermectin against parasitic copepods of Atlantic salmon. *Bull. Eur. Ass. Fish Pathol.*, **7**, 47-54.

4). ULLBERG, S., 1954. Studies on the distriution and fate of 35S-labelled benzyl-penicillin in the body. *Acta radiol. (Stockh.)*, **118**, 1-110.

Discussion

RICHARDS I would have to disagree a little with your conclusion. If we could find a product that had a long active effect, I agree that we would have to have a long withdrawal period, but this may mean that we'd have to use infrequent treatment, so if we had a compound which was used and effective for the first year in the sea, this would dramatically cut down future use of that drug. That of course only as long as the brain effect wasn't toxic to the fish.

HØY Yes of course I agree with you.

MIDTVEDT Did the fish change their behaviour?

HØY No, all fish were given such a very low dose, so we didn't see any toxicity.

SCHNICK What tolerances are established in mammals for ivermectin?

HØY I am not sure.

DUNIER Your model is really nice, but is there any other simpler way to measure residues, may be gas chromatography or enzymatic like cholinesterol?

HØY This is not a residue study, it is just a method to give you an idea of the fate of the drug. The reason why we use it is that we have access to this method in our laboratory. There are other methods, yes.

LAFONT Maybe another drawback to the use of this substance is that excretion is in the unchanged form. It is excreted as theoriginal drug.

HØY Yes.

LAFONT I know that this is a substance which is very active against a wide range of invertebrate animals. Now this raises environmental concerns about the use of this drug in salmon. What do you think about this aspect?

HØY Yes, that could be a problem.

ALDERMAN There is also the problem of waste feed which will produce ivermectin residues in the pile beneath the cage.

JOHNSTON Could you quantify the amount of ivermectin reaching the skin?

HØY No, we didn't try to do that.

ALDERMAN I think that the problem of ivermectin will continue for some time, because of its obvious interest and potential environmental problems.

BIODISPONIBILITE IMMEDIATE DU LEVAMISOLE CHEZ L'ANGUILLE EUROPEENNE (*ANGUILLA ANGUILLA* L.) HOTE DEFINITIF DES NEMATODES *ANGUILLICOLA* YAMAGUTI, 1974

G. Blanc*, S. Loussouarn& L. Pinault****

*Service d'Aquaculture et Pathologie Aquacole.,
**Service de Pharmacie et Toxicologie.
Ecole Nationale Vétérinaire de Nantes, Route de Gachet,
44087 Nantes Cedex 03, France.

Résumé: *L'étude de la pharmacocinétique du lévamisole a été réalisée chez l'anguille européenne après la mise au point et la validation d'une méthode de dosage dans le plasma afin de définir de manière rationnelle le meilleur schéma posologique pour optimiser son efficacité thérapeutique, déjà démontrée sur la base de critères cliniques, de diminuer les risques de persistance de ses résidus dans la chair des animaux traités et de restreindre sa diffusion dans l'environnement.*

La concentration de cet anthelminthique a été déterminée par la méthode du standard interne. La procédure repose sur une extraction chimique puis une quantification par chromatographie en phase gazeuse utilisant un détecteur thermoionique.

L'évaluation de la méthode de dosage portant sur sa fidélité, son exactitude, sa linéarité, sa spécificité et sa sensibilité a permis de vérifier sa validité sur l'intervalle de concentrations de 0,5 à 20 µg/ml.

Les poissons utilisés ont été maintenus en eau douce (valeurs moyennes: température = 21,1°C, pH = 7,5, oxygène dissous = 5,4 mg/l) et avaient un poids moyen de 110g ; ils ont reçu une injection intracardiaque unique de 8 mg de lévamisole par kilogramme de poids vif.

Les concentrations mesurées sont correctement décrites par une équation bi-exponentielle dont les constantes ont été estimées par deux techniques de régression non linéaire basées sur la méthode Damping Gauss-Newton. Les principaux paramètres pharmacocinétiques du lévamisole chez l'anguille sont les suivants : $T_{1/2}(h) = 3{,}99$, $AUC_{0-\infty}$ *(µg.h/ml) = 17,04,* V_{dss} *(l) = 1,67,* Cl_t *(l/h) = 0,47, MRT (h) = 2,16. Ces paramètres indiquent que l'élimination à partir du compartiment central de cette molécule, liposoluble et faiblement basique, est plutôt rapide. Les valeurs faibles de sa demi-vie et de son temps moyen de résidence invitent à une modification du schéma posologique actuellement retenu.*

Summary : **Bioavailability of levamisole in the European eel (*Anguilla anguilla* L.) definitive host of the nematode *Anguillicola* Yamaguti, 1974.**

This paper presents the results of a pharmacokinetic study of levamisole in the European eel. The investigation was designed to determine the most effective therapeutic regime for the use of a product which had already been shown to be clinically effective and to reduce risks of meat residues and adverse environmental effects.

The concentration of the anthelminthic was determined against an internal standard after chemical extraction followed by quantification by gas chromatography with a thermionic detector. The method was demonstrated to be specific and give linear results for levamisole concentrations between 0.5 and 20 μg/ml.

The eels used were held in fresh water at a mean temperature of 21.1°C, pH 7.5 and dissolved oxygen at 5.4 mg/l. The animals were of 110g mean weight and received a single intracardiac injection of 8mg levamisole/kg live weight.

The elimination curves for levamisole gave a good fit to a double exponential curve calculated by two non-linear regression techniques based on the Damping Gauss-Newton method. The principal pharmacokinetic parameters of levamisole in the eel were as follows : $T_{1/2}$*h) = 3.99, AUC0-(ug.h/ml) = 17.04,* V_{dss} *(l) = 1.67,* Cl_t *(l/h) = 0.47, MRT (h) = 2.16> These parameters indicate that the elimination of this liposoluble and weakly basic drug from the central compartment is extremely rapid. The short half life and mean residence time suggests that the currently used treatment regime could well be modified to increase efficacy of treatment.*

Introduction

Jusqu'au début des années 80, les principes actifs utilisés en thérapeutique anti-parasitaire aquacole se sont limités à un petit nombre de composés majeurs: le formaldéhyde, le dichlorvos et le vert malachite (2, 8). La diversification de l'aquaculture, l'apparition de nouvelles espèces parasitaires en Europe ainsi que la diminution des performances de ces molécules ont conduit à la recherche de principes actifs nouveaux. Les dérivés du benzimidazole et de l'imidazothiazole, le lévamisole en particulier , ont fait l'objet depuis quelques années d'une recherche active de leurs potentialités en aquaculture: antiparasitaires (3, 6, 9, 12, 21), antimycosiques (Prost et Sopinska, dans ce volume) et immunomodulatrices (18, 19). La majorité des travaux concernant l' activité thérapeutique du lévamisole repose sur des critères d'efficacité clinique acquis sur de très faibles effectifs.

Si la recherche de molécules nouvelles est nécessaire, elle ne représente qu'une étape primaire dans la mise au point de schémas posologiques rationnels et efficaces. Le besoin de données fondamentales tant pharmacodynamiques que toxicologiques permettant d'optimiser les schémas thérapeutiques et de minimiser les effets indésirables de ces molécules sur l'animal cible, le consommateur et l'environnement a fait l'unanimité des participants au Congrès de l'OIE (Paris, 1991) sur les problèmes posés par la chimiothérapie en aquaculture.

Cette publication présente la première étude sur la détermination des paramètres pharmacocinétiques du lévamisole chez une espèce aquacole d'intérêt économique : l'anguille européenne.

Matériel et méthodes

Conditions expérimentales

Un lot d'anguilles européennes calibrées dont la gamme biométrique correspond à la phase initiale d'élevage qui montre les plus fortes prévalences lors de l'infestation par *Anguillicola crassus* (3) a été sélectionné. A partir de ces anguilles indemnes de pathologie, provenant d'un même biotope et ayant passé deux mois en élevage, ont été échantillonnés 100 individus. Confrontés d'une part aux effets métaboliques que le stress induit (5) chez les espèces aquacoles, susceptibles de modifier le comportement pharmacologique des xénobiotiques, et d'autre part à la petite taille des animaux, nous nous sommes imposés une manipulation rapide des poissons (1 mn) après anesthésie au phénoxy 2-éthanol et une ponction sanguine unique par individu.

Les données biométriques des animaux utilisés sont regroupées sur la Figure 1. Chaque lot de 10 individus représente un point sur la cinétique étudiée. Les animaux ont été répartis dans deux bassins de 100 m^3 divisés chacun en six parties situées au centre des bassins. Les valeurs de la température, de la quantité d'oxygène dissous et du pH sont représentées sur la Figure 2. Les poissons ont été soumis à une photopériode naturelle pendant toute la durée de l'expérimentation. Après anesthésie, le principe actif a été administré à la dose de 8 mg/kg de poids vif (PV) par voie intracardiaque. Les prélèvements sanguins ont été effectués par cette même voie puis conservés dans l'azote liquide.

Protocole analytique

Le détail de la méthode analytique sera présenté dans une autre publication; nous ne décrivons ici que son principe et la validation de la méthode de dosage.

La méthode de dosage fait appel à un étalon interne. Après extraction liquide-solide sur colonnes Bond-Elut, le lévamisole et le standard interne sont

Fig.1: Données biométriques de la population d'anguilles traitées (les moyennes des longueurs, poids, indices de condition, ainsi que les écarts-types sont représentés pour chaque lot).

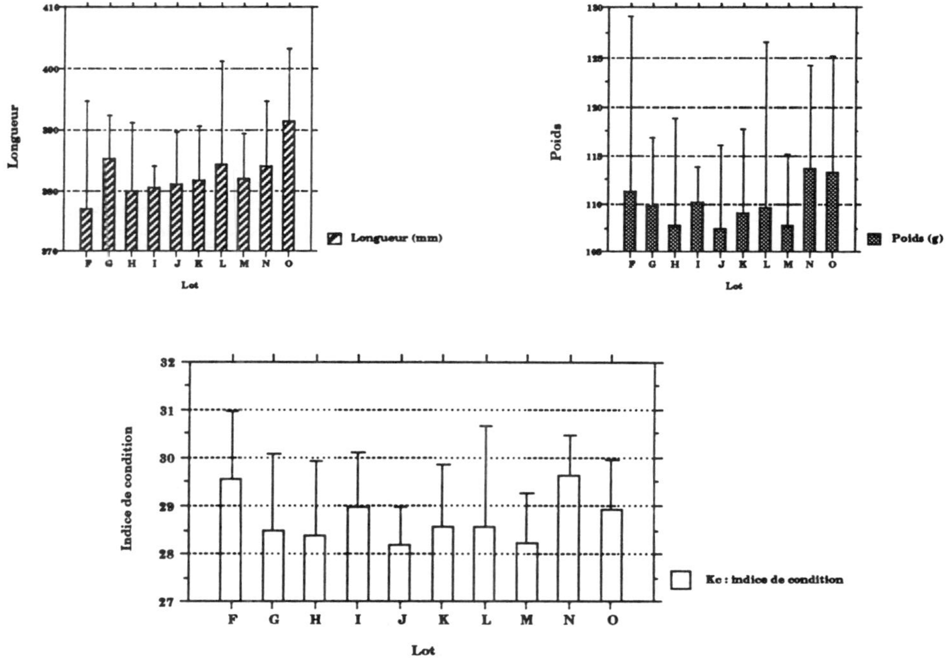

Fig 2: Variations des paramètres de l'élevage dans les bassins expérimentaux (les paramètres statistiques de position et de dispersion ont été calculés sur les valeurs horaires).

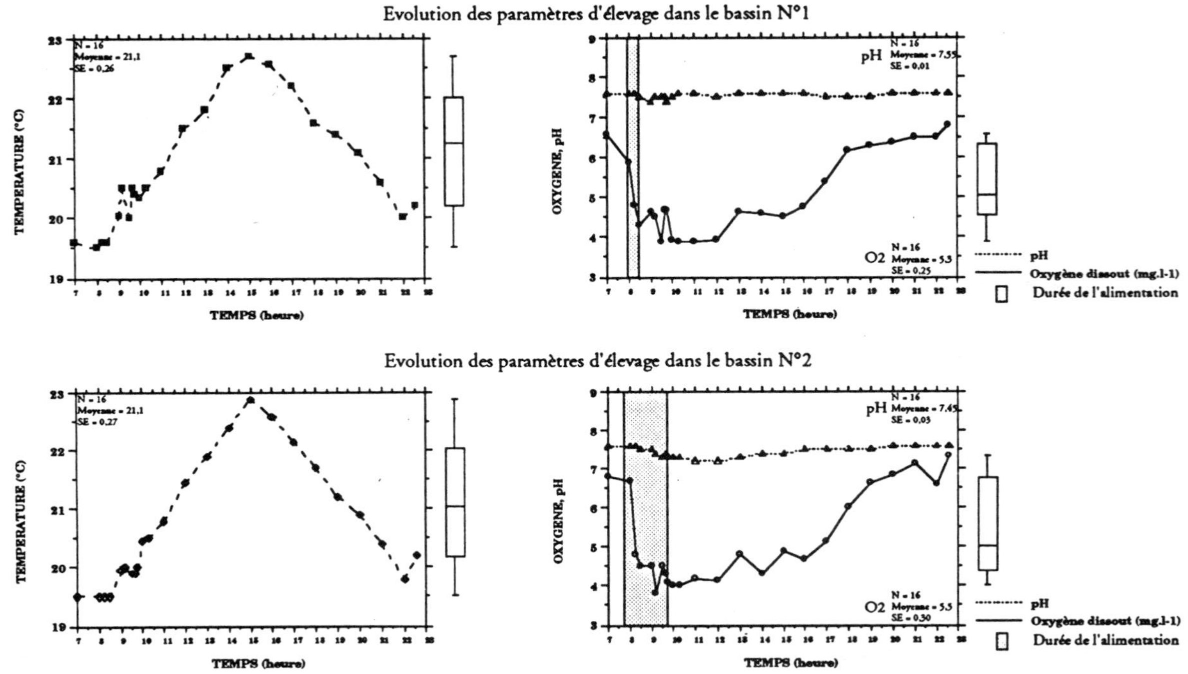

quantifiés par chromatographie en phase gazeuse utilisant un détecteur thermoionique (13).

Les réactifs de qualité analytique proviennent de chez Jannsen Chemica pour le lévamisole (lot N° 0707914) et le standard interne, méthyl-4' tétramisole (lot N° V89723), et de chez Merck pour le n-heptane (qualité uvasol). Les colonnes Bond-Elut C8 proviennent de chez Analytichem International.

L' appareillage chromatographique est un Varian modèle 3700 équipé d'une colonne en verre (200 x 0,2 cm) remplie d'OV 17 à 3% sur supelcoport 80/100 mesh, et le volume de l'injection n'a pas dépassé 2 µl. Les conditions chromatographiques étaient les suivantes: température du détecteur 340°C, température de la colonne 256°C, température de l'injecteur 290°C, gaz vecteur: azote (30 ml/mn), mélange gazeux au niveau du détecteur: hydrogène (4,5 ml/mn) et air (175 ml/mn).

Les aires obtenues ont été calculées par un intégrateur Merck D-2500. La validation de la méthode de dosage du lévamisole a été effectuée sur 5 jours pour 14 niveaux de gamme. Les tests statistiques employés (20) ont été réalisés par les logiciels EXCEL et SYSTAT à partir de feuilles de calculs mises au point par l'auteur senior d'après les définitions données par les normes AFNOR ou par Caporal *et al* (7).

Les données quantifiées ont été ajustées au moyen de deux techniques de régression non-linéaire pondérées par le logiciel BMDP, les paramètres pharmacocinétiques initiaux ayant été calculés par la méthode des résiduelles. Le meilleur modèle est choisi après calcul du critère d'Akaike effectué avec le logiciel PHAR 2.0 du CNEVA, Laboratoire des Médicaments Vétérinaires. Les processus d'élimination ont été supposés suivre une cinétique d'ordre 1. Les micro-constantes ont été calculées d'après les formules données par Gibaldi et Perrier (11), et Shargel et Yu (16).

L'aire sous la courbe et le temps moyen de résidence ont été calculés par la méthode des trapèzes arithmétiques ou logarithmiques en extrapolant à l'infini.

Résultats et discussion

L'hypothèse d'égalité biométrique des différents lots composant la population traitée a été vérifiée à l'aide d'un test de Kruskal-Wallis: au seuil de signification de 5% cette hypothèse est vérifiée pour les longueurs (Hcor = 12,185) et pour les indices de condition (H_{cor} = 15,602), elle ne l'est pas pour les poids (H_{cor} = 21,19), la valeur critique du test Chi-^{2} théorique (9) 0,05 = 16,92.

L'hypothèse d'égalité des conditions expérimentales dans les deux bassins a été vérifiée à l'aide d'un test de comparaison de Kolmogorov-Smirnov: au seuil de signification de 5% cette hypothèse est vérifiée pour la température (N1N2Dobs. = 84), pour l'oxygène dissous (N1N2Dobs. = 84) et pour le pH (N1N2Dobs. = 188); la valeur critique du test N1N2D = 189 au coefficient de sécurité de 95%.

Résultats analytiques

Spécificité:

Elle a été évaluée en comparant des chromatogrammes obtenus après analyse des gammes de concentration en milieu méthanolique et plasmatique supplémentés en lévamisole et standard interne (SI). En solution méthanolique, les pics correspondant au SI et au lévamisole sont bien différenciés, les temps de rétention sont en moyenne de 2,1 mn et de 1,6 mn respectivement. En milieu plasmatique, des pics parasites sont présents, cependant leurs temps de rétention sont suffisamment différents pour ne pas être confondus avec le lévamisole et le SI ; leurs traînées recouvrant partiellement celles de nos analytes, nous avons adopté une procédure d'intégration verticale "vallée-base" privilégiant la mesure de la hauteur à la mesure de l'aire totale du pic. Nous avons noté, par ailleurs, que ces impuretés n'étaient pas présentes quand l'échantillon à doser était dilué, ce qui nous fait penser à la présence d'impuretés dans le plasma.

Tableau I : Fidélité de la méthode d'analyse.

Concentration des standards	Variance de répétabilite	Variance interlaboratoire	Moyenne IC	Répétabilite	Reproductibilité	Valeur stat. (test de Cochran)
0.2/5	4.54E-5	9.98E-4	2.8E-1±0.01	1.91E-2	9.14E-2	7.71E-1*
0.2/10	9.79E-5	1.25E-2	2.7E-1±0.06	2.8E-2	3.18E-1	7.88E-2**
0.5/5	3.09E-5	2.87E-5	5.1E-1 ±.004	1.57E-2	2.18E-2	2.86E-1*
0.5/10	5.66E-3	7.62E-4	5.2E-1±0.04	2.13E-1	2.27E-2	8.35E-1***
1/5	2.52E-4	1.22E-3	1.04±0.02	4.49E-2	1.08E-1	3.72E-1*
1/10	1.42E-3	9.70E-4	1.03±0.03	1.06E-1	1.38E-1	0.42E-1*
2/5	9.88E-4	7.17E-3	2.02±0.04	8.89E-2	2.55E-1	3.85E-1*
2/10	3.21E-3	2.67E-3	2.01±.04	1.60E-1	2.17E-1	5.60E-1*
5/5	2.1E-2	9.43E-3	4.87±0.09	4.10E-1	4.94E-1	7.06E-2*
5/10	4.38E-3	3.99E-3	5.06±0.05	1.87E-1	2.59E-1	3.99E-1*
10/5	1.3E-1	8.65E-1	9.33±0.50	1.02	2.82	4.87E-1*
10/10	5.93E-2	-1.19E-2	10.05±0.12	6.69E-1	6.69E-1	4.26E-1*
20/10	1.07E-1	2.94E-2	20.17±0.2	0.927	1.05	0.63E-1
40/10	7.40E-1	1 .50	33.32±0.78	2.43	2.43	5.64E-1*

Les concentrations des standards sont en µg/l. * : test non significatif, valeurs acceptables pour la concentration considerée. ** : test significatif,valeurs suspectes. *** :test tres significatif, valeurs aberrantes.

Fidélité :

- Répétabilité: ses méthodes de calcul sont classiques, elles supposent que la variance intralaboratoire soit constante, seule la norme ISO 5725 propose un test (variance maximale de Cochran) pour vérifier cette hypothèse..

- Reproductibilité: ses méthodes de calcul diffèrent dans la littérature. La norme ISO 5725 définit la variance de reproductibilité comme étant la somme de la variance interlaboratoires et de la variance moyenne intralaboratoire pour un niveau de gamme donné. Ce mode de calcul prend donc en compte l'influence des facteurs qui sont intégrés dans la définition de la reproductibilité et c'est celui qui a été appliqué ici (Tableau 1). Exceptés les points 0,2/10 et 0,5/10 qui apparaissent douteux et sont retirés de l'intervalle analytique efficace, tous les autres points de la gamme ont une fidélité statistiquement acceptable.

Linéarité:

Les résultats ont été obtenus dans les conditions de validité analytique et selon les gammes précédemment définies. Le traitement statistique a été effectué en trois étapes. Dans un premier temps, nous avons cherché un modèle de régression linéaire sur l'ensemble des points des gammes étudiées par jour d'analyse. La variance totale de la variable dépendante est décomposée en une variance expliquée par la régression et une variance résiduelle elle-même décomposée en une variance due au manque d'ajustement du modèle linéaire et une variance liée à la variance de répétabilité. Un test de Fischer-Snedecor (Fma) permet de conclure en comparant la variance due au manque d'ajustement du modèle linéaire à la variance de répétabilité. Dans le cas où le test indique un manque de linéarité, le point présentant la plus grande variance due au manque d'ajustement est retiré et l'ensemble de ce test est effectué à nouveau de façon à déterminer la plage linéaire de la procédure analytique (Tableau 2). Les gammes sont statistiquement ajustées par un modèle linéaire de 0,2/5 à 5/5 et de 0,2/10 à 20/10. Après vérification de l' hypothèse d'homoscédasticité des variances résiduelles par un test de Bartlett, nous avons effectué une analyse de covariance de façon à s'assurer que les droites de régression obtenues pouvaient être confondues tant au niveau de leur pente que de leur ordonnée à l'origine (Tableau 3). Pour la gamme de 0,2/10 à 20/10 il a fallu supprimer les points 0,2/10 et 0,5/10 ; pour la gamme 0,2/5 à 5/5 seules les trois dernières droites ont été prises en compte car nous avions affaire à deux familles de droites. Enfin, un test de Student significatif a permis de montrer que les pentes communes n'étaient pas différentes de 1.

Tableau II Ajustement linéaire des résponses journalières.

	0.2/5	0.5/5	1/5	2/5	5/5	10/5	R^2	F_{MA}
J1	2.7	1.3	0.8	0.5	3.9	///1.3///	0.998	0.53*
J2	1.9	1.2	0.8	1.4	3.2	6.4	0.998	1.37*
J3	3.1	1.0	1.8	1.4	1.6	3.3	0.998	2.06*
J4	2.3	0.7	1.5	2.0	1.7	///4.4///	0.999	3.65*
J5	2.0	0.9	2.0	1.8	3.6	2.9	0.998	1.65*
Zone linéaire	-----	-----	-----	-----	-----	++++++++		
CV*	5.96	1.40	3.46	4.18	3.51	10.00		

	0.2/10	0.5/10	1/10	2/10	5/10	10/10	20/10	40/10	R^2	F_{MA}
J1	2.7	29.8	5.4	2.1	0.9	2.1	0.5	///2.7///	0.999	0.10*
J2	11.5	11.5	2.5	4.8	0.1	1.0	1.1	///1.9///	0.999	1.228
J3	1.9	1.5	3.6	3.3	1.6	3.5	1.4	///0.6///	0.999	2.22*
J4	3.0	6.2	4.0	1.4	1.3	2.6	2.9	///1.7///	0.999	2.02*
J5	2.3	2.6	1.1	0.5	1.8	2.0	0.9	///4.1///	0.999	2.02*
Zone linéaire	------	------	------	------	------	------	++++			
CV*	37.93	15.10	4.61	3.68	1.74	2.18	1.80	4.27		

///x/// : point supprimé pour obtenir l'ajustment. la zone linéare est representée par des tirets. R^2: est donne quand l'ajustement linéaire est obtenu. *: test non significative, ajustement linéaire accepté. Les valeurs journalières du coefficient de variation sont données, ainsi que la coefficient de variation global CV pour la durée d'évaluation.

Tableau III: Comparaison des droites de régression. Résultats des tests de Bartlett pour l'ensemble des donnée, et ANCOVA pour les gammes 0,2/5 - 5/5 et 1/10 - 20/10.

Gamme 0.2/5 à5/5		
	Pente	Ordonée a l'origine
J1	0.92	0.12
J2	0.94	0.08
J3	0.95	0.05
J4	0.99	0.07
J5	0.99	0.03
Bartlett (Bc)	11.0 Seuill critique (0.01) : 13.28	
Ancova (F)	1.17 Seuill critique (0.05) : 2.61	
Droite commune	0.99	0.00

Gamme 1/10 à20/10		
	Pente	Ordonée a l'origine
J1	1.01	-0.00
J2	1.00	0.02
J3	1.01	-0.06
J4	1.02	-0.01
J5	0.99	0.09
Bartlett (Bc)	0.02 Seuill critique (0.01) : 13.28	
Ancova (F)	2.3 Seuill critique (0.05) : 2.10	
Droite commune	1.00	0.00

Exactitude:

Le biais expérimental ainsi que le biais statistique global (7) de la procédure analytique ont été évalués en utilisant les résultats obtenus le dernier jour de l'étude de fidélité.

Pour les gammes méthanoliques de 0,5/5 à 10/5 puis de 1/10 à 20/10 le biais expérimental moyen est égal à 2,2% et 1,8% et le biais statistique global égal à 1,75% et à 1,27% respectivement.

Pour les gammes plasmatiques de 1/10 à 20/10 et pour 2/5 le biais expérimental moyen est égal à 1,9% et 2,5% et le biais statistique global égal à 1,5% et à 2,1% respectivement.

Sensibilité:

Elle s'exprime classiquement par la pente de la droite de régression, mais aussi par la limite de détection égale à 0,5 µg/ml et par la limite de quantification égale à 0,05 µg/ml.

Rendement d'extraction:

Le rendement d'extraction moyen a été calculé pour tous les points de la gamme 1/10 à 20/10, où il est égal à 100,27%, et pour le point 2/5 de la gamme 0,2/5 à 5/5 où il est égal à 101%.

Analyse pharmacocinétique

Les profils d'évolution de la concentration plasmatique en lévamisole après une administration intracardiaque unique chez l'anguille européenne ainsi que les modèles bi-exponentiels qui s'y rapportent sont représentés sur la Figure 3 . Chaque point obtenu lors de ces dosages correspond ici à un seul individu; ces points sont donc indépendants les uns des autres sur l'ensemble des données. De ce fait, il est possible de définir trois types de réponses pharmacocinétiques différentes caractérisées par les concentrations maximales obtenues (modèle II), les concentrations minimales obtenues (modèle III) et enfin par l'ensemble des concentrations obtenues qui définit une réponse moyenne (modèle I). Ces données ont été ajustées par une régression non-linéaire pondérée. Le choix du

Fig.3: Concentrations plasmatiques en lévamisole mesurées après une injection intra-cardiaque unique de 8 $mg.kg^{-1}$ chez l'anguille européenne, et profils plasmatiques obtenus par différentes méthodes de régression non linéaire.

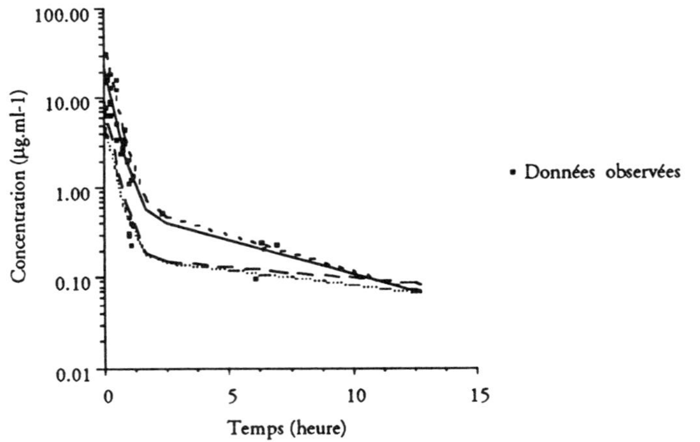

<u>Légende :</u>

—	Régression non linéaire, pondérée par l'inverse de la variance, sur l'ensemble des données quantifiées	$(C = 21,05.e^{-3,08.T} + 0,60.e^{-0,17.T})$
- -	Régression non linéaire, pondérée par l'inverse du carré des concentrations mesurées, sur l'ensemble des données quantifiées	$(C = 9,39.e^{-3,40.T} + 0,17.e^{-0,05.T})$
---	Régression non linéaire, pondérée par l'inverse du carré des concentrations mesurées, sur les données maximales quantifiées	$(C = 43,61.e^{-3,24.T} + 0,75.e^{-0,18.T})$
...	Régression non linéaire, pondérée par l'inverse du carré des concentrations mesurées, sur les données minimales quantifiées	$(C = 5,59.e^{-3,17.T} + 0,17.e^{-0,07.T})$

Fig.4: Etude de la qualité de l'ajustement du modèle I par deux techniques de régression non linéaire.

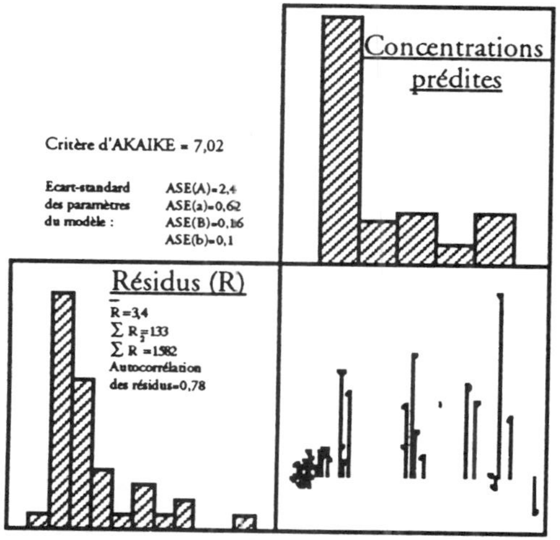

Légende :

1- Régression non linéaire, pondérée par l'inverse du carré des concentrations mesurées, sur l'ensemble des données quantifiées

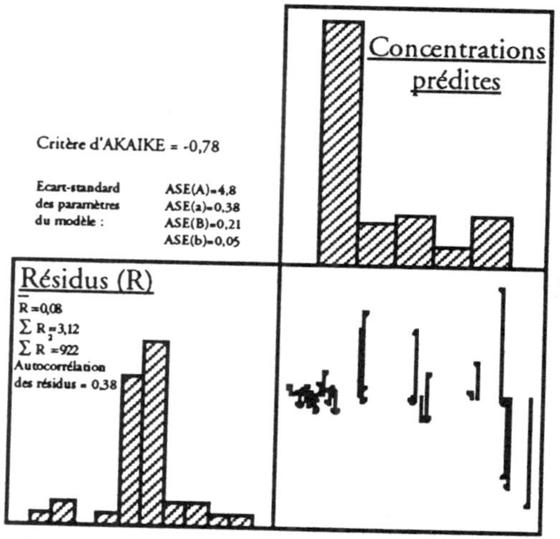

Légende :

2-Régression non linéaire, pondérée par l'inverse de la variance, sur l'ensemble des données quantifiées facteur de pondération est très important car il a une grande incidence sur les estimations des paramètres du modèle pharmacocinétique (14, 15). Nous avons utilisé un facteur de pondération classique égal à l'inverse du carré des concentrations mesurées pour le modèle II et le modèle III. En revanche, dans le cas du modèle I, nous avons testé deux types de facteur de pondération : l'inverse du carré des concentrations mesurées d'une part et l'inverse de leurs variances d'autre part. Ce deuxième choix a été effectué car le coefficient de variation n'était pas constant pour chaque point de la cinétique. Dans le cas de ce dernier modèle nous avons réalisé une analyse des résidus (Figure 4), et il s'avère que ce dernier type de pondération donne les meilleurs résultats. L'analyse des profils plasmatiques recueillis, notamment le critère d'Akaike, permettent de définir deux phases distinctes correspondant aux phases de distribution et d'élimination. Les constantes de vitesse décrivant la diffusion du lévamisole à travers les différents compartiments des modèles obtenus sont regroupées dans le Tableau 4. Les valeurs des autres paramètres pharmacocinétiques du lévamisole après une administration intracardiaque unique chez l'anguille européenne élevée en eau douce à 21°C sont représentées dans le Tableau 5. Au vu des résultats, il semble que nous soyons en présence de deux populations d'anguilles présentant des réponses pharmacocinétiques différentes. En effet, les principaux paramètres pharmacocinétiques fluctuent du simple au double voire au triple selon les modèles obtenus: les paramètres réputés stables (MRT, V_d) varient au-delà de la variation inter-individuelle classiquement admise.

Tableau IV: Constantes de vitesse des modèles pharmacocinétiques retenus (I,II,III) pour décrire la diffusion du lévamisole après administration intracardiaque unique de 8 mg/kg^{-1} chez l'anguille maintenue en eau douce à 20°C.

Parametres	Modele I	Modele II	Modele III
Kel(h-1)	2.1	2.5	1.41
K12(h-1)	0.90	0.65	1.67
K21(h-1)	0.25	0.24	0.17

Légende et signification des abréviations :
K12, K21, constantes de vitesse décrivant la diffusion du lévamisole a partir du compartiment central (1) vers le comparicment périphérique (2) et vice versa,
Kel est la constante de vitesse décrivant l'elimination du lévamisole a partir du compartiment central.
Modele I : micro-constantes obtenues après régression non linéaire, ponderée par l'inverse de la variance, sur l'ensemble de données quantifées.
Modele II : micro-constantes obtenues après régression non linéaire, ponderée par l'inverse des concentrations measurées, sur l'ensemble de données maximales quantifées.

Modele III : micro-constantes obtenues après régression non linéaire, ponderée par l'inverse des concentrations measurées, sur l'ensemble de données minimales quantifées.

Conclusion

Nous avons mis au point une méthode de dosage du lévamisole dans le plasma de l'anguille. Son évaluation dans des conditions d'analyse a permis de définir un intervalle analytique valide de 0,5 µg/ml à 20 µg/ml. Pour l'étude de la cinétique plasmatique du lévamisole cette procédure apparaît satisfaisante. .

La cinétique plasmatique du lévamisole chez l'anguille après injection par voie intracardiaque est bien décrite par un modèle pharmacocinétique à deux compartiments. Le volume du compartiment central ainsi que le volume de distribution à l'équilibre apparaissent très élevés et indiquent une distribution initiale rapide du xénobiotique dans les tissus de l'anguille. Les demi-vies de distribution et d'élimination du lévamisole sont courtes, son temps moyen de résidence faible et une aire sous la courbe peu élevée indiquent de manière globale que l'élimination de cette molécule à partir du compartiment central est rapide; ces caractéristiques permettent de penser que les résidus auront une faible persistance dans les parties consommables. Ces paramètres sont caractéristiques de la distribution d'une molécule liposoluble et faiblement basique.

L'efficacité thérapeutique de l'administration unique du lévamisole selon ce schéma classique (12, 21) est modérée; en revanche, l'administration réitérée du lévamisole proposée par Blanc en 1989 (3) donne de meilleurs résultats.

Tableau V: Paramètres pharmacocinétiques du lévamisole après administration intracardiaque unique de 8 mg.kg^{-1} chez l'anguille maintenue en eau douce à 21°C.

Parametres	Modele I	Modele II	Modele III
$T_{1/2}\alpha$(h)	0.22	0.21	0.22
$T_{1/2}\beta$ (h)	3.99	3.70	0.22
C_0 (µg/ml)	21.6	44.3	5.77
V_c (1)	0.37	0.18	1.38
AUC $_{0-\infty}$(µg^{-1}h^{-1}ml^{-1})	17.04	18.90	4.26
*f*AUC	0.40	0.39	0.92
AUC (math)(µg^{-1}h^{-1}ml^{-1})	10.33	17.47	4.1
V_{dss}(1)	1.67	0.67	15.15

V_{dexp}(1)	13.15	10.69	45.95
{Vd}β(1)	2.70	2.26	25.14
{Vd}β{math}(1)	4.45	2.45	26.15
Clt (l/h)	0.77	0.45	1.95
Clβ (l/h)	0.47	0.42	1.88
MRT (h)	2.16	1.46	7.75

Legende et signification des abreviations :

$T_{1/2}\alpha,\beta$: temps de demi-vie de la drogue pendant les différentes phases définies par le modele pharmacocinétique.
C_0 : concentration plasmatique instannée en lévamisole.
V_c : volume du compartiment central (volume calculé pour un poisson de 1.0kg PV)
$AUC_{0-\infty}$: surface définie sous la courbe d'évolution des concentrations plasmatiques en fonction du temps apres administration intraveineuse unique, calculée par la methode des trapezes.
AUC (math) : surface définie sous la courbe d'évolution des concentrations plasmatiques en fonction du temps apres administration intraveineuse unique, déduite mathématiquement.
*f*AUC : fraction de la dose de lévamisole disponible chez l'Anguille au moment de la deriere ponction sanguine.
V_{dss} : volume de distribution apparent à l'équilibre.
V_{dexp} : volume de distribution extrapolé à l'origine.
{Vd}β : volume de distribution apparent pendant la phase d'élimination.
Clt : clairance totale, elle représente la somme se toutes les clairances chez le poisson.
Clβ : clairance calculée pendant la phase d'élimination.
MRT : temps moyen pendant lequel les molécules de lévamisole séjournent chez l'Anguille.
Modele I : micro-constantes obtenues après régression non linéaire, ponderée par l'inverse de la variance, sur l'ensemble de données quantifées.
Modele II : micro-constantes obtenues après régression non linéaire, ponderée par l'inverse des concentrations measurées, sur l'ensemble de données maximales quantifées.
Modele III : micro-constantes obtenues après régression non linéaire, ponderée par l'inverse des concentrations measurées, sur l'ensemble de données minimales quantifées.

Les schémas thérapeutiques précédents pourront être optimisés en tenant compte d'une part des données pharmacocinétiques présentées dans cet article et d'autre part de la très faible efficacité du lévamisole sur les stades larvaires d'*A.crassus* présents dans l'anguille (9, 12, Blanc, données non publiées).

Remerciements: Les auteurs souhaitent remercier pour leur collaboration Mlle Nathalie Pilet, Aide-technique de laboratoire au Service de Pharmacie et Toxicologie, ENVN, Mr. P. Sanders, CNEVA, Laboratoire des Médicaments Vétérinaires, et le laboratoire VETOQUINOL qui a contribué au financement de cette étude.

Références

1). AFNOR, 1983. Fidélité des méthodes d'essai. Détermination de la répétabilité et de la reproductibilité par des essais interlaboratoires. *Norme NF X 06-041 ISO 5725*, AFNOR Ed., Paris.

2). ALDERMAN D.J., 1991. Malachite green and alternatives. International Conference *Aquaculture Europe 91*, Dublin, Ireland, June 10-12., EAS Special Publication N°14.

3). BLANC G., 1989. L'Anguillicolose: répartition géographique, étude clinique, mise au point d'une méthode de diagnostic par échotomographie, moyens de lutte. *Th. Méd. Vét.*, Ecole Nationale Vétérinaire de Nantes, 194 pp.

4). BLANC G., MARCHAND A.& GERARD J.P., 1990. L'Anguillicolose: un nouveau risque pour la filière Anguille. *Aquarevue*, **28**, 27-32.

5). BRY C., 1988. Le stress chez les poissons d'élevage. Origines possibles, réactions, éléments de prévention. *Rec. Med. Veter.*, **164**, 863-869.

6). BUCHMANN K. & BJERREGAARD J., 1990. Comparative efficacies of commercially available benzimidazoles against *Pseudodactylogyrus* infestations in eels. *Diseases of aquatic organisms*, **9**, 117-120.

7). CAPORAL J., NIVET J.M. & BASCHUNG-BERTRAND M., 1990. Validation analytique: commentaires sur la note explicative européenne et exemples d'application. *S.T.P. Pharma*, **6**, 588-594.

8). DE KINKELIN P., MICHEL C. & GHITTINO P., 1985. *Précis de pathologie des poissons*, INRA-OIE, Paris.

9). FONTAINE Y.A., LE BELLE N., LOPEZ E., QUERAT B., VIDAL B., BARTHELEMY L., SEBERT P., ALINAT J. & PETTER A.J., 1990. Infestation de populations françaises d'anguilles (*Anguilla anguilla* L.) par des nématodes (*Anguillicola crassus*): essais thérapeutiques, évaluation de risques potentiels liés à l'écophysiologie de l'hôte. *Ann. Parasitol. Hum. Comp.*, **65**, 64-68.

10). GALTIER P., 1990. Etablissement de la posologie: cas des antiparasitaires. *Ann. Rech. Vet.*, **21**, suppl. 1, 79S-86S.

11). GIBALDI M. & PERRIER P., 1975. Pharmacokinetics. In, *Drugs and the pharmacokinetic sciences*, vol. 1, J. Swarbrick Ed., New-York.

12). HARTMANN F., 1989. Investigations on the effectiveness of levamisole as a medication against the eel parasite *Anguillicola crassus* (Nematoda). *Diseases of Aquatic Organisms*, 7, 185-190.

13). PATTERSON L. & HOWE R.L., 1978. Thermionic nitrogen-phosphorus detection with an alkali-ceramic bead. *Journal of chromatographic science*, **16**, 275-280.

14). PECK C.C., BEAL S.L., SHEINER L.B. & NICHOLS A. I., 1984. Extended least squares non linear regression: a possible solution to the "choice of weights" problem in analysis of individual pharmacokinetic data. *Journal of Pharmacokinetics and Biopharmaceutics*, **12**, 545-558.

15). PECK C.C., SHEINER L.B. & NICHOLS A.I., 1984. The problem of choosing weights in non linear regression analysis of pharmacokinetic data. *Drug Metabolism Reviews*, **15**, 133-148.

16). SHARGEL L. & YU A.B.C., 1985. *Applied biopharmaceutics and pharmacokinetics*. Appleton-Century-Crofts Ed., Norwalk, Connecticut.

17). SHEINER L.B. & BEAL S.L., 1981. Evaluation of methods for estimating population pharmacokinetic parameters. II - Biexponential model and experimental pharmacokinetic data. *Journal of Pharmacokinetic and Biopharmaceutics*, **9**, 635-651.

18). SIWICKI A.K., 1989. Immunostimulating influence of levamisole on non specific immunity in Carp (*Cyprinus carpio*). *Developmental and Comparative Immunology*, 13, 87-91.

19). SIWICKI A.K., ANDERSON D.P. & DIXON O.W., 1990. *In vitro* immunostimulation of rainbow trout (*Oncorhynchus-mykiss*) spleen cells with levamisole. *Developmental and Comparative Immunology*, **14**, 231-237.

20). SOKAL R.R. & ROHLF F.J., 1981. *Biometry*. Second Edition, W. H. Freeman and Co Eds., New York.

21). TARASCHEWSKI H., RENNER C. & MEHLHORN H., 1988. Treatment of fish parasites. 3 - Effects of levamisole HCL, metrifonate, fenbendazole, mebendazole and ivermectin on *Anguillicola crassus* (nematodes) pathogenic in the air bladder of eels. *Parasitol Res.*, **74**, 281-289.

Discussion

ALDERMAN An interesting contribution, particularly since we have been concentrating on antibiotics and antimicrobials and have not so far considered any of the problems associated with the control of metazoan parasites.

ROTH Do you have any efficacy data on the nematodes?

BLANC A concentration of $8mg^{-1}kg^{-1}$ by means of intra-cardiac or oral administration gave a good efficacy against the adult stage of this parasite. However, the infective larval stage L3, which is in the swim bladder wall, is not haematophagous, and so will not be reached by the drug. Therefore the therapeutic scheme must take into account a combination of pharmacokinetic data and the growth of the third stage larvae.

VINK Have you any information on the occurrence of this parasite in salt water?

BLANC Our epidemiological study, which took place over 18 months in Mediterranean lagoons, has shown that the life cycle of this parasite progress in brackish and saltwater (35 to 40 $^0/_{00}$). In spite of the papers of several authors, it seems that saltwater is not a real barrier to the spread of this parasite.

VINK The reason for asking you is that eel farmers were encountering severe swim bladder parasite problems a few years ago and, as far as I know they were then obtaining their stock entirely from the Ijsselmeer, as wild catch fresh water eel of a weight around 20g. Since about 2 years ago the Dutch eel farmers transferred to the use of glass eel as stocking material and as far as I know the problem is gone.

BLANC It is not possible for glass eel to eat the third stage larvae because the digestive tract is still closed during almost all of the metamorphosis. So I think that is the reason why Dutch eel farmers are restocking with glass eel and no longer with wild eel caught in fresh water. Recently we found that the parasite is able to infest the eel at the very last stages of its metamorphosis at a body length of 67mm. So it is possible for the glass eel to eat the infective stage when they enter estuarine waters or lagoons. Nowadays, for aquacultural purpose, the best precautionary measure to avaoid this parasite is to catch the glass eel before they are able to eat.

THOMEN In Denmark most eel is farmed in recirculated water and there are growing concerns about the use of large amounts of chemicals in such environments. There is a lot of interest in the idea of using UV and ozone treatment of the water to improve the quality. Is there any work on possible side effects of radicals generated by these processes?

BLANC I have no information on this.

CHANGES IN DISTRIBUTION AND ACCUMULATION PATTERNS OF COPPER AND OTHER ESSENTIAL METALS DUE TO COPPER SULFATE TREATMENTS

G. Carbonell & J. V. Tarazona

Department of Animal Health. CIT-INIA
28130 Valdeolmos, Madrid, Spain

Summary *: Copper sulphate has a long history of use to control fish ectoparasites, but can produce marked toxic effects under certain conditions of use. Accumulation in tissue after bath treatment of rainbow trout and the resulting physiological effects have been investigated. Tissue analysis showed no accumulation of copper compared with control fish. In contrast the levels of iron and zinc appeared to be elevated in treated animals resulting in altered Fe/Cu and Zn/Cu ratios. This phenomenon could be caused by an increase in tissue metallothioniens in response to prolonged exposuresince these proteins have a strong affinity for copper. It therefore seems that copper sulphate treatments can lead to a response characterised by a readjustement in mineral equilibria, without a notable accumulation in tissues. Therefore it seems safe to conclude that there is no hazard to consumers from the use of copper sulphate as a fisheries therapeutant.*

Résumé**: Perturbations de la distribution et de l'accumulation du cuivre et d'autres métaux essentiels après traitements par le sulfate de cuivre.**

Le sulfate de cuivre sert classiquement à traiter les ectoparasitoses des poissons, mais peut présenter des effets toxiques marqués dans certaines conditions d'emploi. On s'est intéressé à son accumulation tissulaire et aux perturbations physiologiques pouvant en résulter, après administration soutenue chez la truite arc-en-ciel. L'analyse des tissus n'a révélé aucune accumulation du cuivre par rapport aux poissons témoins (Tableau I). Par contre les taux de fer et de zinc apparaissaient plus élevés chez les animaux traités, induisant une modification des rapports Fe/Cu et Zn/Cu (Tableau III). Ce phénomène pourrait être causé par une augmentation des métallothionéines tissulaire en réponse à l'exposition prolongée, ces protéines présentant une forte affinité pour le cuivre. Il semble donc que les traitements puissent entraîner une réponse caractérisée par un réajustement des équilibres minéraux, sans concentration notable de cuivre dans les tissus. On peut en conclure à l'absence de risque pour les consommateurs.

Introduction

Copper sulfate, is one of the most widely distributed fish farming treatments. It is used to prevent different kinds of processes related directly to the fish or to their environment.

Copper is an essential metal, with recognized biological role (21), and like other heavy metals, it is potentially toxic at high concentrations and it is a widespread water pollutant which can present a risk to aquatic organisms (11,14). The incorrect use of copper sulfate has frequently lead to mortality processes, affecting the production as a whole.

The purpose of the present study, is to study the consequences of the continuous use of copper sulfate in rainbow trout farms relating to copper accumulation in different fish tissues, and changes in copper ratios with other essential metals like iron and zinc.

Material and methods

Two groups of fish farms were chosen. In the first group, fish had been habitually exposed to copper sulfate treatment. In the second group, fish had never been treated with copper sulfate. Water quality was checked to assure that all fish farms were free of external pollution by heavy metals.

Fish were washed in a 6% EDTA solution and deionized water in order to avoid external contamination. Fish were dissected using glass and plastic tools, and fin, gill, opercle, liver, kidney, muscle and bone were obtained. Tissue samples were placed in porcelain crucibles and dried until constant weight at 105ºC and ashed overnight at 500ºC. The ashes were dissolved in 10 ml of 3% HNO_3 (Suprapur^R, Merck) and made up to 25 ml in volumetric flasks with deionized water. Tissues were analyzed for copper, iron and zinc, with acetylene flame by Atomic Absorption Spectrometry (Perkin Elmer, model 3030B).

Statistical analyses of the data were performed with the Student's t-test and one way Analysis of Variance run on an IBM PC computer; the 5% level ($p<0.05$) was considered significant.

Results and discussion

Copper concentrations in the analyzed tissues (fin, operculum, gill, liver, kidney, muscle and bone) were not higher in treated than in non-treated fish.

Copper sulfate treatment, has no apparent effect upon its accumulation in tissues (Table I).

Table I: Student's t test for copper.

ORGAN	t	D.F. .	Prob
Fin	-0.83	26	N.S.
Operculum	0.34	19	N.S.
Gill	-0.45	26	N.S.
Liver	-1.33	28	N.S.
Kidney	-1.12	28	N.S.
Muscle	1.31	27	N.S.
Bone	-1.18	21	N.S.

Statistical differences were observed for fin, gill, liver and kidney in the ANOVA analysis (Table II), showing differences in copper content between different fish farm not related to copper sulfate treatment. These results are in agreement with the lack of correlation between copper concentration in fish tissues and mortality observed several years ago (1,20) and the relatively constant copper concentration in liver found in seasonal studies for copper but not for other metals (15).

TABLE II. Analysis of variance for copper.

ORGAN	F	D.F.	Prob.
Fin	9.72	4 23 27	$p<0.05$
Operculum	1.12	4 16 20	N.S.
Gill	3.98	4 23 27	$p<0.05$
Liver	17.23	4 25 29	$p<0.05$
Kidney	0.88	4 25 29	N.S.
Muscle	2.14	4 24 28	N.S.

However, significant increases in iron and zinc concentrations between copper sulfate treated and non-treated fish were observed in several fish tissues (Table III) and therefore, in some iron/copper and zinc/copper ratios (Table III), changes in zinc/copper ratios could always been explained by changes in total zinc concentrations, but some significant differences were observed for iron/copper ratios without significant changes in iron concentrations. It must be considered that some of these significant differences were observed even when the one way ANOVA grouping by fish farms was not significant. Therefore, such as differences can not be imputed to differences between fish farms.

TABLE III. Analysis of variance and Studen's t test for iron zinc and iron/copper and zinc/copper ratios.

				STUDENT'S t Test		
ORGAN	F	D.F.	Prob	t	D.F.	Prob.
IRON						
Gill	13.80	4 23 27	p<0.05	1.27	26	N.S.
Liver	3.30	4 24 28	p<0.05	1.40	27	N.S.
Kidney	8.75	4 23 27	p<0.05	1.81	26	p<0.05
Muscle	5.80	4 25 29	p<0.05	1.80	28	p<0.05
Bone	16.87	4 23 27	p<0.05	1.31	26	N.S.
ZINC						
Gill	7.29	4 24 28	p<0.05	4.36	27	p<0.05
Liver	6.01	4 24 28	p<0.05	1.89	27	p<0.05
Kidney	10.49	4 23 27	p<0.05	42	26	p<0.05
Muscle	17.14	4 25 29	p<0.05	1.36	28	N.S.
Bone	27.60	4 24 28	p<0.05	2.91	27	p<0.05
Fe/Cu						
Gill	2.39	4 22 26	N.S.	1.87	25	
Liver	3.34	4 25 29	p<0.05	2.71	28	p<0.05
Kidney	3.97	4 25 29	p<0.05	2.94	28	p<0.05
Muscle	1.31	4 23 27	N.S.	0.08	26	N.S.
Bone	3.44	4 18 22	p<0.05	1.47	21	N.S.
Zn/Cu						
Gill	4.37	4 22 26	p<0.05	3.17	25	p<0.05
Liver	2.83	4 25 29	p<0.05	2.06	28	p<0.05
Kidney	2.21	4 25 29	N.S	2.49	28	p<0.05
Muscle	4.13	4 23 27	p<0.05	0.004	26	N.S.
Bone	3.06	4 18 22	p<0.05	3.24	21	p<0.05

One group of compounds that may have a role in such a process are the metallothioneins, a group of low-molecular-weight, cysteine-rich, metal-binding proteins, ubiquitous in the living world (7), isolated from salmonid fish by McCarter *et al.*(12), and proposed as indicators of metal exposure in aquatic organisms (9).

Metallothionein normally occurs in tissues in only trace amounts, however, exposure to sublethal doses of certain metals such as cadmium, copper and zinc induces thionein synthesis and the binding of the apoprotein to the metal, thus forming metallothionein (7, 8, 10)). The affinity of metals for metallothionein is $Zn < Cd < Cu$ been those metals of higher affinity able to replace to those metals of lower affinity (4, 6).

Metallothionein levels were not analyzed in this study, but it can be predicted from current knowledge a significant increase of metallothionein levels in fish periodically treated with copper sulfate. Prasada *et al.*(16) found increase in the levels of copper and zinc in kidney by cadmium administration and of iron by lead and methyl-mercury, always with a several-fold increase in metallothioneins. A similar effect may be considered in order to explain our results, taken into account that tissue distribution of metallothioneins in fish primarily includes the gills, liver and kidney, three organs for which significant differences were found for total Zn and Zn/Cu, Fe/Cu ratios.

Following this hypothesis, the periodic expositions to copper may suppose an increase in metallothionein levels which would lead to changes in Zn, Fe and their copper ratios, while copper concentrations would remained unchanged due to the ability of the fish to increase the elimination rates and maintain homeostatic levels.

The meaning of metal/copper ratios has also been pointed out in other works in our laboratory, i.e. iron/copper and zinc/copper ratios could help to distinguish between survival or dead fish after an acute lethal poisoning by copper sulfate when no differences were observed for copper concentration (3).

From a practical point of view, our results show that periodic treatments with copper sulfate may lead, in any case, to some adjustment in the nutritional requirements of other essential metals, but has not consequences for the human consumer, because of no changes in essential metals or their copper ratios were observed in muscle, that can be considered the edible part of the fish, with the only exception of a low increase in iron levels.

Acknowledgements: This study was supported in part by a grant (INIA Research Project 8569) of the National Institute for Agricultural Researches.

References

1). BRUNGS, W.A., LEONARD, N.E. & McKIM, J.M., 1973. Acute and long term accumulation of copper by the brown bullhead, *Ictalurus nebulosus. J. Fish. Res. Bd. Can.*, **30**, 583-586.

2). BUCKLEY, J.T., ROCH, M., McCARTER, J.M., RENDELL, C.A. & MATHENSON, A.T., 1982. Chronic exposure of coho salmon to sublethal concentrations of copper. I. Effect on growth, on accumulation and distribution of copper and copper tolerance. *Comp. Biochem. Physiol.*, **72C**, 115-119.

3). CARBONELL,G.& TARAZONA,J.V., 1990. Toxicidad por adición de sulfato de cobre en piscifactorías. *Actas tercer Congreso Nacional de Acuicultura*. A. Landin y A. cerviño ed., 917-922.

4). DAY, F.A., FUNK A.E. & BRADY, F.O., 1984. *In vivo* and *ex vivo* displacement of zinc from metallothionein by cadmium and by mercury. *Chem.-Biol. Interact.*, **50**, 159-174.

5). DIXON, D.G. & SPRAGUE, J.B., 1981. Copper bioaccumulation and hepatoprotein synthesis during acclimatation to copper by juvenile rainbow trout. *Comp. Biochem. Toxicol.*, **1**, 69-81.

6). FUNK, A.E. DAY, F.A. & BRADY, F.O. 1987. Displacement of zinc and copper from copper-induced metallothionein by cadmium and by mercury: *in vitro* and *in vivo* studies. *Comp. Biochem. Physiol.*, **86C** (1),1-6.

7). HAMILTON, S.J. & MERHLE, P.M. 1986. Metallothionein in fish: Review of its importance.in assessing stress from metal contaminants. *Trans. American Fish Soc.*, **115**, 596-609.

8). HELD, D.D & HOEKSTRA, W.G., 1984. The effects of zinc deficiency on turnover of cadmium metallothionein in rat liver. *J. Nutr.*, **114**, 2247-2282.

9). HOGSTRAND, C., LITHNER, G. & HAUX, C., 1989. Relationship between Metallotionein, copper and zinc in perch (*Perca fluviatilis*) environmentally exposed to heavy metals. *Marine Environmental Research.*, **28**, 179-182.

10). KÄGI, J.H & KOJIMA, Y., 1987. Chemistry and biochemistry of metallothionein. In, *Metallothionein II*, J.H.R. Kägi & Y. Kojima eds, Birhaüser/Verlag, Basel, pp 25-61.

11). KARIYA, T.Y., HAGA, H. & TSUNDA, T., 1967. Studies on the *post mortem* identification of the pollutant in the fish killed by water pollution.5. Detection of copper in the fish. *Bull. Jap. Soc. Scient. Fish.*, **33**, 818-825.

12). McCARTER, J.A., MATHESON, A.T., ROCH, M., OLAFSON, R.W. & BUCKLEY, J.T., 1982. Chronic exposure of coho salmon to sublethal concentrations of copper. II - Distribution of copper between high- and low-molecular-weight proteins in liver cytosol and the possible role of metallothionein in detoxification. *Com.Biochem. Physiol.*, **72C** (1), 21-26.

13). MOUNT, D.I. & STEPHAN, C.E., 1967. A method for detecting cadmium poisoning in fish. *J. Widl. Manage*, **31**, 168-172.

14). NEFF, J.M., 1985. Use of biochemical measurements to detect pollutant-mediated damage to fish. American Society for Testing and Materials, Special Technical Publication **854**, 155-183.

15). OLSSON, P.E., HAUX, C. & FÖRLIN, L., 1987. Variations in hepatic metallothionein, zinc znd copper levels during an annual reproductive cycle in rainbow trout, *Salmo gairdneri*. *Fish Physiol. and Biochem.*, **3** (1), 39:47.

16). PRASADA, P.V.V. JORDAN, S.A. & BHATNAGAR, M.K., 1989. Combined nephrotoxicity of methylmercury, lead, and cadmium in pekin ducks: Metallothionein, metal interactions, and histopathology. *J. Toxicol. Environ. Health*, **26**, 327-348.

17). PROBST, G.S., 1979. Adsorption and excretion. In, *Cadmium toxicity*, J.H. Mennear ed., Marcel Dekker, New York., pp 40-59.

19). SCOTT, B.J. & BRADWELL, A.R., 1983. Identification of the serum binding proteins for iron, zinc, cadmium, nickel and calcium. *Clin. Chem.*, **29**, 629-633.

20). STOKES, P., 1979. Copper accumulations in freshwater biota. In, *Copper in the environment. Part.I: Ecological Cycling*, Jerome O. Nriagu, ed.

21). SUZUKI, K.T., YAGACHI, K., OHNUKI, R., NISHIKAWA, M., YAMADA, Y.K, 1983. Extent of cadmium accumulation and its effect on essential metals in liver, kidney and body fluids., *J. Toxicol. Environ. Health.*, 11:713-726.

PUBLIC HEALTH ASPECTS OF RESIDUES IN ANIMAL PRODUCTS: FUNDAMENTAL CONSIDERATIONS

M. Yndestad

Department of Food Hygiene, Norwegian College of Veterinary Medicine, P.O.Box 8146 -Dep., 0033 Oslo 1, Norway

Summary: *The world-wide increasing use of veterinary drugs may result in an increasing amount of residues in a large part of foods of animals origin. Public health aspects are associated with toxicological disorders, allergies and bacterial antibiotics resistance problems. Recent toxicological data indicate that there are doubts concerning residues of several drugs such as nitrofurans, sulphamethazine, and chloramphenicol, as well as growth promotors and some antiparasite drugs. Several drugs can induce allergy reactions in humans. This is apparently not a problem associated with low residue level of drugs. It is possible that not all allergic reactions are reported. The most difficult task is to evaluate the relationship between residue levels and resistance problems. Some workers consider that 1/10-1/20 of the MIC value for sensitive intestinal bacteria is a reasonable tolerance level, whereas other scientists regard that level as unrealistic. A maximum residue level (MRL) must be set following a complete evaluation of relevant factors.*

Résumé **: Conséquences pour la santé publique de la présence de résidus thérapeutiques dans les produits d'origine animale.**

Les méthodes d'élevage modernes imposent l'emploi de grandes quantités de médicaments. Ainsi en Norvège en 1990, 50 t de médicaments ont été employées à titre vétérinaire, soit 2 fois plus qu'en médecine humaine. La persistance prévisible de résidus dans les produits alimentaires soulève dès lors des problèmes divers.

- Problèmes toxicologiques: au-delà des délais d'attente, des contaminations à bas niveau excluent-elles les risques d'effets carcinogènes, mutagènes ou tératogènes ? Les doses sans effet observable (DSE) et les doses journalières acceptables (DJA), calculées en appliquant des coefficients de sécurité, n'apportent pas toujours une garantie suffisantes (pour les antibiotiques par exemple: l'oxytétracycline a une DSE de 18 mg/kg!). Pourtant les aspects toxicologiques sont relativement bien étudiés. C'est le cas des furanes, qui ont des effets au niveau des acides nucléiques. Mais ils se dégradent très vite, même après l'abattage, et l'inconnue provient des métabolites. L'aplasie médullaire causée par le chloramphenicol serait également due à des métabolites, l'absence d'effet dose conduisant à en préconiser l'interdiction

pure et simple. Les sulfamides, qui pourraient avoir des effets carcinogénétiques, et certains anthelminthiques sont également suspects.

- Hypersensibilisation: souvent simple intolérance, elle peut se traduire par de véritables allergies et aboutir au choc anaphylactique. Il ne semble pas qu'aux doses où on les détecte les résidus soient allergènes. Seule la réaction de personnes sensibilisées est à craindre. A ce jour les rapports d'accidents liés à la consommation d'aliments n'intéressent que la pénicilline, mais peut-être beaucoup de cas ne sont-ils pas répertoriés?

- Effets microbiologiques: la multiplication des résistances simples ou combinées aux antibiotiques, fréquente dans la nature, se retrouve dans les bactéries de la flore digestive humaine et paraît reflèter l'emploi accru de ces produits. L'augmentation de fréquence des résistances au contact d'animaux d'élevage a plusieurs fois été rapportée à propos des salmonelles. Bien que les perturbations des antibiotiques sur l'équilibre de la flore intestinale ait été très étudiée, le rôle éventuel des résidus demeure obcur. L'idée de fonder les DJA sur les concentrations minimales inhibitrices (CMI) des bactéries sensibles de la flore est assez contestée.

On dispose maintenant de méthodes analytiques très fines, dont les seuils de détection sont parfois proposés pour fixer les valeurs admissibles de résidus. Il en résulte des délais d'attente variables et arbitraires. Logiquement les limites maximales de résidus (LMR) devraient être fondées sur des observations cliniques plutôt qu'analytiques, en tenant compte des consommations alimentaires moyennes. En fait, beaucoup de points (résidus liés, métabolites) demeurent obscurs, et ce sont surtout les DJA qui servent de guide pour les LMR. Des délais d'attente sont alors proposés en fonction des données pharmacocinétiques. Beaucoup de travaux seront encore nécessaires pour conforter cette démarche.

Introduction

In modern livestock farming, including aquaculture, there will always be a need for various types of medicines to prevent or treat different diseases. This is a circumstance which control authorities and consumers both understand and accept. Various drugs are also employed to increase the efficiency of feed utilisation, stimulate growth rate and increase yield (growth promoters). It is the latter use which is the most controversial.

Overall, large quantities of active substances are given to food producing animals (1). In Norway, the quantity of antibiotics used in veterinary medicine, including fish farming, is about double that employed in human medicine (50 tons compared to 25 tons, 1990), even though Norway follows a restrictive policy in this field and the use of antibiotics as growth promotors is normally not allowed.

As veterinary drugs are administered to a large proportion of livestock, in some countries the entire livestock population, for therapeutic, prophylactic or growth promoting reasons, the possibility will always exist that residues will occur in food products of animal origin.

The most important public health aspects connected with such residues will be discussed in the following.

Antibiotics comprise the largest and most important group of drugs employed, and I will therefore primarily address my remarks to these agents. Public health aspects of veterinary drug residues fall naturally into three groups:

- Toxicological aspects
- Drug hypersensitivity
- Microbiological aspects

Toxocological aspects of veterinary drug residues

In most countries, withdrawal times have been laid down for drugs used in animal production. Acute toxicity is therefore hardly relevant in connection with drug residues. The most important aspect is the significance of low level contamination.

According to Raynaud *et al.* (28), the major tests to identify the toxicological properties of residues include:

- Carcinogenicity: irreversible damage to cellular DNA. For this category no residue is permitted; zero tolerance type.
- Mutagenic effect: damage to the genetic components of a cell or organism.
- Teratogenic effect: toxic effect on the embryo or foetus during a critical phase of gestation.

As for any substance, an attempt is made to determine a No Observable Effect Level (N.O.E.L.), and a safety factor (S.F.) is applied in order to extrapolate the N.O.E.L. to human consumers. An Acceptable Daily Intake (ADI) can then be estimated.

As regards drugs such as anthelminthics and non-anti-bacterial growth promoters, it is chiefly toxicological data which must constitute the basis for the determination of maximum residue levels (MRL). However, if only toxicological aspects are taken into account, it would apparently be possible to consume large daily amounts of a number of antibacterial agents. The no-effect level for oxytetracycline has, for example, been found in toxicological studies to be 18 mg/kg body weight per day, or higher (11). In the context of drug residues, toxicological data must therfore always be considered in conjunction with hypersensitivity and the development of bacterial drug resistance. Nevertheless, it is important to note that, in recent years, attention has been focused on several antibacterial agents solely because of adverse toxicological assessments. Some of these substances are still extensively used in veterinary medicine, and the potential health hazards posed by the presence of their residues in food will therefore be briefly described.

Nitrofurans

These substances comprise a group of synthetic antimicrobial compounds based on a furan ring bearing a nitro group at position 5. Studies with ^{14}C-labelled nitrofurans have confirmed that these compounds are able to bind biologically-important molecules such as DNA and RNA (5). They also cause single-strand breaks in DNA and inhibit DNA synthesis (27). In general, most nitrofurans are genotoxic and mutagenic in one or more assay system. They are also proven carcinogens in experimental animals. It is therefore quite clear that residues of these compounds should be present only at the lowest possible levels in edible tissues from animals and fish. A human daily intake of < 0.002 µg/kg b.w. will give a carcinogenic risk of less than $1:10^7$. This would be in the acceptable range (31).

It is important to note that furazolidone degrades quickly in tissues after slaughter. In calves weighing approximately 140 kg which were slaughtered when serum levels after administration of the drug were at their peak (3-4 hours after administration), only traces of the parent drug could be detected two hours after slaughter (26). This shows that it is the <u>metabolites</u> which create the greatest problem, both those which arise due to metabolism of the drug in the live animal and those which appear as a result of degradation after slaughter. A lot of data are still lacking. The European community has been discussing a tolerance level of 1 ppb of the parent drug in edible tissue (26).

Chloramphenicol

Chloramphenicol is another drug against which questions have been raised in the context of residues. The damage to bone marrow cells which can be observed in experimental animals is thought to be due to covalent binding of metabolites, with consequent effect on DNA. This results in aplastic anaemia in predisposed persons, about 1/40,000. There appears to be agreement that this effect is independent of dose (37).

Chloramphenicol has been evaluated by the Joint FAO/WHO Expert Committee on Food Additives (JECFA). The Committee stated among other things:" The mechanism for the pathogenesis of aplastic anaemia is unknown, and no suitable animal model exists. Thus, a no-effect level could not be established and an ADI could not be allocated for chloramphenicol because it was not possible to give an assurance that residues in foods of animal origin would be safe for sensitive subjects. Therefore, the Committee did not recommend an Acceptable Residue Level". The Committee recommended that efforts should be made to replace or prohibit the use of chloramphenicol in food-producing animals, particularly in laying birds and lactating animals where high levels of residues in eggs and milk were major problems (9).

Sulfamethazine

Strong reservations of a public health nature have been raised against this drug, which has been used for years for therapeutic treatment, for improving efficiency of feed utilization, and for growth promotion. The National Center for Toxicological Research in the USA reported that sulfamethazine caused a dose-related increase in follicular cell adenomas of the thyroid gland in rats and mice (24,25). The increase was noted after feeding moderate to high doses of sulfamethazine for 18 to 24 months. It is therefore clearly a serious matter that illegal residues of sulfamethazine (> 0.1 ppm) in pigs in the USA fluctuated between 4% and 5% during the period 1980 to 1987 (3). A screening study carried out in Norway revealed a frequency of pigs with detectable residues of sulfonamides of < 0.01% (Kluge-Berge, Personal communication).

These examples illustrate that toxicological studies are very important with regard to antimicrobial agents when assessing use and acceptable residue levels.

Anthelminthic drugs and growth promotors

These comprise a large group of active substances which deserve attention. They include compounds which exert a teratogenous effect (e.g. albendazol and mebendazol), or a tumour-enhancing effect (e.g. carbadox) (7,11). Several of these drugs are currently being evaluated by the previously mentioned Joint FAO/WHO Expert Committee, and one must therefore await their final opinion. These substances are nevertheless very relevant with regard to public health, among other reasons, because adequate screening methods for the routine determination of residue levels in muscular tissue and organs are lacking. As regards carbadox, JECFA has proposed a maximum residue level of 5 µg/kg, this low figure clearly reflecting the toxicological reservations associated with its use.

Drug hypersensitivity

Generally speaking, hypersensitivity can have an immunological basis (allergy), or a non-immunological basis (intolerance).

Hypersensitivity often manifests itself as a skin reaction (e.g. urticaria and dermatitis), and/or symptoms involving the respiratory system (rhinitis and bronchial asthma). Allergic reactions resulting in anaphylactic shock are more uncommon.

One of the problems associated with drug residues is the possibility of becoming hypersensitive on repeated contact with a drug. Factors which influence sensitisation include intensity, duration and frequency of exposure to the allergen. In this connection, it is important to be aware of the fact that intermittent exposure to low or moderate doses has proven to result in sensitisation more often than prolonged therapy at high dose levels. This is characteristic, for example, of penicillin allergy, and is of significance when considering the issue of drug residues.

As regards hypersensitivity, it seems, nevertheless, that the small amounts found in foods are not likely to induce sensitisation. The significance of drug residues therefore lies in the potential which the drug has to elicit an allergic response in a previously sensitised person.

The route of administration is also of importance for the development of allergy. For example, local application, intramuscular injection, and oral administration of penicillin induces sensitisation in 10-29%, 1-2%, and between 0.1 and 0.5% of patients, respectively (32).

The penicillins constitute the most important group of drugs causing allergy, though it is also well known that, among others, tetracycline, streptomycin, sulfonamides and nitrofuran derivatives may all induce allergy.

In spite of the widespread use of veterinary drugs, there are few published reports describing allergic reactions in humans due to the consumption of foods containing drug residues. Burgat, (6) presented an overview of published accounts of penicillin allergy in humans after intake of foods containing penicillin residues. Milk has most often been associated with cases of allergy, though meat and soft drinks have also been implicated. The incidence of allergy is extremely low, considering the fact that from 1 to 10% of the general human population have a history of adverse reactions to penicillin (1).

This might indicate that such reactions are under-reported. This suggestion is supported by the fact that in Norway, where milk for consumption very seldom contains penicillin (<0.1%), two episodes have occurred during the last ten years in which a number of people developed pruritus, urticaria and itchy-eruption of the face. Milk implicated in one of the outbreaks (5 samples) was analysed at the Department of Food Hygiene, and found to contain from 0. 05 to 0. 6 international units of penicillin per ml. Neither of these outbreaks have been reported in the international literature.

Another reason for the lack of reports may of course be that drug residue levels are so low, that they are unable to cause allergic reactions. As regards penicillins, it has been shown that 3-9 µg penicillin in milk has resulted in allergic reactions in humans (4). The thirty-sixth meeting of the joint FAO/WHO expert committee on food additives recommended that the daily intake of benzylpenicillin from food be kept as low as practicable, and in any case below 30 µg of the parent drug (36).

On the basis of available data, the thirty-sixth JECFA report concludes, among other things, that: hypersensitivity reactions due to the ingestion of food of animal origin containing allergic drug residues are unlikely to be of major health significance (11).

Microbiological aspects

The consumption of veterinary drugs has steadily increased over the years. Data provided by Adkinson (1) give a good idea of just how great this increase has been for antibacterial agents (Table I).

Table 1. Antibiotic production in 1958, 1968, and 1978 (USA).
Amounts given in million kg.
(Values exclude production of sulfonamides)

Year	Total	Medicinal use in humans and animals	Added to animal feed and other applications
1958	1. 9	1.18	0.41
1968	4. 67	2.72	1.95
1978	11.66	6.08	5.58

These figures reveal a considerable increase in the overall consumption for treatment, prophylaxis, and growth-promotion.

A number of observations show that resistance, in the bacterial flora, to antibacterial substances has gradually built up over the years. No-one today is likely to doubt that this is associated with the increasing use of such compounds.

Somewhat simplified, it can be said that there are two types of resistance:

- Natural, or primary resistance
- Acquired, or secondary resistance.

Primary resistance is the term used when bacteria possess an inherent resistance towards certain antibiotics (for example, *E. coli* is naturally resistant to penicillin).

Acquired (secondary) resistance is the most important type with regard to drug residues. This type of resistance can arise spontaneously at a certain rate (mutation). Mutant cells will normally have difficulty in colonising or establishing themselves in a system unless factors are present which suppress the competing flora (e.g. antibiotics). Another type of acquired resistance is transferable resistance associated with bacterial plasmids. Plasmids may contain resistance factors for one or more antibiotics (single or multiple resistance). Transferable resistance can be transferred within the same species, as well as between species, to sensitive cells by resistance plasmid (R-plasmid).

Resistant bacteria in the environment and in foods

The occurrence of multiple-resistant bacteria in the environment is a well known phenomenon. Many investigators believe that also these drug-resistant organisms have become more common recently, due to the extensive use of antibiotics in medicine and agriculture throughout the world (23). Many of these bacteria which possess plasmids exhibit high rates of survival in aquatic environments (18).

It has been documented that bacterial isolates from remote regions such as the Antarctic contain plasmids of one kind or another (42% of sediment isolates and 20% of sea-water isolates). This suggest that there is a certain background level of plasmid-carrying bacteria, which also may carry R-plasmids, in nature (19). Because of their presence in the environment, it is natural that resistant bacteria can occur in drinking water and food. Overall, 33. 9% of 2,653 bacterial isolates from treated drinking water from seven municipal water works showed multiple antibiotic resistance (2). Jayaratne *et al.* isolated nine multiple antibiotic resistant *E. coli* from commercially available meat. All isolates contained plasmids, and were resistant to seven or more antibiotics (17). There is thus no doubt that humans are exposed in some degree to resistant bacteria, both through food and the environment.

Resistant bacteria in animals and man

It is well recognised that the administration of sub-therapeutic amounts of antibiotics leads to an increase in the frequency of occurrence of antibiotic-resistant strains in both animals and man. Levy *et al.* demonstrated that fecal samples from people living on a poultry farm contained more than 80% of tetracycline-resistant bacteria (22). Resistance disappeared, however, when the feeding of the drug to the poultry was discontinued. There are also many other studies which show that resistant bacteria can be found among the intestinal flora of farm-workers who have close and regular contact with animals and their antibiotic-enriched feed.

About three years ago Levy *et al.* made an extensive study of antimicrobial resistance in human fecal flora in the Boston area. In a majority (62.5%) of fecal samples from people without a recent history of taking antibiotics, 10% or more of the total organisms were resistant to at least one antibiotic (21).

Also with regard to pathogenic enteric bacteria such as salmonella, one must anticipate an increase in the incidence of resistant strains in connection with the use of antibiotics. Holmberg *et al.* studied 52 human *Salmonella* outbreaks during the period 1971-83 (16). They showed that the case fatality rate was higher for patients infected with antibiotic-resistant *Salmonella* than for those infected with sensitive pathogens. But no direct evidence of a role of sub-therapeutic use of antimicrobials in animals was presented by the authors. In a second study, Holmberg *et al.* described an outbreak of human salmonellosis caused by *S. newport*. On studying plasmid profiles of human and animal *S. newport* isolates, it was concluded that the antimicrobial-resistant organisms infecting the humans were of animal origin, and that contaminated hamburgers was the probable source (15).

In their excellent review "The use of antimicrobial agents in animal feeds: Implications for human health", Dupont and Steele summarised a number of reported cases of *Salmonella* transfer from animals to man (8). They mention, among other things, that while one study showed that no change has taken place in the resistance pattern in human *Salmonella* isolates during the period 1975-1984 in the USA, another investigations revealed that the frequency with which resistance occurs in certain *Salmonella* types has increased (8).

There is therefore reason to assert that the use of sub-therapeutic concentrations of antibiotics, and contact with antibiotic supplemented feed, seems to encourage the development of resistance, not only in naturally-occurring intestinal bacteria such as *E. coli*, but also in enteric pathogenic bacteria such as *Salmonella*.

There appears to be a high frequency of antimicrobial resistance in human fecal flora from urban and rural communities. There also seems to be a certain normal frequency in nature of plasmids bearing resistance genes (R-plasmids).

It is important to take all these factors into account when considering the significance of residues of antibiotics in foods.

Residues in foods - microbiological significance

According to Somogyi (34), the microbiological issues with regard to drug residues are as follows:

- Can residues exert a selective pressure on the intestinal flora, and thus favour the growth of microorganisms with natural or acquired resistance?
- Can residues give rise, directly or indirectly, to the development of acquired resistance in pathogenic enteric bacteria?

As mentioned previously, it has been documented that there is an increasing frequency of resistant bacteria both in environment, in animals, in food of animal origin and in man. The higher this frequency, the greater is the likelihood that such bacteria under the influence of antibiotics, will be able to establish themselves in man. The problem is then to determine the levels of antibacterial substances which can be considered, with a reasonable degree of certainty, not to exert any adverse effect in this connection.

The human intestinal flora has a complex composition, dominated by various types of anaerobic bacteria (10^9 CFU/g). Enterobacteriacae also make a significant contribution, about $5x10^7$ CFU/g. The intestinal flora otherwise includes a transient flora which may temporarily colonize the gut if protective barriers are perturbed. Savage states, "As long as the communities of anaerobes remain intact in the adult, the population levels of *E. coli* remain low. If the communities of anaerobes are perturbed by antimicrobial drugs, the levels of *E. coli* may rise dramatically" (30).

Bacteria of animal origin usually have difficulty in establishing themselves in human intestines. Such establishment may, however, occur with antibiotic-resistant strains present under the influence of antibiotics (14).

Tancrède failed to observe any changes in the anaerobic flora or in the enterobacteriacae population or the *E. coli* flora during treatment with 2 mg oxytetracycline/day to human volunteers (35). Goldberg *et al.* showed that 10 mg of oxytetracycline given daily for six months resulted in the selection of highly resistant coliform bacteria in some of the volunteers tested (13).

Since the human flora is the hypothetical target of the resistance-inducing action of residues, it has been proposed to use the minimum inhibitory concentration (MIC) of the substance in question as the starting point for the estimation of acceptable residue levels in foods. As test organisms for MIC values, it is natural to use sensitive pathogenic enteric bacteria and *E. coli*.

As concentrations below MIC values may favour the growth of resistant bacteria, it is natural to set the acceptable residue level lower than the MIC value. Based on the studies of Lebek and Egger (20), it is suggested that the residue level should be about 1/10 or 1/20 of the MIC value (33).

Rico states that the French Pharmacopea recommends 1/10 of the model MIC for sensitive *E. coli* populations or the lowest MIC of the most sensitive microorganism in the gastrointestinal flora as the acceptable residue level (29).

Swiss legislation has summarised the problem thus: the average daily intake will be retained as a dose which guarantees the absence of plasmid-selective concentrations in the human intestine (28).

In a very thorough examination of the issues involved, Raynaud *et al.* note that the problems are numerous. "It is difficult to identify a Model MIC value of a sensitive *E. coli* population. It is difficult to determine the lowest MIC value in the anaerobes. How can the intake in ppm in food and the actual level of the antibiotic when in contact with the gastrointestinal flora be correlated?" (28).

The same authors have assessed a proposed MRL for oxytetracycline in muscle of 10 ppb. They conclude as follows: "The rationale of the calculations for no- effect level and microbiological risk is understood, but the use of these factors as coefficients to obtain a tolerance of 10 ppb is not. This seems to be purely theoretical and totally unrealistic". They further state that all things considered and looking at the picture as a whole, a more appropriate tolerance level for muscle and milk would seem to be 0.1 ppm. This is in fact the same tolerance level as proposed by the Joint FAO/WHO Expert Committee on Food Additives (11).

In the above discussion there are clearly some discrepancies in the relationship between antibiotic residue levels in foods and development of bacterial resistance. Further epidemiological data are also needed for a proper evaluation of this problem. The possible synergistic effect of different antibiotics residues has to be considered as well.

Maximum Residue Limits - Withdrawal Times

Modern analytical methods permit most substances to be determined at much lower concentrations than previously. Many countries, including Norway, have general regulations requiring the complete absence of drug residues in foods. In practice, the limit of detection of the method employed is then the maximum level permitted. The maximum residue level is, in these circumstances, not really determined on the basis of scientific evidence, but rather on analytical considerations. The following example with oxolinic acid, one of the most extensively used drugs in fish farming in Norway, will illustrate the point.

Microbiological test systems based on *E. coli* as test organism can detect about 100 ppb in liver. A withdrawal time of about 20 days would then be appropriate. Our HPLC method with a detection limit of 0.5 ppb will give a withdrawal time of more than 80 days.

It is neither logical nor realistic to accept that the question of undesirable drug residues in food should be based on the analytical method available. It is therefore essential to set maximum residue levels (MRLs) in foods. These must be fixed first and foremost on health considerations, not analytical aspects.

In this context, it is also important to take into account:

A. Food factors and intake data
B. Bound residues - bioavailability
C. Metabolites

A: The amount of food consumed is of course of great significance for the setting of an MRL. Codex bases its recommendations on the following daily intake values (10):

- 300 g of meat (as muscle tisssue)
- 100 g of liver, 50 g of kidney and 50 g of tissue fat
- 100 g of egg
- 1.5 litres of milk

B: The use of veterinary drugs in food-producing animals can result in residues that are not extractable from tissues. If bound residues make up a insignificant portion of the total residue, no problem will arise. However, in certain cases, such bound residues will comprise a considerable proportion of the total amount of residue. In such cases, it is important to determine the extent to which the bound residue is releasable in the human gut.

Availability can be estimated in experimental animal feeding trials as, for example, described by Gallo-Torres (12).

C: When considering residues, it is natural to focus attention on the parent drug. In most cases, this is quite in order, as the metabolites are usually less toxic and are excreted at least as rapidly as the active component. Metabolites can arise, not only as a result of metabolism *in vivo*, but also during storage. The significance of metabolites as residues in this connection is an issue of which more knowledge is required.

As a consequence of the different aspects discussed above the Codex Committee on Residues of Veterinary Drugs in Foods has proposed that the MRL should be "based on the type and amount of residue considered to be without toxicological hazard for human health as expressed by Acceptable Daily Intake (ADI) or on the basis of a temporary ADI that utilises an additional safety factor. It (the MRL) also takes into account other relevant public health risks as well as food technological aspects and estimated food intakes" (10).

Once the MRL is fixed, knowledge of the pharmacokinetics of the drug, and results from depletion studies, will allow withdrawal periods to be set for a particular drug in the various animal species. Reliable and safe withdrawal periods are very important for the protection of public health.

Conclusion

The use of veterinary drugs has on a world basis increased steadily during the last decades. There are reasons to believe that more and more foods of animal origin will contain residues. In order to protect the consumer it is important to set maximum residue levels (MRL) for different drugs. The MRL must be based on data from toxicology, allergy and bacteriological resistance studies and not on analytical methods. There are for many drugs still insufficient data, especially concerning bacterial resistance, to set an MRL that is acceptable to safeguard public health.

References

1). ADKINSON F.N., 1980. Immunological consequences of antimicrobials in animal feed. In: *The effects on human health of subtherapeutic use of antimicrobials in animal feed.* Committee to study the subtherapeutic antibiotic use in animals feeds. National Academy of Sciences, Washington DC.

2). ARMSTRONG J.L., SHIGENO D.E., CALOMIRIS J.J. & SEIDLER R.J., 1981. Anibiotic-resistant bacteria in drinking water. *Appl. Environ. Microbiol.*, **42**, 277-283.

3). AUGSBURG J.K., 1989. Sulfa residues in pork: an update. *J. Anim. Sci.*, **67**, 2817-2821.

4). BORRIE P. & BARRETT J., 1961. Dermatitis caused by penicillin in bulked milk supplies. *Brit. Med. J.*, 1267.

5). BOYD M.R., STIKO A.W. & SASAME H.A., 1979. Metabolic activation of nitrofurantoin - possible implications for carcinogenesis. *Biochem. Pharmacol.*, **28**, 601-606.

6). BURGAT, 1986. Le risque allergique des résidus de médicaments vétérinaires. In: *3rd International technical consultation on veterinary drug registration*, Paris.

7). DELATOUR P., LORGUE G., LAPRAS M. & DESCHANEL J.P., 1974. Embryotoxic properties (rat) and residues (sheep and cattle) of three anthelmintics derived from benzimidazole. *Bull. Soc. Sci. Vet. Med. Comp.*, **76**, 147-154.

8). DUPONT H.L. & STEELE J.H., 1987. Use of antimicrobial agents in animal feeds: Implications for human health. *Reviews of infectious diseases*, **9**, 447-459.

9). Evaluation of certain veterinary drug residues in food, 1988. *Thirty second Report of the Joint FAO/WHO Expert Committee on Food Additives*, WHO, Geneva.

10). Evaluation of certain veterinary drug residues in food, 1989. *Thirty fourth Report of the Joint FAO/WHO Expert Committee on Food Additives*, WHO, Geneva.

11). Evaluation of certain veterinary drug residues in food, 1990. *Thirty sixth Report of the Joint FAO/WHO Expert Committee on Food Additives*, WHO, Geneva.

12). GALLO-TORRES H.E., 1977. Methodology for the determination of bioavailability of labeled residues. *Journal of Toxicology and Environmental Health*, **2**, 827-845.

13). GOLDBERG H.S., GOODMAN R.N., LOGUE J.T. & HANDLER F.P., 1961. Longterm, low-level antibiotics and the emergence of antibiotic-resistant bacteria in human volunteers. *Antimicrob. Agents, Chemother.*, **1**, 80-88.

14). HIRSH D.C., BURTON G.S. & BLENDEN D.C., 1974. The effect of tetracycline upon establishment of *Escherichia coli* of bovine origin in the enteric tract of man. *J. Appl. Bacteriol.*, **37**, 327-333.

15). HOLMBERG S.D., OSTERHOLM M.T., SENGER K.A. & COHEN M.L, 1984. Drugresistant *Salmonella* from animals fed antimicrobials. *N. Engl. J. Med.*, **311**, 617-622.

16). HOLMBERG S.D., WELLS J.G. & COHEN M.L, 1984. Animal to man transmission of antimicrobial-resistant *Salmonella*: investigations of U.S. outbreaks, 1971-1983. *Science*, **225**, 883-835.

17). JAYARATNE A.H.G., COLLINS-THOMPSON D.L. & TREVOR J.T., 1987. Plasmid-encoded transferable multiple antibiotic resistance in *Escherichia coli* isolated from meat. *System. Appl. Microbiol.*, **10**, 78-84.

18). KELCH W.J. & LEE J.S., 1978. Antibiotic resistance pattern of gram-negative bacteria isolated from environmental sources. *Appl. Environ. Microbiol.*, **36**, 450-456.

19). KOBORI H., SULLIVAN C.W. & SHIZUYA H., 1984. Bacterial plasmids in antarctic natural microbial assemblages. *Appl. Environ. Microbiol.*, **48**, 515-518.

20). LEBEC G. & EGGER G., 1983. R-selection of subbacteriostatic tetracycline concentrations. *Zbl. Bakt. Hyg. Abt. Orig.*, **A255**, 340-345.

21). LEVY S.B., MARSHALL B., SCHLUEDERBERG S., ROWSE D. & DAVIS J., 1988. High frequency of antimicrobial resistance human fecal coli. *Antimicrob. Agents Chemother.*, **32**, 1801-1806.

22). LEVY S.B., FITZGERALD G.B. & MACONE A.B., 1976. Changes in intestinal flora of farm personnel after introduction of a tetracycline - supplemented feed on a farm. *N. Engl. J. Med.*, **295**, 583-588.

23). MITSUHASHI S., 1977. Epidemiology of bacterial drug resistance. In: *R-factor drug resistance plasmid.* Ed. Mitsuhashi., University Park Press, Baltimore.

24). National Center for Toxicological Research, 1988. - *Chronic toxicity and carcinogenesis study of sulfamethazine in B6C3F, mice*. FDA, Washington DC.

25). National Center for Toxicological Research, 1988. *Chronic toxicity and carcinogenesis study of sulfamethazine in Fisher 344 rats*. FDA, Washington DC.

26). NOUWS J.F.M. & LAURENSEN J., 1990. Postmortal degradation of furazolidone and furaltadone in edible tissues of calves. *The Veterinary Quarterly*, **12**, 56-59.

27). OLIVE P.L., 1978. Macromolecular, antineoplastic and radiosensitization effects of nitrofurans. In: *Carcinogenesis - A comprehensive survey*, Bryan G.T. ed., Vol. 4, 131-169, Raven Press, New York.

28). RAYNAUD J.P., MACKINNON J., KEMP G., BEUKERS R. & METZGER K., 1987. Establishing Tolerance Levels of antimicrobials in human food. *Proc. Bact. Symp. "Rationale view of antimicrobials residues. An assessment of human safety"*, Zürich.

29). RICO A.G., 1986. Médicaments vetérinaires et hygiène publique. In: *Toxicologie et sécurité des aliments.* Ed. Derache, Lavoisier, Paris.

30). SAVAGE D.C., 1980 . Impact of antimicrobials on the microbial ecology of the gut. In: *The effect on human health of subtherapeutic use of antimicrobials in animal feeds.* National Research Council, National Academy of Sciences, Washington DC.

31). SCHLATTER C., 1990. Toxicological assessment of xenobiotics in foods of animal origin. In: *Residues of veterinary drugs in food. Proc. of the EuroResidue conf. Netherlands.* Ed: Haagsma, Ruiter and Czedik-Eysenberg, 65-75.

32). SCHULZ K.H., 1982. Allergy to chemicals: Problems and perspectives. In: *Allergy and hypersensivity to chemicals*, WHO, Copenhagen.

33). SKOVGAARD N., 1986. 3rd International technical consultation on veterinary drug registration, Paris.

34). SOMOGYI A., 1984. Health aspects of veterinary drug residues resistance phenomenae and immunopathological aspects. *Joint FAO/WHO Expert Consultation on Residues of Veterinary Drugs in Feeds.* WHO, Roma.

35). TANCREDE C. & BARAKAT R., 1987. Ecological impact of low doses of oxytetracycline on human intestinal microflora. *Proc. Bact. Symp. "Rationale view of antimicrobials residues. An assessment of human safety"*, Zürich.

36). WHO, 1991. Toxicological evaluation of certain veterinary drug residues in food, 1991. *WHO food additives series*: 27.

37). YUNIS A.A., 1981. Chloramphenicol toxicity and the role of the pNO_2 in aplastic anemia. In: *Safety problems related to chloramphenicol and thiamphenicol therapy*. Ed: Najean, Tognoni and Yunis. Raven Press, New York.

Discussion session for this paper after the following complementary paper by Corpet

RESIDUS ANTIBIOTIQUES ET MICROFLORE DIGESTIVE

D. E. Corpet

Laboratoire des Xénobiotiques, I.N.R.A.
B.P. 3, 31931 Toulouse, France

Résumé: *Les résidus antibiotiques peuvent-ils modifier la microflore de l'homme, et y sélectionner des bactéries résistantes? Chez des volontaires, l'ampicilline ou la tétracycline* per os *(1 ou 2 mg/jour) n'augmentent pas le nombre d'entérobactéries résistantes dans les selles. Un éventuel effet sélectionnant serait cependant masqué par les fluctuations des populations de bactéries résistantes. Des souris hétéroxéniques (souris sans germe, hébergeant la microflore d'un donneur humain), et dixéniques (souris n'hébergeant que deux souches bactériennes identiques, dont l'une porte un plasmide de résistance) sont des modèles d'étude de la flore, vivants et isolés des contaminations. Les doses minimales sélectionnantes dans ces modèles sont comprises entre 0,5 et 10 mg d'antibiotique par litre d'eau de boisson. Ces doses sont supérieures à ce que pourrait ingérer un consommateur dans ses aliments. Enfin les entérobactéries intestinales résistantes de l'intestin de l'homme proviennent directement des aliments, comme le prouve leur disparition des selles de volontaires mangeant des aliments stériles. Les résidus antibiotiques n'ont donc probablement pas d'effet mesurable sur la flore.*

Summary: **Antibiotic residues and digestive microflora**

Can antibiotic residues modify the human gut flora ? Do they select resistant bacteria in vivo ? No increase in the number of resistant fecal enterobacteria was observed in volunteers ingesting a daily dose of ampicillin (1.5 mg/d) or tetracycline (2 mg/d). However, a possible effect of antibiotic residues would be hidden by the natural fluctuations in resistant populations of fecal bacteria. Gnotobiotic mice, harboring either the complex flora from a human donor, or two isogenic E. coli *strains, one of which is carrying a R-plasmid, are living isolated models where the effect of antibiotics can be studied in the absence of contaminations. The minimum antibiotic levels for selecting a resistance plasmid in those gnotobiotic animal models were in the range 0.5-10 mg per liter of drinking water. Extrapolated to human, those concentrations are higher than the level of residues found in food. The resistant enterobacteria found in most human stool samples come from contaminated food, since they are cleared from the gut of volunteers given a sterile diet. Thus antibiotic residues are not likely to have an effect on the gut flora.*

Introduction

Les antibiotiques utilisés lors des traitements vétérinaires, préventifs ou curatifs, sont parfois retrouvés dans les produits animaux. Ces résidus sont rares dans les viandes (moins de 5% des carcasses aux USA), mais plus fréquents dans les laits (plus de 50% à New York), en concentration le plus souvent inférieures à 1 ppm (1 mg/kg de viande). Les molécules incriminées sont le plus souvent les tétracyclines, les sulfamides et les aminosides [2, 18]. Ces faibles doses peuvent-elles avoir un effet sur la microflore des consommateurs? On sait que les antibiotiques à dose thérapeutique détruisent partiellement la flore sensible, permettant l'émergence de bactéries résistantes, parfois pathogènes [12, 17]. Les résidus antibiotiques pourraient détruire certaines bactéries anaérobies strictes de la flore dominante, affaiblir la résistance à la colonisation, permettant l'émergence de bactéries pathogènes. Les résidus pourraient aussi sélectionner des bactéries résistantes aux antibiotiques [10]. Ce dernier effet, *a priori* le plus sensible aux antibiotiques, peut servir de marqueur de l'effet des résidus sur la flore.

Effet de résidus chez des volontaires

Goldberg a administré 10 mg d'oxytétracycline par jour à 50 prisonniers pendant 6 mois. Une élévation sporadique des entérobactéries résistantes fut détectée chez un tiers d'entre eux [11]. Knothe a montré qu'une dose quotidienne de 25 mg d'oxytétracycline pendant plus de 2 semaines augmente la résistance des bactéries fécales de volontaires [13]. Mais ces doses d'antibiotique sont largement supérieures à ce qui peut être ingéré comme résidus.

Pour mimer l'effet des résidus antibiotiques, nous avons donné à des volontaires sains, lors de chaque repas pendant un mois, une petite dose d'ampicilline (1,5 mg/ jour, soit 1000 fois moins que la dose thérapeutique). Au total, comme le montre le Tableau I, le nombre d' *E.coli* résistantes dans les selles de ces volontaires n'a pas augmenté significativement par rapport à une période témoin, contrairement à ce que produit une dose 10 fois plus forte [6].

Tancrède n'observe pas non plus d'effet de doses similaires d'oxytétracycline données à 6 volontaires pendant 7 jours [1]. Dans les deux expériences on observe cependant une augmentation de certaines résistances chez certains volontaires. Mais on observe surtout des fluctuations très importantes dans les populations d'entérobactéries résistantes, d'un jour à l'autre chez la même personne. A cause de ces fluctuations, l'effet sélectionnant des résidus (très petit, s'il existe) est quasi impossible à mesurer, directement chez l'homme.

Tableau I: Excrétion fécale d'*E. coli* résistants aux antibiotiques par des volontaires sains pendant l'administration orale d'ampicilline

Volontaire	dose mg/j	Rapport entre le nombre moyen d'*E. coli* pendant les périodes expérimentales et témoin (sans ampi.), résistants à		
		ampicilline	tétracycline	streptomycine
CP	1.5	63 *	43 *	64 *
FC	1.5	1	0.1	0.1
FL	1.5	3	3	1
JC	1.5	1.5	86 *	34 *
MI	1.5	4	1	3
total (non signif.)		4	5	4
témoin positif DC	30	3600 **	4400 **	700 **

Notes: Rapports calculés à partir des différences des log. des comptages (35 points par volontaire).
Test de Mann Whitney de l'effet de l'antibiotique: ** $p < 0.001$, * $p < 0.01$.

Modèles d'étude *in vitro*

Des modèles d'étude *in vitro* ont été mis au point qui permettent de mesurer l'effet de faibles doses d'antibiotique sur la résistance phénotypique de bactéries [4] ou sur le transfert de plasmide de résistance dans un milieu de culture [3], ou sur la sélection d'une bactérie portant un plasmide dans une culture continue de 3 souches pures en chémostat [14]. Ces systèmes sont séduisants, car très bien contrôlés, mais les bactéries y sont dans des conditions très différentes de celles qui prévalent dans l'intestin [5]. On ne peut donc pas savoir si l'effet sélectionnant des faibles doses d'antibiotique observé *in vitro* est extrapolable à ce qui peut se passer dans l'intestin de l'homme.

Modèle hétéroxénique: souris à flore humaine

Un modèle vivant et isolé de l'écosystème microbien de l'intestin de l'homme peut être construit en donnant à des souris axéniques (souris sans germe, maintenues en isolateur) la microflore fécale d'un donneur humain. La flore de ces souris est alors très semblable à celle du donneur, pendant au moins quelques semaines. Nous avons comparé l'excrétion d'entérobactéries résistantes aux antibiotiques chez ces souris à flore humaine recevant des résidus antibiotiques et chez des souris témoins ayant exactement la même flore mais buvant de l'eau pure. Le Tableau II montre que des doses très faibles d'ampicilline et de chlortétracycline ont sélectionné les entérobactéries résistantes chez les souris traitées [6].

Tableau II: Excrétion fécale d'*E. coli* résistants par des souris hétéroxéniques à flore humaine buvant 0,5 mg d'antibiotiques par litre d'eau.

Antibiotique administré	Rapport entre le nombre moyen d'*E.coli* resistants par des groupes de souris traitées et témoins resistants à :		
	ampicilline	tétracycline	streptomycine
ampicilline	37 *	26 *	5800 **
chlortétracycline	4	24 **	35 **
streptomycine	0.1	0.2	0.1

Notes: Rapports calculés à partir des différences des log. des comptages (2 ou 3 flores testées par antibiotique x 15 à 20 points par groupe de 5 souris).
Test de Wilcoxon de l'effet de l'antibiotique: ** $p < 0.001$, * $p < 0.01$.

Ces doses, extrapolées à l'homme, sont cependant supérieures à ce que pourrait ingérer un consommateur dans ses aliments. Par ailleurs la variabilité de la flore des donneurs rend impossible la duplication d'une expérience, et donc difficile la détermination d'une dose sans effet.

Modèle dixénique: sélection d'un R-plasmide

Un modèle plus simple, et bien reproductible, a été construit en donnant à des souris axéniques deux souches d'*E.coli* isogéniques : l'une était sensible aux antibiotiques, l'autre était identique mais hébergeait un plasmide de résistance à 14 antibiotiques. En l'absence d'antibiotique, la souche sensible "chasse" la souche résistante de l'intestin des souris dixéniques en 2 semaines. Nous avons donné à des groupes de souris dixéniques de faibles doses d'antibiotique via l'eau de boisson [8]. Quand les souris reçoivent une concentration suffisante d'antibiotique, la souche résistante porteuse du plasmide de résistance est sélectionnée, et devient dominante dans l'écosystème digestif en quelques jours ou quelques semaines. A partir des droites de régression entre la dose d'antibiotique et l'effet sélectionnant chez les souris dixéniques, nous avons pu définir une dose minimale sélectionnante (cf. Tableau III).

Pour les antibiotiques que nous avons testés, les doses minimales sélectionnantes données au Tableau III correspondaient à des concentrations supérieures aux résidus des aliments de l'homme [8].

Origine des entérobactéries résistantes de l'homme.

Les travaux exposés plus haut montrent que les résidus antibiotiques ne peuvent sélectionner des entérobactéries résistantes dans l'intestin. Comment expliquer qu'un grand nombre d'individus, non traités, excrètent des bactéries résistantes dans leurs selles [15]? Ces bactéries sont un réservoir pour les plasmides, car elles peuvent transmettre leurs gènes de résistance aux pathogènes: il est donc souhaitable de diminuer leur nombre [16], et donc de connaître leur origine.

Tableau III: Doses minimales sélectionnantes *in vivo*

Antibiotique administré	dose[a] mg/l	Nombre d'expér.[b]	corrélation dose-effet
témoin		10	pas de dose
fluméquine		6	pas d'effet
colistine		7	pas d'effet
tétracycline	6.5	16	0.86
gentamycine	2.3	5	0.99
ampicilline	1	6	0.92
streptomycine	0.9	10	0.91

[a] Dose minimale sélectionnante = concentration dans l'eau de boisson de souris dixéniques, conduisant à une multiplication par 10 en 5 jours de l'inoculum d'*E. coli* porteurs de plasmide de résistance dans l'intestin.
[b] Chaque expérience réalisée sur 5 souris pendant 15 jours.

Nous avons soumis des volontaires à un régime entièrement stérilisé pendant trois semaines: le Tableau IV montre que chez 5 des 6 volontaires, le nombre de bactéries résistantes diminue très fortement pendant le régime stérile.

Tableau IV: Excrétion fécale d'*E. coli* résistants aux antibiotiques par des volontaires sains recevant un régime stérilisé par cuisson.

Volontaire-	Rapport entre le nombre moyen d'*E. coli*pendant les régimes témoin et stérile, résistants à :		
	ampicilline	tétracycline	streptomycine
AT	1.8	22 *	2
DC	81 **	457 **	474 **
FC	26 *	1059 **	183 **
MA	252 **	161 **	1.6
MF	479 **	1702 **	10 *
MI	2432 **	2138 **	1365 **
total	**113 ****	**433 ****	**38 ***

Notes: Les données sont calculées par défaut, puisque les comptages inférieurs à la limite de détection ont été pris par convention égaux à celle-ci, soit 1,4 en log. * : cf. Table I

En fait, les entérobactéries résistantes disparaissent pratiquement en 2 à 3 jours des selles, et ne réapparaissent qu'à l'arrêt du régime stérile [7]. La majorité des entérobactéries résistantes de l'homme proviennent donc d'un apport par les aliments, en particulier ceux qui sont consommés sans cuisson [9].

Conclusion

Les résidus antibiotiques n'ont probablement pas d'effet mesurable sur la flore. Chez l'homme, les nombreuses bactéries intestinales résistantes aux antibiotiques proviennent surtout des aliments.

Références

1). BARAKAT R. & TANCREDE C., 1988. Ecological impact of low doses of oxytetracycline on human intestinal microflora. *Proc. IX Int. Symp. Gnotobiology*, Versailles, France. F. M. Mérieux ed. pp.351-353.

2). BRADY M.S. & KATZ S.E., 1988. Antibiotic antimicrobial residues in milk. *J. Food Protection*. **51**: 8-11.

3). BRADY M.S. & KATZ S.E., 1988. Method to determine effect of antibiotics at residue levels on R-factor transfer. *J. Assoc. Off. Anal. Chem.* **71**: 299-301.

4). BRADY M.S, STROBEL RJ & KATZ S.E., 1988. *In vitro* analytical system for determining the ability of antibiotics at residue levels to select for resistance in bacteria. *J. Assoc. Off. Anal. Chem*. **71**: 295-298.

5). CORPET D.E. & LUMEAU S., 1987. Antibiotic residues and R-plasmid selection : are *in vitro* methods good models ? *Zbl. Bakt. Hyg*. **264 A**: 178-184.

6). CORPET D.E., 1987. Antibiotic residues and drug resistance in human intestinal flora. *Antimicr. Agents Chemother.* **31**: 587-593.

7). CORPET D.E., 1988. Antibiotic resistance from food. *New Engl. J. Med*. **318**: 1206-1207.

8). CORPET D.E., LUMEAU S. & CORPET F., 1989. Minimum antibiotic levels selecting a resistance plasmid in a gnotobiotic animal model. *Antimicrob. Agents Chemother*. **33**: 535-540.

9). CORPET D.E., 1989. Les microbes font de la résistance. *La Recherche* **215:** 1388-1390.

10). FAO/OMS., 1984. Résidus de médicaments vétérinaires dans les aliments. Residues of veterinary drugs in foods. Report of a Joint FAO/WHO Expert consultation. *FAO Food Nutr. Paper*. **32**. Rome, 29 Oct-5 Nov 1984.

11). GOLDBERG H.S., GOODMAN R.N., LOGUE J.T. & HANDLER F.P., 1961. Long-term, low-level antibiotics and the emergence of antibiotic-resistant bacteria in human volonteers. *Antimicr Agents Chemother.* pp.80-87.

12). GUILLOT J.F. & LAFONT J.P., 1989. - Antibiotiques et microflore intestinale. *Rev. Sci. Tech. Off. Int. Epiz*. **8**: 439-452.

13). KNOTHE V.H., 1963. Darmflora und Antibiotika unter besonderer Berucksichtigung der Tetracycline. *Deut. Med. Wochenschr*. **88**: 1469-1477.

14). LEBEK G. & EGGER R., 1983. R-Selection of subbacteriostatic tetracycline concentrations. *Zbl. Bakt. Hyg., I Abt. Orig*. **A 255**: 340-345.

15). LEVY S.B., SCHLUEDERBERG S., ROWSE D. & DAVIS J., 1988. High frequency of antimicrobial resistance in human fecal flora. *Antimicr. Agents Chemother.* **32**: 1801-1806.

16). LEVY S.B., 1990. Starting life resistance-free. *New Eng. J. Med.* **323** : 335-337.

17). TANCREDE C., 1989. Flore intestinale et pathologie infectieuse humaine. *Rev. Sci. Tech. Off. Int. Epiz*. **8**: 405-415.

18). U. S. Department of Agriculture, 1974. Biological residues reports, January to December 1973. U.S.D.A., Washington D.C.

19). VAN KLINGEREN B., 1986. Do residues of antimicrobial drugs constitute a microbiological risk for the consumer ? *Comp. Vet. Pharmacol. Toxicol. Therapy*. **50**: Proc.III Congr. Europ. Assoc. Vet. Pharm. Tox. Ghent, Belgium. A.S. Van Miert *et al* ed. pp.527-538.

Discussion

Joint discussion session with the previous complementary paper by Ynestad.

MIDTVEDT I wish that I could agree with the last two speakers, but I cannot, and I think I have to be the bad boy in the class again! For I don't think that the model they have used is good enough. They are quantitating *E. coli*, and that is one thousandth part of the normal human flora, and as has been shown, it may be a transient part. Therefore another approach to the study of the effect of antibiotics on the intestinal flora may be needed. Instead of studying the number of microbes we have studied some functions and I shall mention one with some bearing on the question of residues. That is the conversion of cholesterol to coprostanol, which is carried out by just one species only, which is an anaerobe *Eubacterium lentum*, found in human as well as in animals. We have followed that parameter and, taking experience from rats, giving rats 0.01 mg^{-1} kg^{-1} rat^{-1} day^{-1} for 5 days, that conversion was completely eradicated for the whole experimental period of 4 weeks. That amount, 0.01 mg^{-1} kg^{-1} rat^{-1}, is far above the residues that can be found in meat. Sure it will depend on the antibiotic, which in this case was clindamycin, so far we have not done that much with the other drug which was tetracycline. Moving then to humans (and we did that experiment with Norwegians, they are cheaper to have as volunteers and there is less paperwork to do in Norway!), we were not able to use such low level of course, so we gave just a therapeutic level and we had a marked effect on intestinal microbial functions for weeks in otherwise healthy Norwegians. Just to sum up then, I don't think that to study *E.coli* as a marker for residues effect is a good model, there might be a better one around.

CORPET I agree that the experiment you give is a very interesting one, and I agree that *E coli* is a very small part of the flora, but I disagree that it is not a significant part. I don't care if my *Eubacterium lentum* is destroyed, because up to now nobody has proved that *Eubacterium lentum* is beneficial in the flora. May be you have a comment, but I do care and I don't want to have too many Enterobacteria which are resistant to drugs because that resistance could be translated to pathogens. Up to now no really useful functions of the flora has been described as being destroyed by antibiotics except for the antagonistic effect against pathogens and against *E coli*.

MIDTVEDT I do care if my normal transformation of cholesterol is altered. I also care (and we have followed several other parameters) if the normal metabolism of bile acids and the metabolism of bilirubin to umbilins is destroyed. I care if the production of short chain fatty acids, the normal breakdown of trypsins and the transformation of mucin are also influenced by the antibiotic. In fact, I do care much more for the function of the flora, than I care about number of *E. coli* present down there.

CORPET Well, most of the functions you describe are thought to be detrimental to the host and an argument for this is that, in most animals that are farmed, when they are given low doses of antibiotics they grow better. I don't want to be supplemented to grow better, nor do I want this for my children. I think that although those parameters are good markers of an effect, and they are very sensitive, and they are very interesting from a scientific point of view, they are not really significant of human health.

MIDTVEDT Time will show.

CORPET I think the normal flora in human beings is very important because of the so called colonisation resistance of the gut. That I support with my Swedish colleague, that is the main point. If that's correct, OK, I'm fine. If not and I don't care that much about *E.coli,* I was convinced already that low level drugs would not induce resistance. I do care a lot about colonisation resistance (which is also called barrier effect in French or antagonistic effect). This is, in my opinion, a major effect and we have to study for example the resistance to a challenge with cholera or with *Salmonella* after supplementing the model with antibiotics. This is a major effect and I think that studying *E coli* is a way to look at colonisation resistance because the normal flora has a great effect against *E coli.*

YNDESTAD Well, I don't say that antibiotics are very good for you, I just say that, in my model, I have no effect. That's it.

MIDTVEDT May I just comment on the colonisation resistance as well. That parameter is a rather rough and rather naive parameter. We also found that, if you are defining colonisation resistance as presence of beta aspartylglycine (we did that), we do find presence of bacterial beta aspartylglycines in rats receiving that low levels of clindamycin. So it was in fact an alteration in colonisation resistance as well.

CORPET I'm sorry to comment further, since this is a very specialized discussion, but, once more, beta aspartylglycine is a marker of the colonisation resistance, it is not the colonisation resistance by itself, but anyway, you're right.

ONARHEIM What about the colonisation factor in the fish when you give the fish antibiotics? What happens to the normal gut flora of fish?

CORPET I am not a specialist in fish flora, but within my knowledge, fish flora is not at all the same as the mammal flora. It is not numerous, there are not a lot of bacteria there and there is almost no colonisation resistance. The fish flora reflects the flora of the water outside, so I would answer that, obviously antibiotics will select, but not with destruction of a colonisation resistance factor which does not exist. It could be different in tropical fish which are in warm water and which harbour many more bacteria, but for trout and salmon, to my knowledge, the flora cannot be considered as an ecosystem.

FIJAN I would like to comment on one of the things mentioned by the first speaker. He mentioned the very important criterion, the mutagenicity of drugs. By grazing through the literature (I am not a specialist in this field), I have noted a title stating that the amount of natural mutagens in water far exceeds the mutagens in heavily polluted waters. Now what would be the stand-point then concerning this small amount of 35 tons of fish drugs in several billions tons of water concerning mutagenicity of a drug used for fish?

YNDESTAD You point to very interesting things. We know that there are a lot of mutagens in drinking water, but how to handle them and what the meaning is, we don't know, therefore I cannot answer your question.

ONARHEIM I just want to correct you, as far as I am concerned it has not been shown that fish do not have an intestinal microflora. That there is no colonisation factor in fish has not been shown.

THOMEN Is there a possibility of introducing probiotics in order to drive away resistant strains?

CORPET Well theoretically, if you can find a probiotic, which is a living bacteria, which will increase the colonisation resistance against the plasmid bearing bacteria, you could do this, but I never saw any paper dealing with this. What I did was to use piglet flora associated mice (gnotoxenic mice harbouring the flora from a pig). Those mice were given bambermycin (which is called in France flavomycin), and this antibiotic at very low level eliminated the plasmid bearing bacteria from the gut of those mice. This is well known in gram negatives, so it is possible to do this. Now I am sure that I was in a parallel way building a resistance in the gram positive flora. I didn't check this, but obviously it is not safe to give an antibiotic to do this kind of thing, but it is possible.

LYDIA BROWN Of the 20,000 samples which you analysed in Norway for residues, do you have any idea generally how many were positive?

YNDESTAD When a sample is positive the fish will not be slaughtered. Concerning the surveillance of slaughtered fish, we have not found a positive at all. In the 20,000 samples, there were many positive, and therefore the farmers received a new withdrawal time. The main point that is that the fish are checked before slaughter. If we find antibiotic residues after slaughter, the whole lot will be condemned without discussion.

HASTEIN Can you give a figure?

YNDESTAD In the slaughtered fish, none were positive. In the pre-slaughter tests there are often positive samples, depending amongst other things on the water temperature. In the winter time we find more positives than in the summer time. Lots of fish where antibiotic residues are found in pre-slaughter samples receive new, prolonged, withdrawal times.

RESIDUE PROBLEMS WITH REGARD TO CULTIVATED FISH

J.H.G. Goebbels

Veterinary Public Health Inspectorate
The Netherlands

Summary : *In recent years aquaculture production in western Europe has increased enormously, increasing risks of environmental pollution and of disease in stressed crowded fish stocks. Consumption of verterinary medicinesin aquaculture has also increased and a number of antimicrobial products have now been licenced s[ecifically for aquaculture use in different countries.*

Lack of detailed knowledge results in uncertainty in areas of pharmacology in fish, particularly in the area of drug residues, for which specific methods are lacking. Moreover the environment is threatened by these treatments and it is becoming urgent that surveillance programmes are developed. These programmes are needed to to prevent other than authorised drugs from being used and to enable proper control throughout the chain of production. Equally, all methods which may reduce the need for drug use must be encouraged.

Résumé **: Le problème des résidus chez les poissons d'élevage.**

La production aquacole s'est énormément accrue ces dernières années en Europe occidentale, augmentant les risques de pollution organique de l'environnement et les risques de maladies pour des élevages régulièrement exposés à des causes d'affaiblissement. La consommation de médicaments a également augmentée, et certains produits antimicrobiens ont reçu des autorisations légales d'emploi dans différents pays (le tableau I répertorie les produits autorisés en Allemagne et en Norvège, et précise les délais d'attente prescrits).

L'état actuel des connaissances implique des incertitudes dans les travaux pharmacologiques relatifs aux poissons, notamment dans le cas des résidus, pour l'analyse desquels nous manquons de méthodes spécifiques. De plus, l'environnement est menacé par les traitements. Il devient urgent de promouvoir des programmes de surveillance, de n'admettre que l'emploi de produits autorisés pour les poissons, ceci sur la base de dossiers étayés par des renseignements précis, d'assurer enfin des contrôles de qualité à tous les niveaux de la chaîne de production. En corollaire, toutes les solutions permettant de restreindre les indications de traitement devront être explorées.

In the last few years the production of farmed fish has increased enormously in different European countries. In Norway for example the production increased from 6,833 x 1000 kg in 1979 to 118,844 x 1000 kg in 1989 (salmon and trout). The present day production in some other European countries is in Denmark 30,000 x 1000 kg (salmon and trout), in France 20,000 x 1000 kg (trout), in the Federal Republic of Germany 60,000 x 1000 kg (trout and carp) and in the Netherlands 1,000 x 1000 kg (trout, catfish and eel).

This large-scale production, as a matter of fact, alters the natural circumstances of the fish. As a consequence of large fish density in rearing ponds the risk for infection diseases has increased. Furthermore the natural selection has decreased and the environment can be polluted by too much organic material, which means that not only the strongest fish will survive. This will also increase the risk for bacterial infections. Finally the need for a rapid weight gain of farmed fish means an additional risk.

As a result, the use of veterinary drugs, for curative as well as for preventive purposes, has increased sharply. In the FRG some antibiotics are already registered for use in aquaculture. In Norway oxytetracycline and sulfadimidine are registered already while oxolinic acid and flumequine are in the registration phase. Withdrawal periods have been set up, depending on the water temperature.

In most of the other countries no drugs are registered special for fish. The major drugs used in the FRG and Norway are with their withdrawal periods given below.

Table I Major drugs used in Norway and FRG

	Withdrawal time in days		
	Norway: advised	Norway: advised	FRG: obligatory
	8-12°C	>12°C	
Furazolidone	90	40	90
Sulfadimidine	40-90	40	30
Oxytetracycline	60-180	60	30
Oxolinic acid	60	40	
Chloramphenicol			30
Trimethoprim			49
Malachite green *			
Dichlorvos	14		
Estomycin sulfate			90
Dimetridazol			30

* Forbidden in FRG but still used

In relation to the above-mentioned concerning approval of drugs the following conclusions can be made:

1. Because as a shortage of knowledge of fish diseases exists, efficiency and dose of the drug must be established empirically. Pharmacokinetic information is only scarcely available.

2. Data with respect to residues data are very often missing. The extrapolation of research data from livestock to fish is impossible as fish is in physiological respect completely different from other production animals. For that reason fish has a totally different drug elimination which is closely related to the water temperature. Therefore the temperature must be taken into account if a withdrawal period is fixed.

3. Analytical methods for residues in fish are only scarcely available. Most of these are developed for mammals.

4. The addition of drugs to the water could have consequences for the environment, especially when drugs are used for the treatment of fish species not farmed in landpens, such as salmon in the Norwegian fjords.

To avoid a large residue problem in future more measures must be taken to restrict the use of veterinary drugs in fish farming.

1. To carry out a surveillance programme obliged in future for EEC countries (directive 86/469/EEC) or alternative programmes for other countries.

2. Use of drugs and vaccines registered for fish only, to achieve the following results:

- Better and more effective treatment resulting in the application of smaller quantities of veterinary drugs
- Residue tolerances can be set up for each edible part of fish and for different kinds of fish
- Routine analytical methods must be given by the pharmaceutical industry
- Studies of the environmental effects of the veterinary drug must be done by the pharmaceutical industry.

3. Integrated quality assurance: The inspection for residues must not only be done after slaughtering of the fish but during the whole production chain: feed - water - fish.

In Norway a similar system is already operational. Inspection takes place both before and after slaughter and a logbook of the veterinary drugs used must be maintained and copies of the prescriptions must be sent to the central authority.

4. In Sweden only a few veterinarians, specialised in aquaculture are allowed to administer drugs to fishes . As long as there are not enough registered drugs for fish, this can be a successful approach. In most of the countries the situation is quite different because all veterinarians are allowed to treat farmed fish.

5. The next years a lot of work has to be done in the fiel of zootechnical research because a good lodging and feeding system will lead to a healthy fish to which less veterinary drugs have to be administered. In this connection the development of a good vaccination programme is also very important.

SURVEILLANCE FOR RESIDUES IN FARMED FISH: PRINCIPLES AND PROBLEMS

D.R. Tennant

Food Safety Directorate, Ministry of Agriculture, Fisheries and Food, 17 Smith Square, London SW1 3JR, UK

Summary : *Surveillance for veterinary residues in farmed fish is undertaken to assess the potential risk to consumers. In order to do this it is necessary to investigate the incidences and concentrations of residues and patterns of consumption. A surveillance programme should be tailored to the purpose of the survey and to the distribution pattern of the particular commodity. Farmed fish samples can be obtained at any point between the farm and the consumer, but sampling should aim to be representative of the whole food supply and must allow for seasonal effects. The choice of tissues, veterinary drug residues and analytical methods should reflect the objectives of the survey and toxicological aspects should be considered when assessing the significance of results.*

Résumé **: Surveillance des résidus dans le poisson d'élevage: principes et problèmes.**

La surveillance des résidus de médicaments vétérinaire dans les produits aquicoles ne diffère pas de celle des autres aliments d'origine animale. Elle vise à connaître les doses moyennes quotidiennement administrées et les concentrations qui en résultent dans les aliments, à déceler l'existence de teneurs particulièrement élevées dans certains tissus, et à prévenir les risques liés aux habitudes alimentaires de certains consommateurs. Il est donc essentiel de déterminer les fréquences et les concentrations des résidus pour chaque produit vétérinaire. Un programme de surveillance de la production nationale et des importations a été mis en place au Royaume-Uni, placé sous l'autorité d'un groupe de travail de composition pluraliste responsable devant l'exécutif gouvernemental, qui en même temps assure l'application de la directive européenne n°86/469 relative aux résidus.

Les critères d'échantillonnage varient selon qu'on cherche à réprimer des fraudes ou à obtenir des informations quantifiées, auquel cas il faut veiller à l'importance et la représentativité des échantillons, à la sensibilité et à la fiabilité des techniques de détection. Les protocoles doivent toujours être standardisés et garantis par une assurance de qualité des analyses. Tout résultat positif notamment, doit être confirmé par une deuxième technique d'analyse. Selon le but recherché les contrôles interviennent à différentes étapes du circuit de commercialisation (fig.1). L'obtention de données objectives et l'identification des sources de résidus imposent de surveiller au moins les exploitations et les points de vente au détail.

La recherche des résidus est orientée vers les molécules autorisées pour les poissons au Royaume-Uni (Tab.I), ou couramment employées à l'étranger. On essaie de détecterles produits caractérisés par une longue persistance (sulfamides, triméthoprime, tétracyclines, acide oxolinique, vert malachite), ou par leur toxicité (dichlorvos). Des enquêtes antérieures (on a montré en 1984 que les tétracyclines étaient les plus fréquemment détectées), peuvent aussi guider les démarches. Il est moins facile, faute de standardisation, de se fier aux données pharmacocinétiques issues d'autres pays. La possibilité pour les vétérinaires de prescrire des formules non autorisées complique la tâche des contrôleurs, de même que la tendance à varier les traitements en réponse à la fréquence des résistances. Enfin, les fluctuations saisonnières s'avèrent parfois équivoques (traitements antimicrobiens davantage employés en saison chaude, mais risques de persistance plus élévés en hiver).

Pour obtenir des données quantitatives, les techniques chimiques et immunologiques, polyvalentes et automatisables, sont préférées aux dosages microbiologiques. Les plus fortes concentrations de résidus se trouvent dans le foie et le rein, mais la chair et la peau font l'objet d'une attention spéciale. Les crustacés, les mollusques ne sont évidemment pas négligés.

Les résultats de ces analyses sont combinés aux statistiques de consommation pour évaluer l'importance des résidus et leurs risques potentiels. Leur communication aux autorités est toujours assortie de publications qui pourraient constituer des documents utiles à la C.E.E. dans l'élaboration de ses futures directives.

Introduction

Food surveillance is undertaken in order to assess the potential risk to consumers of contaminants in food (1). Whilst contamination of farmed fish can arise from several sources, for example water pollution and contaminants in feed, we are presently concerned with assessing the risks from potential exposures to residues of veterinary medicines. This paper thefore will discuss the principles underlying the UK food surveillance philosophy, some of the practical problems to be overcome developing a surveillance strategy, and some ways in which the information obtained can be used.

Principles of Food Surveillance

The way in which surveillance for veterinary drug residues in farmed fish is carried out is fundamentally no different to surveillance for any other chemicals in the UK diet. The questions to be addressed include the following:

'What is the average daily intake of the chemical?'

'What are the concentrations of the chemical in relevant foods in the UK diet?'

'Do any foodstuffs contain specially high concentrations?'

'Does any part of the nation have particularly high intakes of the chemical?'.

To adequately address these questions it is essential to ascertain the concentrations and incidences of the particular residues in foodstuffs. Food surveillance in itself does not provide any assessment of whether the incidence, or concentration, represents a hazard. What it does do is to provide information that allows intakes to be estimated and potential risk to be assessed.

In order to gather information on concentrations and incidences of veterinary drug residues in farmed fish it is necessary to set up a practical surveillance programme. In the UK, this surveillance is co-ordinated by the Working Party on Veterinary Residues in Animal Products (2). This committee consists of representatives from government departments, industry, scientific research, and professional organisations. The Working Party is one of ten that report to the UK Government's Steering Group on Food Surveillance, which reviews the possibility of contamination of any part of the national food supply.

In addition to seeing the results of surveillance for veterinary residues carried out under EC Directive 86/469/EEC (3) (the "Residues Directive"), the Working Party co-ordinates its own surveillance programme. This includes imported and UK-produced food and samples are taken at wholesale and retail levels. Surveillance for veterinary residues in farmed fish is included in the Working Party's surveillance programme.

Sampling strategies for farmed fish

Sampling needs

The first problem when setting up a sampling programme is to decide on the primary objective of the exercise: How are the results to be used? For example, if a scheme is intended to be used to enforce residue limits, it is necessary to plan the sampling in such a way that particular problem areas can be concentrated on and offenders easily identified. The time between sampling and providing a result must be minimal and the result seen to be clearly above, or below, the given limit. There would be no need to quantify all the results provided that residues above the limit were positively identified and confirmed.

By contrast, the criteria for a sampling programme which is intended to collect data about the incidences and concentrations of residues in the food supply with a view to assessing dietary intakes can be quite different. In this case it is necessary to obtain samples that are representative of the whole food supply and being able to identify particular producers is less important. A fast turn-round of data is also less important but it might still be desirable to have a high rate of sample throughput if this means that more data can be produced. If an accurate estimate of intake is to be made it is necessary to avoid false negative and false positive results and provide quantitative data both above and below any action level. The limit of determination should be as low as possible so that the data can be used to investigate trends.

Sampling schemes can be designed to fulfil more than one objective. Under such circumstances compromises can be difficult to avoid. However, the quality of the analytical data should never be allowed to vary. Participation in analytical quality assurance (AQA) programmes is a useful way of maintaining analytical quality. In the UK it is recommended that all positive results obtained in surveillance programmes are confirmed using a different analytical technique, based on different physico-chemical principles and ideally providing information on the chemical structure of the compound as recommended in various EC Directives.

Point of sampling

When developing a sampling strategy one is faced with the problem of choosing which point or points in the distribution chain to obtain samples. The distribution network for farmed fish in the UK is summarised in Figure 1.

Clearly, in order to obtain samples from all UK producers in an enforcement programme it is better to obtain samples directly from farms. However, this introduces practical difficulties, because it is not possible to know which fish on a farm are about to be placed on the market and also medication may not have been fully withdrawn. There are also many fish farms in remote locations which it would be difficult to visit and hence sample. If the scheme is aimed at obtaining surveillance data samples must be obtained later in the food chain so that imports are included and exports excluded in order that sampling is more representative of the product on sale to the consumer. It would also be necessary to include processed products, such as smoked fish, to ensure that the entire food supply is covered.

A better estimate of intakes may be obtained by sampling at the point of consumption. For example, consumers can be asked to produce 'duplicate diets' which can be taken and analysed for the contaminant(s) in question. The problem with such studies is that it is seldom possible to identify the source of any contamination thus making difficult the control of any problems. It is also difficult to include all food consumed, particularly that consumed outside the home (e.g. in restaurants). In practice, particularly in the case of farmed fish, it is most convenient to obtain samples at more than one point in the food chain. To illustrate this, fish markets are used to obtain the majority of UK produced and imported samples and additional samples are taken from stocks on sale at farms, to ensure that all supplies are covered. In this way the origin of each sample can be identified. Further samples can be taken at the retail level to complete the sampling strategy. If information on patterns of consumption of farmed fish is available then estimates of dietary intakes of contaminants can be made.

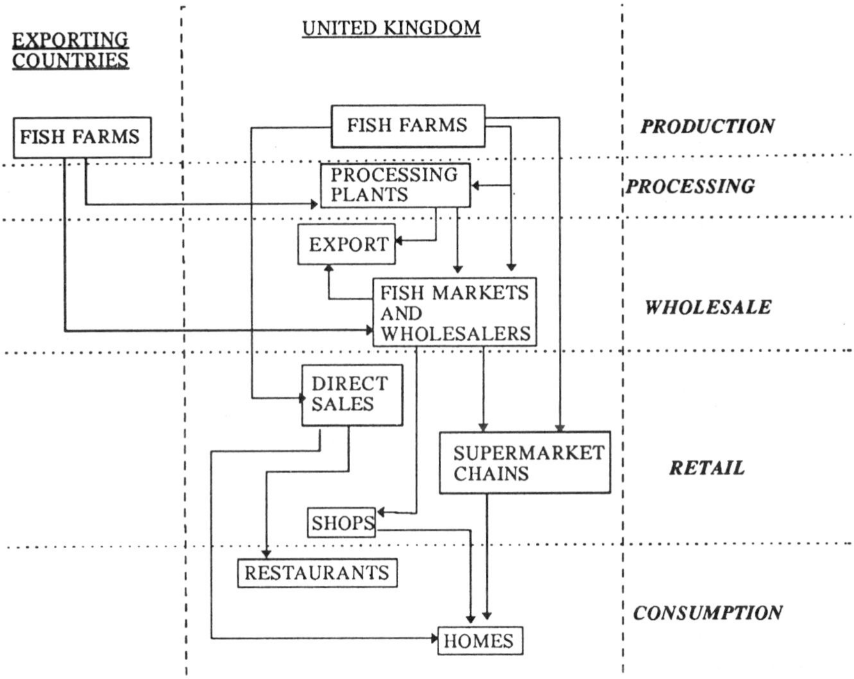

Figure 1 Distribution network of farmed fish in the United Kingdom

Selection of drugs for analysis

In the UK 8 veterinary drugs are known to be regularly used to control diseases in farmed fish (Table I). A further 10 drugs have been notified in countries believed to be exporting to the UK (4). The criterion that the Working Party on Veterinary Residues in Animal Products uses in its selection of analytes is the likelihood that residues will occur. In making its selection it considers patterns of drug use (and abuse), pharmacokinetics, the results of previous surveillance and evidence from data published at home and abroad. The availability of suitable analytical methods is also an important factor.

Table I. Veterinary agents used on farmed fish in the UK and countries exporting farmed fish to the UK(4).

United Kingdom	Exporting Countries
	United Kingdom agents plus:
Amoxycillin	Furazolidone
Chloramine	Nitrofurantoin
Dichlorvos	Sulphamerazine
Formaldehyde	Flumequine
Iodophores	Potassium permanganate
Malachite green	Quart. ammonium salts
Oxolinic acid	Trichlorfon
Oxytetracycline	Vitamin C
Sulphadiazine	Praziquantel
/Trimethoprim	Streptomycin

Published data on pharmacokinetics suggest that whilst residues of furazolidone, trichlorphon, dichlorvos and praziquantel have relatively short half-lives (4,5), residues of sulphonamides, trimethoprim, oxytetracycline, oxolinic acid and malachite green may be longer lived (4,6,7,8). The latter would therefore seem to be more appropriate subjects for residues surveillance since they are longer lived in tissues.

However, other factors such as drug toxicity must also be taken into account. For example, although dichlorvos appears to be rapidly metabolised, it has been speculated that it might have weak mutagenic activity (9). Therefore from the point-of-view of consumer risk assessment it would be desirable to confirm that no residues of dichlorvos are present in the food supply.

During 1984 the Working Party on Veterinary Residues in Animal Products conducted a preliminary survey of residues in 54 samples of farmed trout (2). No residues of any sulphonamide or of furazolidone were detected (limits of determination 0.005 mg/kg and 0.02 mg/kg respectively). Oxytetracycline was detected above 0.005 mg/kg in 7 samples (0.008 to 0.04 mg/kg) and tetracyline was detected in one sample at 0.03 mg/kg.

These preliminary residues data were used in planning subsequent surveillance programmes. The absence of furazolidone residues confirmed that either it was very rapidly metabolised or was not being used in the UK at that time. Although sulphonamide residues were not detected, the relatively slow metabolism of some sulphonamides may make these suitable candidates for future surveillance. Residues of tetracylines suggested that further surveillance to provide better estimates of incidences and concentrations might be worthwhile.

Residues data from countries exporting farmed fish to the UK are more difficult to obtain. Many of the limited data which are available have been produced using microbial inhibition methods and the results are difficult to interpret when planning a food surveillance programme. In a survey of rainbow trout from commercial sources in Italy(10) 2 percent were found to cause antimicrobial inhibition although it was not possible to show which antimicrobial residues were involved. However, 16 percent were positive in a test for flumequine activity(10). A German survey of malachite green in trout reported residues above 0.02 mg/kg in 22 per cent of samples(11). Whilst neither Italy nor Germany are known to export fish to the UK, these kinds of data can be useful since they reflect the patterns of residues that might be expected.

The selection of analytes is further complicated by the use of unlicensed therapeutic agents. Because the number of veterinary medicines that are licensed for use on fish is limited, veterinarians may prescribe other animal medicines. When this is done in the UK a standard withdrawal period of 500° days should be applied(12). This could result in long withdrawal periods during cold weather which some producers may not be prepared to observe. The Working Party aims to obtain information on such extra-licence uses so that its surveillance programme can be extended if necessary.

Seasonal effects

It is widely recognised that the metabolic rate of poikilothermic animals such as fish is dependent upon the temperature of their environment(4). However, very few product data sheets take this into account(13) and thus surveillance needs may vary from season to season. In the colder winter months, recommended withdrawal periods may not be adequate and fish may be slaughtered before metabolism and depuration of residues is complete. This could result in a higher incidence of residues. However, in the summer, higher environmental temperatures may encourage bacterial and fungal diseases so the use of therapeutics might be greater. Although residues should more rapidly deplete in warm weather, the incidence of residues from farmers who disregard withdrawal periods might increase. Furthermore, the increased metabolic rate might make it necessary to administer higher doses during warm weather to ensure that therapeutic doses are achieved. This combination of factors makes it necessary to undertake surveillance throughout the year.

Effects of recurrent drug resistance

Drug resistance is a recurring problem on fish farms. It may develop quickly and occur frequently. Fish farmers and their veterinarians may avoid problems with resistance by making frequent changes in medication. This means that patterns of drug usage on fish farms are not consistent and a wide range of residues could potentially be present.

Analytical approaches

Some of the most important criteria necessary to develop a food surveillance programme relate to analytical methodology. In particular, for food surveillance purposes, high quality results are required so that reliable estimates of intakes can be made. Therefore approaches which rely on the detection of antimicrobial activity are usually unsuitable because they tend to be imprecise, non-specific and liable to produce false positive and false negative results (2). It is also unlikely that they could be adapted to take into account bound residues or toxicologically significant metabolites.

Antimicrobial inhibition methods may be suitable for enforcement programmes where accurate estimates of incidences and concentrations are not required. The lower costs permit relatively high sample throughputs which may allow larger numbers of positive samples to be detected.

The wide range of residues that could be present makes it desirable to employ methods which can detect residues of more than one chemically related compound. The development of such multi-residue techniques is a priority of the Working Party on Veterinary Residues in Animal Products. Chemical methods and immunoassays have been found to be the most appropriate analytical techniques for food surveillance programmes in the UK (2). Automated techniques can increase sample throughput and thus improve the statistical basis of the survey.

Tissues for analysis

Liver and kidney are the principal organs of metabolism and excretion and thus are are likely to contain the highest residues. Most consumers rarely eat these tissues from fish but liver from salmon may be used in the production of pâtès and pastes. Skin and bone may also accumulate residues of antimicrobials (4). Since some consumers eat fish skin and because it is often in intimate contact with flesh during cooking, residues surveys should include samples comprising flesh and skin.

The frequent requirement for several analyses on one sample often necessitates the inclusion of flesh from more than one fish. In such cases, each fish in the sample should come from the same batch so that dissimilarities between individual fish are minimised. Skin and flesh from all fish should be thoroughly homogenised before sub-sampling and analysis.

Sampling should also include farmed crustacea and molluscs. High inputs of antimicrobials may be required to control pathogens when raising shellfish in tropical and sub-tropical climates. Furthermore, antimicrobials may be used to control bacterial growth in the ice used to pack shell-fish during shipping (14). In the UK, imported prawns and shrimps are available in many retail outlets and these may have been produced in aquaculture systems.

Assessment of Surveillance Results

Estimating intakes

The incidences and concentrations of residues can be combined with food consumption data to provide estimates of intakes of residues. These can be compared with toxicologically based acceptable daily intakes (ADIs) to consider whether the food supply is safe. Accurate information on consumption patterns for rarely consumed foodstuffs such as farmed fish is very difficult to obtain. In the UK we presently have data for the general adult population obtained between 1986 and 1987 (Table II). These data were produced by asking 2000 individuals to maintain consumption records.

Table II. Consumption of Salmon and Trout by Adults in the UK, 1986-87 (g/person/day).

Fish	Number of consumers (%)	Mean	Maximum
Salmon	79.3 (3.5)	14.7	87.7
Trout	55.0 (2.5)	22.2	46.3
Total	126.5 (5.7)	8.9	87.7

Reporting and publication of results

Information on veterinary residues in farmed fish gathered by the Working Party on Veterinary Residues in Animal Products is reported back to the Steering Group on Food Surveillance. The Working Party also passes the results of surveillance to the UK Veterinary Products Committee who advise the Government on the licensing of veterinary medicines. This provides a check on the effectiveness of the system of licensing and control, and allows for product licences to be revised or withdrawn if necessary.

After a rigorous assessment of the potential toxicological consequences by consultation with expert advisory committees, the results are published. The most recent data on veterinary residues in farmed fish to be reported by the Working Party were published in a Food Surveillance Paper in 1986 (2).

Future Surveillance

It has been proposed that the Residues Directive (3) should be extended to include residues in farmed fish. Many of the criteria outlined above should be borne in mind when developing plans to enact future EC Directives. However, in the UK residues surveillance of farmed fish directed towards assessing the risk to the consumer will continue to be a high priority.

References

1). Food Surveillance Paper No.24 (1988). "Food Surveillance 1985 to 1988". ISBN 011 2428622, HMSO, London.

2). Food Surveillance Paper No.22 (1986). "Anabolic, Anthelmintic and Antimicrobial Agents". ISBN 011 2428169, HMSO, London.

3). Council Directive 86/469/EEC (1986). *Official Journal of the European Communities*, **L275**, 36.

4). RASMUSSEN, F. (1988) Therapeutics used in Fish Production. Pharmacokinetics, Residues, Withdrawal Periods. European Inland Fisheries Advisory Commission, Fifteenth Session. (EIFAC/XV/88/Inf.13 March 1988.) FAO.

5). MATTSON, N.S., EGIDIUS, E., KRYVI, H. and SOLBAKKEN, J.E. (1987) Fate of radioactivity in Atlantic salmon (*Salmo salar*) following administration of (Methyl-14C)-trichlorphon. *J. Appl. Ichthyol.* **3** 61-67

6). ANHALT, G. Rückstände antibakterieller Therapeutika bei Süßwasserfischen (1978). *Berichte uber Landwirtschafs* **55** (1977/78), 843-854.

7). SALTE, R. and LIESTOL, K. Drug Withdrawal from Farmed Fish (1983). *Acta vet. scand.* **24**, 418-430.

8). ALDERMAN D.J. and CLIFTON-HADLEY, R.S. (1991) Malachite Green: a Pharmacokinetic Study in Rainbow Trout (*Oncorynchus mykiss*) (In press).

9). WRIGHT, A.S., HUTSON, D.H. and WOODER, M.F. (1979) The Chemical and Biochemical Reactivity of Dichlorvos. *Arch. Toxicol.* **42**, 1-18.

10). CORTESI, M L., RIZZO, A.M., BOCHIOCCHIO,M. and SARLI, T.A. (1988) Antibiotic residues in the muscles of commercial trout. *Ind. Aliment. (Pinerola, Italy)* 27(258), 257-159, 262.

11). EDELHÄUSER, M. and KLEIN, E. (1986) Bestimmung von Malachitgrün-Rückständen in Speisefischen. *Deutsche Lebensmittel-Rundschau* **12**, 386-389.

12). GT BRITAIN, PARLIAMENT (1988) - The Medicines Regulations 1989. SI (1988) No. 1009. As amended.

13). NOAH Compendium of Data Sheets for Veterinary Products 1989-90. *Datapharm Publications Ltd.* 1989.

14). Malaysian Food Act 1983. Food Regulations 1985 (P.V.(A)437).

Discussion

ANON What happens with the batches of fishes if you detect residues at levels like processing plants, markets, supermarket chains and shops? Are they confiscated and eliminated and what if for example you detect vitamin C?

TENNANT The surveillance programme of the working party is based on potential hazard to the consumer so we wouldn't look at vitamin C. The object of the exercise is to collect data, its not a control process and so the time that the results are reported back would be long after the food had gone into supply. The purpose of it is to have data available to the authority so that we can have appropriate control procedures enacted. At the present there are no legal limits on veterinary residues in the United Kingdom.

ANON Well but if something is positive and this stuff is then eliminated or deleted or?

TENNANT No, because of the time scale involved, we can't take a sample and hold that particular product until the result is available. It has to go on into the supply chain.

SCHNICK I would like to ask, one of the compounds that is of concern is chloramphenicol and I noticed that it was not on your list of surveillance. Do you plan to add it and if you don't, why not?

TENNANT At the present our information is that chloramphenicol is not widely used in farmed fish production for fish for consumption in the United Kingdom. If it becomes apparent that this isn't the case and chloramphenicol is widely used or is used at all, then we would take action to measure residues.

MATSUSATO Is there a limitation of the use of malachite green in United Kingdom.

TENNANT As I understand the position, malachite green is not licensed for use on fish in the United Kingdom. There is presently no maximum residue in fish tissues.

LA DETERMINATION DES LIMITES MAXIMALES DE RESIDUS DES MEDICAMENTS VETERINAIRES DANS LES ALIMENTS D'ORIGINE ANIMALE: LE POINT DE VUE DU CODEX ALIMENTARIUS

J. Boisseau

CNEVA, Laboratoire des Médicaments Vétérinaires, La Haute Marche, Javené, 35133 Fougères, France

Résumé : *La Commission du Codex Alimentarius a décidé de créer un comité spécifique chargé d'établir des normes en matière de résidus de médicaments vétérinaires dans les aliments. Dès sa première réunion, tenue en 1986, ce comité a établi les critères permettant de définir les médicaments dont les résidus devraient être étudiés en priorité. Depuis, chaque année, ce comité transmet des listes de médicaments prioritaires au Comité mixte FAO/OMS d'experts sur les additifs alimentaires. Ce comité, appelé Jecfa du fait de sa dénomination anglaise (Joint Expert Committee on Food Additives), se réunit annuellement pour fournir au comité du Codex une assistance technique en évaluant scientifiquement la sécurité des résidus de ces médicaments vétérinaires prioritaires. La composition de ce comité diffère de celle du Comité du Jecfa chargé de proposer des normes en matière d'additifs alimentaires. Il associe dans son travail des experts désignés par l'OMS et la FAO.*

Les experts de l'OMS ont, plus particulièrement, la responsabilité d'étudier toutes les informations toxicologiques disponibles concernant, par exemple, la toxicité chronique, la tératogénèse, la mutagénèse, la carcinogénèse, la pouvoir allergisant, les risques microbiologiques. Ils doivent déterminer s'il est possible d'établir, à partir de cet ensemble d'informations, une dose journalière acceptable de résidus, exprimée en mg/kg/j dont l'ingestion n'expose le consommateur à aucun risque.

Les experts FAO ont, de leur côté, la charge de juger du devenir des médicaments vétérinaires dans l'organisme des animaux traités. Ils doivent étudier le métabolisme des médicaments, déterminer l'identité et la cinétique d'élimination des résidus susceptibles d'être retrouvés dans les denrées consommables issues de ces animaux. Au terme de leurs études, ces experts ont pour mission de proposer pour chaque denrée une limite maximale de résidus permettant le respect de la dose journalière acceptable établie par le groupe d'experts OMS. Cette limite maximale est exprimée en mg/kg à l'aide d'un métabolite défini appelé résidu marqueur.

Summary : **The determination of maximum residue limits of veterinary medicines in food of animal origin : the viewpoint of the Codex Alimentarius.**

In 1986 the Codex Alimentarius Commission established a special committee with responsibility for setting standards for residues of veterinary medicines in animal tissues. Since its first meeting in 1986, this committee has established criteria for the selection of those medicines whose residues require priority study. Since then, each year, this Codex Committee has forwarded lists of priority medicines to a combined expert committee of FAO/WHO on food additives. This committee, known as JECFA from its English title (Joint Expert Committee on Food Additives) meets annually to provide the Codex Committee with scientific advice on the evaluation of the safety of residues of the priority veterinary medicines. The composition of this committee differs from that of the JECFA committee charged with establishing standards for food additives. It consists of a combination of experts selected from WHO and FAO.

The WHO experts have primary responsibility for the study of all available toxicological information concerning areas which include chronic toxicity, teratogenicity, mutagenicity, carcinogenicity, potential to cause allergies and microbiological hazards. Using this information, they must determine as far as is practicable an acceptable daily intake (ADI) for residues of veterinary medicines, expressed in mg/kg/day. The ADI is the level of residue, the consumption of which presents no risk to the consumer.

For their part, the other half of the JECFA committee, the FAO experts, are responsible for assessing the behaviour of veterinary medicines in treated animals. They examine studies of the metabolism of these medicines, determine the identity and kinetics of elimination of residues detected in edible tissues of these animals. The outcome of these studies is for these experts to propose for each product a maximum residue limit (MRL). This MRL is the residue level which, if consumed will not exceed the acceptable daily intake (ADI) as established by the group of experts from OMS. This maximum limit is expressed in mg/kg of a defined metabolite, the marker residue.

La plupart des pays se sont dotés de législations pharmaceutiques vétérinaires qui ont pour principal objectif de garantir la qualité hygiénique des aliments d'origine animale et, par conséquent, de protéger efficacement le consommateur vis à vis des résidus médicamenteux qu'il est susceptible d'ingérer. Depuis, le besoin a été ressenti un peu partout de par le monde de protéger la santé publique dans des conditions équivalentes et d'éliminer les entraves sanitaires au commerce international des aliments d'origine animale.

Aussi assiste-t-on depuis quelques années à une évolution rapide en faveur d'une harmonisation internationale en matière de normes hygiéniques.

C'est ainsi que la Commission du Codex Alimentarius a créé, en 1985, un comité spécifique chargé du problème des résidus de médicaments vétérinaires dans les aliments. Le Comité, qui se réunit annuellement aux USA depuis 1986, a, parmi ses missions essentielles, celles

- de sélectionner les médicaments vétérinaires dont il convient d'évaluer en priorité les résidus

- de recommander des limites maximales pour les résidus de ces médicaments

- de déterminer les critères qui doivent guider le choix des méthodes analytiques à utiliser pour détecter les résidus dans les aliments.

Ces listes de médicaments vétérinaires jugés prioritaires sont ensuite soumises à l'examen du Comité mixte FAO/OMS d'experts sur les additifs alimentaires (JECFA) qui se réunit annuellement également, alternativement à Rome et à Genève. Ce Comité scientifique a pour tâche d'évaluer la toxicité des résidus de ces médicaments et de proposer pour ces contaminants, si les données scientifiques nécessaires sont disponibles, des limites maximales qui soient sans risque pour la santé publique. Les experts OMS sont plus particulièrement chargés de l'évaluation toxicologique de ces résidus, les experts FAO de l'étude du métabolisme des médicaments et de la cinétique de leurs résidus.

Evaluation de la toxicité des résidus de médicaments vétérinaires

La première démarche pour protéger la santé publique vis à vis des résidus médicamenteux a consisté tout naturellement à considérer comme inacceptable tout résidu détectable par les méthodes analytiques disponibles. Cette approche, qui avait l'avantage de la simplicité, était applicable dans la mesure où la sensibilité des méthodes analytiques demeurait limitée. Mais l'amélioration rapide de ces méthodes et l'abaissement spectaculaire de leurs limites de sensibilité a eu pour effet d'allonger considérablement la durée des temps d'attente qui devaient être observés pour que de tels résidus ne soient plus détectables. Ces améliorations techniques occasionnaient donc des contraintes de plus en plus handicapantes pour l'utilisation des médicaments vétérinaires sans accroître significativement la protection de la santé publique. Aussi ce concept de tolérance zéro est il considéré désormais comme irréaliste pour des médicaments vétérinaires autorisés. Le Comité du Codex considère ainsi que "les limites maximales de résidus doivent être basées sur le type et la quantité de résidus considérés comme ne présentant pas de risque d'ordre toxicologique pour la santé humaine".

Pour évaluer les potentialités toxicologiques des résidus de médicament, il est nécessaire de disposer des résultats de tout un ensemble de tests toxicologiques. A côté de tests de toxicologie générale, des essais plus spécifiques sont réalisés, en tant que de besoin, pour rechercher des effets mutagènes, carcinogènes, tératogènes, allergiques. La recherche des effets toxiques prend également en compte les effets pharmacologiques qui, s'ils sont souhaitables chez l'animal traité, sont à l'évidence indésirables pour le consommateur d'aliments provenant de ces animaux traités. C'est le cas par exemple des drogues psychotropes et des antibiotiques, ces derniers pouvant affecter la flore bactérienne intestinale du consommateur.

Ces informations toxicologiques peuvent provenir

. d'observations effectuées chez l'homme dans la mesure où le médicament peut également être utilisé en thérapeutique humaine.

. d'expérimentation animale, ce qui est le cas le plus fréquent

. d'expérimentations *in vitro* qui sont plus particulièrement mises à profit pour préciser le mécanisme d'action toxique.

On détermine à partir de cet ensemble d'informations la dose sans effet toxique observée qui correspond à la dose la plus élevée qui n'a permis d'observer le moindre effet toxique chez l'espèce animale la plus sensible. Elle est exprimée en mg/Kg/j.

On extrapole à partir de cette dose sans effet toxique chez l'animal une dose sans effet toxique chez l'homme appelée dose journalière acceptable (DJA) exprimée elle aussi en mg/Kg/j. Cette dose est déterminée en divisant la dose sans effet toxique chez l'animal par un facteur de sécurité. Lorsque les données nécessaires sont disponibles, la valeur habituelle de ce facteur de sécurité est de 100, justifiée par le double postulat que l'homme peut être 10 fois plus sensible que l'animal d'expérience et qu'il peut y avoir, pour un effet toxique donné, une gamme de sensibilité de 1 à 10 parmi la population humaine. La valeur de ce facteur de sécurité peut être supérieure à 100 si les informations disponibles sont insuffisantes ou si certains effets toxiques irréversibles, tératogènes ou carcinogènes, sont à craindre. Dans ces conditions, le facteur de sécurité peut être de 200, 500 voire 1000 ou 2000.

Cette quantité de résidus, exprimée en mg/Kg/j, est multipliée par le poids moyen humain estimé à 60 Kg et ensuite répartie entre les divers aliments d'origine animale qui représentent, classiquement, la base de la nourriture humaine.

La quantité d'aliments consommés joue un rôle important dans l'établissement des limites maximales de résidus dans les divers aliments d'origine animale. Cependant, il est difficile d'obtenir des données exactes sur la consommation alimentaire, notamment à l'échelle internationale, car les habitudes alimentaires sont influencées par des facteurs tels que l'origine ethnique, la religion et le climat. De plus, on manque de connaissances sur le devenir des résidus au cours des opérations de préparation des aliments (par exemple salage, fumage, fermentation, pasteurisation et stérilisation) ainsi qu'au cours de leur préparation domestique.

Pour protéger toutes les catégories de populations, il est raisonnable d'utiliser, dans le calcul des limites, les valeurs les plus élevées des données de consommation alimentaire pour les divers aliments d'origine animale. Le Comité a, en conséquence, retenu les consommations journalières suivantes: 300 g de viande, 100 g de foie, 50 g de rein, 50 g de graisse, 100 g d'oeufs et 1,5 l de lait.

Comme la détermination des limites maximales de résidus comporte un certain nombre de variables, par exemple les facteurs de sécurité utilisés lors de l'établissement des DJA, les temps d'attente et la biodisponibilité des résidus, les défauts de précision au niveau des estimations de consommation alimentaire n'ont, probablement pas une grande importance; il n'est donc pas nécessaire de chercher à affiner davantage cette information.

Etude du métabolisme et de la cinétique tissulaire du médicament vétérinaire

Les experts FAO du JECFA sont plus particulièrement chargés de l'étude

* de la nature des résidus tissulaires issus du métabolisme du médicament vétérinaire

* de la cinétique d'élimination de ces résidus dans les aliments d'origine animale

* des méthodes d'analyse utilisables pour le contrôle du respect des limites maximales de résidus établies.

* Il est en effet d'une importance primordiale de connaître la nature des métabolites médicamenteux susceptibles de contaminer les denrées issues d'animaux traités. Le coût d'une étude métabolique exhaustive conduit, dans la pratique, à limiter ces investigations aux métabolites majeurs bien qu'on ne puisse exclure les potentialités toxiques significatives pour la santé publique de métabolites mineurs. Les intérêts de cette étude sont divers:

- l'analogie des voies métaboliques entre les espèces animales de destination du médicament et les espèces de laboratoire utilisées dans les essais toxicologiques permet de valider ces essais toxicologiques

- l'identification de ces principaux métabolites permet de déterminer ceux qui doivent être pris en compte au vu des conclusions des études toxicologiques. On peut, par exemple, ne pas retenir des composés ayant perdu une structure aromatique dont on sait qu'elle est associée à un effet toxique donné.

A ce sujet, le problème des résidus liés est particulièrement démonstratif. On appelle résidus liés des résidus qui ne sont pas extractibles à partir des tissus et dont la caractérisation est souvent très difficile. Ces résidus peuvent comprendre :

. des résidus de médicaments incorporés par les voies métaboliques normales dans des composés endogènes (par exemple acides aminés, protéines, acides nucléiques). Ces résidus ne posent aucun problème toxicologique.

. les résidus liés chimiquement, issus de l'interaction des résidus de la substance mère ou de ses métabolites avec des macromolécules. Ces résidus peuvent poser un problème toxicologique.

Le Comité estime utile d'associer des techniques *in vivo* et *in vitro* pour l'évaluation des résidus liés.

L'impact biologique de ces résidus liés de médicaments vétérinaires présents dans les aliments dépend en général de leur degré d'absorption après ingestion. En l'absence de données sur les résidus, on supposera qu'ils sont en totalité biodisponibles et que leur activité est égale à celle du constituant le plus toxique.

Pour obtenir les données qui permettront d'évaluer la biodisponibilité de ces résidus, il est en général nécessaire de recourir à des études d'ingestion portant sur les tissus ayant subi les processus d'extraction adéquats et provenant d'animaux d'élevage ayant reçu le médicament sous forme radiomarquée. L'information la plus importante que l'on peut retirer de ces études est la proportion du résidu qui n'est pas biodisponible pour l'espèce animale utilisée comme modèle de "consommateur". Parmi les méthodes *in vivo* connues, celle de Gallo-Torres a été jugée acceptable pour l'étude de la biodisponibilité de ces résidus. Les résidus liés reconnus non biodisponibles par cette méthode peuvent être écartés de l'ensemble des résidus à prendre en compte pour le respect de la DJA.

Le Comité a pensé également que sont utiles des techniques *in vitro*, telles que l'incubation de résidus liés en milieu acide ou en présence d'enzyme appropriée qui pourraient simuler le système digestif humain. Les procédés actuels d'extraction visant à reproduire les processus d'absorption qui ont lieu dans les voies digestives sont intéressants mais sont, cependant, insuffisamment précis pour fournir des données fiables sur la biodisponibilité des composés.

* L'étude des cinétiques tissulaires de ces résidus majeurs apporte également une contribution indispensable à la détermination des limites maximales de résidus. Elle permet en effet de déterminer le résidu marqueur et le tissu cible adéquats qui sont les outils nécessaires aux plans nationaux de surveillance chargés de contrôler le respect des limites maximales de résidus.

On appelle résidu marqueur un résidu dont la teneur dans les tissus, oeufs, ou lait diminue dans un rapport connu par rapport à l'ensemble des résidus .

On appelle tissu cible le tissu dans lequel la cinétique d'élimination du résidu marqueur permet un contrôle pratique du respect des limites maximales établies.

Il a été jugé nécessaire de retenir au moins deux tissus cibles, un pour le contrôle des aliments soumis au commerce international (muscle, graisse) l'autre pour le contrôle des aliments soumis au commerce national (abats).

Cette étude des cinétiques tissulaires permet également de vérifier que les temps d'attente, nécessaires pour que les teneurs tissulaires en résidus soient inférieures à la valeur des limites maximales de résidus, sont réalistes et compatibles avec la bonne utilisation des médicaments vétérinaires.

Un temps d'attente est en effet le délai qui s'écoule entre la dernière administration d'un médicament et la collecte d'aliments à partir de l'animal traité. Son observation permet d'assurer que la teneur en résidus dans ces aliments est conforme à la limite maximale de résidus établie pour ce médicament vétérinaire. En conséquence la détermination d'un temps d'attente irréaliste car trop long, bien que nécessaire pour le respect d'une limite maximale de résidus, conduit à l'interdiction de fait du médicament vétérinaire ou au risque de ne pas voir ce temps d'attente respecté dans la pratique.

Il faut cependant noter à ce sujet que le JECFA n'a pas dans ses attributions de déterminer les temps d'attente adéquats pour les substances qu'il évalue car la durée des temps d'attente est conditionnée par les caractéristiques des spécialités pharmaceutiques vétérinaires et leurs conditions d'utilisation.

Cette étude des cinétiques tissulaires permet enfin d'apprécier si les temps d'attente qui seront respectés dans la pratique, du fait des conditions normales d'utilisation d'un médicament vétérinaire donné, ne conduisent pas à des teneurs en résidus notoirement inférieures aux valeurs des limites maximales de résidus établies au terme de l'évaluation toxicologique. Si c'est le cas, la prise en compte de ces bonnes pratiques peut conduire à réduire la valeur de telles limites maximales de résidus inutilement élevées dans la mesure où des méthodes analytiques pratiques de contrôle de résidus sont disponibles.

* Enfin l'étude des méthodes analytiques destinées à contrôler le respect de ces limites maximales de résidus représente le dernier volet d'activité des experts FAO du JECFA. Il leur appartient en effet de vérifier qu'il existe des méthodes fiables, utilisables en routine, capables de détecter les résidus traceurs dans les tissus cibles déterminés avec la sensibilité requise en égard aux limites maximales de résidus établies.

Conclusion

Les limites maximales de résidus sont déterminées par le Codex et le JECFA au terme d'une démarche scientifique rigoureuse à partir de l'étude des données de toxicologie, de métabolisme, de pharmacocinétique, de l'évaluation des méthodes analytiques, de la prise en compte des bonnes pratiques en matière d'utilisation des médicaments vétérinaires. Cette méthodologie n'en demeure pas moins pragmatique pour s'adapter au cas particulier des médicaments anciens, qui représentent encore l'essentiel de l'arsenal thérapeutique vétérinaire. Elle se doit également d'être flexible pour prendre en compte les acquis récents de la connaissance scientifique dans ce domaine en rapide évolution de la toxicologie alimentaire .

SPOT CHECK FOR RESIDUES IN FARMED FISH IN JAPAN

A. Okamoto

Research Division, Research Department, Fisheries Agency,
Ministry of Agriculture, Forestry and Fisheries,
1-2-1 kasumigaseki Chiyoda-Ku Tokyo, Japan

Summary: *The Japanese Food Sanitation Law prevents people from having antibacterial agents reesidues in food in Japan. The Ministry of Health and Welfare carries out checks to determine whether antibacterial agents remain in fish being sold. Some fish producer organizations and prefecture governments also independantly check drug residues for the consumers who want supply of safe foods. The Fisheries Agency promotes check by them with subsidizing.*

***Résumé*: Contrôle ponctuel des résidus dans les poissons d'élevage au Japon.**

L'importance alimentaire du poisson est élevée au Japon (tableau I), où la consommation annuelle per capita *est estimée à 1.6 kgs (Tableau II). L'évolution des productions aquicoles, passée de 50 000 T en 1965 à 340 000 en 1980 et représentant un chiffre d'affaire de 365 milliards de yens, reflète cette popularité. Les producteurs sont cependant confrontés à des problèmes pathologiques qui tendent à s'accentuer avec l'intensification des élevages. L'impact des pathologies est estimé à près de 20 milliards de yens, tandis que le recours aux thérapeutiques crédite en revanche l'industrie pharmaceutique d'un chiffre d'affaires annuel de 7 milliards.*

Dans ce contexte, la qualité des produits destinés au marché alimentaire est une préoccupation majeure, régie par la loi sur l'hygiène des denrées alimentaires et par la loi sur les questions pharmaceutiques, dont les applications sont placées sous contrôle de l'Agence des Pêches. Le Ministère de la Santé publique a défini des critères proscrivant la présence de résidus pharmaceutiques dans les produits alimentaires. Les méthodes d'inpection sont codifiées et mises à exécution par plus de 6800 agents du Ministère, qui contrôlent aussi bien les produits du marché intérieur que les importations. Les résidus d'agents antimicrobiens font l'objet d'une attention particulière (Tableau III). Certaines organisations professionnelles, ainsi que les stations expérimentales de l'Agence des Pêches réparties dans les préfectures, appuient cette action par des contrôles exécutés à l'échelle locale. Très peu de cas litigieux ont été répertoriés à ce jour, ce qui semble révéler une sensibilisation des producteurs, dont l'encadrement technique demeure cependant un objectif prioritaire.

Introduction

Safety of foods is one of the social issues in Japan today. Some consumers are taking great care of that. Dealing with this issue is one of the reponsibilities of the Fisheries Agency which covers the administration side of aquaculture. The Japanese Food Sanitation Law is intended to prevent people from having antibacterial agents as residues in foods. Additionally, the Pharmaceutical Affairs Law regulates using drugs for fish. Checking drug residues before shipping fish is useful to determine if producers are keeping to these regulations and to prevent fish in which drugs remain from being supplied to consumers. I would like to present the status of drug use for fish and checking drug residues in Japan.

The consumption of farmed fish

Table I shows the production of feeding aquaculture in Japan. The total production increased 7 times from 50,000 tons in 1965 to 340,000 tons in 1988, that is, 19 times increase in value from 19 billion yen in 1965 to 365 billion yen in 1988. Especially the progress of yellowtail, red sea bream, and eel are remarkable. Coho salmon and other marine species have begun to be farmed recently.

Table I: The change of production of feeding aquaculture in Japan (unit: ton).

FISH SPECIES	YEAR		
	1965	1985	1988
Yellowtail	14,799	150,961	165,928
Red sea bream	*	6,990	45,220
Coho salmon	-	28,430	6,496
Carp	7,971	19,105	18,130
Eel	16,017	39,568	39,558
Rainbow trout	*	16,324	15,314
Ayu	*	10,967	13,633
Horse mackerel	*	5,008	6,455
Tilapia	-	4,180	4,760
Flounder	-	1,572	3,090
Kuruma prawn	154	2,151	3,020
Other fish	9,858	8,508	2,048
Total	48,799	293,764	343,652

* included in "other fish"

Table II shows how much Japanese farmed fish and other aquatic animals one Japanese person consumes. The total consumption per one person in 1988 is estimated to 1.6 kg. We do not eat farmed fish with frequency, although the production of feeding aquaculture is increasing.

Use of drugs

Aquaculture is always having problems, fish disease occurrences, in accord with the increase of production of that. Recent damage from disease is estimated to amount 20 billion yen. Improvement and extension of husbandry technique in aquaculture is very important to prevent fish disease occurrences. Chemotherapy is also very efficient to reduce the damage from the disease. Sum of the money currently used for drugs to the disease in aquaculture amounts to 7 billion yen, corresponding to 2% of total production of aquaculture.

Table II.:The consumption of feeding farmed fish per one person in Japan in 1988 (unit: g).

FISH	AMOUNT CONSUMED
Yellowtail	748.8
Red sea bream	148.8
Coho salmon	81.2
Carp	59.5
Eel	243.4
Rainbow trout	75.4
Ayu	72.7
Horse mackerel	29.1
Tilapia	17.6
Flounder	13.9
Kuruma prawn	12.4
Other fishes	54.4
Total	1,556.9

Checking for drug residues by the Ministry of Health and Welfare

The Ministry of Health and Welfare is the authority responsible for the administration of food sanitation. There is a standard established by the Ministry of Health and Welfare based on the Food Sanitation Law. This standard prohibits people from having antibacterial agents residues in foods. The Ministry of Public Welfare appoints check methods for people to inspect if foods are up to the standard. Table III shows check methods selected by the

Ministry of Public Welfare. Food sanitation inspectors who belong to the Ministry of Public Welfare check, if antibacterial agents remain in fish being marketed. There are total of 6,869 food sanitation inspectors including 1,811 whose only job is checking. These inspectors tested 28,360 fresh fish samples between April 1988 and March of 1989. Antibacterial agents were detected in 11 fish as a result of these tests. Fish taken for testing included both domestic Japanese products and imported goods.

Table III.: Detection methods of antibacterial agents

Drugs	Detection method
Antibiotic	Microbioassay
Sulfamonomethoxine	ECD-Gas chromatograph
Sulfadimethoxine	ECD-Gas chromatography
Sulfisozole	ECD-Gas chromatograph
Thiamphenicol	ECD-Gas chromatograph
Sodium nifurstyrenate	Thin layer chromatiography
Oxolinic acid	High performance liquid chromatography
Nalidixic acid	High performance liquid chromatography

Tests for drug residues by producers

Some producer organizations check residues independantly before shipping fish, because they need to deal with the demand from consumers, which is for a safe food supply. The officers belonging to the Fisheries Experimental Stations of prefecture governments also check residues in fish shipped by producers. The Fisheries Agency is subsidizing these tests. Producers check about 1,500 samples every year with this subsidy. The number of samples which the officers of prefectures check a year is about 2,000. We have not any samples in which antibacterial agents remain in these testsso far.

Conclusion

Checking drug residues before shipping fish is useful to prevent fish in which drugs remain from being supplied to consumers. But more important thing is trying to ensure that no drug residues remain in fish. Guiding producers in their proper use is one of the main jobs of the officers of prefecture governments. The Fisheries Agency also subsidizes this work

TRENDS IN INTERNATIONAL COOPERATION FOR AQUACULTURE DRUG REGISTRATION

R.A. Schnick

National Fisheries Research Center, 2630 Fanta Reed Rd., La Crosse, WI 54603-1223, USA

Summary: *Few drugs are registered worldwide for use in aquaculture. Mechanisms must be found to encourage the generation of data that would be acceptable in many countries so that drugs could be marketed worldwide without any repeated studies. Potential mechanisms include universal (1) acceptance of existing guidelines and development of new guidelines for mammalian safety and environmental studies for aquaculture drugs, (2) acceptance of protocols being developed for studies on aquatic species, especially pharmacokinetics and metabolism, (3) willingness to develop appropriate regulations and registration requirements for aquaculture drugs, (4) acceptance of codes developed for aquaculture, and (5) development of an international clearinghouse for information on data generated or in process of being generated throughout the world on aquaculture drugs. With these international and united efforts the aquaculture industry could become attractive to the pharmaceutical industry.*

Résumé **: Tendance à l'harmonisation internationale des règles d'autorisation pour les médicaments d'aquaculture.**

Les compagnies pharmaceutiques se désintéressent actuellement du marché aquicole compte-tenu de son étroitesse, de la protection incertaine des brevets, de la diversité des espèces-cibles et des exigences propres à chaque pays. Or le nombre des produits autorisés est extrêmement restreint. Le besoin d'une harmonisation visant à réduire les dépenses nécessaires aux études pour mettre au point de nouvelles formules est fortement ressenti.

En ce qui concerne les garanties d'innocuité pour les mammifères et l'environnement, déjà des efforts ont été entrepris, par exemple dans la CEE. Les directives de l'OCDE pourraient constituer une base de références pour réajuster les réglementations nationales. La question des résidus est plus complexe, car s'il y a consensus sur la nécessité de définir des délais d'attente, les niveaux retenus sont très variables et la température n'est pas toujours prise en compte. On manque en fait de données pharmacocinétiques relatives aux espèces aquatiques, et il est vraisemblable que des protocoles d'analyse et de contrôle universellement approuvés contribueraient à faire évoluer la situation. Un autre aspect essentiel concerne les modalités de délivrance des autorisations de mise sur le marché. Les règles ne peuvent pas toujours être extrapolées à partir du cas des espèces terrestres, et dans plusieurs pays (Norvège, USA) des réflexions sont en cours pour les adapter au contexte de l'aquaculture.

A l'échelle internationale des comités d'experts ont été mobilisés par des organismes comme la FAO et l'OMS pour travailler à l'élaboration de codes pratiques destinés à normaliser l'emploi des médicaments en aquaculture, à encadrer l'autorisation et la distribution des produits vétérinaires, et à promouvoir des pratiques d'hygiène pour les produits aquacoles. La définition de techniques sandardisées est également à l'étude pour compléter ces documents de référence.

Tous ces efforts risquent cependant de rester vains s'ils ne sont pas doublés par un accès aux données et à l'information. La FAO pourrait jouer le rôle d'une agence de renseignements internationale et coordonner les recherches indispensables. La circulation de l'information peut aussi s'appuyer sur des véhicules comme la Lettre publiée conjointement par l'OIE et l'ITCVDR (International Technical Consultation on Veterinary Drug Registration),qui disposent d'un large réseau de correspondants. Le mouvement d'uniformisation auquel on assiste est en tout cas indispensable pour éviterle gaspillage de temps et d'argent et relancer le dynamisme des firmes pharmaceutiques.

Introduction

Pharmaceutical and chemical companies are inhibited from pursuing the registration of drugs for aquaculture due to the low sales volume, lack of patent protection for existing drug treatments (called "orphan drugs"), tremendous diversity of cultured aquatic species and diseases, and lack of uniform testing guidelines and data requirements from country to country. For a company to make a profit on an aquaculture drug, a worldwide market is needed. However, few data generated on aquaculture drugs are accepted outside of the originating country; and, the required tests are too expensive to repeat from country to country. For example, mammalian safety studies completed on oxolinic acid were acceptable in European countries, but were not acceptable in the United States because they did not meet their guidelines.

Guidelines of mammalian safety and environmental studies

Few drugs known to be effective on diseases of aquatic species have data generated on mammalian safety, environmental fate, residues, and metabolism. These data are also the most expensive to generate, costing a minimum of $3 million to meet those data requirements for a veterinary drug registration in the United States.

It is important to find a mechanism to encourage the generation of data acceptable in many countries. One such mechanism could be guidelines developed by the Organisation for Economic Cooperation and Development (OECD). OECD has developed an array of guidelines for testing chemicals based on Good Laboratory Practices that are acceptable to its 24 member countries, including the United States, Canada, Japan, and several European countries. These guidelines were developed to provide a common basis for the acceptance of data on an international basis, thus, providing "the opportunity to reduce direct and indirect costs to governments and industry associated with testing and assessment of chemicals." Many of the mammalian safety and environmental studies that are required for aquaculture drugs are already part of the OECD testing guidelines. Other guidelines can be added because the OECD has a program for updating and adding tests to their guidelines [7]. Thus, guidelines for residue and metabolism studies could be added to make the system more effective and complete.

The development of new or revised OECD guidelines for drugs would have to be initiated by a member country, extensively reviewed by a panel of experts, endorsed by OECD policy bodies, and finally adopted by the OECD Council. These guidelines are test methods that provide a common basis for data to be acceptable to regulatory authorities internationally. However, they are not a checklist of regulatory requirements. Data requirements will vary from country to country. Core guidelines, however, would be useful for registering aquaculture drugs [10].

Individuals representing the United States, Canada, Europe, and Asia have agreed to contact their respective fishery organizations to pass resolutions advocating the global use of OECD guidelines to register aquaculture drugs. Those resolutions will be sent to the agency or agencies in that country that regulate and register aquaculture drugs. In the United States, the Center for Veterinary Medicine of FDA has been requested to participate in the review and acceptance of OECD guidelines for testing drugs for use in aquaculture. The acceptance of these resolutions by OECD member countries would result in the use of studies from other countries, the reduction of expenses associated with duplicated research, the attraction of pharmaceutical firms to the aquaculture industry, and the overall increase of legal products for aquaculturists throughout the world.

Efforts to develop uniform guidelines for toxicity testing is not a new idea. The regulatory agencies of seven countries (United States, Canada, Japan, France, United Kingdom, West Germany, and Italy) and the European Community collected information on their toxicity protocols to determine if they could be harmonized in testing drugs for human use. They initiated efforts to harmonize several tests so that they would be acceptable to all drug regulatory authorities. This program will reduce or eliminate repeating studies and thus save limited resources [12]. The aquaculture community needs to ensure that we benefit from their efforts.

The Round-Table Consultations on Documentation of Drugs for Use in Aquaculture [6] recommended in June, 1990 that previously approved guidelines for registration of drugs in veterinary use be applied to mammalian safety studies for aquatic species. An additional recommendation would be that countries look at their guidelines to determine if they are the same as the OECD or other international guidelines. If they are slightly different, it would be advantageous to resolve the differences so the guidelines would be consistent from country to country. The same would be true for the environmental testing guidelines that were developed by OECD.

Guidelines for aquatic species studies

As the aquaculture industry grows worldwide, the scrutiny by regulatory agencies increases along with public interest. The major concern are the residues in edible tissues of compounds that are not suitable for human consumption. Currently, no maximum residue levels exist on a worldwide basis.

In 1986, the European Inland Fisheries Advisory Commission recognized that drug residues remain in aquaculture products for varying periods of time necessitating the need for withdrawal times based on accurate data. Withdrawal times are established to ensure that residues of concern are not present in edible food products. Few withdrawal times exist for aquaculture drugs because of the lack of reliable pharmacokinetic data [7].

The establishment of withdrawal times varies greatly from one country to another. Few countries take temperature into consideration. One that does, the United Kingdom, uses degree days (number of days past treatment multiplied by the water temperature). Because so many factors influence the pharmacokinetics of aquaculture drugs and so few data exist, recommendations have been made to have (1) more pharmacokinetic data developed on drugs used on fish, (2) rules for use of chemicals and drugs in fish farming, (3) usable withdrawal times that protect consumers against residues, and (4) rules to protect the environment. Rasmussen [8] suggested that general withdrawal times be established until adequate pharmacokinetic data are available. He recommended 80 days after the last treatment with water temperatures below 10°C and 40 days for water temperatures of 10°C or above. Shepherd [11] recommended a standard withdrawal time of 200 degree days. Thus, an average temperature of 20°C would require a 10-day withdrawal, whereas, 10°C requires a 20-day withdrawal period. In his review of the pharmacokinetic data available on sulfonamides, tetracyclines, nitrofurans, quinolones, and chloramphenicol, Alderman [1] found that technology has allowed for detection of minute levels of a drug in fish tissue. What risks are associated with these levels are difficult to ascertain as well as what would be practical tolerance levels and withdrawal times.

In 1984, the U.S. Food and Drug Administration recognized that the agency did not have much information about the metabolism and disposition of drugs in aquatic species and differences in metabolism and toxicity among species. Therefore, FDA sponsored a workshop on drug disposition and metabolism in aquatic species. The workshop participants concluded that aquatic organisms do not produce unique metabolites [4]. FDA followed that workshop with a roundtable discussion on metabolism of aquatic species at the IR-4/FDA Workshop on Minor Use Drugs: Focus on Aquaculture held October 16, 1990 [9]. The roundtable allowed open dialogue and understanding by both FDA personnel and outside researchers. FDA indicated its willingness to be flexible and agreed to protocols on metabolism studies that are reasonable for aquatic species.

Another group actively involved in the question of residues of veterinary drugs in food is the Federation of Chemical Societies' (FECS) Working Party on Food Chemistry. The FECS met in May, 1990 in Utrecht, The Netherlands, to discuss the increasing demands of regulatory agencies and consumers to guarantee the safety and quality of food in general and food animals (including fish) in particular [5]. The group concluded that quick, easy, reliable, quantitative, and confirming analytical methods are needed for all food-producing industries. The main reason for developing residue data is to ensure safety of the product. A drug should not present any unacceptable risk to either the consumer or the environment.

Regulations and registation requirements

The Norwegian Medicines Control Authority (Statens Lagensmittel Kontrol) initiated a meeting in June, 1990 in Oslo, Norway to develop appropriate protocols and guidelines for testing products to be used in the production of aquatic species. Persons were invited from several European countries, Canada, and the United States. These international participants addressed existing regulations and documentation of pharmaceutical quality, pharmaceutical and clinical trials, safety testing, and environmental impact. The participants recognized the need for separate guidelines, addenda, and regulations for aquaculture drugs that would address the differences between aquacultural practices and terrestrial culture of food animals. In some cases, such as the toxicity testing in target species, the participants recommended the use of OECD guidelines. The results of the conference were presented in report form and also presented at the 5th International Technical Consultation on

Veterinary Drug Registration in The Hague, Netherlands, on October 8-12, 1990 [6]. The participating regulatory officials from countries to the 5th ITCVDR Conference agree with the need for international cooperation in developing guidelines. The conference attendees recommended that regulatory agencies involved in veterinary drug registration take the proper action to ensure (1) that only approved drugs are used in aquaculture according to the approved product information and made available through officially approved distribution channels, (2) coordination of the development of specific guidelines for the demonstration of efficacy, target animal safety, safety of the user, the environment, and consumer, and (3) that withdrawal times be set for veterinary drugs used in aquaculture which reflects the temperature-dependence of the pharmacokinetics of drugs used in aquatic species (B. Fossum, personal communication, The Norwegian Medicines Control Authority, Oslo, Norway).

The U.S. Food and Drug Administration (FDA) is also trying to find mechanisms that will assist the aquaculture community in meeting the data requirements of the agency to ensure the food safety of farmed aquatic animals. These include (1) designation of crop groupings to reduce the number of target animal safety studies, (2) clarification of the regulatory status of water treatment compounds, (3) acceptance of letter petitions for approval of commonly used chemicals that would be placed in the Compliance Policy Guide, (4) review of the minor use guidelines to determine how they can better fit aquaculture, (5) willingness to accept information on various species that might lead to definition of early life stages as non food fish, (6) providing funding for mammalian safety studies, analytical methodology, pharmacokinetics, and special studies, (7) provision for exclusivity if a company has raw data for studies, (8) potential for extra label prescription by veterinarians in conjunction with certified fish pathologists, (9) creation of an Aquaculture Committee and the position of aquaculture coordinator to oversee all activities related to aquaculture, and (10) interact with the U.S. Environmental Protection Agency to determine jurisdiction over chemicals that are used both as drugs and pesticides.

Use codes for aquaculture production

Uniform use regulations throughout the aquaculture producing countries would assure all countries that the food produced is safe for human consumption and would assure the pharmaceutical industry that only properly registered or licensed compounds will be used by the growers. Several codes are under development by the Food and Agriculture Organization in conjuction with the World Health Organization.

The Codex Committee on Residues of Veterinary Drugs in Foods (CCRVDF) with assistance of the Codex Committee on Fish and Fishery Products (CCFFP) are drafting preliminary documents on the control of drug residues in food and the use of veterinary drugs. The documents will eventually become part of the FAO/WHO Codex Alimentarius. The CCRVDF recognizes the necessity for standard methods of analysis and sampling of aquaculture products and the need for a manual of compliance, including sampling and analysis. In the meantime, many countries have set zero tolerances for drugs. The CCFFP recommended such a program because of the human health concerns [2]. A separate Code of Practice for the Use of Veterinary Drugs in Aquaculture is being developed under the lead of the Canadian Bureau of Veterinary drugs because of requests by both codex committees. The CCRVDF is also considering a Code for the Registration and Distribution of Veterinary Drugs. This code would include (1) a registration procedure to objectively evaluate the technical and scientific data with regard to the quality, efficacy, and safety of the veterinary drug, and (2) a distribution system that would ensure that veterinary drugs were made available to farmers in an efficient and safe manner.

One code in draft form is the Code of Hygienic Practice for Products of Aquaculture. This code should assist those engaged in aquaculture to produce high quality, safe fish for human consumption without adversely affecting the environment. Since national regulations on hygienic practices in aquaculture are scare, this code can help all countries raise fish that should be acceptable both in national and international trade. The Code recommends that veterinary drugs be used in compliance with regulations of the official agency having jurisdiction and that no drug residues be above the maximum permissible levels set by that agency [3].

In the United States, a committee named the Working Group on Quality Assurance in Aquaculture Production, was recently formed to develop educational materials for fish farmers about the regulatory status of various chemicals useful in aquaculture management. The committee is to provide recommendations for use of chemicals to assure product quality and safety. The Working Group is coordinating six initiatives that include (1) developing a comprehensive document that summarizes Federal regulation of drugs and chemicals, (2) establishing a central data bank of drugs and hemicals used in aquaculture production, (3) obtaining clarification of the regulatory status of compounds used in aquaculture production, (4) encouraging each aquaculture industry to develop their own quality assurance program, (5) sharing information on production practices with regulatory agencies, and (6) developing species specific producer guides on drug and chemical use in aquaculture production [13].

Information clearinghouse

The value of all these efforts to develop universal regulations, guidelines, and protocols could be diminished if no mechanism were in place to let interested parties know what data are available or in progress on high priority drugs for aquaculture. Additionally, researchers need to know about and use accepted protocols and guidelines for registering an aquaculture drug so their research can be used as part of a registration package.

An initiative is in progress to get the Food and Agriculture Organization (FAO) of the United Nations to act as an international clearinghouse of information on data for aquaculture drugs and protocols for registration-oriented research on aquatic species. FAO could also act as a coordinator of the research on chemicals used as drugs in aquaculture that have no patent protection (i.e., orphan drugs). Potential researchers could check with FAO before they begin the studies so that no research would be duplicated and efforts could be focused on certain priority drugs [9]. A possibility for the dissemination of information regarding aquaculture drugs could be the newsletter on the registration of veterinary drugs that is published jointly by the International Office of Epizootics and the ITCVDR. In fact, the ITCVDR increased its interest in this newletter by providing a larger network of national correspondents at its meetings in October 1990 [2]. OIE and ITCVDR also have the capability to function as a joint clearinghouse, perhaps in place of FAO.

Discussion and conclusions

Mechanisms listed here must be used to address the issues raised here on the regulation and registration of aquaculture drugs. Without international awareness, cooperation, coordination and participation, each country will continue to struggle to meet the chemotherapy needs of the aquaculture industry and at the same time, ensure the safety of the food supply. It is obvious that a start has been made to resolve these challenges.

Pharmaceutical firms need to be assured that regulations, guidelines, and protocols are uniform so that a compound can easily be registered for use in several countries at the same time. Some mechanism needs to be developed that allows parties interested in registering aquaculture drugs to ascertain what data needed for registration are available or what is in progress to prevent duplication and wasted effort.

References

1). ALDERMAN D.J., 1988. Fisheries chemotherapy: a review. In J.F. Muir and R.J. Roberts eds., *Recent advances in aquaculture*, Volume 3, Timber Press, Portland, Oregon, pp. 1-60.

2). CODEX ALIMENTARIUS COMMISSION, 1990. *Report of the Fifth Session of the Codex Committee on Residues of Veterinary Drugs in Foods.* Food and Agriculture Organization and World Health Organization, Rome, Italy, ALINORM 91/31A. 53 pp.

3). FOOD AND AGRICULTURE ORGANIZATION, 1990. *Code of hygienic practice for products of aquaculture.* First draft. Food and Agriculture Organization, Rome, Italy, 48 pp.

4). GUARINO A.M., BATSON D.B. & ZEEMAN M.G., 1986. Workshop on drug disposition and metabolism in aquatic species. *Vet. Hum. Toxicol.*, **28** (Suppl. 1), 1-56.

5). HAAGSMA N., RUITER A. & CZEDIK-EYSENBERG P.B., 1990. *Residues of veterinary drugs in food: Proceedings of the EuroResidue conference.* Noorwijkerhout, The Netherlands, May 21-23, 1990. Rijksuniversitelt Utrecht, Utrecht, The Netherlands.

6). THE NORWEGIAN MEDICINES CONTROL AUTHORITY, 1990. *Round-Table Consultations on Documentation of Drugs for Use in Aquaculture.* Report. The Norwegian Medicines Control Authority, Oslo, Norway.

7). ORGANISATION FOR ECONOMIC COOPERATION AND DEVELOPEMENT (OECD), 1981. *OECD guidelines for testing of chemicals.* OECD, Paris, France. Sections 1-4.

8). RASMUSSEN F., 1988. *Report on the working party on withdrawal period for drugs: Therapeutants used in fish production: pharmacokinetics, residues, withdrawal periods.* European Inland Fisheries Advisory Commission, Food and Agriculture Organization of the United Nations, EIFAC/XV/88/Inf.13. March 1988. 14 pp.

9). SCHNICK R.A., 1991. An overview of current programs on drug development and regulation for aquaculture: U.S. Fish and Wildlife Service Perspective. *Vet. Human Tox.*, **33**, (Suppl. 1), 4-5.

10). SCHNICK R.A., 1991. *Chemicals for worldwide aquaculture.* Proceedings of the Workshop on Fish Disease and Fish Health Managment, Pusan, Korea (in press).

11). SHEPHERD C.J., 1987. How to avoid tissue residues in practice. *Fish Farmer*, **10** (5), 40.

12). SPEID L.H., LUMLEY C.E. & WALKER S.R., 1990 Harmonization of guidelines for toxicity testing of pharmaceuticals by 1992. *Regul. Toxicol. Pharmacol.*, **120**, 179-211.

13). U.S. FOOD AND DRUG ADMINISTRATION, 1991. Quality assurance in aquaculture production. *FDA Veterinarian*, **VI**, (III), 6-7.

CHEMOTHERAPY IN AQUACULTURE : TENTATIVE CONCLUSIONS

J.-P. Lafont

INRA, Station de Pathologie aviaire et de Parasitologie,
Centre de Tours-Nouzilly, 37380 Monnaie, France.

To summarize four days of fruitful work and stimulating discussion presents a real challenge. I should point out at the beginning that my own research does not deal with aquaculture but with chemotherapy and antibiotic resistance in bacteria from warm-blooded animals. This is bound to result in strong personal biases, but it is probably what the organizers of the meeting have been looking for, and the reason why they asked me to present this concluding communication and to be the chairman of the last discussion. They probably expect to find not only the highlights of the session but the provoking comments of an observer exterior to the field of aquaculture as well. That is why I have tried both not to disappoint them and not to betray the opinions expressed by the various participants in this last session.

General comments

1. The growing human population needs fish meat and the meat of other aquatic species as a source of high quality proteins and of diversity in its diet.

2. This need must be fulfilled, at least in part, by farming, since the excessive catching of wild species would rapidly deplete living stocks. Aquaculture accounts for an increasing proportion of the total production of fish and of other aquatic species, and this will probably be a long-lasting trend. Moreover,aquaculture is acquiring a prominent social importance since it is frequently developing in economically less favoured areas.

3. Aquaculture is in a phase of rapid development and growth, and the intensification of fish farming leads to the emergence of infectious and parasitic diseases, as previously observed in other animal production systems undergoing similar changes, such as pig and poultry production.

4. Few vaccines are presently available and their efficacy may be questioned in several cases, at least under practical field conditions. It is unlikely that many new efficient vaccines, convenient for mass administration, will appear in the near future. Furthermore, even when an efficient vaccine is

used, antigenic variation in pathogens is liable to lead to the recurrence of the problem, as observed in vibriosis.

5. Therefore, aquaculture in general and fish farming in particular do need the availability of chemotherapeutic agents to avoid severe economic losses, and chemotherapy will remain one of the main means of controlling transmissible diseases in the foreseable future.

6. The routes of administration of drugs to fish and other aquatic species are inevitably unconventional as compared to treatment of other meat animals. The means of this distribution could represent a hazard to human consumers and to the environment. Regulations are obviously needed in this respect.

Major difficulties

1. Protection of fish and similar species against infectious and parasitic diseases has mainly relied upon chemotherapeutic agents and this has led to the use, overuse, and even occasional misuse of drugs, especially antibiotics.

2. The pharmacokinetics of drugs vary widely according to the species of fish treated, the temperature of the water, the pharmaceutical form etc, and their bioavailability depending on the formulation and the route of administration has been insufficiently investigated. Despite the excellent work of the few research groups involved, many pharmacological data are lacking, and the mechanisms of drug absorption are poorly understood. The pharmacokinetic data available deal mainly with the problems of residues in the meat and less with the efficiency of treatments. Thus, drug usage in aquaculture in the field still appears rather empirical.

3. As a consequence of this intensive use, antibiotic-resistant and multiresistant pathogens have emerged and spread, most prominently in bacteria associated with intensive fish production, such as *Aeromonas salmonicida* and *A.hydrophila*, greatly impairing the efficiency of chemotherapy. Many resistance plasmids, often conjugative, are presently harboured by these pathogens.

4. In the very specific conditions of aquaculture, hygienic procedures efficiently applied to other intensive animal productions (e.g. pigs and poultry) are not relevant or are very difficult to transpose.

5. Due to the rather limited market of aquaculture, to the high costs of research and to the difficulties in complying with the various requirements of national authorities for regulatory approval, the pharmaceutical industry appears reluctant to develop new drugs for this particular market. Registering

a new drug for aquaculture may cost as much, or more, than registering the same product for other meat animals. Thus, some participants consider we are in the presence of a "registration crisis".

6. Another consequence of the intensive use of antibiotics and pesticides is the accumulation of drugs and their metabolites in aquatic environments: accumulation in sediments of drugs administered orally, dispersion in the whole aquatic system of drugs used for bath treatments.

<u>In summary, we are, and will be faced with</u>:

* An increasing number of diseases
* A limited range of approved therapeutic products.
* A pressing need for pharmacological data
* A pressing need for new products in the future.
* An increasing incidence of resistance in pathogens.
* A growing concern about residues in the environment and in fish products.

And, moreover...

"REMEMBER HOW MUCH WE DO NOT KNOW" (M. Yndestad)

What do we need to know ?

Here is a list of what appeared as the most important points and controversial issues:

1. What is a proper treatment, based on pharmacological, rather than empirical, data: dose and dosage schemes, duration, pharmaceutical form, mode of administration ?

2. What is the actual risk, for the human consumer, of selecting antibiotic-resistant bacteria harbouring multi-resistance plasmids in aquatic environments? This risk may differ widely according to the geographical and climatic conditions and to the possible contamination of waters by fecal bacteria of human or warm-blooded animal origin. Obviously, it is not the same in cold waters from subarctic underpopulated regions or in warm waters from tropical, densely populated areas.

3. Is the presence of very low levels of drug residues in the meat of fish of real significance for human consumers ? Do we really need "zero levels" and the very long withdrawal times they imply ? What are acceptable levels of residues to avoid human health concern, and how should we specify withdrawal periods to comply with these ?

4. What are the detrimental effects associated with the persistence of residues in aquatic environments and sediments ?

5. What antibiotics and chemotherapeutic agents do we really need in aquaculture? Are some currently used drugs unnecessary? Could we use more labile molecules less liable to persist for long periods in aquatic environments and sediments? Could we rear aquatic animal species without being compelled to recourse to the use of chemotherapeutic drugs?

Some suggestions: we should...

1. Educate fish farmers about the necessity to base treatment on an accurate professional diagnosis and about the rational use of antibiotics and other therapeutic drugs. This should result in a better protection of people involved in the administration of drugs in the field.

2. Discourage the use of antibiotics bought as chemical-bulk products: prescription of licensed medicines by a properly trained veterinarian, fish pathologist or practitioner, appears as an essential requirement which should be enforced. The professional training of these specialists should concomitantly be improved.

3. Try, whenever possible to "shuttle" between antibiotics, even when resistance to the previously used molecule has not yet appeared. The use of new molecules as first choice treatment is not necessarily the best choice but could be advocated in some instances (*e.g.* fluoroquinolones versus oxolinic acid). This requires a sufficient number of authorized molecules to be available.

4. Rely as much as possible upon accurate microbiological and susceptibility-testing data. It is often impossible to have a sample studied before treatment but a bacteriological study during or even after treatment can be really useful: it never comes too late and could provide useful information about the selective effect of the current treatment, relevant to the possible failure of further treatments.

5. Use more precise susceptibility tests: the usual disc-diffusion test has a good predictive value for the efficacy of an emergency treatment, but precise MIC determinations could be very helpful for future treatments (e.g. evidence of a slight decrease in susceptibility to oxolinic acid).

6. Identify putative new mechanisms of resistance, for epidemiological purposes, and monitor antibiotic resistance in aquatic bacteria potentially pathogenic to Man (*Aeromonas hydrophila* and *salmonicida*, *Edwardsiella*

tarda) and in other pathogens (*Nocardia*, *Yersinia ruckeri*, Mycobacteria...), with the collaboration of medical microbiologists.

7. For the treatment of animals reared in sea-cages, use special devices such as tight enclosures and floating pumps allowing to handle therapeutic solutions with a minimal risk of leakage into the aquatic environment. Such systems are already in use in some countries (*e.g.* Norway) and could probably be further improved.

8. Devise pharmaceutical forms relevant to the treatment of aquatic animal species and less liable to pollute the environment than baths and distribution through the food (encapsulation, coating, *Artemia*...and others?).

9. Develop research on the pharmacokinetics of drugs in fish and if possible in other aquatic species, aimed at defining more informed treatment schemes with a reduced use of active substances.

10. In view of the pressing need for new products in the future, set environmental quality standards and sound regulations for the registration of new medicines, which could be addressed by the pharmaceutical industry with a minimal waste of time and money.

11. Study more extensively the epidemiology of the various diseases, which constitutes the basis for disease control in a majority of cases. This would help to limit control through the use of chemotherapy and vaccines.

12. Reconsider many husbandry practices. Is it possible, in fish farming, to produce and rear, in closed facilities, pathogen-free parental breeding stocks yielding healthy offspring to farmers for growing and fattening as, for instance, in poultry production ? Have all possible hygienic measures been explored ? It seems clear from data obtained with shrimps in South-East Asia, that disease emergence is directly related to animal concentrations...

13. Find new and efficient vaccines and new ways to stimulate the specific and non-specifc defence mechanisms. Several areas of research have been proposed in this respect during the present meeting: recombinant vaccines, stimulation of cytokines, complement, phagocytosis, or of natural killer cells, obtention of transgenic fish with genes encoding interferon or highly efficient hybrid antibodies. Several such research projects are already being carried out and they should be strongly supported by all those involved in aquaculture. The practical consequences of these works cannot be expected on a short term, but they will undoubtedly be essential to the efficient aquaculture of the 21st century.

Brief conclusion

Many problems connected to the use of chemotherapeutic agents in aquaculture are yet to be solved. The rapid growth of this animal production has, in many instances, led to the implementation of inefficient or unsatisfactory procedures, in the absence of relevant scientific information. However it is unquestionable that prominent problems have already been solved, and that solutions are in sight for many other ones. Promising research is under way and the authorities in charge of drug registrations are now taking more openly into consideration the specificities of aquaculture. It can be assumed that the present meeting will significantly contribute to further improvements in the chemotherapy of aquatic animals. Nevertheless, even if appropriate answers are found to many persistent questions about chemotherapy, every participant in the field should never forget that prevention of diseases has to be one of the main concerns of aquaculture.

Acknowledgments: I have introduced into this text the contributions of many participants in the final discussion of the meeting. I very sincerely thank all these participants, especially L. Brown, P. De Kinkelin, N. Fijan, T. Hästein, S.O. Hustvedt, C.S. Lewin, L. Pinault, H. Rodger, R.A. Schnick, K.M. Strub and D.R. Tennant. It is also a pleasure to acknowledge the very kind and efficient help of C. Michel in the preparation of these conclusions.

Index of authors

INDEX OF AUTHORS